Handbook on Alternative, Innovative and Novel Processing Technologies (Milk, Milk Products and other Foods)

THE AUTHOR

Dr. M. K. Srivastava has done his M. Sc (Organic Chemistry) and D. Phil (Micro Analysis of Organic Compounds) from Allahabad University, Allahabad. He has got one year training in Quality Analysis of Milk and Milk Products from National Dairy Development Board (NDDB), Anand and different trainings in analysis of Milk products and Minerals from National Dairy Research Institute (NDRI), Karnal.

He has more than 35 years working experience in Dairy industry and well versed with the Quality and Food Safety Management for Food and Dairy Products and Lab accreditation under NABL. He has good experience in dealing with National (BIS, Agmark, FSSA-2006, NABL etc) and International (Codex, IDF *etc.*) Quality as well as Accreditation agencies.

He has worked as Senior Chemist in Pradeshik Co-operative Dairy Federation Ltd and worked as HOD (Q.A. and R&D) Incharge, Central Lab., Lucknow and undertaken various projects of Council of Science and Technology, U. P. UPDASP etc. He is also academic counsellor in IGNOU for DDT programme and visiting faculty for MSc (Food Processing), Lucknow University. He has published 20 research papers and 3 books. He is an approved Food Safety Lead Auditor.

Handbook on Alternative, Innovative and Novel Processing Technologies (Milk, Milk Products and other Foods)

– Author –

Dr. M. K. Srivastava

M. Sc., D. Phil (A. U.)

Ex Senior Dairy Chemist, PCDF (Parag), Lucknow,
Food Safety Lead Auditor/Consultant (Milk/Food Processing),
Faculty, DDE, IGNOU, Lucknow, Uttar Pradesh
and
Ex Faculty,
MSc (Food Processing and Food Technology,
Lucknow University, Lucknow, Uttar Pradesh

2022

Daya Publishing House®
A Division of

Astral International Pvt. Ltd.
New Delhi – 110 002

ISBN 978-93-5461-007-3 (HB)

Published by : **Daya Publishing House®**
A Division of
Astral International Pvt. Ltd.
– ISO 9001:2015 Certified Company –
4736/23, Ansari Road, Darya Ganj
New Delhi-110 002
Ph. 011-43549197, 8130496929
E-mail: info@astralint.com
Website: www.astralint.com

Digitally Printed at : **Replika Press Pvt. Ltd.**

Dedicated to my Respected

Mother-in-Law
Late Sri Kamdeo Prasad Sinha

Father-in-Law
Late Smt Chintamani Sinha

Eldest Brother
Mr D. K. Srivastava

Bhabhiji
Late Mrs Kanaklata Srivastava

Acknowledgements

The author expresses his gratitude and thanks to all Dairy chemists, Dairy/ Food Scientists, Dairy Institutions, Websites and Manufacturers of Processing/Lab Instruments under references for providing technical information.

The author is also thankful to quality assurance personnel and managers (QA) of PCDF, especially Mrs S. Chopra and Mr H. S. Verma. for supporting and inspiring him for the same.

The author is genuinely indebted to his Mother,Father, Brothers,sisters, wife, kids and other family members for providing continuous moral support without which, it is impossible for author to complete the task.

Last but not Least, author acknowledges his thanks to all his friends and all staff of Central Quality Assurance Lab, PCDF Lucknow.

Dr. M. K. Srivastava

Preface

The Quality of milk and milk products play a very important role as far as consumer acceptability is concerned and if dairy industry/Food industry is not giving emphasis on quality then it is very difficult to survive in present competitive national and international market. Good quality raw materials, infrastructures for processing, packing, quality analysis, plant hygiene *etc.* are the prime requirement for maintaining the quality of milk, milk products and other foods as per the standard quality norms.

The above cited national and international pressures on the quality of milk/milk products, and other food products compel the well equipped dairy industries/Food industries to equip the laboratories also with latest analytical equipments, skilled man power and technical literatures related to the analysis of quality of milk/milk products and other food products because the aforesaid requirements can only make the dairy/food industries able to evaluate the quality of milk, milk products, food products properly and accurately. Realising the importance of compilation of technical literatures related to analysis of milk and milk products, author has made successful attempt to produce a compiled and completed advanced technical information related to the analysis of quality of milk and milk products in the form of **"Handbook on Analysis of Milk"(Chemical and Microbiological Analysis of Liquid Milk)** and **"Handbook on Analysis of Milk Products"(Chemical and Microbiological Analysis of Milk Products**). To provide a compiled and complete advanced technical information related to the instrumental analysis of milk, milk products, other food products *etc.*, author has also written a book entitled **"Instrumentation in Chemical and Microbiological Analysis of Food and Dairy Products (Part-1 and 2)".**

The Novel, Advanced and Modern Processing techniques for Milk, Milk Products and other food products are now available to maintain the quality as well as nutritional values of the foods. Now to provide valuable technical information related to these technologies, author now tries to compile all required technical information as a book **entitled " Handbook on Alternative, Innovative and Novel Processing Technologies (Milk/Milk Products and other Foods) ".** Author expects that the said publication is able to meet the requirements of dairy men, Dairy Educationists and professionals of other food industries.

Dr. M. K. Srivastava

Contents

Section B
Alternate Thermal Processing Technologies

Section C
Other Latest Processing Technologies

Chapter 1

Food Processing–Alternative Technologies

Food processing dates to the prehistoric period. As per archaeological evidences, Food processing was found to be effectively used in the hunter-gather humanities. Initially heating and boiling were found to be effectively used to preserve various foods (meat, fish, vegetables, and roots). Gradually, over time, the need for food preservation became more important and hence some processing techniques such as sun drying, fermentation, cereal grinding, and oven baking became popular [1].

The followings points/things related to consumer's interest have created considerable interest in the development of new food-processing techniques:

1. Consumer demands for high quality foods that are fresh, tasting, crisp texture, natural colour and nutritious. The development of new food processing technologies has increased over the past decade primarily due to aforesaid consumer demand for food.
2. Consumers are also increasingly becoming aware of food-borne disease hazards and are concerned about the safety of their food supply.

Traditional food processing technologies such as freezing, canning, drying *etc.* rely on heat to kill food borne pathogens, (bacteria, viruses, and parasites) to make food safe to eat. For many foods, heating is an effective way to treat foods. However, there are many foods that pose a risk for bacterial or viral food borne disease for which heat is either undesirable or cannot be used. Thermal and Non-thermal processing technologies are now being used to process the foods. Developments in non-thermal technologies have been advanced by both industry and academia in an attempt to meet the challenge of producing safe processed food of a high quality. There is no doubt that high quality food can be produced through the use of non-thermal processing technologies. The safety and microbiological quality of food processed using these technologies, however, needs to be affirmed [2].

Thermal and Non-Thermal Technologies

These technologies are used in the processing and preservation of foods. The earliest modern technique for preservation of food was thermal food processing. These processing techniques provided following desired as well as undesirable changes in foods:

1. **Desirable changes:** These included protein coagulation, starch swelling, textural softening, and formation of aroma components.
2. **Undesirable changes:** Some undesirable changes were also observed. The undesirable changes included the loss of vitamins and minerals, freshness, and flavor.

Consumers became aware of these issues and began searching for foods, which looked fresh and tasted fresh. Hence, the food scientists began considering options for developing new technologies, which would provide a balanced process of preservation [1, 3].

Recently, awareness about good nutrition coupled with the increasing demand for fresher tasting food have paved the way for new food processing technologies [4]. Novel thermal and non-thermal processing technologies based on physical techniques for food preservation have the potential to address the demands of the consumer and deliver high quality processed foods with an extended shelf life that are additive free and have not been subjected to extensive heat treatment. The research and developments in non-thermal or "cold" processing techniques are mainly with the aim that undesirable micro-organisms, allergens and enzymes should be inactivated without damage to nutritional and sensory properties resulting normally from thermal treatment [5]. There has been a consumer demand for minimally processed food that also has an extended shelf-life and is safe to eat [6]. With increasing consumers demand and tightening of food and environmental regulations, traditional food-processing techniques have lost their optimum performance which gave rise to new and powerful technologies.

There is a need for new and improved food processing technologies, which effectively disable the micro-organisms in foods, while at the same time maintaining the quality of food. There is growing interest amongst consumers in foods that are minimally preserved and processed. There is currently much emphasis amongst food processors on the development of novel technologies for minimal processed foods. Conventional heating methods for pasteurizing or sterilizing not only kill or reduce the microbes but also reduce the nutritional and organoleptic quality of foods [7].

Thermal Methods

The thermal processing methods include the followings:

1. Microwave (MW)

2. Radiofrequency (RF)
3. Infrared (IR)
4. Heating, pressure-assisted thermal sterilization (PATS) and
5. Sous-vide processing (SVP).

Non-Thermal Methods

The food industry is generally divided into three main parts: processing, storage and distribution. By the processing of foods, the microbes are eliminated to prevent food from becoming spoiled and cause a disease. It also helps in extension of shelf life, while maintaining the nutritional value of the product. There are quite a few advanced non-thermal Technologies which include the followings:

1. High hydrostatic pressure (HHP) processing
2. Irradiation
3. Ultrasound
4. Pulsed electric field (PEF) and
5. Pulsed light (PL) technologies.

Foods treated with non-thermal process are safer to eat then untreated products but still require refrigeration to delay spoilage. However, like any traditional processing method used to preserve food and keep it safe, a processor must validate that the non-thermal method will work to destroy the specific, targeted pathogen in their specific product. if a non-thermal method is proven effective on one food product it cannot simply be used on another. The same non-thermal method and processing parameters that may work on a fruit puree or salsa may not work on oysters or whole blue berries [6].

Researchers have been studying non-thermal processing methods that will destroy pathogens and keep foods safe to eat, while retaining the sensory attributes and nutrient content similar to raw or fresh products. These alternative processing methods are at various stages of development, and have the potential to destroy pathogens and retain desired food quality [6]. Significant advances have been made in the research, development, and application of these technologies in food processing [8, 9].

The Innovative non-thermal food processing technologies being used/preferred by the food industry in the United States and relationship between processing techniques and different types of foods are well described in following Tables [1]:

Preferred Non-Thermal Food Processing Technologies [1]

Sl. No.	*Processing Techniques*	*Preferred/Used (Per cent)*
1	High Pressure Processing	35.6
2	Pulsed Electric Field	20.0
3	Pulsed Light	13.2

Sl. No.	Processing Techniques	Preferred/Used (Per cent)
4	Irradiation	12.2
5	Ultra Sound	3.4
6	Oscillating Magnetic Fields	7.8
7	Cold Atmospheric	2.0
8	Other	2.0

Relationship between Types of Food and Various Non-thermal Food Processing Technologies [1]

Sl. No.	Foods	Food Processing Techniques (Per cent)						
		HPP	PEF	PL	Irradia-tion	US	OMF	CP
1	Fruits and Vegetables	5.4	2.9	0.5	2.4	0.0	0.0	0.5
2	Meat Products	15.1	3.4	0.5	3.9	0.5	1.5	0.0
3	Seafoods	10.2	2.4	0.0	1.5	0.5	2.9	0.5
4	Dairy	2.4	1.5	3.4	1.5	0.5	1.0	0.0
5	Beverages	2.4	4.0	3.4	1.0	1.0	1.5	0.5
6	Flour and Cereals	8.3	2.0	1.5	3.4	2.0	1.5	0.0
7	Oil and Fat	2.9	3.4	2.9	2.4	0.5	0.0	0.5
8	Liquor	0.5	0.5	2.4	0.0	0.0	0.0	0.5
9	Others	1.5	0.5	2.0	0.0	0.5	0.0	0.5

Conventional thermal processing is a mainstay of the food industry. Alternative processing methods are relatively new in their application but not in their existence. In the quest for better quality, more healthful, minimally processed and safer foods, these technologies have become the subject of intense research, furthering expansion of the knowledge base in this area, catalyzing research and development and ultimately commercialization of certain innovative food processing techniques. Currently, societal driving forces in the market as well as the technological development of alternative thermal and non-thermal preservation methods have provided a fairly well-grounded platform to deliver safer products that also guarantee higher quality.

Non-thermal processes not only do improve the microbiological innocuity and quality of a range of products, but these processes can also be applied in following multiple ways throughout the dairy network [10]:

1. To improve the efficiency and end-quality of processing methods.
2. For physical, nutritional and organoleptic features
3. To preserve unstable bioactive molecules
4. To modulate enzyme activity and
5. To develop ingredients and hypoallergenic products.

Reasons for Selection of Non-Thermal Food Processing Technology (Non-thermal)

The following reasons are responsible for selection of Non-thermal Food Processing Techniques for food processing in United America [1]:

Sl. No.	Reasons for Selection	Per cent
1	Solution for safety problems	25.4
2	Government/regulatory requirements	13.4
3	Enhancing product's shelf life	39.3.
4	Better nutrient/sensory quality	71.1
5	Price (cheaper equipment)	10.5
6	Cost saving (energy, water)	13.4
7	Convenience	11.9

Sources

1. *Harlin Kaur Saroya, Innovative non-thermal food processing technologies used by the food industry in the United States, MSc thesis, 2017, Western Kentucky University, Kentucky.*
2. *Rajan Saha, New and Emerging Non-Thermal Technologies in Food Preservation, April, 2015.*
3. *P. Fellows, Food processing technologies: Principles and practices (3rd ed.), Boca Raton, FL: CRC Press., 2009.*
4. *Hudaa Neetoo (Faculty of Agriculture, University of Mauritius, Réduit, Mauritius) and Haiqiang Chen (Department of Animal and Food Sciences, University of Delaware, Newark, Delaware, US), Alternative Food Processing.*
5. *K. Shiby Technologies.*
6. *Lori F Pivarnik and Randy Worobo, Non-thermal or Alternative Food Processing Methods to Enhance Microbial Safety and Quality, Frequently Asked Questions, (USDA-NIFA, 2011-68003-30005).*
7. *P. Verghese, M. C. Pandey, K. Radhakrishna and A. S. Bawa, Technology, applications and modelling of ohmic heating: a review, J Food Sci Technol. 2014, Oct; 51(10): 2304-2317.*
8. *Ohlsson T, Bengtsson. N, 2002, Minimal processing of foods with non-thermal methods. In: Ohlsson T, Bengtsson N (eds) Minimal Processing Technologies in the Food Industry. Cambridge: Woodhead Publishing, pp. 34-60.*
9. *Senorans FJ, Ibanez E, Cifuentes A, 2011, New trends in food processing. Critical Reviews in Food Science and Nutrition, 507-526.*
10. *Geoffrey W. Smithers, Dairy Product Quality and Non-Thermal Processing Technologies Food Science Australia, Session-4, IDF Dairy Science and Technology Week - Quebec 2008.*

Sustainability and Energy Efficiency of Food Processing Methods [1]

Energy consumption, energy savings, environmental protection, and waste management in the food industry have been in focus for the last 30 years. These have become critical issues in order to lower production costs, maintain economic growth, and improve sustainability in the food industry [1a]. Some insight into the environmental impact of alternative and novel technologies are provided.

Novel thermal technologies such as RF, MW, and ohmic heating (OH) methods and non-thermal methods such as PEF and HHP are continuously being developed and evaluated. Many novel technologies can ensure not only energy savings but also water savings, increased reliability, higher product quality, reduced emissions, and improved productivity [2] and consequently, less impact on the environment. The details related to the energy saving, less gas emission, water saving *etc.* are given below:

1. Energy Savings

An enlightening study was conducted [3], which provided estimates on the potential energy savings of PEF and RF drying systems compared to existing technologies.

a. Pulsed Electric Field (PEF)

- PEF pasteurization was demonstrated to have 100 per cent natural gas savings since thermal input is eliminated.
- The electricity savings of PEF were estimated to be up to 18 per cent, based on the assumed electricity consumption range of the base technology.
- For PEF, the preservation of liquid media results in operation costs about 10-fold higher than those of conventional thermal processing. However, with pulse energy dissipation and the simultaneous resistive heating of the suspending medium, capitalizing on the synergistic effects of mild heat temperatures can render PEF more energy efficient [4].
- The combined synergistic effects of mild heat treatment temperatures and PEF can provide a shorter treatment time with energy recovery and an energy requirement of less than 40 kJ/kg for a reduction of 6 log cycles of *E. coli*. For instance, an energy input of 40 kJ/kg will lead to a temperature increase of 11°C in the case of orange juice, showing that with a maximum temperature of 66°C, the preservation process is still operating at lower maximum temperature and shorter residence times than during conventional heat preservation. Therefore, a reduction of the energy requirements from the original 100 kJ/kg to 40 kJ/kg can be realized, thus rendering PEF energy efficient and able to be easily integrated in existing food processing operations [5].
- Despite the increase of the delivered electrical power, PEF seems less energy intensive than traditional pasteurization methods, leading to

estimated annual savings of 791.2-1055 TJ, while also contributing to reduction of CO_2 emissions [6].

b. Radiofrequency (RF)

- For RF drying applications, the estimated natural gas savings range from 73.8 to 147.7TJ per year, although these savings are masked by the increase in electricity consumption, indispensable to power the RF drying unit.
- Depending on the natural gas consumption of the base tunnel oven, the primary energy savings of RF drying can range from 0 to 73.8 TJ per year.

c. High Hydrostatic Pressure (HHP)

It is one of the most successful developments to date, offering a clean, natural, environment-friendly process.

- With PATS, instantaneous adiabatic compression during pressurization results in a quick increase in the temperature of food products, which is reversed when the pressure is released, providing rapid heating and cooling conditions and hence shorter processing times [7].
- The combined application of high pressure and heat can be utilized to achieve inactivation of spores of *Clostridium* spp. similar to conventional sterilization.
- The specific energy input required for sterilization of cans can be reduced from 300 to 270 kJ/kg when applying the PATS treatment [4].
- Moreover, if the electricity is generated by an environmentally clean, renewable energy source (*e.g.* hydroelectric power), then these processes would effectively contribute to reducing the pollution load, helping to preserve the environment. Furthermore, HHP may partly circumvent the use of cooling systems, which often represents 50 per cent of the total electricity consumption [8].

d. Ohmic Heating

- Studies conducted on meat products at the Agri-Food Canada's Food Research and Development Center (FRDC) have indicated that an Ohmic heating (OH) process could result in energy savings of at least 70 per cent [9].
- In tests conducted at the Louisiana State University Agricultural Center, sweet potato samples were processed using OH (Ohmic heating) prior to freeze drying. OH increased the rate of freeze drying by as much as 25 per cent, which led to significant savings in both processing time and energy use [2, 10].
- In addition, since heat is generated in the bulk fluid, less fouling should take place [11], thus reducing fouling-related costs.

2. Reduced Gas, Effluent Emissions and Water Savings

The main types of gas emissions into the air from food processing operations are related to heat and power production. Traditional thermal methods require large amounts of fossil fuels and water to generate steam. In fact, it is estimated that the food industry's fossil fuel consumption is 57 per cent [12]. Typical boilers may accumulate dirt over time which acts as an insulator, thus reducing the heat transfer rate and wasting energy.

Reduced Gas and, Effluent Emissions

During drying, Ohmic heating, PEF, IR, and RF are allowing precise control of the process temperature and leading to lower energy costs, reduced gas consumption, and fewer combustion related emissions [10, 16, 17].

Water Saving

In contrast, novel food processing systems save water. Some examples are given below:

- ☆ Novel food processing systems powered by electricity may present an environmental advantage. In general, novel processing technologies such as HHP and PEF are considered environment friendly as they may eliminate completely, or at least reduce significantly, the local use of boilers or steam generation systems, and consequently diminish wastewater, thus increasing water savings.
- ☆ Several operations such as peeling, blanching, and drying as used by the food industry often require high water use.

Ohmic Peeling

It, in contrast, reduces environmental problems associated with lye peeling (*e.g.* treatment of wastewater) because it does not use lye in the process [13].

Blanching

It was reported [14] that blanching by OH considerably reduced the extent of solid leaching compared to a hot water process and a short blanching time could be used regardless of the shape and size of the product. For example, blanching of mushrooms using OH was reported to maintain a higher solids content than conventional blanching, while reducing the excessive consumption of water [15].

Drying

Ohmic heating, PEF, IR, and RF can also significantly accelerate drying processes when compared to their conventional counterparts [10, 16, 17].

3. Generation of Solid Waste

Food production and consumption generate solid waste that can be classified as **"food waste" and "non-food waste."** The food waste is any plant or animal tissue in a raw or cooked state that was intended for human consumption but needs disposal as a result of spoilage, expiration, contamination or excess. **The Environmental**

Protection Agency defines food waste as: "Uneaten food and food preparation wastes from residences and commercial establishments such as grocery stores, restaurants, produce stands, institutional cafeterias and kitchens, and industrial sources such as employee lunchrooms" [18].

- ✰ Innovative processing technologies such as HHP and irradiation can dramatically improve the shelf life of raw and processed products with minimal adverse effects on their sensory quality, thereby decreasing food losses and wastage due to microbial, chemical or enzymatic spoilage.
- ✰ Moreover, these technologies (HHP, irradiation *etc.*) can produce valuable compounds that are attractive to consumers, thus **reducing generation of solid waste** such as olive mill wastewater and other agricultural by-products.

Challenges and New Safety Strategies for New Technologies [19]

There are several challenges for new processing technologies and these technologies need new safety strategies to provide safe foods. The Challenges and New Safety Strategies for New Technologies are discussed below [19]:

1. Challenges

Food is deemed unsafe if it constitutes either a physical, chemical or biological hazards to the consumer. Non-thermal technologies have been advanced to replace conventional heat treatments. These new technologies cannot however achieve the broad microbial lethalities that are currently attainable by heat treatment. Current PEF and HPP technologies can only accomplish the equivalent of pasteurization when applied at their maximum lethal doses. The achievement of commercial sterility by these non-thermal technologies is currently not feasible.

Irradiation, ultrasound under pressure, HHP and PEF are effective procedures to inactivate vegetative micro-organisms in foods, **but the high resilience of spores limits their use as a sole method for food preservation**. Therefore, these novel technologies are finding **applications as hurdles** that assure food safety through microbial inactivation in minimally processed high quality products. Current limitations of emerging non-thermal technologies can be overcome when they are combined with conventional preservation methods. For increasing the effectiveness of novel technologies, application of higher or lower temperatures than room temperature is an interesting [19] and give better results.

The application of non-thermal technologies to foods is more likely to result in stress or injury than to cause the death of micro-organisms. An abundance of injured microbial cells in non-thermally-processed food may create new challenges to food processors and regulatory agencies [19].

2. New Safety Strategies

The detection of low levels of pathogens in food is a difficult task particularly when cells are injured. The safety of the non-thermally processed product is compromised if food and storage conditions favour the recovery of injured cells.

Stress of pathogens by non-thermal technologies is a concern and the adaptation of cells to such stress may constitute a microbial hazard. Non-thermal technologies therefore introduce new challenges, and thus warrant the implementation of new safety strategies.

Combinations of Different Non-Thermal Technologies [19]

High pressure, pulsed electric fields, biopreservatives and radiofrequency heating *etc.* are examples of novel technologies, which are currently used or will be used in the near future for preservation of food products. In general these technologies are effective in inactivating vegetative cells of bacteria, yeasts, and fungi. However, inactivation of bacterial spores and for some technologies also enzyme inactivation remains difficult using these technologies alone. **To extend the use of non-thermal processing technologies in the food industry, combinations of these technologies(Hurdle Technology) with traditional or emerging food preservation techniques are interesting.**

Examples of Combinations

Some examples of combination of new technology with traditional technology, are given below:

a. An often-made combination used is cold storage after a (non-thermal) pasteurization treatment.

b. Also other consecutive or simultaneous combinations are interesting to broaden the application range of these technologies:

   1. Combinations of novel technologies with elevated temperature
   2. Lowered temperature and/or
   3. Biopreservatives.

In general, these combinations may result in an enhanced shelf life combined with a good preservation of product quality.

Sources

1. *Hudaa Neetoo (Faculty of Agriculture, University of Mauritius, Réduit, Mauritius) and Haiqiang Chen (Department of Animal and Food Sciences, University of Delaware, Newark, Delaware, US), Alternative Food Processing Technologies.*

1a. *Wang L, 2009, Energy project management in food processing facilities. In: Wang L (ed) Energy Efficiency and Management in Food Processing Facilities. Boca Raton, FL: CRC Press, pp. 57-83.*

2. *Masanet E, Worrell E, Graus W, Galitsky C, 2008, Energy efficiency improvement and cost saving opportunities for the fruit and vegetable processing industry, an energy star guide for energy and plant managers. Technical report: Environmental Energy Technologies Division. www.energystar.gov/ia/business/industry/Food-Guide.pdf, accessed 13 November, 2013.*

3. *Lung R, Masanet E, McKane A, 2006, The role of emerging technologies in improving efficiency: examples from the food processing industry. In: Proceedings of the Industrial Energy Technologies Conference, New Orleans, Louisiana.*

4. *Toepfl S, Mathys A, Heinz V, Knorr D 2006, Potential of high hydrostatic pressure and pulsed electric fields for energy efficient and environmentally friendly food processing. Food Review International 22: 405-423.*

5. *Heinz V, Toepfl S, Knorr D, 2002, Impact of temperature on lethality and energy efficiency of apple juice pasteurization by pulsed electric fields treatment. Innovative Food Science and Emerging Technologies 4: 167-175.*

6. *Lelieveld H, 2005, PEF - a food industry's view. In: Barbosa-Canovas GV, Tapia MS, Cano MP, Martin-Belloso O, Martinez A (eds) Novel Food Processing Technologies. Boca Raton, FL: CRC Press, pp. 145-157.*

7. *Shao YW, Zhu SM, Ramaswamy H, Marcotte M (2008), Compression heating and temperature control for high-pressure destruction of bacterial spores: an experimental method for kinetics evaluation. Food and Bioprocess Technology 3: 71-78.*

8. *Dalsgaard H, Abbots A, 2003, Improving energy efficiency. In: Mattson B, Sonesson U (eds) Environmentally Friendly Food Processing. Boca Raton, FL: CRC Press, pp. 116-130.*

9. *Vicente A, Teixeira J, Castro I, 2006, Ohmic heating for food processing. In: Sun D (ed) Thermal Food Processing: New Technologies and Quality Issues. Boca Raton, FL: Taylor and Francis, pp. 459-501.*

10. *Lima M, Zhong T, Lakkakula NR, 2002, Ohmic Heating: A Value-Added Food Processing Tool. A technical report from Louisiana Agriculture Magazine. <www. lsuagcenter. com/en/communications/publications/agmag/Archive/2002/Fall/Ohmic+ Heating+ A+ Value added+ Food+ Processing+Tool.htm, accessed 11 November 2013.*

11. *Bansal B, Chen X, 2006, A critical review of milk fouling in heat exchangers. Comprehensive Reviews in Food Science and Food Safety 5: 27-33.*

12. *Einstein D, Worrell E, KhrushchM, 2001, Steam systems in industry: energy use and energy efficiency improvement potentials. Lawrence Berkeley National Laboratory. http: //escholarship.org/uc/item/3m1781f1, accessed 13 November 2013.*

13. *Wongsa-Ngasri MSP, 2004, Ohmic heating of biomaterials: peeling and effects of rotating electric field. PhD thesis, Ohio State University, Columbus, OH, pp.1-192.*

14. *Mizrahi S, 1996, Leaching of soluble solids during blanching of vegetables by ohmic heating. Journal of Food Engineering 29: 153-166.*

15. *Sensoy I, Sastry S, 2007, Ohmic heating of mushrooms. Journal of Food Process Engineering* **27***: 1-15.*

16. *Nowak D, Lewicki PP, 2004, Infrared drying of apple slices. Innovative Food Science and Emerging Technologies 5: 353-360.*

17. *Wang WC, 1995, Ohmic heating of foods: physical properties and applications. PhD dissertation, Ohio State University, Columbus, OH, pp.1-157.*

18. *Environmental Protection Agency (EPA), 1997, Terms of Environment: Glossary, Abbreviations and Acronyms. EPA 175-B-97-001. Washington, DC: Environmental Protection Agency.*

19. *Ranjan Saha, New and Emerging Non-Thermal Technologies in Food Preservation April 2015.*

Section-A

Alternate Non-thermal Processing Technologies

Chapter 2

Membrane Technology

The "membrane technology"[1] is the term that is used to collectively represent the separation processes by employing specific semi-permeable membrane filters to concentrate or fractionate a liquid into two liquids of different compositions [1a, 1b] by selectively allowing some compounds to pass while encumbering the others. **The permeates is the liquid that is able to pass the membrane and the "retentate" or "concentrate" is retained liquid.** It covers all engineering approaches for the transport of substances between two fractions with the help of permeable membranes. In general, mechanical separation processes for separating gaseous or liquid streams use membrane technology.

The membrane technology **is a novel non-thermal greener environmental friendly technology** with full of future possibilities that minimizes the following adverse effect of temperature rise:

1. Change in phase
2. Denaturation of proteins
3. Change in sensory attributes of the product.

Membrane separation processes operate without heating and therefore use less energy than conventional thermal separation processes such as distillation, sublimation or crystallization. The separation process is purely physical and both fractions (permeate and retentate) can be used. **Cold separation using membrane technology is widely used in the food technology, biotechnology and pharmaceutical industries.**

The membranes remove unwanted components *e.g.* micro-organisms, drugs or sediments that have a negative impact on product quality, making the final product more attractive in texture and increasing its shelf life [1a]. Membranes enables separations to take place that would be impossible using thermal separation methods. Some examples are given on next page.

1. It is impossible to separate the constituents of azeotropic liquids or solutes which form isomorphic crystals by distillation or recrystallization but such separations can be achieved using membrane technology.
2. Depending on the type of membrane, the selective separation of certain individual substances or substance mixtures is possible.

Important technical applications of membrane technology include the followings:

1. **Drinking water:** The production of drinking water by reverse osmosis.
2. **Petrochemicals:** The recovery of organic vapours such as petro-chemical vapour recovery.
3. **Chlorine Production:** The electrolysis for chlorine production.
4. **Food Processing:** The use of membrane technology as a processing and separation method in food industry is gaining wide application. Membrane separations can be used either as alternatives to conventional techniques or as novel technology for processing new ingredients and foods. Membrane technology is also used for Filtrations in the food industry.
5. **Wastewater Treatment:** In this, membrane technology is becoming increasingly important. With the help of Ultra/Microfiltration it is possible to remove particles, colloids and macromolecules, so that waste-water can be disinfected in this way. This is needed if waste-water is discharged into sensitive waters especially those designated for contact water-sports and recreation [1].
6. **Medical Applications:** About half of the market is in medical applications such as use in artificial kidneys to remove toxic substances by hemodialysis and as artificial lung for bubble-free supply of oxygen in the blood.

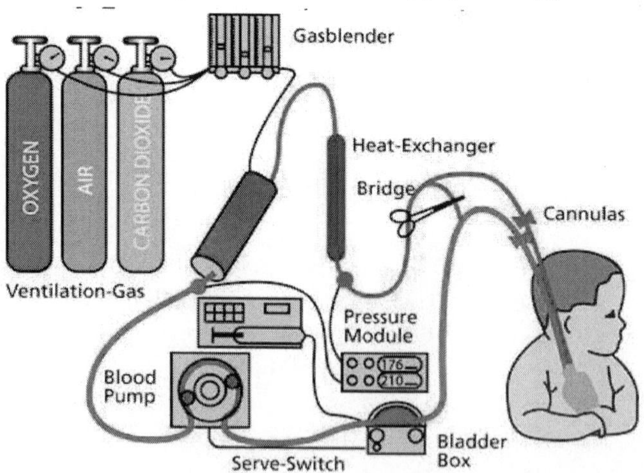

Venous-Arterial Extracorporeal Membrane Oxygenation Scheme [1].

Ultrafiltration for a Swimming Pool [1].

7. **Environment protection:** The importance of membrane technology is growing in the field of environmental protection. Even in modern energy recovery techniques membranes are increasingly used, for example in fuel cells and in osmotic power plants.

The global demand for membrane modules is increasing day by day. With the approximate annual growth of 8 per cent, it was estimated at approximately 15.6 billion USD in 2012 [1], 21.22 billion USD in 2016 and reach 25 billion in 2018 [1c].

Definitions, Terminologies and Flow Geometries [1, 1e]

1. Membrane [1]

It is a selective barrier and it allows some things to pass through but stops others. Such things may be molecules, ions, or other small particles.

Biological Membranes

These include cell membranes (outer coverings of cells or organelles that allow passage of certain constituents) [1f].

Nuclear Membranes

These cover a cell nucleus and tissue membranes, such as mucosae and serosae.

Synthetic Membranes

These are made by humans for use in laboratories and industry.

2. Flow Geometries

The following definitions and terms are important to understand Membrane Technology :

a. Dead-end Filtration [1, 1e]

It is normally applied in laboratory sample preparation and in this dead-end filtration, both pressure vector and feed flow are normal to the membrane [1e]. This flow also offers following advantages and disadvantages [1].

Advantages [1]

Generally, dead-end filtration is used for feasibility studies on a laboratory scale. The dead-end membranes are relatively easy to fabricate which reduces the cost of the separation process. The dead-end membrane separation process is easy to implement and the process is usually cheaper than cross-flow membrane filtration. This filtration process(dead-end) is usually a batch-type process, where the filtering solution is loaded (or slowly fed) into the membrane device, which then allows passage of some particles subject to the driving force.

Disadvantages [1]

The main disadvantage of a dead end filtration is the extensive membrane fouling and concentration polarization. The fouling is usually induced faster at higher driving forces. Membrane fouling and particle retention in a feed solution also builds up a concentration gradients and particle back flow (concentration polarization).

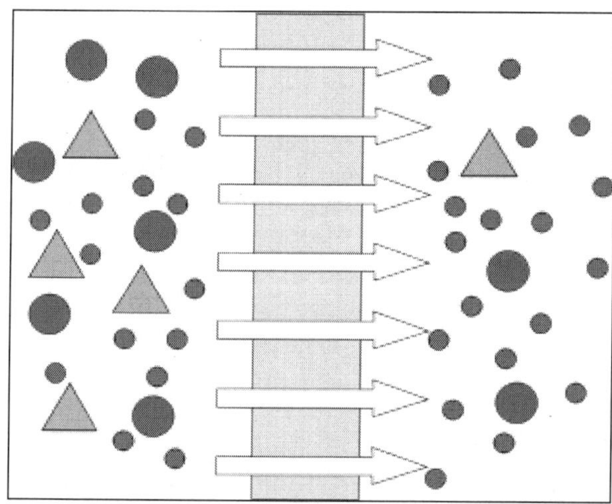

Dead-end Geometry [1e].

b. Crossflow (CF) Filtration

This is also called **tangential** filtration and in this the feed is pumped parallelly with the membrane and it is possible to recirculate the retentate back to the feed flow [1e]. In this type of filtration the feed flow is tangential to the surface of membrane, retentate is removed from the same side further downstream, whereas the permeate flow is tracked on the other side. This flow offers advantages and disadvantages, which are given on next page.

Advantages

1. CF is less susceptible to fouling due to the sweeping effects and high shear rates of the passing flow [1].
2. Compared to dead-end filtration, CF filtration (due to the tangential movement of the feed) is characterized with lower extent of concentration polarization and membrane fouling [1e].

Disadvantages

The tangential flow devices (CF devices) are more cost and labor-intensive [1],

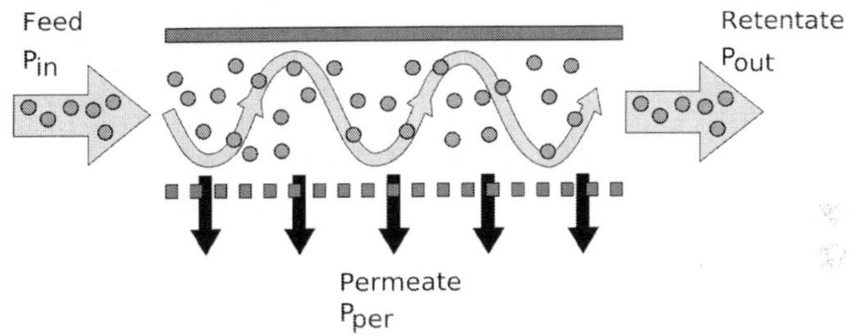

Cross-Flow Geometry [1].

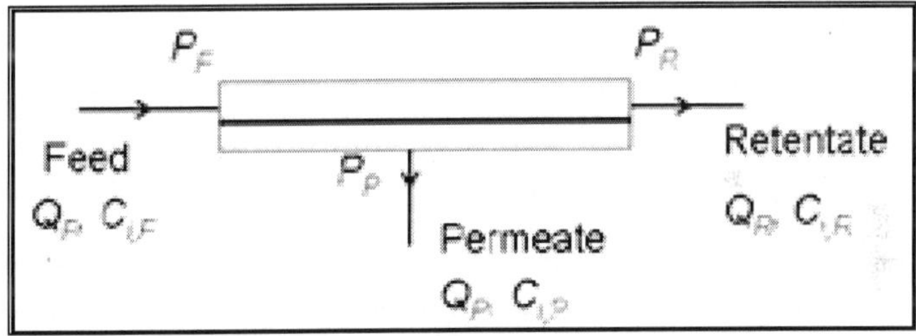

Crossflow Membrane Separation Unit [1e]

3. Concentration Polarization [1e]

It expresses the raise-up in concentration of a macromolecular solute (retained by the membrane) at the surface of the membrane compared to that in the bulk phase (Details given in Chapter-3).

4. Membrane Fouling [1c]

It indicates the deposition of solutes/particles on the surface (**external fouling**) and/or into the membrane pores (**internal fouling**) [2, 2a]. The components causing membrane fouling are called **foulants**. In MF and UF of poly dispersed feeds like milk, fouling can develop very quickly after the start of filtration, leading to a drop in permeate flux. Both the internal and surface foulants act as additional layers of resistance to the transmission of solutes through the membrane (More details given in Chapter-3). The occurrence and extent of concentration polarization and fouling depend on the following factors:

- **i.** Feed properties
- **ii.** Membrane materials and structure
- **iii.** Membrane modules
- **iv.** Operational parameters such as temperatures, TMP values, and CF velocities [1e, 2c, 2d, 2e].

The aforesaid two phenomena (Concentration polarization and Membrane fouling) are two major obstacles causing reduction in membrane separation performance.

5. Diafiltration (DF) [1e]

It is carried out by diluting the retentate with a pure solvent, usually water or a buffer, and re-concentrating the diluted retentate. The main purpose of DF is to increase the purity of the membrane-rejected components (retentate components) and DF can be performed either in batch or in continuous manner.

Mass Transfer [1]

Two basic models can be distinguished for mass transfer through the membrane:

- **i. Solution-diffusion model and**
- **ii. Hydrodynamic model.**

In real membranes, these two transport mechanisms certainly occur side by side, especially during ultra-filtration.

i. Solution-diffusion Model

In the solution-diffusion model, **transport occurs only by diffusion.** The component that needs to be transported must first be dissolved in the membrane. The general approach of the solution-diffusion model is to assume that the chemical potential of the feed and permeate fluids are in equilibrium with the adjacent membrane surfaces such that appropriate expressions for the chemical potential in the fluid and membrane phases can be equated at the solution-membrane interface. This principle is more important for **dense** membranes without natural pores such as those used for reverse osmosis and in fuel cells.

During the filtration process a boundary layer forms on the membrane. This concentration gradient is created by molecules which cannot pass through the membrane. The effect is **referred as concentration polarization** and, occurring during the filtration, leads to a reduced trans-membrane flow (flux). Concentration polarization is, in principle, reversible by cleaning the membrane which results in the initial flux being almost totally restored. Using a tangential flow to the membrane (cross-flow filtration) can also minimize concentration polarization.

ii. Hydrodynamic Model

Transport through pores (in the simplest case) is done convectively. This requires the size of the pores to be smaller than the diameter of the two separate components. Membranes which function according to this principle are used mainly in micro- and ultrafiltration. They are used to separate macromolecules from solutions, colloids from a dispersion or remove bacteria. During this process the not-passing particles or molecules form a pulpy mass (filter cake) on the membrane, and this blockage of the membrane hampers the filtration.

This blockage can be reduced by the use of the cross-flow method (cross-flow filtration). Here, the liquid to be filtered flows along the front of the membrane and is separated by the pressure difference between the front and back of the membrane into retentate (the flowing concentrate) on the front and permeate (filtrate) on the back.

The tangential flow on the front creates a shear stress that cracks the filter cake and reduces the fouling.

Types of Membrane Operations [1]

According to the driving force of the operation it is possible to distinguish following membrane operations :

1. **Pressure driven operations** -Microfiltration, Ultrafiltration, Nanofiltration and Reverse osmosis.
2. **Concentration driven operations-** Dialysis, Pervaporation, Forward osmosis, artificial lung and gas separation.
3. **Operations in an electric potential gradient** -Electrodialysis, Membrane electrolysis *e.g.* chloralkali process, Electrodeionization, Electrofiltration, Fuel cell
4. **Operations in a temperature gradient** - Membrane distillation.

Membrane Shapes and Modules [1, 1e]

The 'module' expresses how membranes are arranged or packed. There are several types of membrane modules, namely plate-and-frame, tubular, spiral-wound, hollow-fiber, and membrane cassette [1e]. Different types of modules are different in packing density, possible applications, and price. The most commonly used synthetic membrane devices (modules) are flat sheets/plates, spiral wounds, and hollow fibers [1]. The details of the some of the modules are given below:

i. Flat Plates/Plate and Frame [1]

These are usually constructed as circular thin flat membrane surfaces to be used in dead-end geometry modules. Plate and frame consist of a series of flat membrane sheets and support plates. The water to be treated passes between the membranes of two adjacent membrane assemblies. The plate supports the membranes and provides a channel for the permeate to flow out of the unit module. The plate-and-frame modules have the **lowest packing density**, which means highest investment for a unit of membrane surface area [1e].

ii. Spiral Wounds [1]

These are constructed from similar flat membranes but in the form of a "pocket" containing two membrane sheets separated by a highly porous support plate [2]. Several such pockets are then wound around a tube to create a tangential flow geometry and to reduce membrane fouling. Spiral-wound modules (diameters in range of 0.1 to 2 mm) are the **most common design for RO and UF membranes [1e].** It provides the **largest packed membrane area per unit volume [1e].**

iii. Hollow Fiber Modules [1]

These modules consist of an assembly of self-supporting fibers with dense skin separation layers, and a more open matrix helping to withstand pressure gradients and maintain structural integrity [2b]. The hollow fiber modules can contain up to 10,000 fibers ranging from 200 to 2500 µm in diameter.

The main advantage of hollow fiber modules is very large surface area within an enclosed volume (It provides the **largest packed membrane area per unit volume [1e]**, increasing the efficiency of the separation process.

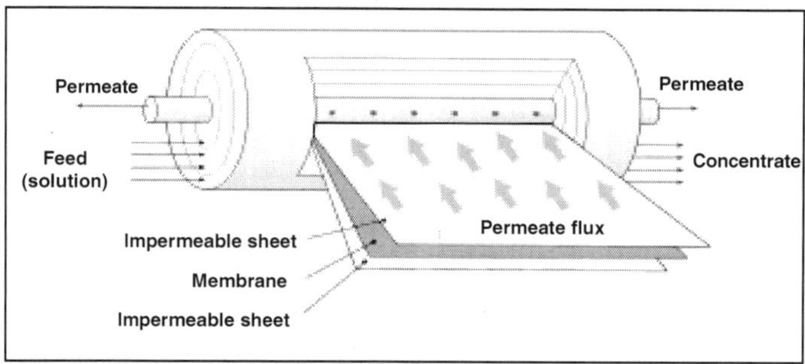

Spiral Wound Membrane Module [1].

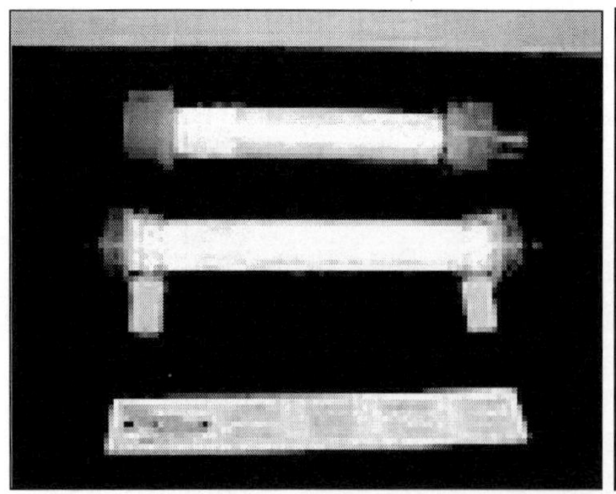
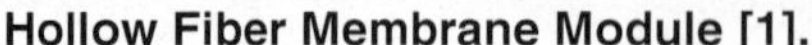

Hollow Fiber Membrane Module [1].

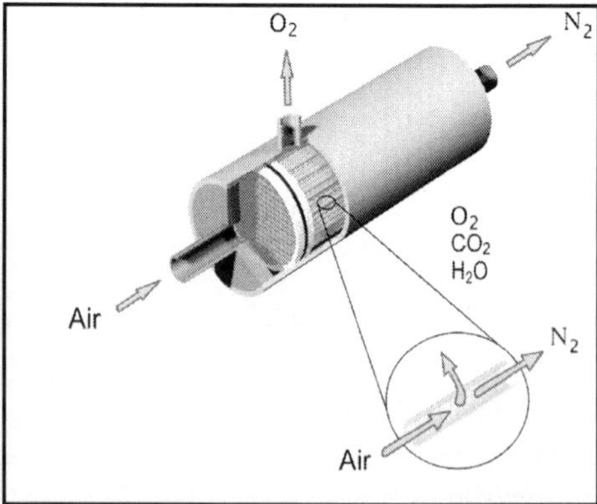

Separation of Air into Oxygen and Nitrogen through a Membrane [1].

Different Membrane Modules.

iv. Tubular Modules (Ceramic membranes) [1e]

With this type of module, the feed channel is wide (*e.g.*, from below 1 cm to 3 cm in diameter) so the feed can be pumped with high CF velocities, which help minimize the formation of concentration polarization and external membrane fouling and hence improve permeate flux and allow the retentate to be concentrated to a high level (higher VCR). The tubular modules have the **lowest packing density,** which means highest investment for a unit of membrane surface area [1e]. These are used widely in MF of skim milk (Pl see in Chapter-03).

Membrane Performance and Governing Equations [1]

a. Membrane Performance

The selectivity of membranes is very high due to the unique mechanisms of action such as ion exchange, solution diffusion, *etc.* The membranes are suitable to different types of plant design and expansion due to their compact design and need very low maintenance. The operations of the membranes are very simple, competitive and do not required any specialized knowledge to handle or operate them [1a]. **To get the better and desired effect, it is sometimes necessary to use a combination of membranes rather than single membrane [1a].**

b. Selection of Membranes

The selection of synthetic membranes for a targeted separation process is usually based on following few requirements:

1. Membranes have to provide enough mass transfer area to process large amounts of feed stream.
2. The selected membrane has to have high selectivity (rejection) properties for certain particles. It has to resist fouling and to have high mechanical stability.
3. It also needs to be reproducible and to have low manufacturing costs.

c. Governing Equations

The main modeling equation for the dead-end filtration at constant pressure drop is represented by **Darcy's law [2b]:**

$$\frac{dV_p}{dt} = Q = \frac{\Delta p}{\mu} A\left(\frac{1}{R_m + R}\right)$$

where, V_p and Q are the volume of the permeate and its volumetric flow rate respectively (proportional to same characteristics of the feed flow), μ is dynamic viscosity of permeating fluid, A is membrane area, Rm and R are the respective resistances of membrane and growing deposit of the foulants. Rm can be interpreted as a membrane resistance to the solvent (water) permeation. This resistance is a membrane intrinsic property and is expected to be fairly constant and independent of the driving force, Δp. R is related to the type of membrane foulant, its concentration in the filtering solution, and the nature of foulant-membrane interactions.

Darcy's law allows for calculation of the membrane area for a targeted separation at given conditions. The solute sieving coefficient is defined by the equation [2b]:

$$S = \frac{C_p}{C_f}$$

where, C_f and C_p are the solute concentrations in feed and permeate respectively.

Hydraulic permeability is defined as the inverse of resistance and is represented by the equation [2b]:

$$L_p = \frac{J}{\Delta_p}$$

where, J is the permeate flux which is the volumetric flow rate per unit of membrane area.

The solute sieving coefficient and hydraulic permeability allow the quick assessment of the synthetic membrane performance.

Other Definitions

Permeate flux, Volume concentration ratio (VCR), Average trans-membrane pressure (TMP), Transmission of component, Apparent rejection which represents separation efficiency, Selectivity of the membrane between two components along with the calculation formulas are also given below [1e]:

i. Permeate Flux

It is represented by the given formula :

$$J = \frac{Qp}{Ae} \quad (L/m^2\, h \text{ or } m^3\, m^{-2}\, s^{-1})$$

where, Ae is effective surface area of the membrane.

ii. Volume Concentration Ratio (VCR)

It is represented by the given formula :

$$VCR = \frac{V_{Fi}\ V_{Fi}}{V_{Rt}\ V_{Fi}\ V_{pt}}$$

where, V_{FI} - Feed volume at the beginning of operation, V_{Rt} - The retentate volume at time *t* and V_{pt} - Permeate volume at time *t*

iii. Average Trans-membrane Pressure (TMP)

It is represented by the given formula :

$$TMP = \frac{P_F\ P_R}{2} P_p$$

where, P_F, P_R and P_p are pressures at the feed inlet, retentate outlet, and permeate outlet, respectively.

iv. Transmission of Component i

It is represented by the given formula:

$$Tr_i(\%) = \frac{C_{ip}}{C_{iR}} \times 100$$

Where, C_{iR} and, C ip are the concentration of component *i* in the retentate and at the outlet of the membrane pores, respectively.

Membrane Separation Processes [1]

Membrane separation processes have a very important role in the separation industry. These membrane separation processes were not considered technically important till mid 1970. Membrane separation processes differ based on separation mechanisms and size of the separated particles. The details related to the widely used membrane separation processes are given below:

i. Microfiltration

Microfiltration (MF) is a membrane separation **process similar to ultrafiltration (UF)** but with even larger membrane pore size allowing particles in the range of 0.2 to 2 micrometers to pass through. The pressure used is generally lower than that of UF process [1a]. **MF is normally used for separation of suspended particles and micro-organisms from soluble components in feed** [1e]. More details are given in Chapter 3.

ii. Ultrafiltration

UF is a variety of membrane filtration in which hydrostatic pressure forces a liquid against a semi-permeable membrane. Suspended solids and solutes of high molecular weight are retained, while water and low molecular weight solutes pass through the membrane. It is a separation/fractionation process using a 10,000 MW cutoff, 40 psig, and temperatures of 50 to 60°C with polysulfone membranes [1a]. UF can be applied to separate soluble macromolecules such as proteins and peptides [1e]. Diafiltration (DF) is a specialized type of ultrafiltration process in which the retentate is diluted with water and re-ultrafiltered, to reduce the concentration of soluble permeate components and to further increase the concentration of retained components [1a]. More details are given in Chapter 3.

iii. Nanofiltration

Nanofiltration (NF) is **a type of reverse osmosis** where the membrane has a slightly more open structure allowing monovalent ions to pass through the membrane and rejecting the divalent ions to a great extent. NF is applied for partial demineralization and, at the same time, concentration. Basically, NF allows monovalent salts to pass through while it retains multivalent salts [1e].

iv. Reverse Osmosis

RO is a high pressure-driven membrane filtration process which is based on a very dense membrane. The Reverse osmosis (RO) membrane pore size is very small allowing only small amounts of very low molecular weight solutes to pass through the membranes. It is a concentration process using a 100 MWCO (molecular weight cutoff), 700 psig, temperature lower than 40°C with cellulose acetate membranes and 70-80°C with composite membranes [1a].

v. Electrolysis
vi. Dialysis
vii. Electrodialysis
viii. Gas separation
ix. Vapor permeation
x. Pervaporation
xi. Membrane distillation, and
xii. Membrane contactors [3].

Characteristics of Membrane Processors

Several hundreds of thousand square meters of membrane (Ultrafiltration-400,000, Nanofiltration- 300,000, Reverse osmosis- 100,000, Microfiltration- 50,000) are currently operating [1d]. Some characteristics of aforesaid processes are mentioned below:

1. All processes except for pervaporation involve no phase change.
2. All processes except (electro) dialysis are pressure driven.
3. Microfltration and ultrafiltration is widely used in food and beverage processing (beer microfiltration, apple juice ultrafiltration), biotechnological applications and pharmaceutical industry (antibiotic production, protein purification), water purification and wastewater treatment, the microelectronics industry, and others.
4. Nanofiltration and reverse osmosis membranes are mainly used for water purification purposes.
5. Dense membranes are utilized for gas separations (removal of CO_2 from natural gas, separating N_2 from air, organic vapor removal from air or a nitrogen stream) and sometimes in membrane distillation. The later process helps in the separation of azeotropic compositions reducing the costs of distillation processes.

Pore Size and Selectivity [1]

The degree of selectivity of a membrane depends on the membrane pore size. Depending on the pore size, they can be classified as microfiltration (MF), ultrafiltration (UF), nanofiltration (NF) and reverse osmosis (RO) membranes. The pore sizes of technical membranes are specified differently depending on the manufacturer. One common distinction is **by nominal pore size**. It describes the maximum pore size distribution [4] and gives only vague information about the retention capacity of a membrane. The exclusion limit or "cut-off" of the membrane is usually specified in the form of NMWC(nominal molecular weight cut-off, or MWCO, molecular weight cut off, with units in Dalton). It is defined as the minimum molecular weight of a globular molecule that is retained to 90 per cent by the membrane. The cut-off, depending on the method, can by converted to so-called D_{90}, which is then expressed in a metric unit. In practice the MWCO of the membrane should be at least 20 per cent lower than the molecular weight of the molecule that is to be separated.

Generally speaking, the pore sizes (or MWCO-molecular weight cut-off in cases of NF and RO) of membranes decrease in the order from MF to RO. Operational trans membrane pressure (TMP) values increase from MF to RO. For example, in MF applications the applied TMP is rarely higher than 3 bar. While those for UF are normally in range of 3-7 bar and for NF 10-30 bar. RO membranes, theoretically, allow only water to permeate. This means that RO processes works against chemical potential difference, namely osmotic pressure. For this reason, the operational TMP applied in RO is normally much higher (*e.g.*, 10-75 bar) than in other pressure-driven separation processes [1e].

Classes of Filter Membrane [1]

Filter membranes are divided into four classes according to pore size:

Sl. No.	*Pore Size*	*Molecular Mass*	*Process*	*Filtration*	*Removal of*
1	> 10	-	"Classic" filter	-	:
2	> 0.1 μm	> 5000 kDa	microfiltration	< 2 bar	larger bacteria, yeast, particles.
3	100-2 nm	5-5000 kDa	ultrafiltration	1-10 bar	bacteria, macro-molecule, proteins, larger viruses.
4	2-1 nm	0.1-5 kDa	nanofiltration	3-20 bar	viruses, 2-valent ions [5].
5	< 1 nm	< 100 Da	reverse osmosis	10-80 bar	salts, small organic molecules

The form and shape of the membrane pores are highly dependent on the manufacturing process and are often difficult to specify. Therefore, for characterization, test filtrations are carried out and the pore diameter refers to the diameter of the smallest particles which could not pass through the membrane.

Rejection and Pore Size Determination [1]

The rejection can be determined in various ways and provides an indirect measurement of the pore size.

1. One possibility is the filtration of macromolecules (often dextran, polyethylene glycol or albumin), another is measurement of the cut-off by gel permeation chromatography. These methods are used mainly to measure membranes for ultrafiltration applications.
2. Another testing method is the filtration of particles with defined size and their measurement with a **particle sizer or by laser induced breakdown spectroscopy (LIBS).**

3. A vivid characterization is to measure the rejection of dextran blue or other colored molecules.
4. The retention of bacteriophage and bacteria, the so-called "**bacteria challenge test**", can also provide information about the pore size.

Sl. No.	*Nominal Pore Size*	*Micro-organism*	*ATCC Root Number*
1	0.1 μm	*Acholeplasma laidlawii*	23206
2	0.3 μm	*Bacillus subtilis spores*	82
3	0.5 μm	*Pseudomonas diminuta*	19146
4	0.45 μm	*Serratia marcescens*	14756
5	0.65 μm	*Lactobacillus brevis*	:

6. Physical Methods

To determine the pore diameter, physical methods such as **porosimetry** (mercury, liquid-liquid porosimetry and **Bubble Point Test**) are also used, but a certain form of the pores (such as cylindrical or concatenated spherical holes) is assumed. Such methods are used for membranes whose pore geometry does not match the ideal, and we get "nominal" pore diameter, which characterizes the membrane, but does not necessarily reflect its actual filtration behavior and selectivity.

Selection of Membrane [1]

In addition to the pore size, the selectivity is highly dependent on the followings:

1. Sparation process
2. The composition of the membrane and its electrochemical properties.

With high selectivity, isotopes can be enriched (Uranium enrichment) in nuclear engineering or industrial gases like nitrogen can be recovered (gas separation). Ideally, even racemics can be enriched with a suitable membrane.

When choosing membranes selectivity has priority over a high permeability, as low flows can easily be offset by increasing the filter surface with a modular structure. In gas phase filtration different deposition mechanisms are operative, so that particles having sizes below the pore size of the membrane can be retained as well.

Membrane Performance Improvement [1e]

The following conditions are helpful in minimizing the membrane fouling and improving the membrane performance :

1. Operational conditions should be optimized. It is not always true that increasing TMP results in increased permeate flux.
2. The 'best' filtering conditions are the settings at which all sections (along the length) of the membrane work at the critical flux, the flux at which fouling starts to occur.

3. The uniform trans membrane pressure (UTMP) configuration for tubular modules (Patented by Alfa-Laval AB in Sweden) is a 'breakthrough.

There are **several techniques which can be applied to increase the membrane separation performance** (improve permeate flux or prevent fast drop of permeate flux and maintain membrane selectivity). The details of some of the these techniques are given below [1e]:

1. Permeate Backflow Techniques

The backflow techniques help lengthen the filtration time before the permeate flux decreases to an unaccepted level and as such they reduce the frequency of chemical washings. These include the followings:

a. **Backwashing [2, 11]-** Backwashing involves using a third solution (*e.g.*, water) as a backflow while others use the filtrate of the process.
b. **Backflushing [12, 13].**
c. **Backpulsing [14].**
d. **Backshock [15, 16].**

An extra pump (hence require extra energy, more complicate control system) in the permeate pipe is needed to convert the TMP periodically during the operation and by that the backflow of the permeate lifts up the foulants which are then swept away by the crossflow of the retentate. The main differences of these methods are the applied frequency (*e.g.*, backflow is applied periodically after a set time period during filtering), pressure discrepancy, and length of time for the permeate backflow.

2. Use of Dynamic Membranes [17] or Application of External Forces

The use of dynamic membranes [17] or application of external forces like electric field and ultrasound, application of Gas/air sparging [18] are other techniques for enhancement of membrane separation performance.

3. The use of spacers (in organic polymeric membranes) and turbulence promoters (in tubular ceramic membranes to create turbulent flow at the surface of the membrane or to increase shear) also prove their efficiency in improvement of membrane separation.

Advantages and Disadvantages of Membrane Technology [19]

Advantages

The followings are well-known advantages gained from the application of membranes:

1. Lower energy consumption
2. No thermal stress acting on the feed constituents
3. Recovery of fresh water
4. Reduced waste treatment costs
5. Relatively low space and capital requirements,

Disadvantages

A number of disadvantages and problems associated with the application of membrane are as follows:

1. The restricted membrane operating life
2. The attack of chemicals and pH changes on the membrane material
3. The fouling and scaling phenomena
4. The limited range of operating pressure
5. The product safety as well the need for additional permission documents.

Factors Responsible for Increasing Trend in Applications of Membrane Technologies

The details related to the factors/reasons for aforesaid trend are given below [1d]:

1. **Green Technology**: Membrane separation processes are green technology due to their higher efficiency in energy usage and the winging away from using chemicals and additives, which is both better for environment and human health.
2. **Economic production**: Membrane separation processes can be used as alternatives to conventional processing methods in a way that the formers assist more economic production and better quality products in terms of both technological functionalities and nutrition value.
3. **Recovery of Valuable components**: These technologies create the possibilities to recover valuable components in diluted effluents, by-products, wastewater and bring more benefits to the producers.
4. **Wastewater treatment**: Membrane separations are a very good technology for, especially permeated water from certain membrane filtrations can be reused in production activities
5. **Products with favorable characteristics**: The membrane techniques can create products, ingredients with favorable characteristics that conventional techniques are unable to offer.

Sources

1. *Wikipedia.*

1a. *Pavan Kumar, Neelesh Sharma, Rajeev Ranjan, Sunil Kumar, Z. F. Bhat, and Dong Kee Jeong, Perspective of Membrane Technology in Dairy Industry-A Review, Asian-Australas J Anim Sci., 2013, Sep; 26(9): 1347-1358.*

1b. *Winston Ho WS, Sirkar KK. In: Membrane Handbook. Winston Ho WS, Sirkar KK, editors. Van Nostrand Reinhold; 1992. pp. 3-16.*

1c. *Osada, Y. and Nakagawa, T., Membrane Science and Technology, New York: Marcel Dekker, Inc, 1992.*

1d. *Saxena A, Tripathi BP, Kumar M, Shahi VK. Membrane-based techniques for the separation and purification of proteins: An overview. Adv Colloid Interface Sci. 2009; 145: 1-22.*

1e. *Dhinesh Kumar, V and Ramasamy, D, Review on Membrane Technology Applications in Food and Dairy Processing, Journal of Applied Biotechnology and Bioengineering, Volume 3, Issue 5, 2017.*

1f. *Cheryan, M 1998, Ultrafiltration and Microfiltration Handbook. Lancaster, PA.: echonomic Publishing Co., Inc.*

2. *Hong S, Krishna P, Hobbs C, Kim D, Cho J, 2005, Variations in backwash efficiency during colloidal filtration of hollow-fiber microfiltration membranes. Desalination 173(3): 257-268.*

2a. *Fillaudeau L, Lalande M, 1998, A practical method to predict steady-state flux and fouling in the crossflow microfiltration of rough beer with 1.40 [mu]m Tubular Ceramic Membranes. Food and Bioproducts Processing 76(4): 217-223.*

2b. *Zeman, Leos J., ydney, Andrew L., Microfiltration and Ultrafitration, Principles and Applications, New York: Marcel Dekker, Inc, 1996.*

2c. *Dewettinck K, Le TT, 2011, Membrane separations in food processing. In: Proctor A (Ed.), Alternatives to conventional food processing. RSC, Royal Society of Chemistry, UK, pp. 184-253.*

2d. *Lamminen MO, Walker HW, Weavers LK, 2006, Effect of fouling conditions and cake layer structure on the ultrasonic cleaning of ceramic membranes. Separation Science and Technology 41(16): 3569-3584.*

2e. *Lee DN, Merson RL, 1975, Examination of cottage cheese whey proteins by scanning electron microscopy: relationship to membrane fouling during ultrafiltration. J Dairy Sci 58(10): 1423-1432.*

3. *Mulder M., Basic Principles of Membrane Technology, Kluwer Academic Publishers, Netherlands, 1996.*

4. *Jornitz, Maik W., Sterile Filtration, Springer, Germany, 2006.*

5. *Van Reis R., Zydney A. Bioprocess membrane technology. J Mem Sci. 297, 2007: 16-50.*

6. *Templin T., Johnston D., Singh V., Tumbleson M. E., Belyea R. L. Rausch K. D. Membrane separation of solids from corn processing streams. Biores Tech. 97, 2006, 1536-1545.*

7. *Ripperger S., Schulz G. Microporous membranes in biotechnical applications, Bioprocess Eng. 1, 1986, 43-49.*

8. *Thomas Melin, Robert Rautenbach, Membran-verfahren, Springer, Germany, 2007, ISBN 3-540-00071-2.*

9. *Munir Cheryan, Handbuch Ultrafiltration, Behr, 1990, ISBN 3-925673-87-3.*

10. *Eberhard Staude, Membranen und Membranprozesse, VCH, 1992, ISBN 3-527-28041-3.*

11. *Katsoufidou K, Yiantsios SG, Karabelas AJ, 2007, Experimental study of ultrafiltration membrane fouling by sodium alginate and flux recovery by back-washing. Journal of Membrane Science 300, 137- 146.*

12. *Kim HG, Park C, Yang J, Lee B, Kim SS, et al., 2007, Optimization of back-flushing conditions for ceramic ultrafiltration membrane of disperse dye solutions. Desalination 202: 150-155.*

13. *Kuberkar VT, Davis RH, 2001, Microfiltration of protein-cell mixtures with crossflushing or backflushing. Journal of Membrane Science 183: 1-14.*

14. *Wen E, Cinelli LD, Murray D, Lander RJ, Sagar SL, et al., 2005, Purification of a poly-saccharide conjugate vaccine using microfiltration membranes in backpulsing mode. Journal of Membrane Science 258(1-2): 23-34.*

15. *Guerra A, Jonsson G, Rasmussen A, Nielsen EW, Edelsten D, 1997, Low cross-flow velocity micro-filtration of skim milk for removal of bacterial spores. International Dairy Journal 7(12): 849-861.*

16. *Gan Q, 2001, Beer clarification by cross-flow microfiltration-effect of surface hydrodynamics and reversed membrane morphology. Chemical Engineering and Processing 40: 413-419.*

17. *Jaffrin MY, 2008, Dynamic shear-enhanced membrane filtration: A review of rotating disks, rotating membranes and vibrating systems. Journal of Membrane Science 324: 7-25.*

18. *Li ZY, Kittikun AH, Youravong W, 2008, Separation of protease from yellowfin tuna spleen extract by ultrafiltration: Effect of hydrodynamics and gas sparging on flux enhancement and selectivity. Journal of Membrane Science 311(1-2): 104-1113*

19. *Josef Robert, Rolf Kuemmel, and Hans Fahlenkamp, Application of Membrane Technologies in Dairy Industries, Proc. Intern'l Conf. on Wastewater Treatment and Reuse Adapted to Mediterranean Area, 25-28 October 2000, Tunisia (WATRAMA'2000), p.272-275.*

Applications of Membrane Technologies in Food/Dairy Industries

The use of membrane technology in food and beverage industries has become increasingly important. These applications range from well-established pressure-driven processes for concentration and fractionation of dairy products, clarification of fruit juice, wine and vinegar, degumming of vegetable oils to emerging technologies such as forward osmosis and membrane distillation. Some applications of Membrane technologies in Food/Dairy Industries are given below:

A. Applications in Dairy Industry [1]

Milk, by nature, is a complicated and poly-dispersed system with suspended particles and soluble molecules. The largest natural particles of milk are fat globules and the following are casein micelles with average diameter of 0.4µm. As such, the use of any membrane types for processing of milk or dairy processing-derived streams becomes possible [1d].

- ✰ Membrane technology has been applied in the dairy industry since the **early 1960s** and now comes the second behind the water treatment technology and serves as viable alternatives for more traditional dairy processes like distillation, evaporation or extraction [1a].
- ✰ In the modern dairy processing, membranes play a major role in the followings- **i-**clarification of the milk, **ii-** increase the concentration of the selected components and **iii-** separation of the specific valuable components from milk or dairy by-products *etc.* [1a].
- ✰ The dairy processing industry is known for high energy and water consumption, in part due to product diversification. The energy consumption in some dairy companies has been significantly reduced through the coupling of heat and power generation. Although the consumption of water and the generation of wastewater has been reduced over the past years, it has not been decreased sufficiently [1]. The **membrane technology improves the economics of dairy by reducing the cost of production as well generating new revenue resources** [1b].
- ✰ Membrane technology proves a suitable and economical alternative to many important processing stage of milk in dairy industry such as centrifugation, bactofugation, evaporation, demineralization of whey, *etc.* [1c].
- ✰ The following membrane processes are being useds in dairy industry for different works [1]:
  - i. **Reverse osmosis (RO)**: Water recycling.
  - ii. **Nanofiltration (NF)**: Whey treatment, Demineralization, Fractionation of milk and whey.
  - iii. **Ultrafiltration (UF)**: Whey treatment, Fractionation of milk and whey.
  - iv. **Microfiltration (MF)**: Bacteria removal.
  - v. **Electrodialysis (ED)**: Fractionation of milk and whey.
  - vi. **Membrane reactions (MR).**
- ✰ The details of some applications of Membrane Technology in dairy are given below:

1. Lactose Separation/Whey Processing [1, 1a]

Whey is a dairy by-product which is obtained during the preparation of milk products *viz.* cheese, paneer and casein. By application of different membrane filtration technology, the nutrients in whey are concentrated, fractionalized or purified into valuable products such as whey protein concentrate/isolates, α-lactalbumin, β-lactoglobulin, lactose and salts [1a].

The membrane separation technology can be successfully applied to the separation of the disaccharide lactose from process waters/whey of the dairy industry. The comparative status of conventional and new innovative treatment process is tabulated on next page [1, 3].

<table>
<tr><th>Conventional Process</th><th>Innovative Treatment Process</th></tr>
<tr><td>1-Processing Line-This starts from aqueous whey, which after pasteurization and separation passes a UF unit where a high-molecular protein concentrate is separated. This concentrate is then transformed by successive evaporation and flash drying steps into protein powder. The UF permeate which contains approximately 50 g/l lactose is further concentrated in a multistage vaporizer and finally transferred to a crystallization vessel where solid lactose is caused to precipitate by spontaneous crystallization or addition of seeding crystals. Separation of the crystals, conveyor drying and, if necessary, recrystallization to achieve pharmaceutical quality are the final steps of the process.</td><td>It has been designed which differs from the conventional one by introducing a nano-filtration unit capable of improving the separation of organic and inorganic constituents. NF membranes operating at relatively high pressures show rejection rates of clearly more than 99 per cent with respect to lactose while lactic acid and citric acid as well as mono-mono-valent electrolytes are allowed to permeate through the membranes. At the same time, due to an extremely extended induction period prior to crystallization, the lactose concentration can be easily increased up to a degree of super-saturation of more than 200 per cent without immediate crystallization. In a special crystallizer, the addition of seeding crystals initiates the mass crystallization of lactose crystals which can be separated and collected by a decanter system.</td></tr>
<tr><td colspan="2">Advantages of NF
1. The thermal stress acting on the organic constituents is much lower since maximum temperatures of only 40 °C are applied. Therefore, Maillard-type decomposition reactions leading to high-molecular and colored substances can be completely prevented,
2. The product purity is high, whereas electrolyte contamination is considerably lower.
3. The energy requirement reduces by about 50 per cent, the yield of lactose increases from 70 to 93 per cent
4. The amount of COD released with the process water reduces to nearly a third of the original values.</td></tr>
</table>

Benefits

1. About 2/3 of the membrane area installed in the dairy industry is used for the treatment of whey and about 1/3 for milk [5].
2. Up to 60 per cent more saving on fuel has been reported in whey concentration by applying RO over the traditional evaporation methods.
3. By application of UF and DF, the protein content of the whey protein concentrate can be increased by 35 to 85 per cent of the total solids.
4. by removing bacteria and fat by passing whey through MF, the protein content of whey protein isolates can be increased to 90 per cent of the total solid content [1a, 11].

2. Wastewater Management

The wastewater of a composite dairy manufacturing different dairy products *e.g.* Butter, Paneer, Cheeses *etc.*, contains high concentrations of organic contents and because of this it has BOD and COD approx. 32,000 mg/l and approx. 60,000 mg/l respectively. Before discharging it to river or pond or agricultural land, it was treated by conventional effluent treatment plant (ETP) which is cost intensive.

It was studied that by means of innovative combination of different membrane processes the load of harmful substances in the water can be reduced significantly (COD reduction by 50 to 70 per cent). In conventional processes, approximately one ton of steam is needed for the precipitation/extraction of proteins out of ten cubic meters of whey. As the amount of whey is reduced by the factor of 20, steam/heating oil, can be saved per day through the utilization of membrane technology [1, 2, 4].

3. Application in Extending Shelf Life of Milk [1a]

Extended shelf life (ESL) milk products are products that have been treated in a manner to reduce the microbial count beyond normal pasteurization, packaged under extreme hygienic conditions, and have a prolonged shelf life under refrigerated conditions [6]. The **MF** may be used to reduce the microbial load in liquid milk and increase the shelf life without any changes in its composition and sensorial qualities [6a, 6b]. Using modified membrane structures, microbial load can be reduced significantly without affecting the milk composition [6c].

1. ESL milk has been prepared by removing bacteria from milk by MF without causing any compositional change or negligible decrease in the total protein by 0.02 to 0.03 per cent [7].
2. A thermal treatment or combination of heat treatment and membrane filtration are involved in the production of ESL milk [8]. It was reported the extension of shelf life of refrigerated milk from 12 d to 18 d at 8°C by the combined application of MF and HHT without compromising sensory attributes by reducing the number of psychrotrops [9].
3. The somatic cell count can be reduced up to 100 per cent by the application of combined processes of MF (microfiltration) followed by HHT (high heat treatment) [10].
4. A study was conducted [6d], where process was developed for extended shelf life (ESL) of milk for 33 days, using a combination of microfiltration and thermal treatment. There is least change in the main composition of the ESL milk compared to raw untreated milk. For microfiltration processing, low fat milk is more preferred, though slight change in protein, calcium and lactose was observed after the treatment [6c]
5. **Membrane CF MF** offers an efficient alternative to heat treatment for removal of bacteria from liquid food. Actually, removal of bacteria and spores from skim milk is one of the most important applications of MF in dairy industry. This technique is referred as **cold pasteurization [1d]**. Based on total microbial counts, the shelf life of this micro-filtered pasteurized milk when stored at 4.2 C was beyond 92 days but the real shelf life was limited to 42 days due to proteolysis [1d]. This is still much longer than the shelf life of normal pasteurized milk, which is about a week at refrigeration condition. High temperature, short-time (HTST) treated milk can last only up to 14 days due to spoilage bacteria. Ultra-pasteurization can extend the shelf life to 45 days under refrigeration condition, however, this heat treatment already exert distinct cooked flavor.

6. Microfiltration is a decisive step in the **milk sterilization process**. This method mechanically eliminates a good part of the bacteria in milk but the morphology of micro-organisms limits its efficiency. Combined with thermal treatment, micro-filtration extends the shelf life and ensures the safety of milk. However enzymatic reactions can limit the shelf life of microfiltered and pasteurized milk. These occur when raw milk is of lesser microbial and enzymatic quality [1f].

4. Application in Cheese Industry [1a]

Membrane technology has successfully used in cheese industry for better compositional control, improving the nutritive quality and yield of cheese by increasing the total solid content, utilization of whey during preparation of cheese, comparatively reducing the requirement of rennet and starter culture.

- ☆ UF can be used to standardize/concentrate milk for continuous production of cheeses. Not like MF, UF retains both caseins and whey proteins and that it increases cheese yield. It also reduces total processing time due to shorter clotting/coagulation time and elimination of several processing steps (such as curd cutting and whey drainage); and reduces the use of salt, starter culture, rennet and calcium chloride [1d].
- ☆ UF can be applied for production of cottage cheese, soft cheese varieties, some hard cheeses like Cheddar and Parmesan; the types of cheeses in which the presence of high concentrations of whey proteins does not create an issue in term of sensorial properties [1d].
- ☆ The UF and MF are most commonly used in the cheese industry. The details related to rthe application of UF and MF in cheese industry are tabulated below [1a]:

Sl. No.	*Ultrafiltration(UF)*	*Microfiltration(MF)*
1	It is highly remunerative by concentrating milk by a factor of 1.2 to 2.0 times [14]and increase in the casein: protein ratio [15] leading to fewer requirements of processing equipments, better compositional and quality control as well as improving the yield of cheese.	The MF casein concentrated milk is very suitable for various cheese preparations due to its superior microbial quality achieved after removing bacteria and spores from milk, as well as optimization of the different major milk components.
2	After concentration of milk by membrane filtration technology, it is possible to make cheese without whey, thus do not require the use of cheese vat and avoid cumbersome processing steps of removal and draining of whey.	UHT milk need higher amount of rennet for coagulation. By application of UF, UHT milk is coagulated with lesser amount of rennet by compensating the hindrance to coagulation arising from the formation of complexes between k-casein and β-lactoglobulin by reduction of the zeta potential of the casein micelle [17].
3	UF is called as complementary process of cheese making [11].	The MF has future possibilities in standardization of protein in cheese milk and fortification with casein micelle powder [17].

Sl. No.	Ultrafiltration(UF)	Microfiltration(MF)
4	UF have not given good results in making of fresh cheese from acidified curd as it is difficult to process high acidic milk by these traditional membranes. With the introduction of advanced ceramic and polysulfone membranes, it is possible to apply UF to acidic milk with pH 4.4 to 4.6 in the production of better quality of fresh cheese [16].	The MF followed to HTT (high temperature treatment) is more efficient than single stage bactofugation in terms of microbial quality. But this treatment of MF followed by HTT modifies rennet coagulation and increases the water holding capacity, necessitating adjustments in the parameters during cheese preparation [18].

Benefits of UF

Some of the benefits of using UF in cheese-making are given below [1f]:

1. Plant space is used better
2. The Lower amount (some time) of enzyme (rennet) is required
3. There is a reduced volume of milk to handle
4. Fewer cheese-making vats are required
5. There is little or no whey production because most of the water and lactose has already been removed
6. An increase in yield of 10-30 per cent with soft and semi-soft cheeses due to the inclusion of the whey proteins, was observed.

5. Application in Milk Fat Processing [1a]

In addition to the traditionally energy intensive centrifugation, the cream separation from whole milk is also possible by energy saving membranes technology producing skim milk with good storage quality and improved sensory attributes of cream without causing any damage to fat globular membranes. The cream with smaller fat globules have fine texture and improved flavour than cream with large fat globules. Technically, it is possible to use MF to separate fat globules from whole milk instead of using a cream separator [1e]. However according to the available literature, there seems to be no industrial application of MF for this purpose yet [1d].

6. Application in Desalting or Demineralization [1a]

The cheese whey is rich in salt and acids and the reduction or demineralization of whey is essential before its use and to alleviate the environmental hazards. The demineralization is done by **electrodialysis and ion exchange process [19]** in the dairy industry to get up to 60 per cent reduction of minerals [20]. The NF membranes with molecular weight cutoff (MWCO) ranging between 200 to 1,000 Daltons are best suited for demineralization of whey as these membranes are permeable to salt and monovalent ions but impermeable to organic compounds. The NF is more economical than electro-dialysis and a method of choice for partial desalting of whey. NF membranes are highly permeable to water and monovalent ions. The mineral contents of the whey is reduced by 35 per cent and ash content by 3 to 4 times in addition to increase concentration of whey by applying nanofiltration (NF).

7. Application in Milk Protein Processing [1a]

The membrane technologies are being used for **fractionation of milk proteins for making cheeses, caseins and whey proteins and for milk protein standardization.** Membrane technology has revolutionized the casein industry. Application of different membrane types with desired pore size either alone or in combination with enzymatic action, chromatography lead to the concentration, fractionalization and purification of milk proteins [1a].

A. *Microfiltration*

MF membranes with pore size 0.1-0.2m can be used to fractionate skimmed milk into the following fractions [1d]:

i. *Retentate Fraction*

This fraction is enriched with micellar caseins. This casein fraction can be used for vatless production of semi-hard and hard cheeses or for production of casein-enriched ingredients without using acids to precipitate caseins and then other chemicals to neutralize the products [22, 22a, 22b]. The volume of the retentate from the MF is much lower than that of the original milk and that the production of cheese from the retentate requires less quantity of rennet and additives, reduces total production time, increases capacity of the plant. MF production of cheese also allows flexible changing of the casein/whey protein ratio or the casein/fat ratio for desired structural and sensorial properties of final products.

ii. *Sterile Permeate*

It is containing native whey proteins. MF permeate (micro-filtrate), considered as sweet whey, contains the major whey proteins in highly functional (undenatured) state. Compared to classic cheese whey, the microfiltrate contains no rennet, bacteria, and starter culture, contains extremely low lipid content (*i.e.*, <0.01 per cent) and is more consistent in composition among production batches and different companies [22a]. This whey can be directly subjected to UF for manufacture of functionality-enhanced **whey protein concentrate (WPC) and whey proteins isolate (WPI).**

The membrane separation of bacteria and spores also leads to production of high **quality whey protein concentrate (WPC) and whey protein isolates (WPI**) with inherent functional properties from whey [11] as it avoid the denaturation of whey or serum proteins. These whey proteins are used in food industry for their high biological value as well as their ability to improve the functional properties (emulsifying, foaming and gelling) of the food products [12]. With the use of membrane technology, it is possible to concentrate and separate whey proteins in their un-denatured form with high functional property in native whey as compared to traditional sweet cheese whey. **Native whey protein concentrates (NWPC) and native whey protein isolates (NWPI) are produced by UF concentrating the native whey [13]**. At present, these native whey proteins are utilized extensively in human nutrition as an integral component of weight balancing products and in baby foods [1a].

Whey proteins are obtained by concentrating and purification from whey. Whey protein from skimmed milk was separated by polymeric microfiltration in permeate flux [22c]. The **β-lactoglobulin** is separated by the membrane process followed by heat treatment. The β lactoglobulin can be fractionized from defatted whey proteins by using MF and centrifugation followed by purification by UF with electrodialysis (ED) or DF.β-lactoglobulin is purified by diafiltration of supernatant. The **α-lactalbumin** is separated and purified by applying UF on solubilised MF retentate at neutral pH.

B. *Ultrafiltration (UF)*

By application of ultra filtration, the level of protein content in milk is adjusted without adding any extraneous protein source by removing the water. The milk protein concentrate (MPC) containing 50 to 58 per cent proteins of good functionality are prepared by application of MF, UF and diafiltration (DF) technologies either alone or in combinations.

C. *DF*

Upto 90 per cent casein concentration is possible by DF and it can be used for industrial, pharmaceutical as well as edible purposes [21].

8. Reduction in Transport Cost

The transport cost of milk is reduced by membrane technology by pre-concentrating it, removing the water and lactose or dewatering it at farm level and thus decreasing the volume of transported milk without damaging its sensory attributes [23].

Sources

1. *Josef Robert, Rolf Kuemmel, and Hans Fahlenkamp, Application of Membrane Technologies in Dairy Industries, Proc. Intern'l Conf. on Wastewater Treatment and Reuse Adapted to Mediterranean Area, 25-28 October 2000, Tunisia (WATRAMA'2000), p.272-275.*

1a. *Pavan Kumar, Neelesh Sharma, Rajeev Ranjan, Sunil Kumar, Z. F. Bhat, and Dong Kee Jeong, Perspective of Membrane Technology in Dairy Industry-A Review, Asian-Australas J Anim Sci. 2013 Sep; 26(9): 1347-1358.*

1b. *Siebert JW, Alejandro L, Sung-Yong K. The commercial potential of new dairy products from membrane technology. J Food Distrib Res. 2001; 3: 24-33.*

1c. *Pouliot Y. Membrane processes in dairy technology-From a simple idea to worldwide panacea. Int Dairy J. 2008; 18: 735-740.*

1d. *Dhinesh Kumar, V and Ramasamy, D, Review on Membrane Technology Applications in Food and Dairy Processing, Journal of Applied Biotechnology and Bioengineering, Volume 3 Issue 5 -2017.*

1e. *Glimenius AR, Jansson GT, Kemi KWH, Sandblom RM (1979) Filtering method for separating skim milk from milk products. Alfa-Laval AB, Sweden.*

1f. *M. Cheryan, Membrane Separations, Food Technology, University of Illinois, Urbana, Illinois, USA(Academic Press.*

1g. *Veronika Kaufmann, Microfiltration and Thermal Treatment for Extending the Shelf Life of Milk: Factors Affecting Product Quality and Stability, Munich Technical University, Germany, Session 4, IDF Dairy Science and Technology Week - Quebec 2008.*

2. *Rautenbach, R., Albrecht, R.: Membrantrenn-verfahren. Salleund Sauerländer, Frankfurt, Berlin, München, Aarau, 1981.*

3. *Kümmel, R.; Robert, J.: Application of Membrane Processes in Food Technologies. In: Belafi-Bako et al., Integration of Membrane Processes into Bioconversions, Kluwer Academic Plenum Publishers, New York 2000, p.143-154.*

4. *Scott, K., Handbook of Industrial Membranes. Elsevier Advanced Technology, Oxford 1997.*

5. *Saxena A, Tripathi BP, Kumar M, Shahi VK. Membrane-based techniques for the separation and purification of proteins: An overview. Adv Colloid Interface Sci. 2009, 145: 1-22.*

6. *Rysstad R, Kolstad J. Extended shelf life milk-advances in technology. Int J Dairy Technol. 2006, 59: 85-96.*

6a. *Soumitra Banerjee and Shanker Lal Shrivastava, Recent Trends in Milk Processing, Approches in poultry, dairy and Veterinary sciences 2015, Crimson Publishers, ISSN: 2576-9162.*

6b. *Pafylias I, Cheryan M, Mehaiab MA, Saglam N, 1996, Microfiltration of milk with ceramic membranes. Food Research International 29(2): 141- 146.*

6c. *Hoffmann W, Kiesner C, Clawin R, Martin D, Einhoff K, 2006, Processing of extended shelf life milk using microfiltration. International journal of dairy technology 59(4): 229-235.*

6d. *García LF, Rodríguez FR, 2014, Combination of microfiltration and heat treatment for ESL milk production: Impact on shelf life. Journal of Food Engineering 128: 1-9.*

7. *Hoffmann W, Kiesner C, Clawinradecker I, Martin D, Einhoff K, Lorenzen PC, Meisel H, Hammer P, Suhren G, Teufel P. Processing of extended shelf life milk using microfiltration. Int J Dairy Technol. 2006; 59: 229-235.*

8. *Lorenzen PC, Decker IC, Einhoff K, Hammer P, Hartmann R, Hoffmann W, Martin D, Molkentin J, Walte HG, Devrese M. A survey of the quality of extended shelf life (ESL) milk in relation to HTST and UHT milk. Int J Dairy Technol. 2011, 64: 166-178.*

9. *Damerow G. Die anwendung der mikrofiltration fur die konsummilch, kessel-milch, Molke. Dtsch Molkerei Zeitung. 1989, 110: 1602-1608.*

10. *Pedersen PJ. New applications of membrane processes. Brussels, Belgium: 1992. pp. 33-50. IDF Spl Iss-9201.*

11. *Lipnizki F. Cross-flow membrane applications in the food industry. In: Peinemann Klaus-Viktor, Nunes Suzana Pereira, Giorno Lidietta, editors. Membrane Technology,*

Vol.3: Membranes for Food Applications, WILEY-VCH Verlag GmbH and Co. KGaA; Weinheim: 2010.

12. *Harper WJ. New applications of membrane processes- In IDF Spl. Iss. 9201. Brussels, Belgium: 1992, pp. 77-108.*
13. *Maubois JL, Fauquant J, Famelart MH, Caussin F. Milk microfiltrate a convenient starting material for fractionation of whey proteins and derivatives. Proc. 3rd Int Whey Conf. The importance of whey and whey components in food and nutrition; München. Germany. 2001, pp. 59-72.*
14. *Rosenberg M. Current and future applications of membrane processes in the dairy industry. Trends Food Sci Technol. 1995, 6: 12-19.*
15. *Guinee TP, Kennedy BT, Kelly PM. Effect of milk protein standardization using different methods on the composition and yields of cheddar cheese. J Dairy Sci. 2006, 89: 468-482.*
16. *Jelen P, Renz-Schauen A. Quarg manufacturing innovations and their effects on quality, nutritive value, and consumer acceptance. Food Technol. 1989, 43: 74-81.*
17. *Mistry VV, Maubois JL. Application of membrane separation technology to cheese production. In: Fox PF, editor, Cheese: Chemistry, Physics and Microbiology. Chapman and Hall; London: 1993, pp. 493-522.*
18. *Pedersen PJ, Ottosen N. New Applications of Membrane Processes. Brussels, Belgium: 1992, pp. 67-76, IDF Spl Iss 9201.*
19. *Greiter M, Novalin S, Wendland M, Kulbe KD, Fischer J. Desalination of whey by electrodialysis and ion exchange resins: analysis of both processes with regard to sustainability by calculating their cumulative energy demand. J Memb Sci. 2002, 210: 91-102.*
20. *Kelly PM, Horton BS, Burling H. New applications of membrane processes. IDF Bulletin 9201. Brussels, Belgium: 1991. Partial demineralization of whey by nanofiltration; pp. 130-140.*
21. *Maubois JL, Ollivier G. New Applications of Membrane Processes. Brussels, Belgium: 1992, pp.15-22. IDF Spl. Iss. 9201.*
22. *Lawrence ND, Kentish S, O'Connor A, Stevens G, Barber A., 2006, Microfiltration of skim milk for casein concentrate manufacture. Desalination 200: 305-306.*

22a. *Britten M, Pouliot Y, 1996, Characterization of whey protein isolate obtained from milk microfiltration permeate. Lait 76(3): 255-265.*

22b. *Espina VS, Jaffrin MY, Ding LH, 2009, Comparison of rotating ceramic membranes and polymeric membranes in fractionation of milk proteins by microfiltration. Desalination 245(2/3): 714-722.*

22c. *Govindasamy-Lucey S, Jaeggi JJ, Johnson ME, Wang T, Lucey JA. Use of cold microfiltration retentates produced with polymeric membranes for standardization of milks for manufacture of pizza cheese. J Dairy Sci. 2007, 90: 4552-4568.*

23. *Rattray W and Jelen P. Freezing point and sensory quality of skim milk as affected by addition of ultrafiltration permeates for protein standardization. Int Dairy J. 1996, 6: 569-579.*

B. Applications in Food Industry [1]

The food industry is one of the largest users of membrane technology. With new developments in low fouling membranes and modules, and membranes stable in organic solvents, the applicability of this technology will widen considerably in the industry, especially in the production of **'neutraceuticals'** and **grain processing**. Food industry applications of membrane technology are tabulated below [1, 2]:

Sl. No.	*Foods*		*Membrane Technology*	*Applications*
1	Oilseeds, cereals, legumes	Soybean processing	UF.	Protein concentrates and isolates.
			CMR.	Protein hydrolysates.
			UF, NF.	Oil degumming and refining.
			UF, RO.	Recovery of soy whey proteins.
		Corn refining	MF, UF, RO.	Steepwater clarification and concentration.
			RO.	Light-middlings treatment- water recycle.
			CMR.	Saccharification of liquefied starch.
			MF, UF.	Purification of dextrose streams.
			CMR.	Fermentation of glucose to ethanol.
			MF, UF, NF, RO, ED, PV.	Downstream processing of fermentation broths.
			MF, UF, NF, RO.	Wastewater treatment.
2	Fruits and vegetables	Juices- Apple	UF,RO.	-
		Apricot, citrus	MF, UF, RO, ED.	-
		Cranberry, grape	UF, RO.	-
		Kiwi, peach	UF, RO.	-
		Pear, pineapple	MF, UF, RO.	-
		Vegetables- Tomato	RO.	-
		Pigments-Anthocyanins, Betanins.	UF, RO.	-
		Wastewater- Apple, pine apple, potato	UF, RO.	-
3	Beverages	Wine, beer, vinegar	MF, UF.	Clarification
		Low-alcohol beer	RO	-
4	Sugar	Beet and cane extracts, maple syrup, candy wastewaters	MF, UF	Clarification
			ED	Desalting
			RO	Preconcentration

Sl. No.	Foods		Membrane Technology	Applications
5	Animal products	Gelatin:	UF	concentration and de-ashing
		Eggs:	UF, RO.	concentration and reduction of glucose
		Animal by-products	UF	blood, wastewater treatment

1. Processing of Fruit Juices [1]

There are following three areas where membranes can be applied in the processing of fruit juices:

a. Clarification [1]

The process of fruit juice preparation is usually labour- and-time-consuming. Membrane filtration can replace the holding, filtration and decantation steps. The largest application is apple juice but the following have also received considerable attention:

- ☆ Apricot, Carrot, Cherry, Cranberry (which was one of the earliest applications of ceramic membranes), Blackcurrant, Grape, Guava, Kiwi, Lemon, Lime, Maple Sap, Melon, Orange, Passion fruit, Peach, Pear, Pineapple, Plum, Raspberry, Strawberry and Tomato.
- ☆ Citrus juices (grapefruit, orange, lemon, lime) are being upgraded by combining UF and adsorbent resin technology to remove bitter compounds such as limonin, naringin, hesperidin, polyphenols and many other off-Savour compounds.

b. Concentration [1, 3]

RO with the appropriate polyamide composite membrane can concentrate juice without a significant loss of aroma, sugar or acids. Due to limitation, RO can be used as a preconcentration step. The RO is used for production of fruit juice concentrates of greater than 42 Brix. The Fresh Note process by Du Pont and FMC allowed the production of highly concentrated (42-70Brix) fruit juices using a combination of high and low retention RO membranes. UF is first used to separate the pulpy solids from the serum. Concentration of tomato juice presents a difficult problem, because it has a high pulp content (25 per cent Fibre) and a high viscosity (which behaves in a non- Newtonian manner). Because of this, tubular modules are probably best. The colour of the Final tomato concentrate is very good, and shows none of the browning normally associated with evaporation.

c. De-acidification [1]

The electrodialysis or nanofiltration are being used to reduce the acidity in citrus juices.

2. Vegetable Oils Processing [1]

In place of physical or chemical refining, it is possible to use UF membranes for the degumming step, thereby producing a substantially oil-free lecithin. In membrane refining, the crude oil is treated with a solvent such as methanol to extract the free fatty acids. After phase separation, the methanol layer is subjected to nanofiltration to recycle the methanol while producing a free fatty acid concentrate. Dewaxing with microfiltration membranes can also be done. The application of membrane technology in the edible oil industry is expected to have following advantages:

1. Reduce energy consumption
2. Reduce losses of oil
3. Reduce the usage of chemicals and water
4. Reduce the discharge of contaminated effluents.

3. Application in Sugar Refining [1]

The membranes are used for clarification and purification of the extracted juices.

- ☆ By using the UF or MF to remove the colloidal and macromolecular impurities, a clear decolorized thin juice can be obtained with little or no need for addition of lime, carbon dioxide or sulfite. The lime could be eliminated completely, if ion exchange is done immediately after UF.
- ☆ Membrane technology reduces macromolecules and reduced lime levels and hence fouling and scaling of the evaporator is reduced, which in turn reduces down time and cleaning costs. Higher yield of sugar and better crystallization are also possible. A high quality soft cane molasses is obtained and this can go directly to chromatographic separation to recover sucrose and fructose.
- ☆ UF or MF is used in sugar industry to clarify thick juice, reducing bacterial counts and storage losses.

4. Vegetable Proteins Processing [1]

UF has been successfully used to upgrade the quality of the soy protein(obtained from Soyabean after removing the oil) by selectively removing undesirable components such as oligosaccharides (implicated in gastrointestinal stress when consuming soya beans) and phytic acid (which forms insoluble chelates with minerals and can form complexes with proteins that reduce their bioavailability). The UF technique usually results in higher yields because of the inclusion of soy whey proteins that are normally lost in conventional manufacturing methods.

Sources

1. *M. Cheryan, Membrane Separations, Food Technology, University of Illinois, Urbana, Illinois, USA(Academic Press).*
2. *Cheryan, M. and Alvarez, J. 1995, Membranes in food processing, Membrane Separations. Technology, Principles and Applications, p. 415. Amsterdam: Elsevier.*
3. *Cheryan, M., 1992, Concentration of liquid foods by reverse osmosis, Handbook of Food Engineering, p. 393. New York: Marcel Dekker.*

Chapter 3

Ultrafiltration (UF) and Microfiltration (MF)

1. Ultrafiltration (UF) [1]

It (UF) is a variety of membrane filtration in which forces like pressure or concentration gradients (hydrostatic pressure forces) lead to a separation through a semipermeable membrane. Suspended solids and solutes of high molecular weight are retained in the retentate, while water and low molecular weight solutes pass through the membrane in the permeate. Ultrafiltration (UF) is a pressure-driven barrier to suspended solids, bacteria, viruses, endotoxins and other pathogens to produce water with very high purity and low silt density. The pore size of the ultrafiltration membrane is usually 10^3 - 10^6 Daltons. This separation process is used in industry and research for purifying and concentrating macromolecular (10^3- 10^6 Da) solutions, especially protein solutions.

Ultrafiltration is not fundamentally different from reverse osmosis, microfiltration or nanofiltration, except in terms of the size of the molecules it retains. Both of these separate based on size exclusion or particle capture. It is fundamentally different from membrane gas separation, which separate based on different amounts of absorption and different rates of diffusion. Ultrafiltration membranes are defined by the molecular weight cut-off (MWCO) of the membrane used. Ultrafiltration is applied in cross-flow or dead-end mode. A membrane or, more properly, a semi permeable membrane, is a thin layer of material capable of separating substances when a driving force is applied across the membrane.

UF is ideal for the removal of colloids, proteins, bacteria, pyrogens and macromolecules larger than the membrane pore size from water. Once considered a viable technology only for desalination, membrane processes are increasingly employed for removal of bacteria and other micro-organisms, particulate material, and natural organic material, which can impart color, tastes, and odors to the water and react with disinfectants to form disinfection byproducts (DBP) [2].

UF processes are currently preferred over traditional treatment methods for the following reasons [1, 2]:

1. No need for chemicals (coagulants, flocculates, disinfectants, pH adjustment, aside from cleaning)
2. Size-exclusion filtration as opposed to media depth filtration
3. Good and constant quality of the treated water/treated product in terms of particle and microbial removal, regardless of feed quality
4. Process and plant compactness
5. Simple automation
6. Environmently friendly
7. Capable of exceeding regulatory standards of water quality, achieving 90-100 per cent pathogen removal [3].

UF processes are currently limited by the high cost incurred due to membrane fouling and replacement [4]. Additional pretreatment of feed water is required to prevent excessive damage to the membrane units. In many cases, UF is used for pre filtration in reverse osmosis (RO) plants to protect the RO membranes.

Sources

1. *Wikipedia.*
2. *Crystal quest.com.*
3. *Water treatment membrane processes, New York, McGraw Hill. 1996.*
4. *Edwards, David; Donn, Alasdair; Meadowcroft, Charlotte, May 2001, "Membrane solution to a "significant risk" Cryptosporidium groundwater source". Desalination 137 (1-3): 193-198.*

Principles [1]

The basic operating principle of ultrafiltration uses a pressure induced separation of solutes from a solvent through a semi permeable membrane. The relationship between the applied pressure on the solution to be separated and the flux through the membrane is most commonly described by the Darcy equation:

$$J = \frac{TMP}{\mu R_t}$$

where, J is the flux (flow rate per membrane area), TMP is the transmembrane pressure (pressure difference between feed and permeate stream), μ is solvent viscosity and R_t is the total resistance (sum of membrane and fouling resistance).

Membrane Fouling [1]

Concentration Polarization (CP)

When filtration occurs the local concentration of rejected material at the membrane surface increases and can become saturated. In UF, increased ion concentration can develop an osmotic pressure on the feed side of the membrane. This reduces the effective TMP of the system, therefore reducing permeation rate. The increase in concentrated layer at the membrane wall decreases the permeate flux, due to increase in resistance which reduces the driving force for solvent to transport through membrane surface. CP affects almost all the available membrane separation process.

1. In RO, the solutes retained at the membrane layer results in higher osmotic pressure in comparison to the bulk stream concentration. So the higher pressures are required to overcome this osmotic pressure.
2. CP plays a dominant role in ultrafiltration as compared to microfiltration because of the small pore size membrane [2].

It must be noted that concentration polarization differs from fouling as it has no lasting effects on the membrane itself and can be reversed by relieving the TMP. It does however have a significant effect on many types of fouling [3].

Types of Fouling [1]

The following types of fouling are described here:

1. Particulate Deposition

The following models describe the mechanisms of particulate deposition on the membrane surface and in the pores:

a. **Standard blocking**-Macromolecules are uniformly deposited on pore walls.
b. **Complete blocking**-Membrane pore is completely sealed by a macromolecule.
c. **Cake filtration**- Accumulated particles or macromolecules form a fouling layer on the membrane surface, in UF this is also known as a gel layer.
d. **Intermediate blocking**-when macromolecules deposit into pores or onto already blocked pores, contributing to cake formation [4].

2. Scaling

As a result of concentration polarization at the membrane surface, increased ion concentrations may exceed solubility thresholds and precipitate on the membrane surface. These inorganic salt deposits can block pores causing flux decline, membrane degradation and loss of production. The formation of scale is highly dependent on factors affecting both solubility and concentration polarization including pH, temperature, flow velocity and permeation rate [5].

3. Biofouling

Micro-organisms will adhere to the membrane surface forming a gel layer known as biofilm [6]. The film increases the resistance to flow, acting as an additional barrier to permeation. In spiral-wound modules, blockages formed by biofilm can lead to uneven flow distribution and thus increase the effects of concentration polarization [7].

Sources

1. *Wikipedia.*
2. *Brian, P. L., 1965, Concentration polarization in reverse osmosis desalination with variable flux and incomplete salt rejection, Ind. Eng. Chem. Fund. 4: 439-445.*
3. *Rizvi, edited by Anil Kumar Pabby, Ana Maria Sastre, Syed S. H., Pabby, Anil Kumar, Rizvi, Syed S. H.; Sastre, Ana Maria, 2007, Handbook of membrane separations: chemical, pharmaceutical, and bio-technological applications. Boca Raton, Fla.: CRC Press.*
4. *Bruijn, J P F, Salazar, F N, Borquez, R, 2005, "Membrane blocking in ultrafiltration: A new approach to fouling". Food and Bioproducts Processing 83 (3): 211-219.*
5. *Antony, Alice, Low, Jor How, Gray, Stephen; Childress, Amy E., Le-Clech, Pierre, Leslie, Greg, Nov.-2011, "Scale formation and control in high pressure membrane water treatment systems: A review". Journal of Membrane Science 383 (1-2): 1-16.*
6. *Flemming, H. C., Schaule, G., Griebe, T.; Schmitt, J., Tamachkiarowa, A., Nov.-1997, "Biofouling -the Achilles heel of membrane processes". Desalination 113 (2-3): 215-22.*
7. *Baker, J. S., Dudley, L. Y., Sept.-1998, "Biofouling in membrane systems-A review". Desalination 18 (1-3): 81-89.*

Membrane Arrangements [1]

Some information related to membrane arrangement are also discussed in Chapter-2 under Membrane Technology. Depending on the shape and material of the membrane, different modules can be used for ultrafiltration process [4]. Commercially available designs in ultrafiltration modules vary according to the required hydrodynamic and economic constraints as well as the mechanical stability of the system under particular operating pressures [5]. The main modules used in industry include the followings:

1. Tubular Modules

The tubular module design uses polymeric membranes cast on the inside of plastic or porous paper components with diameters typically in the range of 5 - 25 mm with lengths from 0.6 - 6.4 m [2]. Multiple tubes are housed in a PVC or steel shell. The feed of the module is passed through the tubes, accommodating radial transfer of permeate to the shell side. This design allows for easy cleaning however the main drawback is its low permeability, high volume hold-up within the membrane and low packing density [2, 5].

2. Hollow Fibre

This design is conceptually similar to the tubular module with a shell and tube arrangement. A single module can consist of 50 to thousands of hollow fibres and therefore are self-supporting unlike the tubular design. The diameter of each fibre ranges from 0.2 – 3 mm with the feed flowing in the tube and the product permeate collected radially on the outside. The advantage of having self-supporting membranes as is the ease at which it can be cleaned due to its ability to be back flushed. Replacement costs however are high, as one faulty fibre will require the whole bundle to be replaced. Considering the tubes are of small diameter, using this design also makes the system prone to blockage [3].

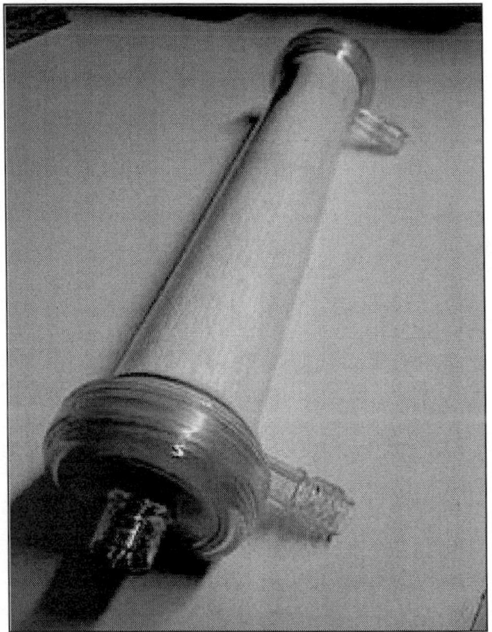

Hollow Fibre Module [1].

3. Spiral-wound Modules

These are composed of a combination of flat membrane sheets separated by a thin meshed spacer material which serves as a porous plastic screen support. These sheets are rolled around a central perforated tube and fitted into a tubular steel pressure vessel casing. The feed solution passes over the membrane surface and the permeate spirals into the central collection tube. Spiral-wound modules are a compact and cheap alternative in ultrafiltration design, offer a high volumetric throughput and can also be easily cleaned [5]. However it is limited by the thin channels where feed solutions with suspended solids can result in partial blockage of the membrane pores [3].

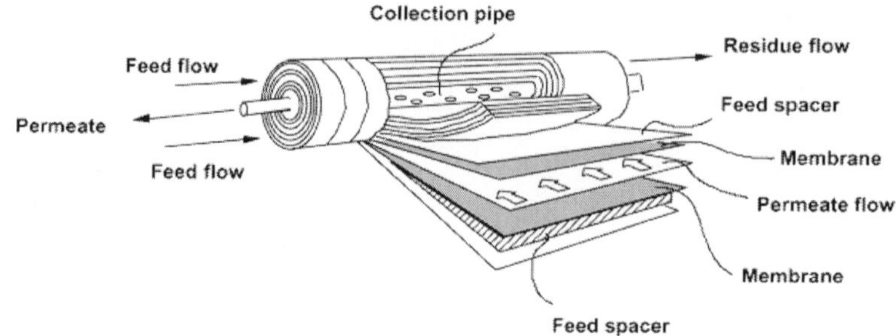

Spiral-Wound Membrane Module [1].

4. Plate and Frame

This uses a membrane placed on a flat plate separated by a mesh like material. The feed is passed through the system from which permeate is separated and collected from the edge of the plate. Channel length can range from 10 - 60 cm and channel heights from 0.5-1 mm [3]. This module provides low volume hold-up, relatively easy replacement of the membrane and the ability to feed viscous solutions because of the low channel height, unique to this particular design [5].

Sources

1. *Wikipedia.*
2. *Tamime, A. Y. Membrane Processing Dairy and Beverage Applications. Chicester: Wiley.*
3. *Cheryan, Munir, 1998, Ultrafiltration and Micro-filtration Handbook. CRC Press.*
4. *Futselaar, Harry; Weijenberg, Dick C. September 1998)."System design for large-scale ultrafiltration applications". Desalination 119 (1-3): 217-224.*
5. *Belfort, Georges (1 February 1988). "Membrane modules: comparison of different configurations using fluid mechanics". Journal of Membrane Science 35 (3): 245-270*

Process Characteristics [1]

The process characteristics of a UF system are highly dependent on the type of membrane used and its application. Manufacturers' specifications of the membrane tend to limit the process to the following typical specifications [2-5]:

Sl. No.	*Parameters*	*Hollow Fibre*	*Spiral-wound*	*Ceramic Tubular*
1	pH	2-13	2-11	3-7
2	Feed Pressure (psi)	9-15	<30-120	60-100
3	Backwash Pressure (psi)	9-15	20-40	10-30
4	Temperature (°C)	5-30	5-45	5-400
5	Total Dissolved Solids (mg/L)	<1000	<600	<500
6	Total Suspended Solids (mg/L)	<500	<450	<300
7	Turbidity (NTU)	<15	<1	<10

Sl. No.	*Parameters*	*Hollow Fibre*	*Spiral-wound*	*Ceramic Tubular*
8	Iron (mg/L)	<5	<5	<5
9	Oils and Greases (mg/L)	<0.1	<0.1	<0.1
10	Solvents, phenols (mg/L)	<0.1	<0.1	<0.1

Sources

1. *Wikipedia.*
2. *"Membrane Products". Koch Membrane Systems. Retrieved 9 October 2013 .*
3. *"Water Treatment Primer for Communities in Need" (PDF). US Department of the Interior Bureau of Reclamation. Retrieved 11 October 2013.*
4. *"Operation and Maintenance Manual-UF-6-HF Ultrafiltration System" (PDF). Con-Serv Manufacturing. Retrieved 10 October 2013.*
5. *Laîné, prepared by Joseph G. Jacangelo, Samer Adham, Jean-Michel, 1997, Membrane filtration for microbial removal. Denver, CO: AWWA Research Foundation and American Water Works Association. ISBN 0898678943.*

Process Design Considerations [1]

There are many factors which must be considered while designing a new membrane separation facility or considering its integration into an existing plant. A composed approach can be applied to determine many of these characteristics to simplify the design process for most applications. Some design areas include the followings:

a. Pre-treatment

Treatment of feed (water, Food *etc.*) prior to the membrane is essential to prevent damage to the membrane and minimize the effects of fouling which greatly reduce the efficiency of the separation. Types of pre-treatment required to be given to feed, are often dependent on the type of feed and its quality. Some examples of pretreatment are given below:

1. In wastewater treatment, household waste and other particulates are screened.
2. Other types of pre-treatment common to many UF processes include pH balancing and coagulation [2, 3].
3. Appropriate sequencing of each pre-treatment phase is crucial in preventing damage to subsequent stages. Pre-treatment can even be employed simply using dosing points.

b. Membrane Specifications

1. **Material-**Most UF membranes use polymer materials (polysulfone, polypropylene, cellulose acetate, polylactic acid). Ceramic membranes are used for high temperature applications.

2. **Pore size-**A general rule for choice of pore size in a UF system is to use a membrane with a pore size one tenth that of the particle size to be separated. This limits the number of smaller particles entering the pores and adsorbing to the pore surface. Instead they block the entrance to the pores allowing simple adjustments of cross-flow velocity to dislodge them [9].

c. Operation Strategy

The details related to the operation strategies (Flow type, flow velocity, flow temperature, pressure, post treatment, cleaning *etc.*) of UF are given below [1]:

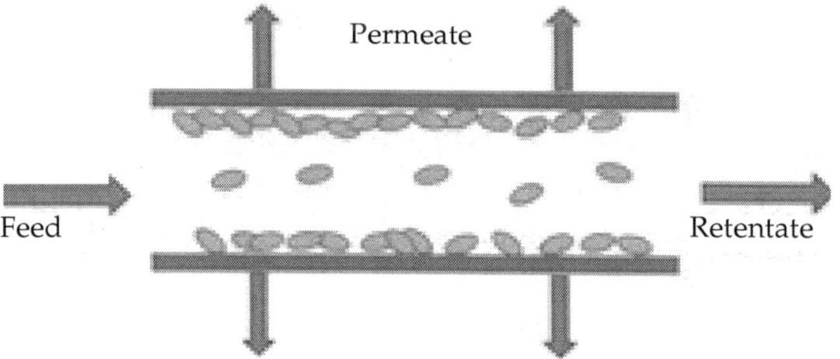

Schematic of cross flow operation.

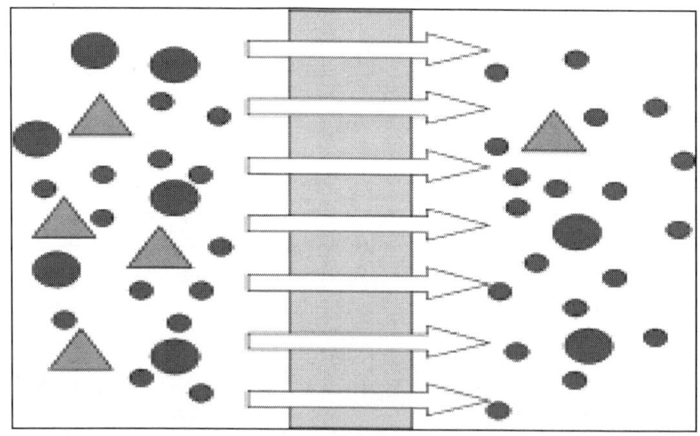

Schematic of Dead-end Operation

i. Flow Type

UF systems can either operate with cross-flow or dead-end flow. In dead-end filtration the flow of the feed solution is perpendicular to the membrane surface. On the other hand in cross flow systems the flow passes parallel to the membrane surface [4]. Dead-end configurations are more suited to batch processes with low suspended solids as solids accumulate at the membrane surface therefore requiring frequent back flushes and cleaning to maintain high flux. Cross-flow configurations are preferred in continuous operations as solids are continuously flushed from the membrane surface resulting in a thinner cake layer and lower resistance to permeation.

ii. ***Flow Velocity***

Flow velocity is especially critical for hard water or liquids containing suspensions in preventing excessive fouling. Higher cross-flow velocities can be used to enhance the sweeping effect across the membrane surface therefore preventing deposition of macromolecules and colloidal material and reducing the effects of concentration polarization. Expensive pumps are however required to achieve these conditions.

iii. ***Flow Temperature***

It is recommended to operate a plant at the temperature specified by the membrane manufacturer so that excessive damage to the membrane can be avoided. In some instances however temperatures beyond the recommended region are required to minimise the effects of fouling [3]. Economic analysis of the process is required to find a compromise between the increased cost of membrane replacement and productivity of the separation.

iv. ***Pressure***

Pressure drops over multi-stage separation can result in a drastic decline in flux performance. This can be improved using booster pumps to increase the TMP in the final stages. This will incur a greater capital and energy cost which will be offset by the improved productivity of the process [3]. With a multi-stage operation, retentate streams from each stage are recycled through the previous stage to improve their separation efficiency.

v. ***Multi-Stage, Multi-Module***

Multiple stages in series can be applied to achieve higher purity permeate streams. Due to the modular nature of membrane processes, multiple modules can be arranged in parallel to treat greater volumes [5].

vi. ***Post-Treatment***

Post-treatment of the product streams is dependent on the followings:

a. Composition of the permeate and retentate
b. Its end-use or government regulation.

In cases such as milk separation both streams (milk and whey) can be collected and made into useful products. Additional drying of the retentate will produce whey powder. In the paper mill industry, the retentate (non-biodegradable organic material) is incinerated to recover energy and permeate (purified water) is discharged into waterways. It is essential for the permeate water to be pH balanced and cooled to avoid thermal pollution of waterways and altering its pH.

j. Cleaning

Extremely fine membrane filters of Ultrafiltration systems need to be properly cleaned. Cleaning of the membrane is done regularly to prevent the accumulation of foulants and reverse the degrading effects of fouling on permeability and selectivity. Regular backwashing is often conducted every 10 min for some processes to remove cake layers formed on the membrane surface [9]. By pressurising the permeate stream and forcing it back through the membrane, accumulated particles can be dislodged, improving the flux of the process. Backwashing is limited in its ability to remove more complex forms of fouling such as biofouling, scaling or adsorption to pore walls [24]. These types of foulants require chemical cleaning to be removed.

The cleaning process used depends on whether a UF system is being used to remove organic or inorganic contaminants, or even both. The common types of chemicals used for cleaning of membranes are given below [6, 7]:

a. Acidic Solutions for the Control of Inorganic Scale Deposits

To remove **inorganic contaminants** the best treatment is with citric acid at a maximum concentration of 3.0 per cent. The acid should circulate for 1 to 3 hours. Hydrochloric acid can also be used.

b. Alkali Solutions for Removal of Organic Compounds

To remove organic contaminants the general cleaning protocol for the cleaning of tubular membranes is to use a low foam, medium alkaline detergent at 0.6 per cent to 1 per cent for a maximum of 40 to 60 minutes.

c. Biocides. When bio-fouling is evident.

During designing a cleaning protocol, following factors are to be considered essentially:

1. **Cleaning time**: Adequate time must be allowed for chemicals to interact with foulants and permeate into the membrane pores. However if the process is extended beyond its optimum duration it can lead to denaturation of the membrane and deposition of removed foulants [6]. The complete cleaning cycle including rinses between stages may take as long as 2 hours to complete [8].
2. **Aggressiveness of chemical treatment**: With a high degree of fouling it may be necessary to employ aggressive cleaning solutions to remove fouling material. This may not be suitable in some applications, if the membrane material is sensitive, leading to enhanced membrane ageing.
3. **Disposal of cleaning effluent**: The release of some chemicals into wastewater systems may be prohibited or regulated therefore this must be considered. For example the use of phosphoric acid may result in high levels of phosphates entering water ways and must be monitored and controlled to prevent eutrophication.

Summary of common types of fouling and their respective chemical treatments [9]

Sl. No.	Foulant	Reagent	Time and Temp.	Mode of Action
1	Fats and oils, proteins, polysaccharides, bacteria	0.5M NaOH with 200 ppm Cl_2	30-60 min 25-55 °C	Hydrolysis and oxidation
2	DNA, mineral salts	0.1M – 0.5M acid (acetic, citric, nitric)	30-60 min 25-35 °C	Solubilization
3	Fats, oils, biopolymers, proteins	0.1 per cent SDS, 0.1 per cent Triton X-100	30 min – overnight 25-55 °C	Wetting, emulsifying, suspending, dispersing
4	Cell fragments, fats, oils, proteins	Enzyme detergents	30 min – over night 30 - 40 °C	Catalytic breakdown
5	DNA	0.5 per cent DNAase	30 min – over night 20 – 40 °C	Enzyme hydrolysis

Sources

1. *Wikipedia.*
2. *Rosehill Recycled Water Scheme-Fairfield Recycled Water Plant"(PDF). Sydney Water.*
3. *Nordin, Anna-Karin; Jönsson, Ann-Sofi (1 Nov. 2006). "Case study of an ultrafiltration plant treating bleach plant effluent from a pulp and paper mill". Desalination 201(1-3): 277-289.*
4. *Farahbakhsh, Khosrow; Adham, Samer S.; Smith, Daniel W. (June 2003). "Monitoring the Integrity of Low-Pressure Membranes". Journal AWWA: 95-107.*
5. *Water treatment membrane processes. New York [u.a.]: McGraw Hill. 1996. ISBN 0070015597.*
6. *Cui, edited by Z. F.; Muralidhara, H. S., 2010, Membrane technology: a practical guide to membrane technology and applications in food and bioprocessing (1st ed.). Amsterdam: Butterworth-Heinemann. pp. 213-254. ISBN 978-1-85617-632-3.*
7. *Gao, Wei; Liang, Heng; Ma, Jun; Han, Mei; Chen, Zhong-lin; Han, Zheng-shuang; Li, Gui-bai (1 May 2011). "Membrane fouling control in ultrafiltration technology for drinking water production: A review". Desalination 272 (1-3): 1-8.*
8. *Wallberg, Ola; Jönsson, Ann-Sofi; Wickström, Peter (1 December 2001). "Membrane cleaning - a case study in a sulphite pulp mill bleach plant". Desalination 141 (3): 259–268.*
9. *Cheryan, Munir, 1998, Ultra-filtration and Microfiltration Handbook. CRC Press,*

New Developments in Membrane and UF [1]

Some information related to the new developments in Membrane and UF Technology are mentioned below:

- In order to increase the life-cycle of membrane filtration systems, energy efficient membranes are being developed in membrane bioreactor systems.

Technology has been introduced which allows the power required to aerate the membrane for cleaning to be reduced whilst still maintaining a high flux level.

- ✰ Mechanical cleaning processes have also been adopted using granulates as an alternative to conventional forms of cleaning, **this reduces energy consumption and also reduces the area required for filtration tanks [2].**
- ✰ Membrane properties have also been enhanced to reduce fouling tendencies by modifying surface properties. This can be noted in the biotechnology industry where membrane surfaces have been altered in order to reduce the amount of protein binding [3].
- ✰ Ultrafiltration modules have also been improved to allow for more membrane for a given area without increasing its risk of fouling by designing more efficient module internals. The current pre-treatment of seawater desulphonation uses ultrafiltration modules that have been designed to withstand high temperatures and pressures whilst occupying a smaller footprint. Each module vessel is self supported and resistant to corrosion and accommodates easy removal and replacement of the module without the cost of replacing the vessel itself [2].

Sources

1. *Wikipedia.*
2. *Bennett, Anthony, November 2012."Membrane technology - Developments in ultrafiltration technologies". Filtration + Separation 49 (6): 28-33.*
3. *Ag, S, September 2012, "Energy-efficient membrane is designed for MBR systems". Membrane Technology 2012 (9): 4.*

Applications of Ultrafiltration [1]

Ultrafiltration is employed in order to recycle flow or add value to later products in chemical and pharmaceutical manufacturing industries, food and beverage processing industries and wastewater treatment. Blood dialysis also utilizes ultrafiltration. The details related to the applications of UF in Milk Protein processing, Lactose separation, Sugar refining, Vegetable protein and oil processing, Cheese manufacture, Fruit juices processing *etc.* are given in Chapter-2. Other important applications are discussed here.

1. Drinking Water

UF can be used for the removal of particulates and macro-molecules from raw water to produce potable water. It has been used to either replace existing secondary (coagulation, flocculation, sedimentation) and tertiary filtration (sand filtration and chlorination) systems employed in water treatment plants or as standalone systems in isolated regions with growing populations [2]. When treating water with high suspended solids, UF is often integrated into the process, utilising primary (screening, flotation, filtration) and some secondary treatments as pre-treatment stages [3].

Drinking Water Treatment 300 m³/h using Ultrafiltration in Grundmühle Waterworks (Germany).

2. Dairy Industry

a. Protein Concentration

UF is used extensively in the dairy industry; particularly in the processing of cheese whey to obtain whey protein concentrate (WPC) and lactose-rich permeate [4, 5]. In a single stage, a UF process is able to concentrate the whey 10-30 times the feed [6]. The original alternative to membrane filtration of whey was using steam heating followed by drum drying or spray drying. The product of these methods had limited applications due to its granulated texture and insolubility. Existing methods also had inconsistent product composition, high capital and operating costs and due to the excessive heat used in drying would often denature some of the proteins [4].

Comparison between Traditional Methods and UF Processes

UF Processors used for this application [4, 6]:

1. Are more energy efficient
2. Have consistent product quality, 35-80 per cent protein product depending on operating conditions
3. Do not denature proteins as they use moderate operating conditions.

Cheese whey contains high concentrations of calcium phosphate which can potentially lead to scale deposits on the membrane surface. As a result substantial pretreatment must be implemented to balance pH and temperature of the feed to maintain solubility of calcium salts [6].

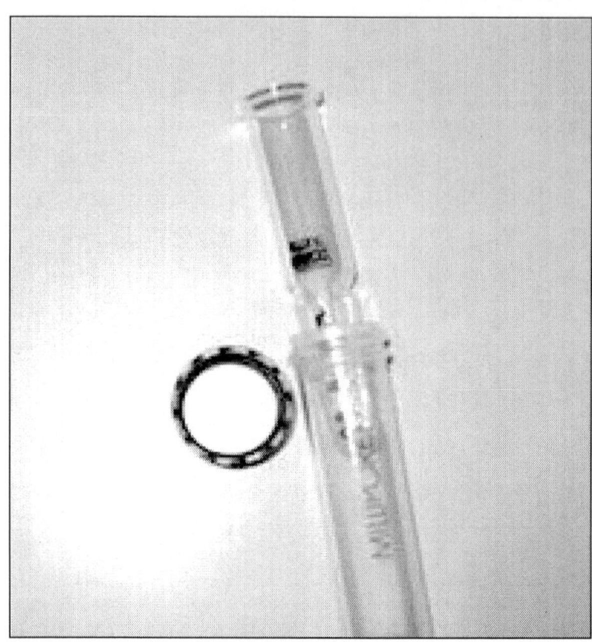

A selectively permeable membrane can be mounted in a centrifuge tube. The buffer is forced through the membrane by centrifugation, leaving the protein in the upper chamber.

b. Electrodialysis with Ultrafiltration Membrane (EDUFM)

EDUFM is a new technique which isolates fractions enriched with acidic or basic bioactive peptides in a very selective manner. EDUFM has been used to concentrate the sequence of β-lg 142-148 which has anti-hypertensive properties. Results of various tests have shown that increasing the effective membrane surface and the strength of the electric field up to a certain threshold makes it possible to quadruple the concentration ratio of the peptidic sequence. However, increasing flow had no effect on the efficiency of the process but modified the selectivity of the peptides [7].

3. Other Applications

UF is also being used for the following purposes:

1. Filtration of effluent from paper pulp mill.
2. Cheese manufacture.
3. Removal of pathogens from milk.
4. Process and wastewater treatment,
5. Enzyme recovery.
6. Fruit juice concentration and clarification [8]- UF is a m**ethod of producing sterile and concentrated juices with improved flavor and reduced acid. UF** is employed to preferentially pass an ultrafiltration (UF) permeate containing flavor and aroma components while retaining spoilage micro-

organisms in a UF retentate. The UF retentate is then treated to inactivate a sufficient number of spoilage micro-organisms to inhibit spoilage of the juice under storage conditions. A UF permeate containing flavor and aroma components is then recombined with the UF retentate to provide a food juice suitable for storage while avoiding the loss or alteration of the desirable flavor and aroma components of fresh juice encountered in conventional juice processing. Further, the UF permeate can be further treated by reverse osmosis to concentrate flavor and aroma components in an RO retentate.

7. Dialysis and other blood treatments,
8. Desalting and solvent-exchange of proteins (via diafiltration),
9. Laboratory grade manufacturing.

Sources

1. *Wikipedia.*
2. *Clever, M.; Jordt, F.; Knauf, R.; Räbiger, N.; Rüdebusch, M.; Hilker-Scheibel, R. (1 December 2000). "Process water production from river water by ultrafiltration and reverse osmosis". Desalination 131 (1-3): 325-336.*
3. *Laîné, J.-M.; Vial, D.; Moulart, Pierre (1 December 2000). "Status after 10 years of operation - overview of UF technology today". Desalination 131 (1-3): 17-25.*
4. *Tamime, A. Y. Membrane Processing Dairy and Beverage Applications. Chicester: Wiley.*
5. *Nigam, Mayank Omprakash; Bansal, Bipan; Chen, Xiao Dong (1 January 2008). "Fouling and cleaning of whey protein concentrate fouled ultrafiltration membranes". Desalination 218 (1-3): 313-322.*
6. *Cheryan, Munir, 1998, Ultrafiltration and Microfiltration Handbook. CRC Press.*
7. *Laurent Bazinet, Electrodialysis with Ultrafiltration Membrane: A New Way to Fractionate Peptides STELA Research Centre/Nutraceutical and Functional Foods Institute (INAF), Laval University, Quebec, Canada, Session-4, IDF Dairy Science and Technology Week - Quebec 2008.*
8. *Ranjan Saha, New and Emerging Non-Thermal Technologies in Food Preservation, April 2015.*

2. Microfiltration (MF) [1]

Membrane microfiltration (MF) is a pressure-driven physical separation process suitable for separation of particles, somatic cells, bacteria and spores from liquid foods. According to its mode of application, MF can be classified into following two systems:

i. Dead-end System

Dead-end is one of the simplest modes, where feed flows perpendicular to the membrane and after some time, the permeate flux decrease, as a result of cake formation due to increase in volume of retained particles over the membrane [2].

ii. Cross-flow System

This is also called **tangential** filtration and in this the feed is pumped parallelly with the membrane and it is possible to recirculate the retentate back to the feed flow. To improve the efficiency of MF filtration, in 1907 Bechhold found an increase in permeate flow when feed flowed parallel to the membrane, and this laid the foundation of cross-flow MF. Feed flows parallel to the membrane surface, and the shear exerted by the feed solution flowing parallel to the membrane surface can sweep the deposited particles towards the retentate side; thereby reducing the formation of the cake layer, and the cake remains relatively thin [4].

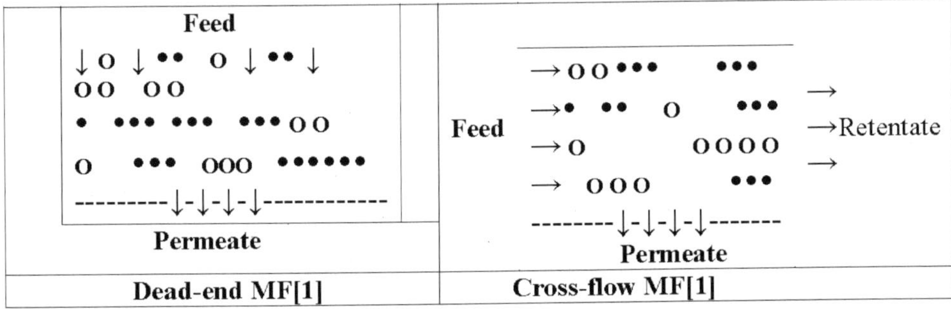

Dead-end MF[1] | Cross-flow MF[1]

It is one of the oldest membrane technologies, the first commercial membrane was available in the 1920s, and was mainly used for bacteriological analysis of water [3].

In MF the membrane has a microporous structure and separation of particles takes place according to particle size exclusion using a semi-permeable membrane. The following basic requirement for the membrane have been described as follows [5]:

1. The membrane should have high mechanical strength with chemical and thermal stability
2. Have high resistivity towards different kinds of chemicals including detergents, acid, base used for cleaning purposes
3. Also resistivity towards microbial action.

In MF, solvent and smaller particles pass through the pores in the membrane due to a pressure gradient, leaving large molecules unfiltered.

1. **Permeate:** The portion passing through the membrane is known as permeate,
2. **Retentate:** while the unfiltered portion is rich in retained macromolecules, usually larger than the pore size and is smaller in volume compared with feed and permeate, thus it is known as retentate [2].

MF processing can be carried out at low temperature to remove bacterial cells, spores and somatic cells, so, cooked flavour cause by thermal treatment can be avoided.

Types of Membrane [1]

According to Berk (2013), membranes used for commercial application can be classified into the followings [5]:

1. **Organic polymer**: It includes cellulose and its derivatives (mainly cellulose acetate), polyolefines, polysulfones, polyamides, chlorine and fluorine substituted hydrocarbons.
2. **Inorganic (ceramic) membranes**: These are based on oxides of zirconium, titanium, silicium and aluminum. Membranes made of cellulose acetate are among the most popular in the dairy industry due to their low cost and low fouling characters. On the other hand, ceramic membranes are well known for their longer operational life and resistance to cleaning chemicals and high temperature [2].

Effect of MF on Physico-chemical Properties [1]

Very little research has been done in the area of MF processing of milk to investigate physico-chemical properties. Use of microfiltration also limits the application of heat treatment and results in less damage to the nutritional and physico-chemical properties of the raw material, so MF can play an important role to preserve liquid foods with natural properties, whilst ensuring food safety [5]. It has been observed that pH and acidity of permeate obtained after MF 1.4 μm was the same as raw skim milk.

Sources

1. *Dipendra Khanal, Non-thermal processing of skim milk: Impact on microbial reduction, physico-chemical properties and quality of Brie type cheese, 2014, AThesis presented to The University of Guelph, Ontario, Canada.*
2. *Cheryan M.1998, Ultrafiltration and microfiltration handbook. Technomic Publishing Company, Inc.xvii+527pp.*
3. *Huisman IH.2000. Membrane Separations micro-filtration, Wilson ID, editor. In: Encyclopedia of separation science. Oxford: Academic Press.1764 p.*
4. *Ripperger S and Altmann J. 2002. Crossflow microfiltration – state of the art. Separation and Purification Technology 26(1): 19-31.*
5. *Berk Z. 2013. Chapter 10 - membrane processes. Berk Z, editor. In: Food process engineering and technology (second edition). San Diego: Academic Press. 259 p.*
6. *G Csan-Guiziou G. 2010. Removal of bacteria, spores and somatic cells from milk by centrifugation and microfiltration techniques. Griffiths MW, editor. Improving the Safety and Quality of Milk, Volume 1: Milk Production and Processing: 349-372.*

Application of MF

Membranes are already well established and are gaining popularity in the dairy industry [8], with the total MF area used in the dairy industry increasing to 15,000 m^2 in 2010 from less than 750 m^2 in the 1990s [6]. The details related to the applications of MF in Milk Protein processing, Milk fat separation, Sugar refining,

Cheese manufacture, Fruit juices processing, microbial removal *etc.* are given in Chapter 2. Other applications are discussed here.

In Dairy Industry

MF is being used for the processing of milk and milk products.

a. Milk Procesing [1]

The use of MF in the dairy industry for fat separation, bacterial removal, spore removal and casein concentration (milk fractionation) were described [2]. This technology can be used in the food industry to clarify the food material and separate the suspended particles in the range of 0.10 - 5 µm [2, 3]. MF is described as a non-thermal technology, which can be used to extend the shelf life of milk by reducing the microbial load and remove spores from the milk, while keeping the organoleptic quality as before. The limitation of MF was pointed out [4], as only applicable for skim milk processing due to clogging caused by fat globules.

Moreover, MF treated milk has high quality and extended shelf life, in comparison with heat processing, as, unlike MF, the latter does not remove the dead cell debris along with potentially active enzymes such as thermoduric proteases [5].

Membrane processing was first used in the dairy industry to separate milk components in the late 1960s, and now is widely used for whey and cheese processing [6]. The application of MF in the dairy industry have recently reviewed [7] and described that membrane processing was first used in the dairy industry to separate **cream and skim milk** using polymeric filters with pore sizes of 0.2 - 10 µm. Later on 2 µm ceramic membranes were successfully used to obtain skim milk virtually free of fat. **With the development of ceramic membranes in the 1980s**, the use of MF became more widely used in the dairy industry and its major application was bacterial removal from skim milk.

Microbial Inactivation [1]

The first commercial application of MF in the dairy industry was in Sweden, where it was used to remove spores of *Bacillus* spp. from milk. This resulted in an extension of shelf life to 16 - 21 days from 6 - 8 days, as well as producing milk with better quality than heat-treated product [8]. Membrane microfiltration can reduce the microbial load in milk by 4 - 5 log10 cycles, but to meet the regulations which states that microbial reduction must be from application of heat, MF can be only used along with thermal pasteurization to extend the shelf life of milk. In the UK, the shelf life of milk has been extended up to 23 days by using MF and market share for this type of milk is up to 11 per cent [6]. Some dairy companies in Canada, UK, and Scandinavia are using MF to process skim milk obtained after separation of milk. The permeate is mixed with separately heated cream (95°C, 20 s), and finally this mixture is HTST pasteurized (72°C, 15 s) before aseptic filling. A summary on microbial removal by membrane micro-filtration is tabulated on next page [1]:

Sl. No.	Micro-organism	Treatment Media	Processing Conditions	Log_{10} Reduction
1	Total bacterial load [10]	Skim Milk	MF 1.4 µm, ceramic membrane, Tp=50 °C	5.18
2	Total bacterial load [8]		MF 1.4 µm	2.1-3.1
3	Total bacterial load [6]		MF-1.4µm, ceramic membrane	2-3
			MF1.4 µm, Sterilox® membrane	3-4
4	Spore forming bacteria [11]		MF-1.4µm	>4.5
5	Spores [12]		MF-1.4 µm, ceramic membrane, Tp= 55 °C	>4.5
6	*Bacillus cereus* Spores [13]		MF-1.4 µm, ceramic membrane	3.5

b. Milk Products Processing

Cheese

There is an economic benefit to using MF for cheese making as it increases production plant capacity and the amount of cheese production per day but without an increase in cheese yield [9]. In the process of cheese making, a ceramic membrane of 1.4 µm pore size was used [10] to process raw milk at 50 °C and found a 5.18 log10 cycle reduction of total bacterial count.

Some researchers have pointed out the benefits of cheese made for MF treated milk, and suggest they are of superior quality due to the absence of pathogenic as well as spoilage micro-organisms. Researcher [6] has also reported that application of MF to produce Camembert type soft cheese resulted in a product comparable with traditionally prepared cheese in terms of flavour and texture. PEF and MF treated milk was used for raw milk cheese., Cheddar cheese, Brie-type cheese mould ripened soft cheese.

Sources

1. *Dipendra Khanal, Non-thermal processing of skim milk: Impact on microbial reduction, physico-chemical properties and quality of Brie type cheese, 2014, A Thesis presented to The University of Guelph, Ontario, Canada.*
2. *Cheryan M. 1998, Ultrafiltration and microfiltration handbook. Technomic Publishing Company, Inc. xvii+527pp.*
3. *Ripperger S and Altmann J. 2002. Crossflow microfiltration - state of the art. Separation and Purification Technology 26(1): 19-31.*
4. *Guerra A, Jonsson G, Rasmussen A, Nielsen EW, Edelsten D. 1997. Low cross-flow velocity microfiltration of skim milk for removal of bacterial spores. International Dairy Journal 7(12): 849-861.*
5. *Olesen N and Jensen F. 1989. Microfiltration - the influence of operation parameters on the process. Milchwissenschaft-Milk Science International 44(8): 476-479.*

6. GCsan-Guiziou G. 2010. *Removal of bacteria, spores and somatic cells from milk by centrifugation and microfiltration techniques. Griffiths MW, editor. Improving the Safety and Quality of Milk, Volume 1: Milk Production and Processing: 349-372.*

7. *Fernandez Garcia L, Alvarez Blanco S, Riera Rodriguez FA. 2013. Microfiltration applied to dairy streams: Removal of bacteria. Journal of the Science of Food and Agriculture 93(2): 187-196.*

8. *Daufin G, Escudier JP, Carrere H, Berot S, Fillaudeau L, Decloux M. 2001. Recent and emerging applications of membrane processes in the food and dairy industry. Food and Bioproducts Processing 79(C2), 89-102.*

9. *Papadatos A, Neocleous M, Berger AM, Barbano DM. 2003. Economic feasibility evaluation of microfiltration of milk prior to cheese making. Journal of Dairy Science 86(5): 1564-1577.*

10. *Awad S, Ahmed N, El Soda M. 2010. Influence of microfiltration and adjunct culture on quality of domiati cheese. Journal of Dairy Science 93(5): 1807-1814.*

11. *Trouve E, Maubois JL, Piot M, Madec MN, Fauquant J, Rouault A, Tabard J, Brinkman G.1991. Retention of various microbial species during milk epuration by cross-flow microfiltration. Lait 71(1): 1-13.*

12. *Maubois JL.2002. Membrane microfiltration: A tool for a new approach in dairy technology. Australian Journal of Dairy Technology 57(2): 92-96.*

13. *Olesen N and Jensen F. 1989. Microfiltration - the influence of operation parameters on the process. Milchwissenschaft-Milk Science International 44(8): 476-479.*

Combined Technologies for Food Processing (Hurdle Technology) [1]

The combination of more than one preservation technique to create a barrier for the micro-organism, which makes them unable to survive and prevents their growth, is known as **hurdle technology** [2]. In hurdle preservation, microbial stability and safety can be achieved at low intensity of each preservation technique, thus, minimizing the negative impact on quality of foods, resulting in improved nutritional quality and sensory properties [3]. Some examples are given below:

i. PEF and MF [1]

PEF is a proven technology for microbial inactivation, but in some cases, very high electric field with large number of pulses are required to achieve the desired degree of inactivation, which may sometimes produce technical as well as quality problems, so combinations of PEF, temperature and MF can be used as hurdles to ensure food safety.

ii. PEF and Temperature

As a non-thermal technology, microbial inactivation by PEF in milk can be achieved at low (nonlethal) temperature, however, studies have reported the synergistic effect of temperature and PEF for microbial inactivation. An increase in bacterial inactivation with increase in temperature was evident due to an increase in the susceptibility of bacterial cells to electroporation at higher temperatures. The synergistic effect of PEF and temperature in apple juice processing was also reported.

iii

The non-thermal hurdle-processing approaches also include the combination of PEF and ultrasonication [4], combination of PEF and HHP [5].

Sources

1. *Dipendra Khanal, Non-thermal processing of skim milk: Impact on microbial reduction, physico-chemical properties and quality of Brie type cheese, 2014, A thesis presented to The University of Guelph, Ontario, Canada.*
2. *Leistner L and Gorris LGM. 1995. Food preservation by hurdle technology. Trends in Food Science and Technology 6(2): 41-46.*
3. *Leistner L. 2000. Basic aspects of food preservation by hurdle technology. International Journal of Food Microbiology 55(1-3): 181-186.*
4. *Palgan I, Munoz A, Noci F, Whyte P, Morgan DJ, Cronin DA, Lyng JG.2012. Effectiveness of combined pulsed electric field (PEF) and manothermosonication (MTS) for the control of Listeria innocua in a smoothie type beverage. Food Control 25(2): 621-625.*
5. *Huang E, Mittal GS, Griffiths MW. 2006. Inactivation of Salmonella enteritidis in liquid whole egg using combination treatments of pulsed electric field, high pressure and ultrasound. Biosystems Engineering 94(3): 403-413.*

Chapter 4

Pulsed Electric Field for Food/Milk Processing

The food industry is continuously investing for new preservation techniques to replace conventional food preservation techniques owing to increasing consumer demand for foods which are natural in taste, nutrient-rich and safe for consumption **[1, 1a]**. Pulsed electric field (PEF) is one of the most promising emerging technologies for the replacement of traditional thermal pasteurization among non-thermal processes, apart from microfiltration eventually combined with moderate heat treatment [1d]. Very little heat is generated during PEF processing due to the short treatment time. Application of PEF technology in food processing may avoid the detrimental effects on flavor, freshness, nutrition and bioactive functional components [1, 2]. **Besides PEF, other promising non-thermal processing techniques includes ionizing radiation, high intensity light pulses, high pressure processing (HPP), electric or magnetic fields, high power ultrasound, oscillating magnetic field, high voltage arc discharge,** *etc.*

PEF is an emerging non-thermal technology that is gaining in popularity, since it preserves food by inactivation of micro-organisms using low energy inputs and reduces the time for processing. This novel non-thermal technology can be used as a food preservation technology, which inactivates spoilage and pathogenic micro-organism and enzymes and ensures extended shelf life and food safety **[1, 1a]**.

PEF technology has considerable potential for use in the dairy industry. Its major application is non-thermal inactivation of vegetative micro-organisms in foods. It causes little or no change in color, flavor and nutritive value as opposed to the impact by thermal treatments. Other applications such as inactivation of enzymes may show promise for improving the product quality. **PEF is effective for extending the shelf life of milk. PEF treatment is largely restricted to liquid foods that can endure high electric fields and is not suitable for fluids containing**

particulate or gases, without pretreatment. PEF may be useful for treatment of products containing heat sensitive components such as whey proteins and immunoglobulins [1a].

The pasteurizing effects of electric fields in foods were first observed in the early 1900s. Pulsed electric fields (PEF) technology in the pasteurization of liquid food products has shown positive results. Processing times can be reduced considerably, which in turn reduces the loss of nutrients and offers important savings in energy. PEF has been used successfully to pasteurize some liquid foods, but it is still not used commercially in milk pasteurization, although several trials have shown the positive effects of PEF milk pasteurization, which could allow for its future use at the industrial level [1b].

Pulsed Electric Field and other Thermal Processes

Pulsed Electric Field (PEF) is an innovative non-thermal technology which could be used as an alternative to the traditional thermal process to inactivate the micro-organisms and enzymes in liquid foods such as milk [3]. Thermal processes alter the nutritional and sensory qualities of the food. The conventional HTST processing technique can affect the organoleptic and nutritional properties of milk to varying degrees [4, 5].

It is revealed from the research studies conducted on PEF technology, that this non-thermal technology can be used as complementary or to replace traditional thermal processing, as a means to avoid the negative effect of thermal technology such as color alterations, flavor damage, and vitamin and nutritional losses to produce minimally processed and fresh-like food products. So, PEF processed foods are considered to be superior in terms of quality attributes, in comparison with those processed using traditional thermal technologies [5a, 6].

As compared to thermal processing, the PEF process is considered more energy efficient as the microbial or enzymatic inactivation is achieved at ambient or mild temperatures by the application of short bursts of high intensity electric fields to liquid food flowing between two electrodes for a couple of microseconds [6, 7, 8].

Principle of PEF [1a, 1c, 5a]

The following studies help in understanding the principle of PEF:

1. A German engineer, Heinz Doevenspeck, in the 1960s, showed that microbial inactivation in food can be achieved by the application of pulsed electric fields.
2. After this, researchers [5b] in the late 1960s worked on PEF and found that "loss in the semi permeability properties" of cell membranes was the cause of microbial death [5a].
3. Researchers [5b] pioneered the study of microbial inactivation by PEF and showed that electric field strength and treatment time are the important factors for microbial inactivation,

The precise mechanisms by which the micro-organisms are destroyed by electric fields are not fully understood although it is thought that cell inactivation occurs by several mechanisms, discussed below.

1. The formation of pores in cell membranes [8a]
2. Formation of electrolytic products or highly reactive free radicals
3. Oxidation and reduction reactions within the cell structure that disrupt metabolic processes
4. Disruption of internal organelles and structural changes [8b]
5. Production of heat produced by transformation of induced electrical energy.

The basic principle of the PEF technology is the application of short pulses of high electric fields with duration of micro seconds to milliseconds and intensity in the order of 10- 80 kV/cm. The processing time is calculated by multiplying the number of pulses times with effective pulse duration. The applied high voltage results in an electric field that causes microbial inactivation.

Mathematical Expression of Microbial Inactivation [5a]

Mathematical expression of microbial inactivation is essential to predict the desired level of microbial destruction by PEF processing. Lack of knowledge of inactivation kinetics fails to establish appropriate PEF processing conditions, which ultimately results either in under or over processing, leading to food safety problems and economic loss [5a, 5f]. Some of the kinetic models used in PEF processing are described below:

i. First Order Kinetics for Microbial Inactivation

After observing the linear destruction of micro-organisms beyond the limit of applied electric field strength, researchers [5g] modeled microbial inactivation based on first order kinetics with respect to electric field strength and treatment time and this can be expressed as:

Electric field strength $\log \frac{N}{N_0} = \ln(S) = -b_E (E - E_C)$

Treatment time $\ln(S) = -b_E \ln(t/t_c)$

where, N = survivor fraction, No = initial population, E = applied electric field (kV/cm), Ec = critical electric field strength (no inactivation below this value), *t*= treatment time (pulse number × time constant; μs), *tc* = critical treatment time to reach Ec (no inactivation below this value), b_E = coefficient of regression, depends on E and *n*, *n* = number of pulses.

After combining aforesaid two equations, the following model for microbial inactivation [5g] was derived :

$$S = \frac{t^{-(E-E_C)/k}}{t_C}$$

where, k = specific rate constant, value of *Ec, tc,* and k depends on microbial species.

Critical electric field strength (E*c*) is related to cell size, the bigger the cell, the smaller the threshold strength required for inactivation. With the advancement of technology and knowledge it became apparent that a linear relationship of microbial inactivation with respect to applied electric field and treatment time does not always apply and the relation also differs with types of micro-organism. Various critical factors, large range of operational conditions and different types of equipment used in PEF processing makes the task even more complex to compare the result of different types of microbial inactivation [5g].

ii. Double Logarithmic Model for Microbial Inactivation

On the basis of the Fermi equation, Peleg (1995) proposed a double logarithmic model to describe the sigmoidal shaped curve of microbial inactivation with respect to applied electric field [5h]:

$$S = \frac{100}{1 + \exp\left[\frac{E - E_C(n)}{K(n)}\right]}$$

where, S = percent surviving micro-organisms, E = applied electric field (kV/cm), Ec = critical electric field strength (value of E, when S = 50 per cent), k = steepness of the survival curve around Ec (kV/cm), n = number of pulses.

iii. Other Models

Some of the other kinetic models generally used in PEF inactivation includes **log-logistic kinetic model** and a **Weibull distribution based model [5a].**

Mechanisms of Microbial Inactivation in PEF Processing [1c, 1e]

Several theories have been proposed to explain electroporation and microbial inactivation. The following two mechanisms have been proposed for the mode of PEF action on microbial membrane:

i. Electroporation

Formation of pores (termed electroporation) in the cell membrane and leakage of cell content into the environment is the widely accepted theory to explain the mechanism of cell death. This pore formation is due to structural changes in the microbial cells and membranes of micro-organisms, developed by applied electric field. Cell exposed to high voltage electric field pulses temporarily destabilizes the lipid bilayer and proteins of cell membranes [1c]. Electroporation is a process of temporary destabilization of protein and the lipid bilayer in a microbial cell under

PEF treatment In this process, the focus is to increase the membrane absorptivity since there is membrane compression and poration [1e, 1g].

The theory given by Zimmermann and others [1h, 5a] has been accepted for the mechanism of electro-permeabilisation and microbial inactivation. This theory assumes the microbial cell is a capacitor filled with dielectric material of low dielectric constant, surrounded by an environment of high electrical conductivity. The applied electric field increases trans-membrane potential, due to accumulation of free charges with opposite polarity on both sides of the membrane surface. These charges are the result of differences in dielectric constant, which attract each other resulting in cell compression. **When the applied electric force equals the critical electric field (electric field at which membrane breakdown occurs), reversible pore (electroporation) formation in the membrane occurs, and a further increase in applied electric field results in electro-permeabilisation,** *i.e.*, leakage of cellular contents into the extracellular environment, which brings about cell death. The formation of pores in the cell membrane leads to reversible (electrical) or irreversible (mechanical) breakdown leading to cell death and depends on the magnitude of the trans-membrane potential [1i, 5a]. The formation of large pores in the membrane results in an irreversible breakdown, and pore size can be enlarged by increasing the applied electric field strength and treatment intensity. Nevertheless, greater trans-membrane potential is induced in large cells compared to small cells and in both of these cases, makes the cells more susceptible to PEF treatment, resulting in mechanical destruction leading to cell death [5a].

ii. Electrical Breakdown

Normal resisting potential difference across the bacterial membrane is 10 mV which leads to the build-up of a membrane potential difference due to charge separation across the membrane [1c]. In this process, the PEF needs to obtain an electric field strength that is higher than the critical value, which can provoke a trans-membrane potential greater than 1volt, to most of the vegetative cells. This will lead to an immediate discharge and decomposition of the membrane [1e, 1f].

In both cases, the phenomena starts by electroporation, by which the cell wall is perforated and cytoplasm contents leak out resulting in cell death [10, 11]. In the pulsed electric field food processing methodology, the inactivation of vegetative cells is greater when high electric field intensities are used with an increased duration of pulses [1e, 1g].

Arcing or Dielectric Breakdown [5i]

PEF causes undesirable arcing or dielectric breakdown in a material. Arcing occurs when the applied field strength becomes equal to the dielectric strength of a material. When a liquid food is subjected to PEF, vapor bubbles present cause arcing. Gases or vapors have a much lower dielectric strength than do pure liquids. Roughness of the electrode surface also causes dielectric breakdown of food materials. In order to avoid the occurrence of arcing, the following points are recommended for consideration [5j]:

1. Use of smooth electrodes.
2. Use of carefully designed treatment chambers to provide uniform electric field strength
3. Degassing
4. Pressurizing the liquid in the treatment chamber to prevent bubble formation.

Types of Pulse Shapes [5a]

Food materials contain ions that make them good conductors of electricity and when an electrical field is applied, electrical current flows into the liquid food and is transferred to each point in the liquid because of the charged molecules present [9]. The electric current is allowed to flow through the food material for an extremely short period of time. A capacitor is therefore needed to generate pulses. The capacitor slowly charges and then quickly discharges its stored electrical energy. The electric field may be applied in the form of exponentially decaying, square wave, bipolar, or oscillatory pulses and at ambient, sub-ambient, or slightly above-ambient temperature. After the treatment, the food is packaged aseptically and stored under refrigeration. The following two types of pulse shape (*i.e.*, exponential decay and square wave) are normally used in PEF processing:

i. A Square Wave Pulse

It is generally regarded as superior over exponential decay pulse for microbial inactivation [5c]. In the case of a square pulse, the voltage increases and decreases quickly but remains constant for some time. The voltage increases instantaneously to a peak value, where it is held for some time before decreasing to zero almost instantly. A square wave pulse needs a pulse-forming network consisting of inducers and capacitors and is more complex and challenging to generate than an exponential decay pulse [5d]. Square pulses can maintain their peak voltage for a longer time than can exponential pulses. Square pulses in addition generate less heat than do exponential pulses. Although more complex circuits are required for the generation of square pulses, they are preferred for their advantages in food applications [5i].

ii. Exponential Decay Pulse

In this, the voltage is unidirectional and increases rapidly to a maximum level then decreases to ground level slowly. Exponential decay pulses are easy to generate.

A Bipolar Pulse

Bipolar forms are also in use depending on circuit design. It consists of one positive and another negative pulse, while monopolar has either all positive or negative pulses. Bipolar pulses are more efficient for microbial inactivation than mono polar pulses (square or exponential decay), due to alternate movement of charged groups of the cell membrane causing extra stress, which results in greater susceptibility to electric field strength [5e].

The types of pulses generated during PEF are illustrated on next page.

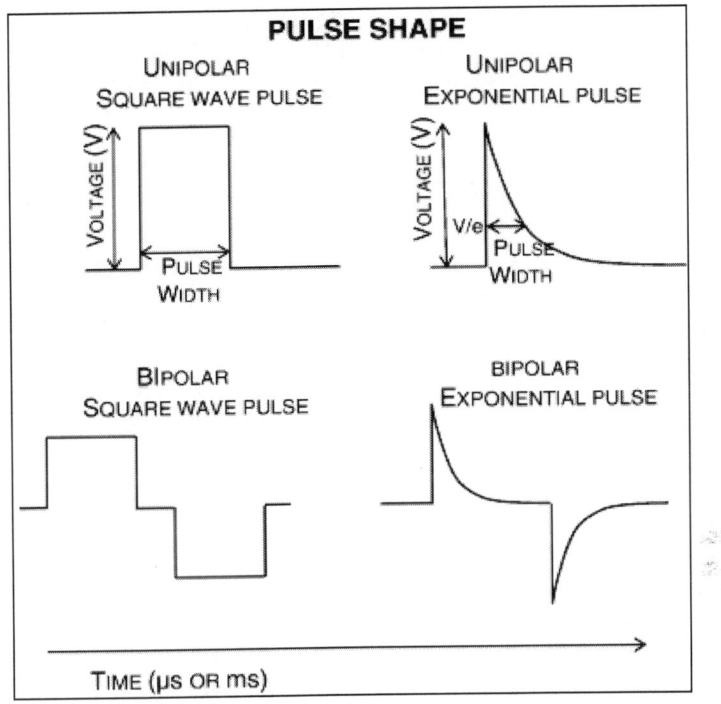

Different Types of Pulse Shape Used in PEF Treatment [5k].

Working

Presently PEF technology has had an increasing interest for liquid food purification, and for refining mass transfer operations in the food industry [1e]. Pulsed electric field (PEF) processing involves the application of high-voltage pulses to foods located between a series of electrode pairs. The electrical fields (generally at 20-80 kV/cm) are achieved through capacitors that store electrical energy from DC power supplies [1c, 1d]. When a short electric pulse (1-100 µsec) is applied to food, there is a pronounced lethal effect on micro-organisms [1c]. The handling time was analyzed by multiplying the number of pulses with the actual pulse duration [1e]. The electric field developed due to this applied high voltage results in microbial inactivation.

Sources

1. *Yeom HW, Streaker CB, Zhang QH and Min DB, 2000, Effects of pulsed electric fields on the quality of orange juice and comparison with heat pasteurization. J Agric Food Chem, 48: 4597-4605.*

1a. *Arjun Chaudhari, A. H. Jana and J. P. Prajapati, Application of Pulsed Electric Field for Milk Processing, Department of Dairy Technology, SMC College of Dairy Science, AAU, Anand.*

1b. *Bermúdez-Aguirre D, Fernández S, Esquivel H, Dunne PC, Barbosa-Cánovas GV, Milk processed by pulsed electric fields: evaluation of microbial quality, physico-chemical characteristics and selected nutrients at different storage conditions, Food Sci. 2011 Jun-Jul; 76(5): S289-99.*

1c. *Hudaa Neetoo (Faculty of Agriculture, University of Mauritius, Réduit, Mauritius) and Haiqiang Chen (Department of Animal and Food Sciences, University of Delaware, Newark, Delaware, US), Alternative Food Processing Technologies.*

1d. *Juliane Floury, Noël Grosset, Nadine Leconte, Maryvonne Pasco, Marie-Noëlle madec, Romain JEANTET, Continuous raw skim milk processing by pulsed electric field at non-lethal temperature: effect on microbial inactivation and functional properties, Lait 86, 2006, 43-57.*

1e. *Harlin Kaur Saroya, Innovative non-thermal food processing technologies used by the food industry in the United States, MSc thesis, 2017, Western Kentucky University, Kentucky.*

1f. *Griffiths, M., and Walking-Riberio, M., Pulsed electric field processing of liquid food and beverages, Emerging Technologies of Food Processing, 115-145, 2014.*

1g. *Novickij, V., Grainys, A., Lastausk ienë, E., Kananavièiûtë, R., Pamedytytë, D., Kalëdienë, L., Miklavèiè, D. (n.d.). Pulsed electromagnetic field assisted in vitro electroporation: A pilot study. Scientific Reports, 6, 33537.*

1h. *Zimmermann U, Pilwat G, Riemann F. 1974. Dielectric breakdown of cell membranes. Biophysical Journal, 14, 881-899.*

1i. *Pagan R, Condon S, Raso J. 2005. Microbial inactivation by pulsed electric fields. Novel Food Processing Technologies 141: 45-68.*

2. *Li SQ, Zhang QH, Lee YZ and Pham TV, 2003, Effetcs of pulsed electric field and thermal processing on the stability of bovine IgG in enriched soymilk. J Food Sci, 68: 1201-1207.*

3. *Shamsi K., Wan J., Versteeg C and Sherkat F, 2009, Effects of pulsed electric field treatment on the rheological and textural properties of rennet-induced curds of bovine milk. Poster presented at the 4th Innovative Foods Centre Conference, 17-18 September, 2008, Brisbane, Australia.*

4. *Steele J., 2000. History, trends and extent of pasteurization. J Am Vet Medical Assoc, 217: 175-178.*

5. *Mertens B and Knorr D., 1992, Developments of non-thermal processes of food preservation. Food Technol, 5: 124-133.*

5a. *Dipendra Khanal, Non-thermal processing of skim milk: Impact on microbial reduction, physico-chemical properties and quality of Brie type cheese, 2014, A Thesis presented to The University of Guelph, Ontario, Canada.*

5b. *Sale AJH and Hamilton WA., 1967, Effects of high electric fields on micro-organisms: I. killing of bacteria and yeasts. Biochimica Et Biophysica Acta (BBA) - General Subjects 148(3): 781-788.*

5c. *Qin BL, Zhang QH, Barbosacanovas GV, Swanson BG, Pedrow PD. 1994. Inactivation of micro-organisms by pulsed electric-fields of different voltage wave-forms. IEEE Transactions on Dielectrics and Electrical Insulation 1(6): 1047-1057.*

5d. *Evrendilek GA and Zhang QH. 2005. Effects of pulse polarity and pulse delaying time on pulsed electric fields-induced pasteurization of E. coli O157: H7. Journal of Food Engineering 68(2): 271-276.*

5e. *Barbosa-Canovas GV and Sepulveda D. 2005. Present status and the future of PEF technology. BarbosaCanovas, GV Tapia, MS Cano, MP, editor.44 p.*

5f. *Wouters PC, Alvarez I, Raso J. 2001. Critical factors determining inactivation kinetics by pulsed electric field food processing. Trends in Food Science and Technology 12(3-4): 112-121.*

5g. *Hulsheger H, Potel J, Niemann e.g. 1981. Killing of bacteria with electric pulses of high-field strength. Radiation and Environmental Biophysics 20(1): 53-65.*

5h. *Peleg M. 1995. A model of microbial survival after exposure to pulsed electric-fields. Journal of the Science of Food and Agriculture 67(1): 93-99.*

5i. *Singh R. Paul, Professor of Food Engineering Department of Biological and Agricultural Engineering, Department of Food Science and Technology University of California Davis, FAO-2001, Technical Elements of New and Emerging Non-Thermal Food Technologies.*

5j. *Sale, A. J. H. and Hamilton, W. A.1967. In J. Biochim. Biophys. Acta 148: 781788.*

5k. *J. Raso, W. Frey, G. Ferrari, G. Pataro, D. Knorr, J. Teissie and D. Miklavcic, Recommendations guidelines on the key information to be reported in studies of application of PEF technology in food and biotechnological processes, Innovative Food Science and Emerging Technologies, Volume 37, Part C, October 2016, Pages 312-321.*

6. *Barbosa-Canovas GV, Pierson MD, Zhang QH and Schaffner DW, 2000, Pulsed electric fields. Special supplement: kinetics of microbial inactivation for alternative food processing technologies. J Food Sci, 65: 65-79.*

7. *Gaudreau M, Hawkey T, Petry J and Kempkes M, 2008, Pulsed electric field processing for food and waste streams. Food Australia, 60: 323-325.*

8. *Jose A, Guerrero-Beltran, Sepulveda DR, Maria M, Gongora-Nieto MM, Swanson B and Barbosa-Canovas GV, 2010, Milk thermization by pulsed electric fields and electrically induced heat. J Food Engg, 100: 56-60.*

8a. *Toepfl S, Heinz V, Knorr D, 2007, High intensity pulsed electric fields applied for food preservation. Chemical Engineering and Processing 46: 537-546.*

8b. *Barbosa-Canovas GV, Gongora-Nieto MM, Pothakamury UR, Swanson BG, 1999, Preservation of Foods with Pulsed Electric Fields. San Diego, CA: Academic Press.*

9. *Zhang QH, Barbosa-Cánovas GV and Swanson BG, 1995, Engineering aspects of pulsed electric field pasteurization. J Food Engg, 25: 261-281.*

10. *Tsong T., 1990, Reviews on electroporation of cell membranes and some related phenomena. Bioelectro-chem Bioenergetics, 24: 271.*

11. *Martin-Belloso O, Elez-Martinez P., 2005, Food safety aspects of pulsed electric field. In Emerging Technologies for Food Processing. Da-Wen Sun (Ed.), Elsevier Academic Press, London, pp.184-217.*

Pulsed Electric Field Equipment [1, 1a]

There are following two kinds of equipment available for the operation of PEF technique:

1. **Batch equipment**
2. **Continuous equipment.**

Commercially, the continuous process is preferred because the process does not have to be stopped. This results in a continuous flow.

There are a few important components, which play a major role in the equipment.

The basic-set up of the PEF unit consists of the following components:

1. Repetitive a high-voltage DC source also known as a pulse generator
2. Inductors and a capacitor bank
3. A discharge switch to discharge energy to electrodes around a treatment chamber [2]/high voltage control unit (HV switch)
4. Liquid handling system
5. A treatment chamber.

The main purpose of the pulse generator is to receive low voltage as an alternating current (AC) from a power supply and to convert it to high voltage (HV), which is then supplied to PEF chambers [7a]. Capacitors are charged by direct current (DC) power obtained from an amplified and rectified AC source. The low level voltage is collected and stored in a capacitor and this releases the energy to the food material through a HV discharge switch [7b]. The devices normally used as switches include Ignitron, Gas Spark Gap, Trigatron, or Thyratron [7c].

More than one capacitor can be combined to store more energy.

i. Energy Stored in the Capacitor [7]

The energy stored in the capacitor can be expressed as [7, 7a]:

$$\text{Energy} = Q = \frac{1}{2}CV^2$$

where, C = capacitance (F), V = voltage (V).

2. Electric Field Strength Developed [7]

The electric field strength (E; kV/cm) developed in the treatment chamber by supplying the electric energy stored in capacitor can be expressed as [4]:

a. For Parallel Plate Electrodes

$$\text{Electric Field Strength} = E = \frac{V}{D}$$

b. For Coaxial Electrodes

$$\text{Electric Field Strength} = E = \frac{V}{r / n \frac{R_1}{R_2}}$$

where, V = electric potential difference between two points (V), D = distance between two electrodes (cm), r = radius of electric field measurement (cm), R_1 and R_2 = radii of the inner and outer electrodes (cm).

Once the food is treated, it is then passed through a sterile packaging line. Also, there are supervising and control equipment that include temperature sensors made of optic fibers, electric monitors, data procurement, and regulatory micro-processors [1a, 1b].

The switching mechanism modulates several processing parameters, including the voltage pulse frequency, the duration of the treatment and the waveform [2]. The voltage and current applied, as well as the strength of the electric field, can be measured using an oscilloscope. Since the process of subjecting food to PEF unavoidably leads to the generation of heat, the treatment chamber is coupled to a cooling system to avoid overheating of the sample [3].

Electrodes

Electrodes used for food treatment should be of food grade material and sterilizable, and made up of electro-chemically inert materials. The material for construction of electrodes is usually stainless steel but gold, platinum, carbon and metal oxides can also be used as an alternative [5c, 7].

Parallel plate electrode configuration is the simplest in design and consists of a rectangular parallel piped shape channel of insulating material with two electrodes on opposite sides. In the co-linear electrode configuration, couples of tubular electrodes are spaced with insulator spacer tubes [3b].

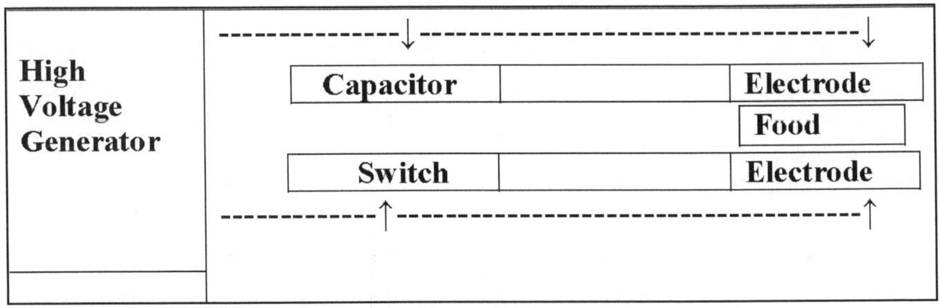

A Simplified General Design of a Pulsed Electric Field Apparatus [3a].

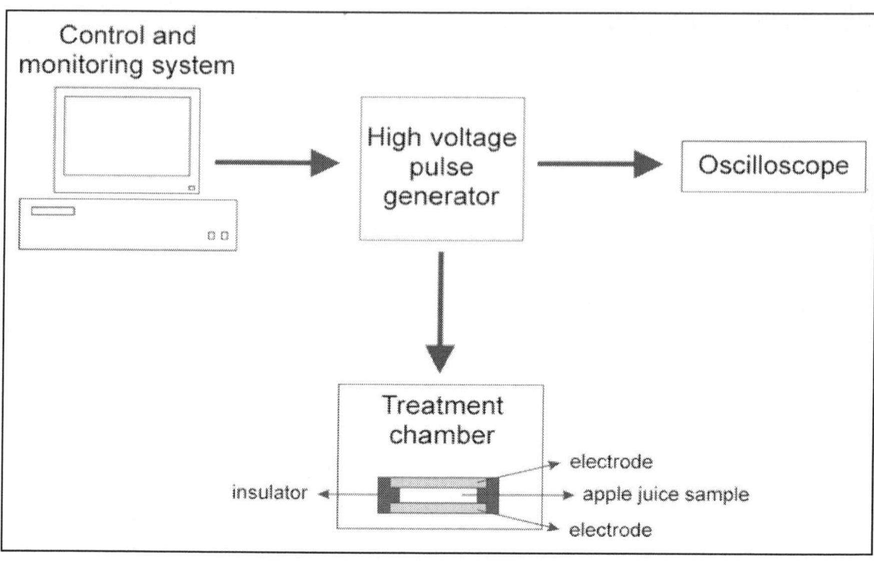

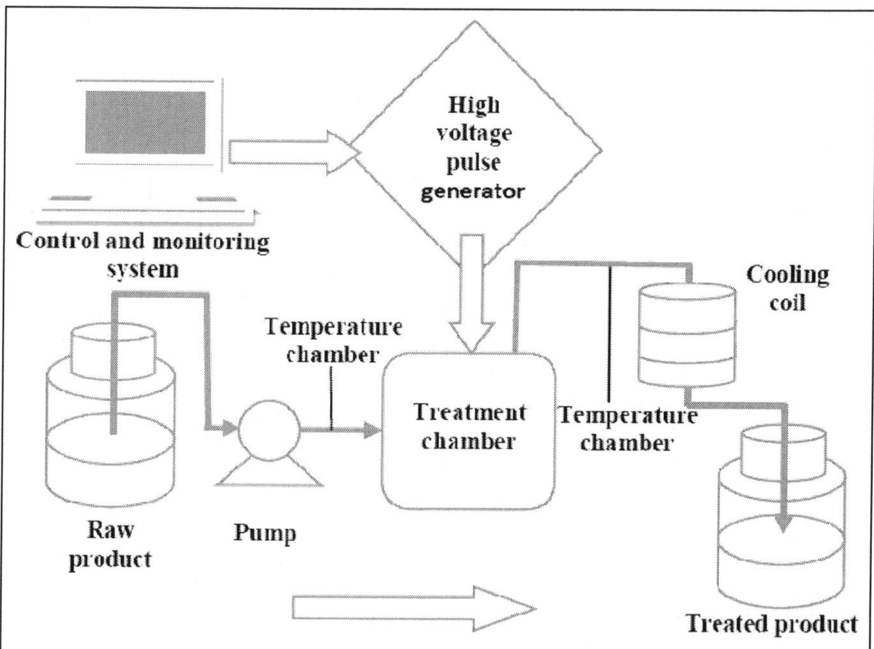

Flow Chart of a PEF Food Processing System [3c].

Types of Chambers

In PEF, food samples may be processed in following types of chamber:

i. A Static Chamber (Batch mode)

In the static system, a unit block of food material is placed in the chamber, where it is treated with applied electric field strength and is replaced with another food block after a definite time [7], Static PEF treatment chambers typically comprise two electrodes held in position by insulating materials that also enclose the food materials [3]. Uniform electric fields can be achieved using parallel plate electrodes with a gap sufficiently smaller than the electrode surface dimension [3].

a. A static chamber of Washington State University contains disk-shaped electrodes (area 27 cm^2) made of stainless steel polished to a mirror-like surface, with a gap that could be set at either 9.5 or 5.1 mm. Electric field strengths of up to 70 kV/cm can be used. Electrodes contain built-in jackets that allow the circulation of water in order to maintain low temperatures [3a].

ii. Pumped through a Chamber (Continuous mode)

In the continuous system the food is pumped through the chamber and is cooled before collection. The continuous system is an updated and modified form of the static system and is suitable for pumpable liquid foods. Due to greater treatment uniformity, the continuous system is more effective for microbial inactivation than the static system [7].

a. An example of a continuous chamber is a flow-through unit using low flow rates developed at Washington State University [3]. The chamber consists of two electrodes, a spacer, and two lids. The parallel-plate stainless steel electrodes have baffled flow channels between the electrodes to eliminate dead corners and ensure uniform treatment [3]. The two stainless steel electrodes are separated by a polysulfone spacer. Once treated, the pumpable food is then filled into individual consumer packages or stored using aseptic packaging equipment [3].

b. The continuous PEF equipment (Europulse, Cressensac, France) [5] uses an original pressurized spark gap switching technology (dry air) with a high repetitive rate, connected to a pulse forming line consisting of a coaxial cable and lumped elements [6]. It is designed to allow a widely adjustable operating pulse width (from 50 up to 3000 ns), electric field strength (from 30 up to 80 $kV{\cdot}cm^{-1}$) and pulse frequency (from 1 up to 815 Hz), and to produce a volumetric flow rate (from 1up to 10 $L{\cdot}h^{-1}$). This equipment, including a 2-kW high-voltage power supply charging capacitors and an interactive computer control (Labview software), generates square waveform pulses.

Arrangement of Chambers [7]

Parallel and coaxial chamber configurations are commonly used for processing foods by PEF [5b]. The applied high voltage results in an electric field that causes microbial inactivation.

i. Parallel Chamber Configuration

Parallel plates are normally used with a batch system [5c]. Parallel plate configuration is easy to design and produces uniform electric field distribution; however, electric intensity is greatly reduced in the wall of the electrodes [5d].

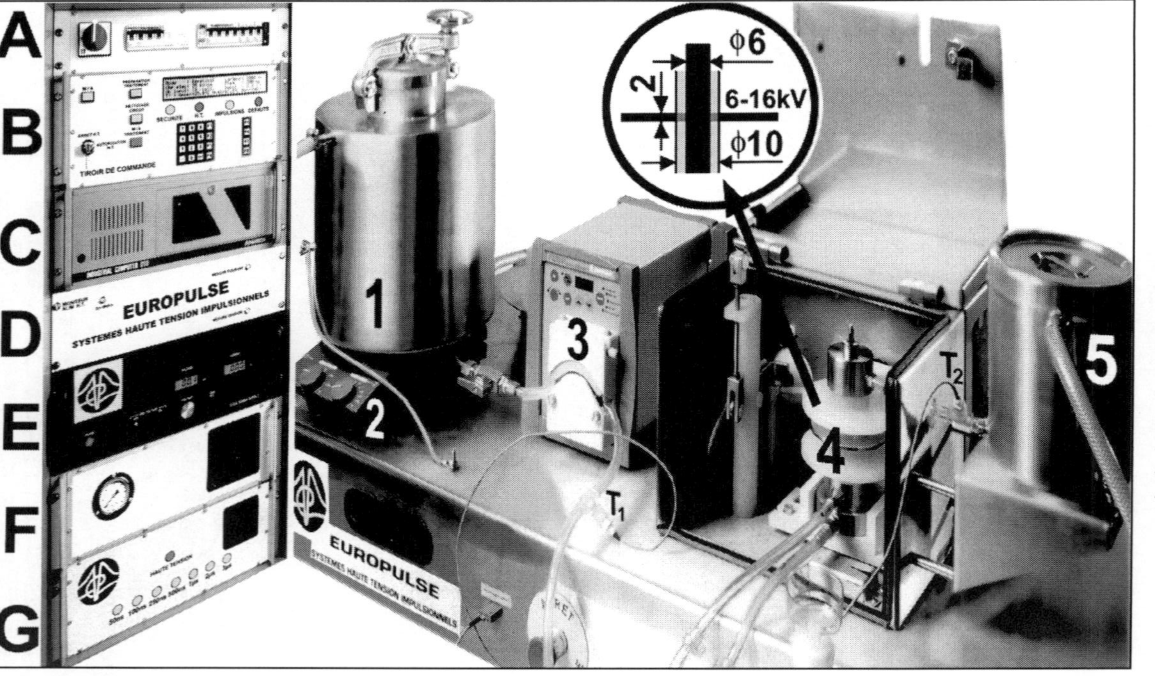

Control panel: A- power supply, B-programing and control drawer, C- computer, D and E-high voltage monitor and power supply (50 kV, 2 kW), F- spark gap switch (0–815 Hz); G: high voltage energy storage (50 ns, 100 ns, 250 ns, 500 ns, 1 μs, 2 μs and 3 μs).

Hydraulic line: 1. Supply tank (12 L), 2. magnetic stirrer, 3. peristaltic pump (0–25 $L \cdot h^{-1}$), 4. PEF treatment chamber, 5. tubular heat exchanger, T1, T2- thermocouples.

Equipment for Continuous PEF Treatment of Liquid Products [5].

ii. Coaxial Chamber Configuration

Coaxial systems are used for continuous operation [5c]. The coaxial plate is best known for smooth and uniform product flow and is suitable for large industrial applications [5d].

Integrated and Commercial PEF System [5a]

i. Integrated PEF System

An integrated PEF system consists of a fluid handling unit, high voltage pulse generator, PEF treatment chambers, and packaging machine. The fluid handling unit delivers stable, uniform flow with sterilize-in-place (SIP) and clean-in-place (CIP) functions. The pulse generator supplies high voltage pulses into foods flowing through PEF treatment chambers. Treated foods are packaged continuously [5a].

ii. Commercial PEF System

In the US, the first commercial scale continuous PEF system is installed at Department of Food Science and Technology, Ohio State University. This PEF system is part of a new food treatment system assembled by a DoD sponsored, University directed industry consortium. Diversified Technologies Inc., Bedford, MA, builds commercial PEF systems of processing volumes ranging from 500 to 2,000 liters per hour, with The Ohio State University supplying the PEF treatment chambers [5a].

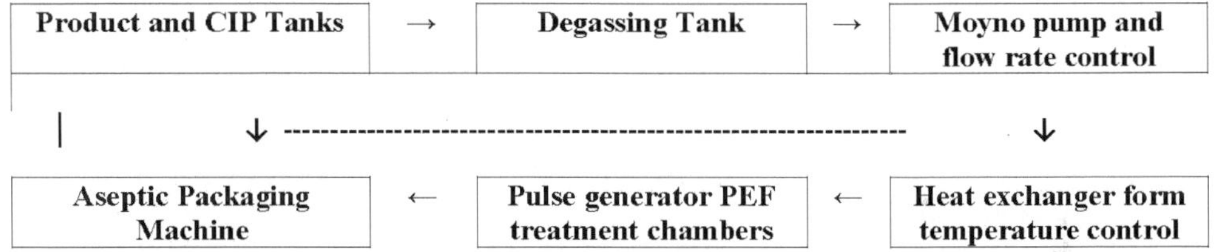

An Integrated Pulsed Electric Field Set up and Aseptic System for Juice Processing [3a].

Sources

1. *Hudaa Neetoo (Faculty of Agriculture, University of Mauritius, Réduit, Mauritius) and Haiqiang Chen (Department of Animal and Food Sciences, University of Delaware, Newark, Delaware, US), Alternative Food Processing Technologies.*

1a. *Harlin Kaur Saroya, Innovative non-thermal food processing technologies used by the food industry in the United States, MSc thesis, 2017, Western Kentucky University, Kentucky.*

1b. *Knoerzer, K., Innovative food processing technologies: Advances in multiphysica simulation. Chichester, U. K: Wiley- Blackwell, 2011.*

2. *Ohlsson T, Bengtsson N, 2002, Minimal processing of foods with non-thermal methods. In: Ohlsson T, Bengtsson N (eds) Minimal Processing Technologies in the Food Industry. Cambridge: Woodhead Publishing, pp. 34-60.*

3. *Barbosa-Canovas GV, Pothakamury UR, Palou E, Swanson BG, 1998, Light-pulse sterilization of foods and packaging. In: Barbosa-Canovas GV, Pothakamury UR, Palou E, Swanson BG (eds) Non-thermal Preservation of Foods. New York: Marcel Dekker, pp. 139-159.*

3a. *Singh R. Paul, Professor of Food Engineering Department of Biological and Agricultural Engineering, Department of Food Science and Technology University of California Davis, FAO-2001, Technical Elements of New and Emerging Non-Thermal Food Technologies.*

3b. *J. Raso, W. Frey, G. Ferrari, G. Pataro, D. Knorr, J. Teissie and D. Miklavcic, Recommendations guidelines on the key information to be reported in studies of application of PEF technology in food and biotechnological processes, Innovative Food Science and Emerging Technologies, Volume 37, Part C, October 2016, Pages 312-321.*

3c. *M. E. A. Mohamad and A. H. Amer Eissa, Pulse Electric Fields for food processing Technology, Structure and Function of Food Engineering, DOI: 105772/48678.*

4. *Barbosa-Canovas GV and Sepulveda D. 2005. Present status and the future of PEF technology. BarbosaCanovas, GV Tapia, MS Cano, MP, editor.44 p..*

5. *Juliane Floury, Noël Grosset, Nadine Leconte, Maryvonne Pasco, Marie-Noëlle madec, Romain JEANTET, Continuous raw skim milk processing by pulsed electric field at non-lethal temperature: effect on microbial inactivation and functional properties, Lait 86, 2006, 43-57.*

5a. *Ranjan Saha, New and Emerging Non-Thermal Technologies in Food Preservation, April 6, 2015.*

5b. *Motarjemi Y, Moy GG, Jooste PJ, Anelich LE. 2014. Chapter 5 - milk and dairy products. Motarjemi Y and Lelieveld H, editors. In: Food safety management. San Diego: Academic Press. 83 p..*

5c. *Toepfl S, Heinz V, Knorr D. 2005, 4 - Overview of pulsed electric field processing for food. Sun D, editor. In: Emerging technologies for food processing. London: Academic Press. 69 p.*

5d. *Jeyamkondan S, Jayas DS, Holley RA. 1999. Pulsed electric field processing of foods: A review. Journal of Food Protection 62(9): 1088-1096.*

6. *Jeantet R., Carballeira Fernández J. A., Roignant M., Cochet M. F., Baron F., Korolczuk J., Brion J. C., Performances d'un nouvel appareil de décontamination microbienne de produits liquides en continu par champs électriques pulsés, Ind. Aliment. Agric. 4, 2003, 9-14.*

7. *Dipendra Khanal, Non-thermal processing of skim milk: Impact on microbial reduction, physico-chemical properties and quality of Brie type cheese, 2014, A Thesis presented to The University of Guelph, Ontario, Canada.*

7a. *Zhang Q, Barbosa-Cánovas GV, Swanson BG. 1995. Engineering aspects of pulsed electric field pasteurization. Journal of Food Engineering 25(2): 261-281.*

7b. *Min S, Evrendilek GA, Zhang HQ. 2007. Pulsed electric fields: Processing system, microbial and enzyme inhibition, and shelf life extension of foods. IEEE Transactions on Plasma Science 35(1): 59-73.*

7c. *Ho S and Mittal GS. 2000. High voltage pulsed electrical field for liquid food pasteurization. Food Reviews International 16(4): 395-434.*

Shelf Life and Storage of PEF Treated Food [1]

i. Shelf Life

In general, the shelf-life of PEF-treated and thermally pasteurized foods is comparable. PEF pasteurization kills micro-organisms and inactivates some enzymes and, unless the product is acidic, it requires refrigerated storage. For heat-sensitive liquid foods where thermal pasteurization is not an option (due to flavor, texture, or color changes), PEF treatment would be advantageous.

ii. Storage

PEF pasteurized products currently are stored refrigerated. In some cases (for example, milk), this is necessary for safety (to prevent the growth of spores in low-acid foods). For acid foods, refrigeration is not necessary for microbial stability, but is used to preserve flavor quality for extended periods of time.

Sources

1. *Ranjan Saha, New and Emerging Non-Thermal Technologies in Food Preservation, April 6, 2015.*

Precaution and Limitations [1]

Details related to the precaution to be taken while operating PEF equipment and limitations of PEF are given below:

Precaution

Since this equipment utilizes high electricity, the protection of the operators becomes of prime concern. Hence, it is placed in a restricted area with interlocked gates. Also, the connections to the chambers are isolated and earthed to prevent leakage. This process uses ordinary electricity. The **facility meets electrical safety standards and no harmful environmental by-products are produced**.

Limitations

The PEF Technology has the following limitations:

1. PEF technology is restricted to food products that can withstand high electric fields, have low electrical conductivity and do not contain or form bubbles(food products with no air bubbles).
2. The particle size of the food in both static and flow treatment modes is a limitation. The maximum particle size in the liquid must be smaller than the gap of the treatment region in the chamber in order to ensure proper treatment.
3. PEF is a continuous processing method, which is not suitable for solid food products that are not pumpable.

4. In addition, there are other major cost drivers that depend on the electrical properties of the food being processed.
5. Intensive research conducted on PEF has brought the technology to the brink of commercial uptake, although PEF energy requirements are too considerable and costly to make it a good choice for most food applications [2].

The main limitations [3] are summarised below:

1. This process is restricted to liquid or small particles foods.
2. This process can only be implemented to foods, which can withstand high electric fields.
3. If a food requires salt to be added, it can only be done after PEF processing.
4. There cannot be any bubbles. The bubbles must be removed since they can cause safety issues, or can lead to a non-uniform treatment.

Sources

1. *Harlin Kaur Saroya, Innovative non-thermal food processing technologies used by the food industry in the United States, MSc thesis, 2017, Western Kentucky University, Kentucky.*
2. *Hudaa Neetoo (Faculty of Agriculture, University of Mauritius, Réduit, Mauritius) and Haiqiang Chen (Department of Animal and Food Sciences, University of Delaware, Newark, Delaware, US), Alternative Food Processing Technologies*
3. *Knoerzer, K., Innovative food processing technologies: Advances in multiphysica simulation. Chichester, U. K: Wiley-Blackwell, 2011.*

Parameters Affecting Microbial Inactivation by PEF [1, 1a]

The success of PEF technology to pasteurize food largely depends on its ability to inactivate different types of micro-organisms, including spoilage and pathogenic bacteria. In electroporation process, microbial inactivation occurs. The following various parameters that can affect the microbial inactivation by PEF are as follows:

a. Process Parameters (Parameters Related to Pulsed Electric Field *etc.*)

The antimicrobial efficacy of the PEF process varies as a function of numerous processing parameters. There are numerous treatment factors that can affect microbial inactivation, including the electric field strength, number of pulses, pulse duration(PEF treatment time), pulse shape, pulse width and processing temperature (inlet and outlet), flow rate, frequency, chamber configurations *etc.* [1, 1f]. The increased microbial inactivation with increase in electrical field intensity is related to the electroporation theory of microbial inactivation, which states that the induced potential difference across the cell membrane is proportional to the applied electric field intensity [1g]. The details related to the some of the aforesaid parameters/factors are given on next page.

i. Electric Field Intensity/Strength

It refers to the field strength locally present in the treatment chamber during the sample treatment, and depends on the voltage applied between the electrodes, geometry of the treatment chamber, and the spatial distribution of dielectric properties of the material between the electrodes [2a]. Microbial inactivation increases when electric field strength is increased above a certain value, the critical electric field strength. This critical electric field strength depends on microbial types and morphology, and most of the micro-organisms are not inactivated by electric field strengths of less than 4-8 kV/cm. To achieve microbial inactivation, the applied field needs to be greater than the critical electric field for particular micro-organisms [3]. When the cells are exposed to an electric field of sufficient strength, the field induces the accumulation of electric charges at the nonconductive microbial membranes. In the pulsed electric field food processing methodology, the inactivation of vegetative cells is greater when high electric field intensities are used with an increased duration of pulses [1a]. *Listeria innocua* (smaller cell size) requires a minimum dose of 15 kV/cm to become inactivated, while the larger cells of *S. cerevisiae* are affected at field strength as low as 2-4 kV/cm.

ii. Pulse Wave Shape

Exponentially decaying and square wave pulses (bipolar or mono polar form) are the most common wave shapes used in PEF systems. An exponentially decaying wave is a unidirectional voltage that rises rapidly to a maximum value and decays slowly to zero. Square wave pulses maintain a peak voltage above the critical field strength for a longer period of time and are more lethal than exponentially decaying pulses. Bipolar pulses are reported to be more lethal than mono polar pulses [4].

Pulse duration, or *pulse width,* for a square pulse is the time that the voltage is kept at the maximum value (peak voltage). In the case of exponential decay pulses, the pulse width is defined as the time needed to decrease the voltage to 37 per cent of its peak value [2a].

iii. Treatment Temperature

When the electric field is applied to the flowing liquid between the electrodes, its temperature rises. Therefore, proper cooling of the product (by using cooling water bath) between the treatment chambers is necessary to avoid heat damage and to limit microbial inactivation only to the PEF effect.

PEF treatment at moderate temperatures (50 to 60°C) exhibits additive effects on the inactivation of micro-organisms and their spores [4]. Researchers observed higher microbial inactivation when inlet and/or outlet temperature of product was increased in PEF treatment [4].

iv. Treatment Time and Frequency

Treatment time refers to the number of pulses applied multiplied by the pulse width (or pulse duration) [2a]. The inactivation of micro-organisms increases with an increase in the pulse number up to a certain extent [1e]. The electric field pulses between 1 and 5 µs produced the best results for microbial inactivation [5]. Treatment time is another factor which affects microbial inactivation, and is the effective time during which product is subjected to the field strength. It is calculated as a product of number of pulses and the width of the pulses applied. Longer pulse width results in higher inactivation and increased pulse width gives longer treatment time [6]

b. Parameters Related to Micro-organism (Microbial Parameters)

Physiological state of the bacteria, type of micro-organisms *etc.*

i. Type of Micro-organisms

Microbial cell size, species, strain, Gram type, and growth stage have been found to affect the inactivation of micro-organisms in food by PEF treatment [1g].

- ✰ Larger micro-organisms are more sensitive to electric field strength than those of smaller size. Small cells develop lower trans-membrane potential in an electric field than larger cells and develop greater resistance, so, larger cells are more easily permeabilized than small cells [1g].

  Trans-membrane potential (TMP) and cell size are related and the relation can be presented by the following equation [1h]:

  $$TMP = kErcos(\varphi)$$

  where, k = form factor (3/2 for spherical cells), E = electric field strength, r = cell radius φ= angle between the electric field and the point of interest on the membrane (1 at the poles)
- ✰ Gram-positive bacteria are more resistant to PEF treatment than Gram-negative ones [1e]. Gram-negative bacteria are more easily inactivated by PEF treatment than gram-positive bacteria, and spores are most resistant.
- ✰ Logarithmic phase cells are more sensitive to stress than lag and stationary phase cells. Microbial growth in logarithmic phase is more susceptible to the applied electric fields, inactivation effects in the logarithmic phase were 30 per cent greater than in stationary phase [2].
- ✰ Researchers [4] also reported that yeasts are more sensitive to applied electric field than bacteria due to their larger size, although at low electric fields they were more resistant than Gram-negative bacteria.

c. Product Parameters (Parameters Related to the Condition of Food Sample)

PEF treatment has been applied to different types of food products and physical and chemical properties of different products have been shown to influence microbial inactivation [1g]. Other factors that also influence the degree of inactivation include the temperature of the food, pH, ionic strength, water activity, density, viscosity and electrical conductivity [1b].

i. Electrical Conductivity of a Food

The electrical conductivity of a food depends on its composition and presence of ions, salts and acids. Products of low conductivity include beer (0.143 Siemens/m) and black coffee (0.182 Siemens/m), while products of relatively higher conductivity include vegetable juice cocktail (1.556 Siemens/m), and tomato juice (1.697 Siemens/m). The study [1c] stated that lowering the conductivity of the treated product increases the difference in the ionic concentration between the cytoplasm and the suspending fluid, which can in turn weaken the membrane structure by increasing the flow of ionic substances across the membrane. It was observed [1d] that the foods with high conductivities are difficult to work with because they generate smaller peak electric field strengths across the treatment chamber. A higher inactivation of *Lactobacillus plantarum* was achieved in low conductivity medium (4 mS/cm buffer solution) than in higher conductivity medium (16 mS/cm buffer medium) when all other parameters were kept constant [1f].

ii. pH of Food

Researchers [1f] reported in a review that some studies did not find any effect of pH on microbial inactivation and concluded that the effect of pH depends on microbial type.

iii. Water Activity

Low water activity food makes micro-organisms more resistant to PEF, resulting in lower microbial inactivation [1g].

Sources

1. *Arjun Chaudhari, A. H. Jana and J. P. Prajapati, Application of Pulsed Electric Field for Milk Processing, Department of Dairy Technology, SMCCollege of Dairy Science, AAU, Anand-388110.*

1a. *Harlin Kaur Saroya, Innovative non-thermal food processing technologies used by the food industry in the United States, MSc thesis, 2017, Western Kentucky University, Kentucky.*

1b. *Hudaa Neetoo (Faculty of Agriculture, University of Mauritius, Réduit, Mauritius) and Haiqiang Chen (Department of Animal and Food Sciences, University of Delaware, Newark, Delaware, US), Alternative Food Processing Technologies.*

1c. *Jeyamkondan S, Jayas DS, Holley RA, 1999, Pulsed electric field processing of foods: a review. Journal of Food Protection 62: 1088-1096.*

1d. *Barbosa-Canovas GV, Gongora-Nieto MM, Pothakamury UR, Swanson BG, 1999, Preservation of Foods with Pulsed Electric Fields. San Diego, CA: Academic Press.*

1e. *Hulsheger H, Potel J and Niemann EG., 1983, Electric field effects on bacteria and yeast cells, Radiation Environ. Biophysics, 22: 149-162.*

1f. *Wouters PC, Alvarez I, Raso J. 2001. Critical factors determining inactivation kinetics by pulsed electric field food processing. Trends in Food Science and Technology 12(3-4): 112-121.*

1g. *Dipendra Khanal, Non-thermal processing of skim milk: Impact on microbial reduction, physico-chemical properties and quality of Brie type cheese, 2014, A Thesis presented to The University of Guelph, Ontario, Canada.*

1h. *Nuumann E. 1992. Membrane electroporation and direct gene transfer. Bioelectrochemistry and Bioenergetics, 28 (1-2), 247-267.*

2. *Qin BL, Barbosa-Cánovas GV and Swanson BG, Pedrow PD and Olsen RG, 1995, Inactivating micro-organisms using pulsed electric fields continuous treatment system, IEEE Transactions Dielectrics Electrical Insulation, 34: 43-50.*

2a. *J. Raso, W. Frey, G. Ferrari, G. Pataro, D. Knorr, J. Teissie and D. Miklavcic, Recommendations guidelines on the key information to be reported in studies of application of PEF technology in food and biotechnological processes, Innovative Food Science and Emerging Technologies, Volume 37, Part C, October 2016, Pages 312-321.*

3. *Castro A, Barbosa-Canovas GV and Swanson BG, 1993, Microbial inactivation of foods by pulsed electric fields. J Food Proc Preserv,* **17:** *47-73.*

4. *Shamsi K., Wan J., Versteeg C and Sherkat F., 2009, Effects of pulsed electric field treatment on the rheological and textural properties of rennet-induced curds of bovine milk. Poster presented at the 4th Innovative Foods Centre Conference, 17-18 September 2008, Brisbane, Australia.*

5. *Sepulveda-Ahumada DR, 2003, Preservation of fluid foods by pulse electric fields in combination with mild thermal treatments, Ph. D Thesis. Pullman WA: Washington State University.*

6. *Min S, Evrendilek GA, Zhang HQ. 2007. Pulsed electric fields: Processing system, microbial and enzyme inhibition, and shelf life extension of foods. IEEE Transactions on Plasma Science 35(1): 59-73.*

Benefits, and Shortcomings of Alternative Non-thermal Processing Method PEF [1]

Process considerations, benefits and shortcomings of alternative non-thermal processing method-PEF are summarized in following table:

Sl. No.	*Process Considerations*	*Benefits*	*Shortcomings*	*Examples of Applications*
1	Electric field intensity	Effective against vegetative bacteria	Not suitable for non-liquid foods	Fruit juices
2	Chamber design	Relatively short processing time	Post process recontamination possible	Milk
3	Electrodes design	Suitable for pumpable foods	Less effective against enzymes and spores	Whole liquid egg
4	Pulse width	Minimal impact on nutrients, flavor or pigment compounds	Adverse electrolytic reactions could occur. Potential health risk (electrolytic reaction).	Soups
5	Treatment time Temperature	No evidence of toxicity	i. Not currently energy efficient. ii. Restricted to foods with low electrical conductivity	Heat-sensitive foods

Sl. No.	*Process Considerations*	*Benefits*	*Shortcomings*	*Examples of Applications*
6	Microbial species	—	Not suitable for product that contain bubbles	—
7	Microbial load	—	Scaling up of process difficult	—
8	Physiological age of organisms, Product acidity, Product conductivity, Presence of antimicrobials	—	—	—

Opportunities for PEF Processing [2]

The followings are the opportunities for PEF:

1. Continuous process.
2. Fresh-like quality
3. Retention of nutrients
4. Application for acid food
5. Low acid food: combined process for spore inactivation = under investigation.

Drawbacks [2]

There are following **drawbacks** of the PEF technology:

1. Ohmic heating occurs during the PEF discharge, which cause the temperature of the sample to rise, and hence a cooling system has to be in place in order to maintain a low temperature of the liquids. Therefore, a significant amount of energy is wasted in both unwanted heating up and cooling down the liquids.
2. Since the electrodes have to be immersed in the liquid, they are regarded as major contamination sources to the liquid due to the erosion of electrodes during discharge. Quality losses such as color deterioration have been observed. PEF technology is not being used to preserve foods commercially at present.
3. The combinations of PEF with other preservation technologies are being investigated to increase the lethal effect of this non-thermal process and to extend its application to different liquid foods.

Sources

1. *Hudaa Neetoo (Faculty of Agriculture, University of Mauritius, Réduit, Mauritius) and Haiqiang Chen (Department of Animal and Food Sciences, University of Delaware, Newark, Delaware, US), Alternative Food Processing Technologies.*
2. *Ranjan Saha, New and Emerging Non-Thermal Technologies in Food Preservation, April 6, 2015.*

1. Application of PEF in Milk/Food Processing

Pulsed electric field processing treatment has gained its popularity over the last two to three decades in the food processing industry. Major changes observed in this technique include: inactivation of micro-organisms, removal, pressing, osmotic dryness, and freezing. Application of PEF technology has been successfully demonstrated for the pasteurization of foods such as juices, milk, yogurt, soups, and liquid eggs. PEF is also applied to enhance extraction of sugars and other cellular content from plant cells, such as sugar beets. PEF also found application in reducing the solid volume (sludge) of wastewater.

Thre following technical issues that are important in the industrial application of PEF have been noted [11b]:

1. Determining the optimum electric field strength for inactivating bacteria
2. Provision to cool the food material that heats up due to Joule heating effects
3. Dielectric breakdown in foods
4. Proper selection of power and flow rates
5. Operational safety issues.

Due to the recent progress in the pulsed electric field processing, the commercial market for this process has increased [11].

A. Application of PEF in Milk Processing [1]

Freshly drawn raw milk has its own anti-microbial defence mechanism *i.e.* lactoferrin, lactoperoxidase, lysozyme, and possibly N-acetyl-β-D-glucosaminidase (NAGase) [1f]. But this anti-microbial defence mechanism of milk is temporary, which with time gets weaker, making milk prone to microbial spoilages.

Bovine milk contains easily digestible proteins, fats, along with lactose, minerals, vitamins, and other components, which makes it nature's single most complete food. The high water activity (0.995), and water content (~88 per cent), almost neutral pH and an ample supply of essential nutrients in milk make it a suitable growth medium for numerous spoilage and pathogenic micro-organisms. These properties make raw milk a highly perishable food and easily subject to microbial contamination [8h]. Consumption of raw milk, or products made from raw milk, has been associated with outbreaks of foodborne illness associated with several pathogen [8h]. Milk being a fluid food, is a good product eligible for PEF treatment [1b].

1. The occurrence of fewer undesirable changes in milk properties, undergoing PEF treatment, with reduction of bacterial load present in milk, was reported [1c].
2. Effects of PEF treatment was found to increase with increase in treatment temperature [1d].
3. Researchers [1e] reported microbial stability of PEF treated milk similar to thermally treated pasteurized milk, but without any thermal induced damages.

4. It was observed [1a] that raw skim milk when subjected to PEF treatment (intensity of 35 kV/cm; pulse with width 3 µs; and time of 9 µs) did not show any significant difference in color, pH, proteins, moisture and particle size. Application of PEF treatment (35 kV/cm; 2.3 µs of pulse width at 65°C for <10 s immediately after HTST pasteurization extended the shelf life of milk to 78 days at 4°C [2]. PEF treatment of bovine immunoglobulin enriched soymilk at dose of 41 kV/cm for 54 µs did not cause any significant change in bovine IgG activity but resulted in a 5.3 log reduction of initial microbial flora [3].

Microbial Inactivation and Shelf Life Enhancement of Milk [1]

Significant reduction in the count of spore forming bacteria in milk is possible with PEF treatment. PEF treatment results in less flavor degradation of milk than equivalent heat treatments, and does not exert any chemical or physical changes in milk fat, protein integrity, and casein structure. Hence, PEF may be used to preserve heat-sensitive dairy products such as whey protein concentrates. An extension of raw milk shelf life to 2 weeks at refrigeration temperature using PEF (two steps of seven pulses, and one step of six pulses at 40 kV/cm) was noted. There was no difference in the sensory quality of PEF treated and heat pasteurized milk. High intensity PEF (HIPEF) treatment of whole milk at 35.5 kV for 1000 µs with 7 µs bipolar pulses at 111 Hz reduced the mesophilic aerobic count from 3.2 log to 1 log cycle ensuring microbial stability for 5 days at 4°C without significant change in acidity, pH and FFA; proteolysis and lipolysis was not observed [4]. It was reported [5] that combination of PEF and heat treatment reduced the *Salmonella enteritidis* count in skim milk by 2.3 log cycle compared to 1.2 log cycle obtained by PEF treatment alone.

The shelf life of skim milk subjected to HIPEF treatment (40 kV/cm, 60 µs; 36 kV/cm, 84 µs) was enhanced to 14 days when stored at 4°C [6]. It was reported [4] that whole milk processed with HIPEF (35.5 kV/cm, 1000 µs) had a shelf life of 5 days at 4°C and change in acidity was not observed during storage. Subjecting chocolate milk to HIPEF (30 kV/cm, 45 µs) prior to heating at 105°C and 112°C for 31 s resulted in product having shelf life of 119 days at ambient (37°C) storage temperature.

The effectiveness of high pulsed electric field treatment of milk (Different types of milk) on bacterial inactivation is tabulated below [8g]:

Sl. No.	*Types of Milk*	*Bacteria*	*PEF Treatment Conditions (E, nb, duration and shape of the pulses, treatment time)*	*Mode*	*Temperature (°C)*		*Decimal Reduction No. (log N_0/N)*
					Inlet	*Outlet*	
1	Whole milk	*L. mono cytogenes* [6a]	25 to 35 kV·cm^{-1}; square waveform; 1.5 µs/pulse at1700 Hz; 100 to 600 µs.	Conti-nuous	25 50	n.d	1.5 to 4
		L. innocua [6b]	28 to 29 kV·cm^{-1}; 545 pulses maxi; series of 16 to 32 pulses at 1.1 or 100 Hz; waveform; 30 to 400 µs.	Batch	21.5	Esti-mated 25 to 45	0.2 to 2.5

Sl. No.	Types of Milk	Bacteria	PEF Treatment Conditions (E, nb, duration and shape of the pulses, treatment time)	Mode	Temperature (°C)		Decimal Reduction No. (log N_0/N)
					Inlet	Outlet	
2	2 per cent fat milk	*L.monocyto genes* [6a]	25 to 35 kV·cm⁻¹; square waveform; 1.5 µs/pulse at 1700 Hz; 100 to 600 µs	Conti-nuous	25 50	Not Deter-rmined	1 to 2.5
3	Skim milk	*L.monocyto genes* [6c]	15 to 30 kV·cm⁻¹; 5 to 50 pulses;waveform; 3.25 µs/pulse, 1pulse/min	Batch	35 0 50 55	Not Deter-rmined	0.03 to 0.87 0.3 (± 0.7) 1.5 (± 1.0) 4.5 (± 2.3)
		L. innocua [6b]	28 to 29 kV·cm⁻¹; 545 maxi; 16 to 32 pulses at 1.1 or 100 Hz; waveform; 30 to 400 µs	Batch	21.5	Esti-mated 25 to 45	0.2 to 1.5
		Staphyloccocus aureus [6d]	35 kV·cm⁻¹; 3.7 µs/ pulse at 250 Hz; bipolar square wave; 450 µs	Conti-nuous	7	< 40	3.7
		P. fluorescens	35 kV·cm⁻¹; 64 pulses; bipolar square wave; 47 µs (1 pass) to 188 µs (4 passes)	Conti-nuous	22		1 to 4 passes.
		Lactococcus lactis [6e]					0.5 to 2.2
		Bacillus cereus					0.1 to 0.3
		Total flora [6e]					0.6 to 3

n.d.: Not Deterrmined.

B. Application of PEF in Food Processing [7]

PEF research has mostly focused on the inactivation of microbes suspended in various pumpable non-particulate foods with free flowing characteristics including fruit juices, liquid eggs, milk, and pea soup [7a].

PEF alone in Food Processing

The continuous flow PEF system was used with electric field strength of 20 kv/ cm and flexible frequencies in apple, orange, and watermelon juices.

1. **PEF Treatment of Apple juice and tissue** -PEF has shown promise for liquid food processing by reducing/inactivating total counts, *L. monocytogenes*, *Salmonella* spp, *E. coli* O157: H7 and *S. aureus* [8h].
   a. **Apple juice:** A 6-log reduction of S. cerevisiae in PEF-treated apple juice after 10 pulses of 35 kV/cm at 22–34 °C was reported [7b],
   b. **Apple Tissues [11, 11b]:** A study [11b], related to the effects of PEF on vacuum freeze-drying process in apple tissues was conducted.

The apple slices were subjected to a PFF treatment at an electric field strength of E=800V/cm for various values of disintegration index Z. The vacuum cooling was related to lower the temperature to sub-zero level. This is when the freeze-drying experiment was performed at a pressure of 10m bar. It was found the PEF treatment resulted in the preservation of the dried samples, and it increased the tissue pores.

2. It was showed [7c] that the total aerobic counts of reconstituted orange juice were reduced by 3-4-log cycles when treated with a PEF unit operating at <32 kV/cm,
3. The fresh apple juice inoculated with *E. coli* O157: H7 and *E. coli* 8739 was treated with PEF and reported that PEF achieved a 5-log reduction with a treatment temperature of 35°C [7d],
4. A PEF unit for treatment of fresh orange juice at a pilot scale was described [7h]. It was concluded that the PEF treatment inactivated 99.9 percent of microbial flora, with the square waves being most effective. Compared with heat pasteurization, the PEF-treated orange juice retained more vitamin C and flavor [7h]. In orange juice, *Leuconostoc mesenteroides, E. coli,* and *L. innocua* were inactivated by as much as 5-log cycles at 50 kV/cm and 50°C [7e].
5. **Extraction of anthocyanin**-The study on application of PEF processing (mashed cabbage in batch chamber was exposed to PEF) on the extraction of anthocyanin from red cabbage, using water as a diluter, was carried out. The anthocyanin extraction was enhanced by 2.15 times [11].
6. The effect of PEF treatment (electric field strength of 5 kv/cm along with 3.6 kj/kg of energy was applied) on grape was studied [11, 11a]. It was found when wine phenolic content was increased, the effects lasted for a longer period of time.
7. **PEF processing of Cane sugar juice** [11]–PEF was applied to fresh sugarcane juice using the static treatment chamber with and without lemon and ginger, respectively.
   a. **Fresh juice without lemon and ginger**- The samples were subjected to different field strengths (from 30 KV cm{[1] and 50 KV cm{[1]) with restive pulse numbers (150 and 300). These were investigated at a room temperature (31°C) and refrigeration temperature(4°C) for 30 days and results were then compared to untreated samples at room temperature. It was found that the samples treated at fields strength 30 KV cm^{-1}, 150 pulses were more stable when compared to the untreated samples.
   b. **Fresh sugar cane juice with lemon and ginger**- The samples were subjected to the PFF process(10 KV cm^{-1}, 20 KV cm^{-1}, and 30 KV cm^{-1} with 150 pulses) for 14 days and were refrigerated at 4°C. The results indicated the best reduction of microbes was achieved at 20 KV cm^{-1} for 150 pulses [11].

Application of PEF with other Effects (Combined Effect)

Some authors have also reported the synergistic inactivation upon combination of several hurdles such as PEF, pH, water activity (aw), salt strength, temperature and the presence of antimicrobial compounds such as nisin and lysozyme. For instance:

1. The combined effect of temperature, acidity and a number of pulses on microbial inactivation in orange juice was studied [7f] and reported a 6-log reduction in the natural microbiota following application of 20 pulses of an electric field of 80 kV/cm at a temperature of 44°C and nisin concentration of 100 IU/mL.
2. It was shown [7g] that the application of PEF at moderately high temperatures of 60°C in combination with the antimicrobial compounds nisin and lysozyme resulted in higher inactivation ratios. It is thought that the presence of antimicrobials sensitizes microbial cells to injury and ultimately loss of viability.

2. Effect of PEF Treatment on Sensory and Nutritional Quality of Milk/Food

A. Effect of PEF Treatment on Sensory and Nutritional Quality of Milk [1]

1. The vitamin content including thiamine (B1), riboflavin (B2), cholecalciferol (D), and tocopherol (E) of PEF treated milk remained unchanged after treatment of up to 400μs at field strength of 18.3 to 27.1 kV/cm for 400μs; 93 per cent of the ascorbic acid was retained. This is significantly higher retention of vitamins as compared to LTLT and HTST pasteurized milks.
2. The pH, titratable acidity, fat content, fat and protein integrity, starter growth, rennet clotting time, yield, cheese production, calcium distribution, color, moisture and particle sizes remained unaffected by the PEF treatment at field strength of 20-80 kV/cm.
3. It was showed [8] that heat treatment of milk at a temperature > 65°C adversely affected its rennet coagulation property.
4. PEF treatment up to 35 kV/cm for 73μs did not exert any significant effect on the bovine IgG in a protein-enriched soymilk. Flavored milk, yogurt, and yogurt drink treated with PEF coupled with heat had lower yeast and mold count, showed increased shelf life and did not exhibit any change in the sensory score of resultant product [8a].
5. Since the viscosity of PEF-treated milk decreased and coagulation properties were enhanced for high field levels (45-55 $kV \cdot cm^{-1}$) with 2.1 to 3.5 μs cumulated treatment time and square waves [8g].
6. Any significant change in physical and chemical properties of PEF treated skim milk and the product was fresh-like with natural nutritional characteristics was not observed [8h].

B. Effect of PEF Treatment on Sensory and Nutritional Quality of Food [7]

In addition to dairy products, significant research has been also conducted to determine the effect of PEF on the sensory and nutritional quality of fruit juices.

1. The benefit of PEF over heat treatments in the retention of flavor in fruit juices was demonstrated [7f]. The application of combined PEF (80 kV/cm) and thermal treatment of up to 50°C was reported to have little to no effect on the vitamin C content of orange juice compared to heat-treated samples. PEF-treated apple juice and cider also had higher vitamin C retention [8b].
2. PEF-processed tomato juice presented better rheological properties following PEF processing, marked by smaller and more uniform-sized particles compared to their heat-treated counterparts. The overall higher acceptability of PEF-treated tomato juice was observed [8c], due to the reduced extent and rate of browning after treatment and during subsequent storage.
3. Although some researchers reported color fading of PEF-treated apple juice and cider after multiple pulses at 50-466 kV/cm [8d], other researchers have reported virtually no color loss after 94 µsec at 35 kV/cm.

3. Inactivation of Enzymes by PEF [1]

i. **Alkaline phosphatase enzyme**: The alkaline phosphatase enzyme activity in simulated milk ultrafiltrate (SMUF) subjected to electric field strength of 22 kV/cm and seventy pulses using static chamber was reduced by 65 per cent [8e].

ii. **Plasmin activity**: A 90 per cent reduction in plasmin activity in SMUF was attained by use of PEF at strength of 30 kV/cm. Destruction of plasmin by non-thermal PEF is of significance since its activity can cause bitterness and gelation in UHT milk.

iii. **Lipase activity**: PEF treatment at a dose of 21.5 kV/cm achieved 60 per cent reduction in the milk lipase activity too.

iv. **Protease and lipases Inactivation**: Inactivation of protease and lipases produced by psychrotropic micro-organism such as Bacillus and Pseudomonas by use of PEF is of significance in dairy industry since these enzymes are heat stable and some remain active even after high-temperature treatment. The inactivation of enzymes increased with field strength, treatment time, input energy and pulse frequency and was dependent on the composition of milk.

v. **The inactivation of enzymes achieved in skim milk was higher than that of whole milk:** It was reported [8f] a 90 per cent reduction in plasmin activity in SMUF at 30 kV/cm after 50 pulses. PEF treatment at 21.5 kV/cm and high energy input reduced the lipase and peroxidase activity of raw milk by 65 per cent and 25 per cent respectively with negligible effect on alkaline phosphatase activity [9].

4. Milk Processed by Pulsed Electric Fields

Evaluation of microbial quality, physico-chemical characteristics, and selected nutrients at different storage conditions [10]-Pulsed electric fields (PEF) technology was used to pasteurize raw milk under selected treatments. Processing conditions were as follow:

i. **For skim milk (SM):** Temperature 20, 30, and 40 °C, electric field 30.76 to 53.84 kV/cm, and pulse numbers 12, 24, and 30.

ii. **For whole milk (WM)):** 12, 21, and 30 (2 µs pulse width, monopolar).

Physico-chemical parameters (pH, electrical conductivity, density, color, solids nonfat [SNF]) and composition (protein and fat content) were measured after processing. Shelf life of SM and WM was assessed after processing at 46.15 kV/cm, combined with temperature (20 to 60 °C) and 30 pulses. Mesophilic and psychrophilic loads and pH were evaluated during storage at 4 and 21°C.

Results showed following trends:

1. Minor variations in physico-chemical properties after processing. **There was an interesting trend in SM in SNF, which decreased as treatment became stronger.**
2. Similar behavior was observed for fat and protein, showing a 0.18 per cent and 0.17 per cent decrease, respectively, under the strongest conditions.
3. Protein and fat content decreased in WM samples treated at 40°C, showing a decrease in protein (0.11 per cent), and an even higher decrease in fat content.
4. During storage, PEF-treated milk samples showed higher stability at 4 °C with minor variations in pH; after 33 d, pH was higher than 6. However samples at 21 °C showed faster spoilage and pH dropped to 4 after 5 d.
5. Growth of mesophilic bacteria was delayed in both milks after PEF processing, showing a 6- and 7-log cycles for SM and WM respectively, after day 25 (4°C); however, psychrophilic bacteria grew faster in both cases.

Sources

1. *Arjun Chaudhari, A. H. Jana and J. P. Prajapati, Application of Pulsed Electric Field for Milk Processing, Department of Dairy Technology, SMC College of Dairy Science, AAU, Anand-388110.*

1a. *Michalac SL, Alvarez VB and Zhang QH., 2003, Inactivation of selected micro-organisms and properties of pulsed electric field processed milk. J Food Process Preser, 27: 137-151.*

1b. *Soumitra Banerjee and Shanker Lal Shrivastav, Recent Trends in Milk Processing, Approches in poultry, dairy and Veterinary sciences 2015, Crimson Publishers, ISSN: 2576-9162.*

1c. Bendicho S, Barbosa CGV, Martýìn O (2002) Milk processing by high intensity pulsed electric fields. Trends in Food Science and Technology 13(6): 195-204.

1d . Sharma P, Bremer P, Oey I, Everett DW (2014) Bacterial inactivation in whole milk using pulsed electric field processing. International Dairy Journal 35(1): 49-56.

1e. Sharma P, Oey I, Bremer P, Everett DW 2017, Microbiological and enzymatic activity of bovine whole milk treated by pulsed electric fields. International Journal of Dairy Technology, doi: 10.1111/1471- 0307.12379.

1f. Ekstrand B, 1989, Antimicrobial factors in milk-a review. Food Biotechnology 3(2): 105-126.

2. Sepulveda-Ahumada DR, 2003, Preservation of fluid foods by pulse electric fields in combination with mild thermal treatments. Ph. D Thesis. Pullman WA: Washington State University.

3. Li SQ, Zhang QH, Lee YZ and Pham TV, 2003, Effects of pulsed electric field and thermal processing on the stability of bovine IgG in enriched soymilk. J Food Sci, 68: 1201-1207.

4. Odriozola-serrano I, Bendicho-porta S and Martin Belloso O., 2006, Comparative study on shelf life of whole milk processed by high intensity PEF or heat treatment. J Dairy Sci, 89: 905-911.

5. Floury J, Grosset N, Lesne E and Jeantet R., 2006, Continuous processing of skim milk by a combination of pulsed electric field and conventional heat treatments: does a synergetic effect on microbial inactivation exist? Lait, 86: 203-211.

6. Fernandez-Molina JJ, Fernandez-Gutierrez SA, Altunakar B, Bermudez-Aguirre, Swanson, B G and Barbosa-Canovas GV, 2005, The combined effect of pulsed electric fields and conventional heating on the microbial quality and shelf-life of skim milk. J Food Process Preserv, 29: 390-406.

6a. Reina L. D., Jin Z. T., Zhang Q. H., Yousef A. E., Inactivation of Listeria monocytogenes in milk by pulsed electric fields, J. Food Prot. 61, 1998, 1203-1206.

6b. Picart L., Dumay E., Cheftel J. C., Inactivation of Listeria innocua in dairy fluids by pulsed electric fields: influence of electric parameters and food composition, Innov. Food Sci. Emerg. Technol. 3, 2002, 357-369.

6c. Fleischman G. J., Ravishankar S., Balasub-ramaniam V. M., The inactivation of Listeria mono-cytogenes by pulsed electric field (PEF) treatment in a static chamber, Food Microbiol.21, 2004, 91-95.

6d. Evrendilek G. A., Zhang Q. H., Richter E. R, a pplication of pulsed electric fields to skim milk inoculated with Staphylococcus aureus, Biosyst. Eng. 87, 2004, 137-144.

6e. Michalac S., Alvarez V., Ji T., Zhang Q. H., Inactivation of selected micro-organisms and properties of pulsed electric field processed milk, J. Food Proc. Preserv, 27, 2003, 137-151.

7. Hudaa Neetoo (Faculty of Agriculture, University of Mauritius, Réduit, Mauritius) and Haiqiang Chen (Department of Animal and Food Sciences, University of Delaware, Newark, Delaware, US), Alternative Food Processing Technologies.

7a. *Vega-Mercado H, Gongora-Nieto MM, Barbosa-Canovas GV, Swanson BG, (1999), Non-thermal preservation of liquid foods using pulsed electric fields. In: Rahman MS (ed) Handbook of Food Preservation. New York: Marcel Dekker, pp. 487-520.*

7b. *Qin BL, Zhang Q, Barbosa-Canovas GV, Swanson BG, Pedrow PD, 1995, Pulsed electric field treatment chamber design for liquid food pasteurization using a finite element method. Transactions of the American Society of Agricultural Engineers 38: 557-562.*

7c. *Zhang QH, Qiu X, Sharma SK, 1997 Recent development in pulsed electric field processing. In: New Technologies Yearbook. Washington, DC: National Food Processors Association, pp. 31-42.*

7d. *Evrendilek GA, Zhang QH, Richter ER, 1999, Inactivation of Escherichia coli O157: H7 and Escherichia coli 8739 in apple juice by pulsed electric fields. Journal of Food Protection 62: 793-796.*

7e. *McDonald CJ, 2000, Effects of pulsed electric fields on micro-organisms in orange juice using electric field strengths of 30 and 50 kV/cm. Journal of Food Science 65: 984-9870*

7f. *Hodgins AM, Mittal GS, Griffiths MW, 2002, Pasteurization of fresh orange juice using low-energy pulsed electric field. Journal of Food Science 67: 2294-2299.*

7g. *Liang Z, Mittal GS, Griffiths MW, 2002, Inactivation of Salmonella typhimurium in orange juice containing antimicrobial agents by pulsed electric field. Journal of Food Protection 65: 1081-1085.*

7h. *Kalchayanand, N, Sikes, T., Dunne, C. P. and Ray, B. 1994. Hydrostatic pressure and electroporation have increased bactericidal efficiency in combination with bacteriocins. In J. Appl. Environ. Microbiol. 60: 4174-4177.*

8. *Fox PF, Guinee TP, Cogan TM, and Mc Sweeney PLH, 2000, Fundamentals of Cheese Science. Aspen Publishers, Gaithersburg, MD.*

8a. *Yeom HW, Evrendilek GA, Jin ZT and Zhang QH., 2007, Processing of yoghurt based products with pulsed electric field: microbial, sensory and physical evaluations. J Food Process Preserv, 28: 161-178.*

8b. *Evrendilek GA, Jin ZT, Ruhlman KT, Qiu X, Zhang QH, Richter ER, 2000, Microbial safety and shelf-life of apple juice and cider processed by bench and pilot-scale PEF systems. Innovative Food Science and Emerging Technologies 1: 77-86.*

8c. *Min Z, Zhang QH, (2003), Effects of commercial-scale pulsed electric field processing on flavor and color of tomato juice. Journal of Food Science 68: 1600-1606.*

8d. *Ortega-Rivas E, Zarat-Rodriguez E, Barbosa-Canovas GV, 1998, Apple juice pasteurization using ultra-filtration and pulsed electric fields. Food and Bioprocess Technology 76: 193-198.*

8e. *Castro AJ.m, 1994, Pulsed electric field modification of activity and denaturation of alkaline phosphate. Ph. D Dissertation, Washington State University, Washington.*

8f. *Vega HM, Powers JR, Barbosa-Canovas GV, Swanson BG, and Luedecke L., 1995, inactivation of protease from Pseudomonas fluorescens M3/6 using high voltage pulse electric fields. IFT 1995. Annual Meeting. Book of Abstracts, paper no. 89, pp. 267-278.*

8g. *Juliane Floury, Noël Grosset, Nadine Leconte, Maryvonne Pasco, Marie-Noëlle madec, Romain JEANTET, Continuous raw skim milk processing by pulsed electric field at non-lethal temperature: effect on microbial inactivation and functional properties, Lait 86, 2006, 43-57.*

8h. *Dipendra Khanal, Non-thermal processing of skim milk: Impact on microbial reduction, physico-chemical properties and quality of Brie type cheese, 2014, A Thesis presented to The University of Guelph, Ontario, Canada.*

9. *Grahl T. and Märkl H, 1996, Killing of micro-organisms by pulsed electric fields. Applied Microb. Biotec., 45: 148-157.*

10. *Bermúdez-Aguirre D, Fernández S, Esquivel H, Dunne PC, Barbosa-Cánovas GV., Milk processed by pulsed electric fields, Food Sci. 2011 Jun-Jul; 76(5): S289-99.*

11. *Harlin Kaur Saroya, Innovative non-thermal food processing technologies used by the food industry in the United States, MSc thesis, 2017, Western Kentucky University, Kentuckyand.*

11a. *Puértolas, E., Saldaña, G., Alvarez, I., and Raso, J., Effect of pulsed electric field processing of red grapes on wine chromatic and phenolic characteristics during aging in oak barrels, Journal of Argiculture and Food Chemistry, 58, 2351-2357, 2010.*

11b. *Sale, A. J. H. and Hamilton, W. A.1967. In J. Biochim. Biophys. Acta 148: 781788.*

11c. *Singh R. Paul, Professor of Food Engineering Department of Biological and Agricultural Engineering, Department of Food Science and Technology University of California Davis, FAO-2001, Technical Elements of New and Emerging Non-Thermal Food Technologies.*

Chapter 5

Oscillating Magnetic Fields

Preservation of foods with high energy Oscillating Magnetic Fields (OMF) is a potential, excellent alternative to conventional methods, as it inactivates micro-organisms and denatures enzymes in fresh or prepared foods. Static (SMF) and oscillating (OMF) magnetic fields have been explored for their potential as microbial inactivation methods. OMF has the potential to improve the quality and shelf life compared to other conventional process. OMF effects the micro-organisms, cell membranes, malignant cells and the potential application of magnetic field for food preservation. For SMF, the magnetic field intensity is constant with time, while an OMF is applied in the form of constant amplitude or decaying amplitude sinusoidal waves [1a]. Exposure to high intensity oscillating magnetic fields has been reported to have a variety of effects on biological systems ranging from selective inactivation of malignant cells to the inactivation of bacteria on packaging materials and in foods. Treatment times are very short, typically from 25 ms to a few milliseconds, and field strengths are very high, typically from 2 Tessla to about 100 Tessla at frequencies between about 5-500 kHz. Efficacies of treatments did not exceed about 100-fold reductions in numbers of vegetative micro-organisms inoculated into milk *(Streptococcus thermophilus),* orange juice (*Saccharomyces* spp.), bread rolls (**mould spores**) and no inactivation of bacterial spores has been reported, so the practical potential for the technique, as it has been developed so far, appears to be limited [1c]. The magnetic field can be either of the followings [1, 1a]:

Homogenous Methodology

In a homogeneous magnetic field, the field intensity B is uniform in the area enclosed by the magnetic field coil.

Heterogeneous Methodology

On the other hand, in a heterogeneous field, B is non uniform, with the intensities decreasing as distances from the center of the coil increases.

Oscillating Magnetic Fields (OMF) [1] are known for their microbial inactivation methods. The OMF are applied in the form of decaying or constant amplitude sinusoidal waves. OMF applied in the form of pulses reverses the charge for each pulse, and the intensity of each pulse decreases with time to about 10 per cent of the initial intensity [1b]. OMF are applied as pulses, thereby, reversing the charge of each pulse. The intensity of each pulse decreases with time to 10 per cent of the initial intensity. For future development, there is a requirement of systems, where power source, number pulses, and frequencies applied to the food can be monitored [2].

The **advantages of OMF** include the followings:

1. Aseptic treatment of solid and liquid foods inside a flexible package
2. Generation of minimal heat within the food
2. Reduction of energy requirements [2].

Sources

1. *Harlin Kaur Saroya, Innovative non-thermal food processing technologies used by the food industry in the United States, MSc thesis, 2017, Western Kentucky University, Kentucky.*

1a. *U. S. Food and Drug Administration (USDA) Center for Food Safety and Applied Nutrition June 2, 2000, Kinetics of Microbial Inactivation for Alternative Food Processing Technologies Oscillating Magnetic Fields.*

1b. *Pothakamury, U. R., Barbosa-Cánovas, G. V., and Swanson, B. G.1993, Magnetic-field inactivation of micro-organisms and generation of biological changes. Food Technol.47(12): 85-93.*

1c. *Ranjan Saha, New and Emerging Non-Thermal Technologies in Food Preservation, April 2015.*

2. *Barbosa Canovas, G., Schaffner, D., Pierson, M., and Zhang, Q., Oscillating magnetic fields. Journal of Food and Science, 65, 86-89, 2000.*

Oscillating Magnetic Field Working and Equipment [1, 1a]

Preservation of foods with OMF involves sealing food in a plastic bag and subjecting it to 1 to 100 pulses in an OMF with a frequency between 5 to 500 kHz at temperatures in the range of 0 to 50 °C for a total exposure time ranging from 25 to 100 ms [1a].

The oscillating magnetic fields are produced by passing a fluctuating current by the electromagnets [2]. The field strengths are high when compared with the earth magnetic field, which would approximately be 5 to 100 tesla. Frequencies ranging between 5-500 kHz are applied for around 25µs to a few milliseconds. This helps in the inactivation of vegetative cells. However, if the frequencies are higher than 500 kHz, the food gets heated, and the inactivation process is not effective. In recent studies, it has been observed there are not much of an effect on the spores or enzymes. In addition, it can lead to the growth of vegetative cells. For instance, when an oscillating magnetic field strength of 12.0 T was applied to milk at a frequency of 6000 Hz, cells were reduced from 25,000 to 970ml^{-1} [3]. Magnetic field treatments are carried out at atmospheric pressure and at moderate temperatures. The temperature of the food increases 2-5 °C.

The region in which a magnetic body is capable of magnetizing the particles around is called the magnetic field. Isotropic susceptibility and anisotropic susceptibility of magnetization are there. Magnetic field is measured in terms of magnetic intensity B. Magnetic fields are homogeneous and heterogeneous in nature [4]. Magnetic flux density is a force that an electromagnetic source exerts on charged particles. Magnetic flux density is measured in Telsa (1 Telsa =104 gauss) [4].

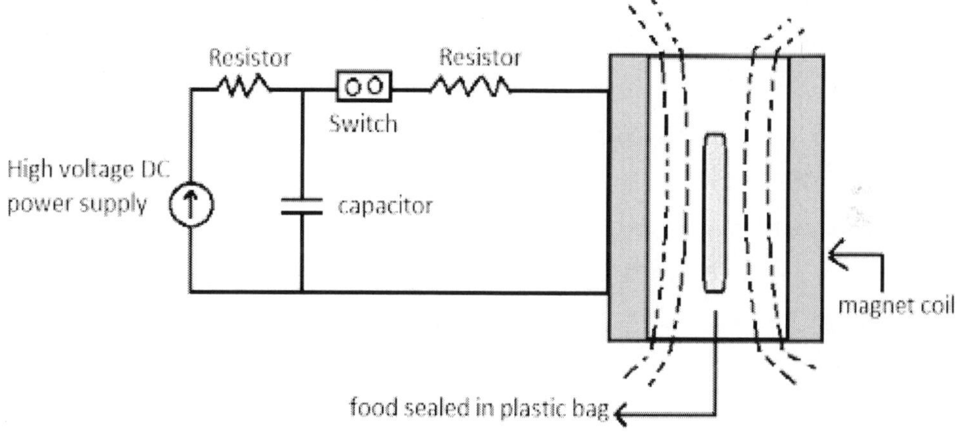

Circuit for OMF System [5].

Sources

1. *Harlin Kaur Saroya, Innovative non-thermal food processing technologies used by the food industry in the United States, MSc thesis, 2017, Western Kentucky University, Kentucky.*

1a. *U. S. Food and Drug Administration (USDA) Center for Food Safety and Applied Nutrition June 2, 2000, Kinetics of Microbial Inactivation for Alternative Food Processing Technologies Oscillating Magnetic Fields.*

2. *Barbosa Canovas, G., Schaffner, D., Pierson, M., and Zhang, Q., Oscillating magnetic fields. Journal of Food and Science, 65, 86-89, 2000.*

3. *Fellows, P., Food processing technologies: Princilpes and practices (3rd ed ed.). Boca Raton, FL: CRC Press, 2009.*

4. *Maya Sharma, Ph. D. Scholar at CTAE, Udaipur, Oscillating magnetic field for food processing.*

5. *Praveen Kumar, OMF for food processing, IIT, Kharagpur.*

Process Control and Monitoring [1]

The following controls are required:

1. Data acquisition systems must be installed in the processing area to monitor and control the power source, number of pulses, and frequencies applied to the food.
2. Food composition, temperature, size of unit, among other factors also would require control and monitoring to assure constant treatments.
3. Any deviation from the specified conditions such as temperature changes must be continuously recorded and appropriate responses taken.
4. If the system shuts down or fails to deliver the described treatment during processing, the food must be reprocessed to assure quality and safety [1].

Inhibition of Micro-organism and Mechanisms of Microbial Inactivation [1]

Inhibition of Micro-organism [1a]

According to the study [2] exposure to magnetic fields causes inhibition in the growth and reproduction of micro-organisms. OMF of intensity (magnetic flux densities) of 5 to 50 telsa (T) and frequency of 5 to 500 kHz was applied and reduced the number of micro-organisms by at least 2-log cycles. Within the magnetic field of 5-50 T, the amount of energy per oscillation coupled to 1 dipole in the DNA is 10^{-2} to 10^{-3} eV [2].

OMF of this intensity can be generated by using the following methods:

1. Superconducting coils
2. Coils which produce DC fields, or
3. Coils energized by the discharge of energy stored in a capacitor [2a].

Inhibition or stimulation of the growth of micro-organisms exposed to magnetic fields may be a result of the magnetic fields themselves or the induced electric fields. The latter is measured in terms of induced electric field strength and induced current density. To differentiate between electric field and magnetic field effects, a cylindrical enclosure containing cells and a medium that can be adapted to in vitro studies employing uniform, single-phase, extremely low frequency (ELF) magnetic fields is recommended.

Mechanisms of Microbial Inactivation [1]

SMF or OMF have some potential to inactivate micro-organisms in food. The following two theories were reported [1a] to explain the inactivation mechanisms for cells placed in SMF or OMF:

i. The First Theory

It stated that a "weak" OMF could loosen the bonds between ions and proteins. Many proteins vital to the cell metabolism contain ions. In the presence of a steady background magnetic field (such as that of the earth), the biological effects of OMF are more pronounced around particular frequencies, the cyclotron resonance frequency of ions.

An ion entering a magnetic field B at velocity v experiences a force F given by F = qv * B, it is known as ICR model. When v and B are parallel, F is zero. When v is normal to B, the ion moves in a circular path. For other orientations between v and B, the ions move in a helical path. The frequency at which the ions revolve in the magnetic field is known as the ion's gyro frequency, denoted by 'n'. Which depends on the charge/mass ratio of the ion and the magnetic field intensity [1b].

ii. Second Theory

It considers the effect of SMF and OMF on calcium ions bound in calcium-binding proteins, such as calmodulin. The calcium ions continually vibrate about an equilibrium position in the binding site of calmodulin. A steady magnetic field to calmodulin causes the plane of vibration to rotate, or proceed in the direction of magnetic field at a frequency that is exactly equal of the cyclotron frequency of the bound calcium. Adding a "wobbling" magnetic field at the cyclotron frequency disturbs the precision to such an extent that it loosens the bond between the calcium ion and the calmodulin [1a].

It was suggested [2] that the inactivation of micro-organisms may be based on the theory that the OMF may couple energy into the magnetically active parts of large critical molecules such as DNA. Within 5-50 T range, the amount of energy per oscillation coupled to 1 dipole in the DNA is 10^{-2} to 10^{-3} eV. Several oscillations and collective assembly of enough local activation may result in the breakdown of covalent bonds in the DNA molecule and inhibition of the growth of micro-organisms [1a].

The effects of magnetic fields on microbial growth and reproduction are classified in following three types [3]:

1. Inhibitory
2. Stimulatory
3. None observable.

The inactivation of micro-organisms with OMF in milk, yogurt, orange juice, and bread roll dough was reported [2]. According to this report [2] only 1 pulse of OMF was adequate to reduce the bacterial population between 10^2 and 10^3 cfu/g. The intensity of the magnetic field required to achieve these effects varied between 2-25 T and a frequency range from 5-500 Hz. The effect of magnetic fields on micro-organisms are summarised in table [1a] given on next page.

Sl. No.	*Micro-organism*	*Type of Magnetic Filed*	*Field Strength T*	*Frequency of pulse Hz*	*Effect*
1	*Streptococcus themophilus* in milk [4]	Omagnetic field	12.0	6,000 (1 pulse)	Cell population reduced from 25,000 cells/ml to 970.
2	*Escherichia coli* [4]	Smagnetic field	0.3	0	Growth simulated.
3	*E. coli* [4]	Omagnetic field	0.15	0.05	Inactivation of cells when concentration was 100 cells/ml.
4	*Saccharomyces* in yogurt [2]	Omagnetic field	40.0	416,000 (10 pulses)	Cell population reduced from 3, 500 cells/ml to 25.
5	*Saccharomyces* in orange juice [2]	Omagnetic field	40.0	416,000 (1 pulse)	Cell population reduced from 25,000 cells/ml to 6.

Note: S magnetic field = Static magnetic field; O magnetic field = Oscillating magnetic field.

Sources

1. *U. S. Food and Drug Administration (USDA) Center for Food Safety and Applied Nutrition June 2, 2000, Kinetics of Microbial Inactivation for Alternative Food Processing Technologies Oscillating Magnetic Fields.*

1a. *Pothakamury, U. R., Barbosa-Cánovas, G. V.and Swanson, B. G. (1993). Magnetic-field inactivation of micro-organisms and generation of biological changes. Food Technol. 47(12): 85-93.*

1b. *Maya Sharma, Ph. D. Scholar at CTAE, Udaipur, Oscillating magnetic field for food processing.*

2. *Hofmann, G. A. 1985. Deactivation of micro-organisms by an oscillating magnetic field. U. S. Patent 4, 524,079.*

2a. *Gersdorf, R., deBoer, F. R., Wolfrat, J. C. Muller, F. A., Roeland, L. W. 1983. The high magnetic facility of the University of Amsterdam, high field magnetism. Proceedings International symposium on High Field Magnetism. Osaka, Japan. 277-287.*

3. *Yoshimura, N. 1989. Application of magnetic action for sterilization of food. Shokukin Kihatsu 24(3): 46-48.*

4. *Moore, R. L.1979. Biological effects of magnetic fields. Studies with micro-organisms. Can. J. Microbiol., 25: 1145-1151.*

Effects of Oscillating Magnetic Fields in Food Processing [1]

Some factors *e.g.* magnetic field intensity, electrical resistivity, and microbial growth stage are believed to influence microbial inactivation. The details related to the effects of some factors on OMF are tabulated on next page.

Sl. No.	Factors	Effect
1	Electrical Resistivity	Foods need to have a high electrical resistivity (greater than 10 to 25 ohms-cm) for micro-organisms to be inactivated by OMF. The applied magnetic field intensity depends on the electrical resistivities and thickness of the food being magnetized, with larger magnetic fields intensities used with products with large resistivity and thickness.
2	Magnetic Field	Exposure to a magnetic field may stimulate or inhibit the growth and reproduction of micro-organisms. A single pulse of intensity of 5 to 50 T and frequency of 5 to 500 kHz generally reduces the number of micro-organisms by at least 2-log cycles [2]. High intensity magnetic fields can affect membrane fluidity and other properties of cells.
3	Microbial Growth Stage	Working with homogeneous (7 T) and inhomogeneous (5.2 to 6.1 T and 3.2 to 6.7 T) magnetic fields, a growth stage dependent response of Escherichia coli bacterial cultures was found [3]. The ratio of cells under magnetic field to cells under geomagnetic field was less than 1 during the first 6 h of treatment and greater than 1 after 24 h. It was also found that cell survival was greater under inhomogeneous compared with homogeneous fields.

Sources

1. *U. S. Food and Drug Administration (USDA) Center for Food Safety and Applied Nutrition June 2, 2000, Kinetics of Microbial Inactivation for Alternative Food Processing Technologies Oscillating Magnetic Fields.*
2. *Hofmann, G. A. 1985. Deactivation of micro-organisms by an oscillating magnetic field. U. S. Patent 4, 524,079.*
3. *Tsuchiya, K., Nakamura, K., Okuno, K., Ano, T. and Shoda, M. 1996. Effect of homogeneous and inhomogeneous high magnetic fields on the growth of* Escherichia coli*. J Ferment Bioeng 81(4): 343-346.*

Applications of Oscillating Magnetic Fields in Food Processing

Some examples related to the application of OMF in food processing are given below:

i. Processing of Crab Sticks [1]

The effects of OMF on crab sticks were analyzed [2]. The crab sticks were frozen both with and without oscillating magnetic field application. The entire process was observed for approximately 12 months and during this process, various quality attributes were assessed. The results revealed that the oscillating magnetic field had no effect on the crab sticks. The drip loss, water-holding capacity, toughness, and whiteness remained the same. Additionally, the quality of the frozen samples also remained the same. It is important to note during this experiment; the strength of the oscillating magnetic field was lower than 2mT. This is only two orders of magnitude more than earth's natural magnetic field. The frequency range was between 6 to 59Hz.

ii. Processing of Garlic Bulbs [1]

Researchers [3] conducted a study to find out the reaction of garlic bulbs when subjected to oscillating magnetic fields, as compared to freezing them under normal conditions. The results of this study showed substantial cooling in the garlic bulbs, during some of the freezing trials. However, freezing stage the oscillating magnetic fields had a significant effect on the garlic bulbs over the normal freezing method. Also, the researchers concluded super-cooling was more effective with garlic bulbs if they were frozen at an initial ambient 21 ± 1°C rather than 4 ± 5 °C.

iii. Effect of OMF Pulses on the Survival of Selected Micro-organisms [4]

The antimicrobial effect of oscillating magnetic field pulses at the 20T amplitude, 250 s period and a full time of duration below 1 ms, on micro-organism (Ervinia carotovora, Streptomyces scabies and Alternaria solani, taken from paralysed potatoes), was studied.

OMF System Used Study [4]

Oscillating pulsing magnetic field that has an induction up to several score T and growing time of orders of magnitude ms can result from discharging a **high voltage condenser through a suitable solenoid**.

1. The following parameters of the condenser and the solenoid as well as the way the power connection has been constructed:

   Parameters of the Condenser - Capacity, Induction, Charge, and Voltage

   A battery of high-voltage condensers (C) with a total capacity of 1.5 mF is charged up to the required voltage through a high voltage supply, with a maximum working voltage of 5 kV and a working current up to 100 mA.

   Parameters of the Solenoid -Induction, Geometry of winding.

   The solenoids (L) producing magnetic field are single-layer coils which have been wound with a copper wire of 20-40 mm^2 surface and a rectangular section. Epoxy-glass composite material has served as the internal and external coil isolation. The *ca.* 17mm internal diameter of the solenoids has been chosen to enable placing standard test-tubes with analyzed material inside. The coils were placed inside a special holder which provides the appropriate mechanical resistance. It was rarely possible to use the same coil more than 10 times at the highest applied voltage.

2. The discharge of the condenser batteries went through the spark gap which has an electrode gap precisely regulated by a step engine. In that system, there is another alternative discharging system built on the basis of high-tension thyristors, which allows getting single unidirectional magnetic field pulse.
3. Directing the process of charging and pulse discharging is controlled remotely from the control panel placed, for the safety reasons, in a different room.

4. The parameters of the oscillating magnetic field were accurately measured by a induction probe placed for that very purpose inside the solenoid and connected with the **digital oscilloscope Voltcraft DSO2100**.
5. The magnetic field generated in the solenoids had features of strongly suppressed oscillation vibrations with the growing magnetic induction of over 100 kT s-1 of speed. The oscillation periods were from 250 to 350 s depending on the coils used. The results of the research were obtained while using a 250 s oscillation period coil.
6. A full pulse would last below 4 periods which means that its duration is *ca.* 1 ms. An example (for a coil producing pulse of 250 s period and 20 T amplitude) of oscillating magnetic field pulse route with a marked period, amplitude and the total duration read out from the digital oscilloscope Voltcraft DSO 2100, was used in study.
7. The Kruscal-Wallis and LSD test were used for statistical analysis of differences in reaction of the tested micro-organisms treated with pulses of oscillating magnetic field.

Observation [4, 5, 6]

The tested microbiological material had different reactions when treated with the magnetic field. The tested bacteria *Ervinia carotovora* and *Streptomyces scabies* were more sensitive to the oscillating magnetic field pulses effects. The tested *Alternaria solani* fungus turned out to be the least sensitive to the analyzed magnetic field effects. Treatment with pulses at the maximum amplitude of 20T caused the reduction of the surviving bacteria to *ca.* 4000 times in the case of **Ervinia carotovora,** *ca.* 1000 times in the case *Streptomyces scabies,* and only 30 times for the fungus Alternaria solani. On the basis of observations, following conclusions were derived:

1. The micro-organisms used in the study are good as indicators in testing the lethal activity of oscillating magnetic field.
2. A varied oscillating magnetic field pulses effect has been shown on the tested micro-organisms. The treatment with the oscillating magnetic field pulses reduces the number of tested bacteria in liquid nutrient medium 4 000 times and the number of tested fungus 30 times.

Sources

1. *Harlin Kaur Saroya, Innovative non-thermal food processing technologies used by the food industry in the United States, MSc thesis, 2017, Western Kentucky University, Kentucky.*
2. *Otero, L., Pérez-Mateos, M., Rodríguez, A. C., and Sanz, P. D., Electro-magnetic freezing: Effects of weak oscillating magnetic fields on crab sticks. Journal of Food Engineering, 200, 87-94, 2017.*
3. *James, C., Reitz, B., and James, S., The freezing characteristics of garlic bulbs(Allium sativum L.) frozen conventionally or with the assistance of an oscillatingweak magnetic field. Food and Bioprocess Technology, 8(3), 702-708, 2014.*

4. *J. Lipiec, P. Janas, and W. Barabasz, Effect of oscillating magnetic field pulses on the survival of selected micro-organisms, Int. Agrophysics, 2004, 18, 325-328.*

5. *Pothakamary U. R., Barbosa-Canovas G. V., and Swanson B. G., 1993, Magnetic-field inactivation of micro-organisms and generation of biological changes. Food Technology, 47, 85-88.*

6. *Ray B., 2001. Fundamental Food Micro-biology. Second Edition. CRC Press, Boca Raton, London, New York, Washington.*

Chapter 6

High Hydrostatic Pressure Technology

High pressure processing (HPP) [1, 2], is also described as **high hydrostatic pressure (HHP), or ultra high pressure (UHP) processing** in literature [1a, 1b]. The application of high hydrostatic pressure (HHP) in food preservation has received particular attention as a viable alternative (economically and technologically) to thermal processes [2, 3]. High pressure processing has been demonstrated as a non-thermal technique to eliminate pathogenic and spoilage bacteria as well as inhibit several enzymes bring about to enhance safety and shelf-life extension for foods. Furthermore, this technique can be also preserved color, flavor and nutritional values, including anti-oxidative components and properties of fruit and vegetable products.

High Pressure Processing (HPP) is a way to process foods without using heat. It has the potential to produce high-quality, fresh, nutritious, safe-to-eat foods without using chemical preservatives. Unlike thermal treatment, HPP does not significantly affect the flavour and nutritional value of foods. Therefore, with HPP it is possible to produce safe fresh foods with enhanced organoleptic (smell, taste *etc.*), nutritional and functional properties with an extended shelf life. The successful development of HPP in Australia would make it possible to commercialise high quality foods such as meats, seafood, and horticultural and dairy products, which will contribute to greater financial returns to the Australian agrifood industry [1c].

Developmental Steps for HHP Technology

It has been the subject of intense research effort over the last 15-20 years. The details related to the development of HHP technology are given below:

1. Among the array of non-thermal processing technologies, HHP has garnered the most attention since the early 1990s.

2. HPP was first used in 1899 in the United States and towards 1990, the research began when better equipment was developed, and products like fruits, meats, and juices were being made in the USA [3a].
3. The first research on the effect of high pressure on food was first carried out in the nineteenth century [4] describing an increase in shelf-life for products such as milk, fruit and other foods, but the scientific development, its application in the food industry, and the foodstuff marketing are much more recent and have taken place in the past two decades [5, 6].
4. For the last 20 years, HHP has been explored by food research institutions as well as the food industry with the goal of enhancing the safety, quality, nutritional, and functional properties of a wide variety of foods with minimal deleterious effects on their nutritional and organoleptic characteristics [1, 7, 8].
5. During commercial HHP processing, foods are exposed to pressures of the order of 200-1000 MPa for a few minutes using a suitable pressure-transmitting fluid such as water.

This non-thermal process is currently being used commercially in the food industry to target specific pathogens in specific food products such as Vibrio para-haemolyticus and *Vibrio vulnificus* in oysters, and *Listeria monocytogenes* as post-process treatment on sliced deli meats, guacamole and juices [9]. **One of the principal advantages of the HHP process is the expanded shelf-life and improvement of food safety** due to the inactivation affected in the microbial population. The loss of viability of micro-organisms through HHP is probably the result of a combination of injuries in the cell. The resistance of micro-organisms is highly variable, depending mainly on the type of organism and the food matrix involved, *e.g.* spores show great resistance to inactivation [2]. **Food products pasteurized by high pressure are now commercially available in a number of countries including Japan, France, Spain, North America, and the UK [1].** HHP-processed products experience a significant extension of shelf-life. Nonetheless, HHP-processed foods are usually non sterile and therefore must be refrigerated to maintain their sensory characteristics and microbiological stability [2].

While planning an HHP treatment to ensure the microbiological stability of food and its safety, it is important to consider the utilization of fractional factor designs as a reliable and cost effective methodology in experiment planning. **The synergistic effect** of pre- and post-processing factors ("hurdle" effect) should be explored, since microbial inactivation rates can be further enhanced, as compared to HHP processing alone [2].

Sources

1. *Hudaa Neetoo (Faculty of Agriculture, University of Mauritius, Réduit, Mauritius) and Haiqiang Chen (Department of Animal and Food Sciences, University of Delaware, Newark, Delaware, US), Alternative Food Processing Technologies.*

1a. *FDA, 2014, Kinetics of Microbial Inactivation for Alternative Food Processing Technologies - High Pressure Processing. U. S. Food and Drug Administration.*

1b. *Harlin Kaur Saroya, Innovative non-thermal food processing technologies used by the food industry in the United States, MSc thesis, 2017, Western Kentucky University, Kentucky.*

1c. *Ranjan Saha, New and Emerging Non-Thermal Technologies in Food Preservation, April 2015.*

2. *E. Rendueles, M. K. Omer, O. Alvseike, C. Alonso-Calleja, R. Capita and M. Prieto, Microbiological food safety assessment of high hydrostatic pressure processing: A review, journal homepage: www.elsevier.com/locate/lwt.*

3. *Patterson, M. F, 2005, Microbiology of pressure-treated foods. Journal of Applied Microbiology, 98, 1400-1409.*

3a. *Tewari, G., Jayas, D., and Holley, R. (1999). High pressure processing of foods: An overview. Science Aliments, 19, 619-661.*

4. *Hite, B. H, 1899, The effect of pressure in the preservation of milk. Bulletin of West Virginia University Agricultural Experiment Station, 58, 15-35.*

5. *Farr, D, 1990, High pressure technology in the food industry. Trends in Food Science and Technology, 1, 14-16.*

6. *Considine, K. M., Kelly, A. L., Fitzgerald, G. F., Hill, C., and Sleator, R. D, 2008, High pressure processing effects on microbial food safety and food quality. FEMS Microbiology Letters, 281(1), 1-9.*

7. *Jung S, Tonello-Samson C, Lamballerie-Anton MD, 2011, High hydrostatic pressure food processing. In: Proctor A (ed) Alternatives to Conventional Food Processing. London: RSC Publishing, pp. 254-306.*

8. *Tewari G, 2007, Microwave and radiofrequency heating. In: Tewari G, Juneja V K (eds) Advances in Thermal and Non-Thermal Food Preservation. Ames, IA: Blackwell Publishing, pp.91-98,.*

9. *Lori F Pivarnik and Randy Worobo, Non-thermal or Alternative Food Processing Methods to Enhance Microbial Safety and Quality, Frequently Asked Questions, (USDA-NIFA #2011-68003-30005).*

Process and Principle

The HPP process included immersing the packages of food in liquid, and then suddenly releasing the pressure uniformly throughout the food. By doing so, the pressure is applied evenly, and all parts of the food receive the same treatment. At present, industrial application is widespread for a range of pressures between 100 and 800 MPa, depending on the desired objective. The HPP process is known to be a non-thermal process. However, when pressure is applied to the foods, the temperature in then rise (approximately 3 °C per100 MPa, depending on the composition of the food) due to adiabatic heating. This is generated by the density of water and food components [1]. This can go higher if the foods contain fat. Once the depressurization is complete, the temperature falls back due to adiabatic cooling [2].

HHP relies on the **isostatic** principle (meaning that the food product is compressed instantaneously and uniformly from every direction and returns to its original shape when the pressure is removed**) and adiabatic**, which means that no matter the food shape or size, there is little variation in temperature with increasing pressure [3, 4]. This prevents the food from being deformed or heated, which would modify its organoleptic properties. This hydrostatic compression is capable of inactivating micro-organisms. HHP can also be combined with heat, where the process temperatures during treatment can vary from subzero temperatures to above the boiling point of water (100°C) [5].

The basic principles of high pressure technology determine the behavior of foods under pressure. These basic principles include [1, 6] the following important principles:

I. **Le Chatelier's Principle**- Any reaction that results in decrease in volume are enhanced whereas, reactions that increases the volume are repressed.

II. **Isostatic Principle**- A consistent pressure is applied on the foods from every direction. This process compresses the foods and once the pressure is released, they return to their original shape.

III. **Principle of Microscopic Ordering**- At a contact temperature, if the pressure is increased, the degree of ordering of molecules of a given substance also increases

If the food product contains enough moisture within itself, the pressure will not cause any damage at the macroscopic level if the pressure is being applied uniformly [7].

In a high-pressure process, the pressure vessel is filled with a food product and pressurized for a desired time, following which it is depressurized. The time required to pressurize the vessel is influenced by the compressibility of the pressure medium and the nature of the food material. If water is used as the pressure-transmitting medium for most food materials, compressibility is similar to that of the pressure medium. Typically, the pressurization time for foods is independent of the quantity of food placed in the pressure vessel. The presence of air in the food increases the pressurization time, since air is considerably more compressible than water. After pressurization, the food is kept under high pressure for the required process time, which may be for several minutes. Upon completion of the pressure exposure, depressurization can be done quite rapidly [8].

Sources

1. *Harlin Kaur Saroya, Innovative non-thermal food processing technologies used by the food industry in the United States, MSc thesis, 2017, Western Kentucky University, Kentucky.*

2. *P. Fellows, Food processing technologies: Princilpes and practices (3rd ed.), Boca Raton, FL: CRC Press., 2009.*

3. *J. P. Smelt, Recent advances in the microbiology of high pressure processing. Trends in Food Science and Technology, 9, 152-158, 1998.*

4. *Wilson, D. R., Dabrowski, L., Stringer, S., Moezelaar, R. and Brocklehurst, T. F, 2008, High pressure in combination with elevated temperature as a method for the sterilisation of food. Trends in Food Science and Technology, 19, 289-299.*

5. *Caner C, Hernandez RJ, Harte BR, 2004, High-pressure processing effects on the mechanical, barrier and mass transfer properties of food packaging flexible structures: a critical review. Packaging Technology and Science 17: 23-29.*

6. *C. Blany, and P. Masson, Effects of high pressure on proteins. Food Reviews International, 9(4), 611-628, 1993.*

7. *Crawford, Y. J., Murano E, A., Olson, D., and Shenoy, K., Use of high hydrostatic pressure and irradation to eliminate clostridium sporogenes in chicken breast, Journal of Food Protection, 59, 711-715, 1996.*

8. *Ranjan Saha, New and Emerging Non-Thermal Technologies in Food Preservation, April 2015.*

High Hydrostatic Pressure Equipment [1]

HHP equipment (huge and bulky) consists of the following components [9, 3]:

1. A pressurized vessel - A high-pressure vessel along with its end closures and a way to confine the end closures [2, 3].
2. A low-pressure pump [2]
3. A pressure generation system-An intensifier to generate higher pressures [2, 3]
4. System control - A data collection system, along with controls and some instrumentation [2, 3]
5. A temperature control device [3]
6. A material handling system [3].

A Pressurized Vessel

The pressure vessel is a very important component of this equipment in which food materials are subjected to high pressure. In the case of food applications, a unique requirement for the high-pressure vessel is that it must undergo several thousand processing cycles per year to process large volumes of foods. The large number of required pressurized and depressurized cycles increases metal fatigue and reduces the life of the vessel. Furthermore, the vessel itself must be protected from any corrosion due either to the food material itself or to any liquids used for cleaning, and must be easy to clean [3a].

There are quite a few aspects one should keep in mind when designing the vessel. For instance, it has to be dimensionally stable in a safe-fail way. That means, if the vessel is failing, it should breakdown with a leak. There are pressure-conducting fluids in the vessel that allow the pressure to be distributed uniformly to the product sample [9].

The construction of vessels depends on one of three common approaches, depending on the vessel operating pressure and diameter. The details related to the three cylindrical vessel designs [4] are given below:

i. **A single forged monolithic chamber-** The monolithic vessel or "monoblock" is commonly fabricated for operating pressures of 400-600 MPa for small vessels with internal diameters not exceeding 15 cm.
ii. **Prestressed multilayered (multiwall)**-a series of concentric tubes shrunk fit on each other to form a multiwall chamber. Used for high pressure and larger dimensions.
iii. **Wire-wound vessels**-a stainless steel core tube compressed by wire winding [4]. Used for high pressure and larger dimensions.

Closing and Opening Systems HHP Vessels

These vessels require rapid closing and opening systems that allow fast loading and unloading of the product. Small-diameter or lower pressure vessels typically use threaded- or beech-type closures or in some cases a pin closure [4]. At higher pressures and larger diameters, closure loads become so large that a secondary structure such as a yoke or external frame is required to hold the end closures in place [4].

Volume of Vessels

The volume of the vessel may vary from less than 1 L (for laboratory-scale applications operating at <1000 MPa) to up to 500 L for processing units operating at 600 MPa.

Working [1, 9]

The foods to be processed should be packaged flexibly when they are loaded into the high-pressure chamber. A disposable liner is injected into the stainless-steel cylinder and filtered water is used as the isostatic compression fluid. Finally, the pressure vessels are sealed. When operating the unit, a pressure-transmitting fluid such as potable tap water is pumped from the reservoir into the pressure vessel and pressure is generated either directly or indirectly by compression.

Pressurizing Systems [3a]

The following two types of pressurization systems defined as indirect and direct, are commonly employed in the industry:

Indirect Compression

In the indirect compression method, an intensifier is used, which pumps the fluid directly into the vessel until the desired pressure is reached [1, 9]. The intensifier is a high-pressure pump used to increase the pressure to desired levels. The intensifier is separate from the high-pressure vessel. This system requires high-pressure tubing and appropriate fittings to convey the pressurized medium to the pressure vessel [3a].

Direct Compression

In a direct system, the pressure intensifier is located within the pressure vessel. In this system, both pressure intensifier and the vessel are fabricated as a single unit, and the total size of the vessel can be quite large [3a]. This uses a piston to compress the hydrostatic medium in the vessel [3]. In the direct method, the fluid is compressed in the vessel by moving a piston using hydraulic pressure, thereby reducing the volume [9].

Temperature control before and during processing can be achieved by pumping a heating/cooling medium around the pressure vessel in the larger or more modern systems [5]. The various components of the production systems have been shown in following figure:

A Horizontal HPP Production System [9, 10].

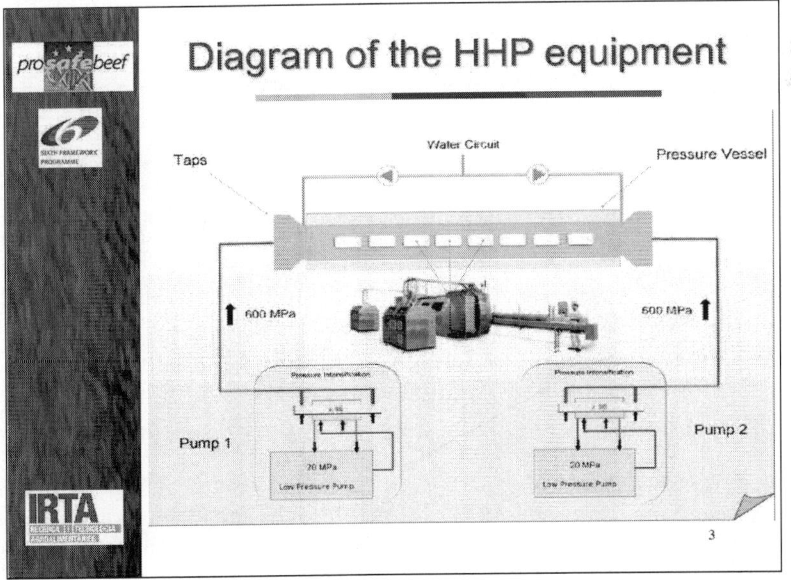

Other Design for HHP [9a].

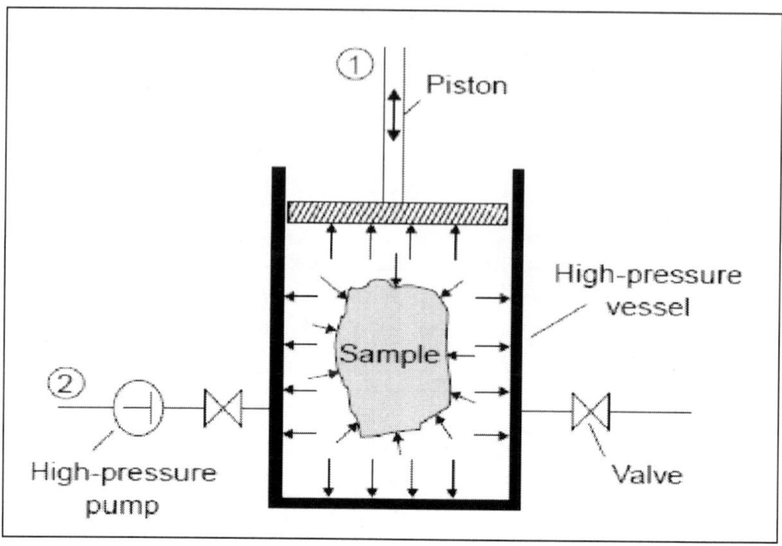

Other Design for HHP [9b].

Mode of Operation/Arrangements of HPP System

In a high-pressure process, the pressure vessel is filled with a food product and pressurized for a desired time, following which it is depressurized. A simplified mode of operation used in a high-pressure processing system is shown below [3a]:

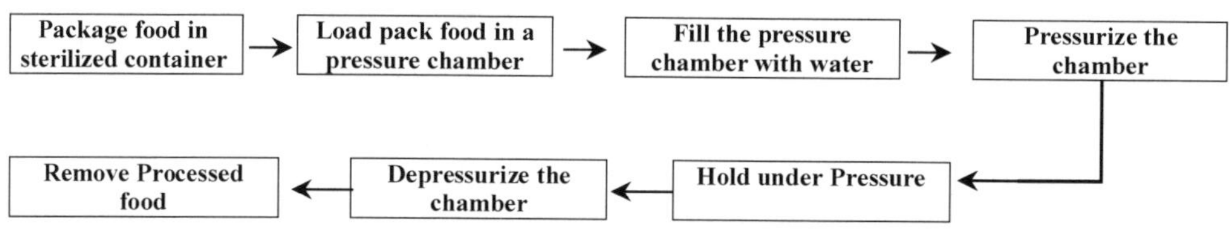

A Schematic Flow Diagram of High-pressure Processing.

The time required to pressurize the vessel is influenced by the compressibility of the pressure medium and the nature of the food material [3a].

- ✰ If water is used as the pressure transmitting medium for most food materials, compressibility is similar to that of the pressure medium.
- ✰ Typically, the pressurization time for foods is independent of the quantity of food placed in the pressure vessel.
- ✰ The presence of air in the food increases the pressurization time, since air is considerably more compressible than water.

After pressurization, the food is kept under high pressure for the required process time, which may be for several minutes. Upon completion of the pressure exposure, depressurization can be done quite rapidly.

Conceptually, HPP system can be arranged in to following two ways [1, 3a, 9]:

1. A Semi-continuous Bulk Mode

This HHP system treats unpackaged pumpable liquids or slurries in a semi-continuous bulk mode [1, 9]. This system involves a combination of multiple pressure vessels that are sequenced to provide a continuous flow. In this while one vessel is being pressurized, another may be in a decompression mode [3a]. The reduction in cost of a semicontinuous process is about 27 percent over a batch process for 500 litre per hour production [3b].

2. Batch Mode

This HHP system treats packaged food in a batch mode [9]. Most of the high-pressure equipment currently used is operated in the batch mode. Since pressurizing and depressurizing steps can be rapidly accomplished, the low efficiency associated with batch processing may be minimized. Rapid pressurizing and depressurizing cycles also can cause metal fatigue and reduce the life of equipment. Above 4 000 kg/cm^2, the weight of equipment increases significantly, as does its cost [3a]. Batch mode offers following advantages [1, 3a, 9]:

i. A major advantage of the batch over the continuous process is that it reduces the risk of large quantities of product becoming contaminated by the lubricants or wear particles of the machinery [6] and potential post processing contamination. Different types of foods can be processed without cross-contamination, there is no need for clean up between runs, the equipment is relatively simple, and there is no risk of large quantities of foods becoming contaminated in case of equipment malfunction [3a].

ii. In addition, the productivity of the batch processes can be increased by operating the pressure vessels in tandem, with no lag in the processing times, so that the system operates sequentially [6]. Several pressure vessels may be operated in a controlled sequence to minimize any time lag associated with the time required for the pressurization of vessels [3a].

iii. Examples of products suitable for batch processing are cooked meats, stews, guacamole, fruits in juice, shellfish, and ready-meal entrées [7, 8].

With HHP, equipment reliability has traditionally been an issue but with advances in instrumentation and experience gained by equipment manufacturers in a number of installations for high-pressure pasteurization, the issue seems to have been resolved. Although the capital cost of HHP equipment remains an issue, the concept of treating a variety of foods in flexible containers that are scalable to larger containers or larger pasteurization high-pressure vessels offers a lot of opportunity to the food industry.

Wet/Dry Bag Configuration [3a]

a. Wet Bag Configuration

This is more suitable for food processing and in this, a mould is first filled with the material outside the pressure vessel. The filled mould is then moved into the pressure vessel containing a pressure medium. With cold isostatic pressure, water is used as the pressure medium.

b. Dry Bag Configuration

In this, the mould is fixed in place within the pressure vessel. The material to be treated is filled into the mould. The mould remains separated from the pressure medium by an elastomer tool.

In high-pressure processing, inert gases or water are the most commonly used pressure media. The relative incompressibility of water compared with gases makes it the preferred pressure medium in many applications. The decrease in volume of water is about 5 percent when its pressure is increased from 0 to 4000 kg/cm^2 at 22° C. This volume reduction is much smaller compared with inert gases, where high volume reductions can make operations more hazardous. When water is used as a pressure medium when subjecting food materials to high pressure, there is instantaneous and uniform transmission of the pressure throughout the product being treated. Typically, small amounts of oil may be added into the water for anticorrosive and lubricant purposes [3a].

Sources

1. *Hudaa Neetoo (Faculty of Agriculture, University of Mauritius, Réduit, Mauritius) and Haiqiang Chen (Department of Animal and Food Sciences, University of Delaware, Newark, Delaware, US), Alternative Food Processing Technologies.*

2. *Guan D, Hoover DG, 2005, Novel non-thermal treatments. In: Sapers GM, Gorny JR Yousef AE (eds) Microbiology of Fruits And Vegetables. Boca Raton, FL: CRC Press, pp. 497-523.*

3. *Mertens B, 1995, New Methods of Food Preservation. New York: Blackie Academic and Professional, pp. 135.*

3a. *Singh R. Paul, Professor of Food Engineering Department of Biological and Agricultural Engineering, Department of Food Science and Technology University of California Davis, FAO-2001, Technical Elements of New and Emerging Non-Thermal Food Technologies.*

3b. *Besser, R. E., Lett, S. M., Weber, J. T., Doyle, M. P., Barrett, T. J., Wells, J. G. and Griffin, P. M. 1993. An outbreak of diarrhea and hemolytic uremic syndrome from Escherichia coli O157: H7 in fresh-pressed apple cider. In J. Amer. Med. Assoc. 269: 2217-2220.*

4. *Ting E, 2010, High pressure processing equipment fundamentals. In: Zhang HQ, Barbosa-Cánovas GV, Balasubramaniam B, Dunne CP, Farkas DF, Yuan JT (eds) Non-thermal Processing Technologies for Food. Oxford: Wiley-Blackwell.*

5. *Fonberg-Broczek M, Arabas J, Kostrzewa E et al. (1999), High-pressure treatment of fruit, meat and cheese products -equipment, methods and results. In: Oliveira FAR, Oliveira JC (eds) Processing Foods: Quality Optimization and Process Assessment. Boca Raton, FL: CRC Press, pp. 282-298.*

6. *Palou E, Lopez-Malo A, Barbosa-Canovas GV, Swanson BG, 2007, High-pressure treatment in food preservation. In: Rahman MS (ed) Handbook of Food Preservation. New York: CRC Press, pp. 815-853.*

7. *Jung S, Tonello-Samson C, Lamballerie-Anton MD, 2011, High hydrostatic pressure food processing. In: Proctor A (ed) Alternatives to Conventional Food Processing. London: RSC Publishing, pp. 254-306.*

8. *Patterson MF, Ledward DA, Rogers N, 2006, High-pressure processing. In: Brennan JG (ed) Food Processing Handbook. Wokingham: Wiley-Vch, pp.173-197b.*

9. *Harlin Kaur Saroya, Innovative non-thermal food processing technologies used by the food industry in the United States, MSc thesis, 2017, Western Kentucky University, Kentucky.*

9a. *Gopal jee Chourasia, Application of High hydro pressure technology, Agricultural and Food Engineering Deptt., IIT, Kharagpur.*

9b. *A. Goyal, Ravinder Kaushik, Vivek Sharma, M. K. Sihag, High Pressure Processing and Its Impact on Milk Proteins: A Review, Researchgate.*

10. *Huang, H., Wu, S., Lu, J., Shyu, Y., and Wang, C., Current status and future trends of high-pressure processing in food industry. Food Control, 72, 1-8, 2017.*

Commercial and Pilot Scale HPP System [1]

Some examples of commercial HPP system are given below:

1. Commercial HHP System

i. The Wakayama plant in Japan commercially used this system for the treatment of tangerine juice. This system involves a combination of multiple pressure vessels that are sequenced to provide a continuous flow. In this while one vessel is being pressurized, another may be in a decompression mode. In this process, three 50 litre pressure cells are sequenced to achieve a production rate of 4 000 litre per hour [2].

ii. The pressure system, known as ACB high-pressure liquid processor (GEC Alsthom ACB, France) is equipped with a chamber having an internal volume of 4 litres. The compression process is done with water up to a maximum pressure of 400 MPa. Programmable pressure controllers are used to adjust pressurization and decompression rates. This unit has been used in selected processing steps in wine production [1].

iii. **QUINTUS Food Press**- This is a equipment developed specifically for food processing. The pressure vessel is pre-stressed using a spring steel wire and remains in a pre-stressed state even under pressure. A replaceable liner is inserted into the cylinder for additional safety of operation. The press uses a retractable frame fabricated of pre-stressed wire winding in order to keep the top and bottom closures safely in place. The press is pressurized with an external pressure intensifier. As a total system, the QUINTUS Press may be incorporated in a bulk processing line, where the product is kept within large bulk containers during processing and storage [1].

2. Pilot Scale

A pilot-scale ultra-high-pressure food processor manufactured by Flow International Corporation (Kent, Washington) is used. This unit was developed for use with pumpable products. The food is pumped into the pressure vessel, the pressure is raised and held for the required time and the food is discharged into filling containers by a computer-controlled process. An ultra-high-pressure pump (FLOW WaterNifeâ) is used to pump the food into the chamber. In a multi-chamber system, filling, pressurizing and discharging operations can be appropriately sequenced to achieve maximum production rates [1].

Sources

1. *Singh R. Paul, Professor of Food Engineering Department of Biological and Agricultural Engineering, Department of Food Science and Technology University of California Davis, FAO-2001, Technical Elements of New and Emerging Non-Thermal Food Technologies.*
2. *Deuchi, T. Hayashi, R. 1992. In High Pressure and Biotechnology, Eds. C. Balny, R. Hayashi, K. Heremans and P. Masson. 353-355. Montroge: INSERM/John Libbey Eurotext Ltd, England.*

Relation between Pressure and Temperature [1]

The relationship between pressure and temperature in a typical HHP was studied and described [2]. The molecules with a lower molecular weight have lesser effect on them through the HPP process. Therefore, this helped with the retention of vitamins, flavor compounds, and pigments in the processed food. At the same time, there are some compounds, which change during the HPP process. For instance, gelatinization of carbohydrates is achieved once the pressure is increased and not when the temperature increases. Additionally, proteins can be altered from their natural quality with an increase in temperature [2]. The relation between pressure and temperature in HPP can be easily understood by the following figure [3]:

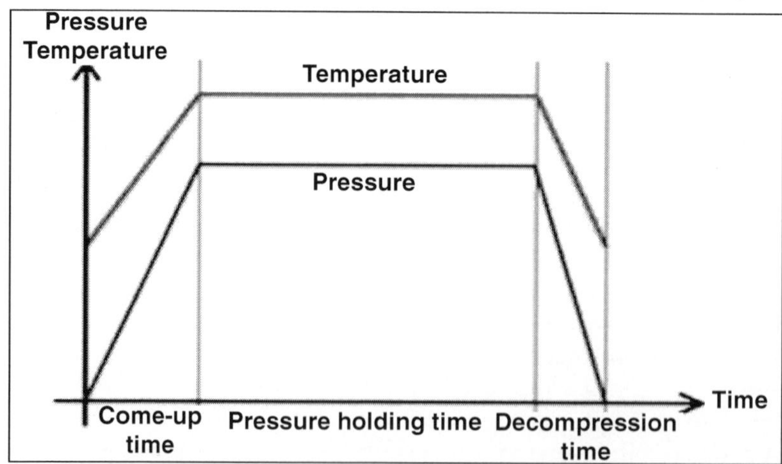

Pressure, Temperature, and Time during a HPP Process.

Sources

1. *Harlin Kaur Saroya, Innovative non-thermal food processing technologies used by the food industry in the United States, MSc thesis, 2017, Western Kentucky University, Kentucky.*

2. *Muntean, M., Marian, O., Barbieru, V., Catunescu, G. M., Ranta, O., Drocas, I., and Terhes, S., High pressure processing in food industry - Characteristics and Applications. Agriculture and Agricultural Science Procedia, 10, 377-383, 2016 and 3-Ferstl, C., and P, F., Insights on technology and regulatory requirements, The NFLWhite Paper Series, 10, 1-6, 2013.*

Effects of HHP on Molecules, Structures and Microbiota

The details related to the effect of HHP on molecular structure and microbiota are given below:

A. Effects of HHP on Molecules and Structures [1]

The effects of HHP on molecule and structure are summarized below:

i. High hydrostatic pressure does not alter the low-energy, covalent bonds, which have low compressibility and do not break within the ranges of pressures normally used in food. Therefore the primary structure of molecules such as proteins or fatty acids remains intact [2].

ii. The ionic bonds and hydrophobic interactions, responsible for maintaining the secondary and tertiary structure of proteins, are disrupted and this event is associated with decreases in volume [2, 3, 4].

iii. The secondary, tertiary or quaternary structure of large molecules and complex organized structures such as membranes are altered, due to the rupture of ionic bonds, some hydrogen bonds and hydrophobic and electrostatic interactions.

iv. Molecules such as vitamins, amino acids, flavour molecules or other low-molecular-weight compounds are hardly affected and as a result, the organoleptic and nutritional properties are slightly modified [3, 5].

v. The different chemical composition and structural properties of the cell membrane in Gram-positive and Gram-negative micro-organisms result in differences in resistance to HHP. **Gram-positive bacteria are generally more resistant compared to Gram-negative [6].**

vi. Fatty acids(FA) are partially responsible for resistance to HHP lethal effects [7].

vii. There are other components and cellular functions sensitive to high pressures that are altered or inhibited such as the ribosome [8], the protein synthesis and enzyme activity [2, 9] and the structure of DNA-enzyme complexes, as the DNA can be degraded due to the action of endonucleases not normally in contact with DNA [10].

viii. A series of morphological and structural changes in the cell, such as the separation of the membrane from the cell wall, the lengthening of the cell, the compression of gas vacuoles [12] and the condensation of nuclear material [9, 13] are also described.

Sources

1. *E. Rendueles, M. K. Omer, O. Alvseike, C. Alonso-Calleja, R. Capita, M. Prieto, Microbiological food safety assessment of high hydrostatic pressure processing: A review, journal homepage: www.elsevier.com/locate/lwt.*

2. *Considine, K. M., Kelly, A. L., Fitzgerald, G. F., Hill, C., and Sleator, R. D. (2008). High pressure processing effects on microbial food safety and food quality. FEMS Microbiology Letters, 281(1), 1-9.*

3. *Aymerich, T., Picouet, P. A., and Monfort, J. M, 2008, Decontamination technologies for meat products. Meat Science, 78(1-2), 114-129.*

4. *Ross, A. I. V., Griffiths, M. W., Mittal, G. S., and Deeth, H. C, 2003, Combining non-thermal technologies to control food borne micro-organisms. International Journal of Food Microbiology, 89(2e3), 125-138.*

5. *Farkas, D., and Hoover, D. G, 2000, High pressure processing. Journal of Food Science, Supplement Kinetics of Microbial Inactivation for Alternative Food ProcessingTechnologies, 65, 47-64.*

6. *Shigehisa, T., Ohmori, T., Saito, A., Taji, S., and Hayashi, R, 1991, Effects of high hydrostatic pressure on characteristics of pork slurries and inactivation of micro-organisms associated with meat and meat products. International Journal of Food Microbiology, 12(2-3), 207-215.*

7. *Casadei, M. A., Manas, P., Niven, G., Needs, E., and Mackey, B. M, 2002, Role of membrane fluidity in pressure resistance of Escherichia coli NCTC 8164. Applied and Environmental Microbiology, 68(12), 5965-5972.*

8. *Kaletunc, G., Lee, J., Alpas, H., and Bozoglu, F, 2004, Evaluation of structural changes induced by high hydrostatic pressure in Leuconostoc mesenteroides. Applied and Environmental Microbiology, 70(2), 1116-1122.*

9. *Wouters, P. C., Glaasker, E. and Smelt, J. P, 1998, Effects of high pressure on inactivation kinetics and events related to proton efflux in Lactobacillus plantarum. Applied and Environmental Microbiology, 64(2), 509-514.*

10. *Chilton, P., Isaacs, N. S., Mackey, B., and Stenning, R, 1997, The effects of high hydrostatic pressure on bacteria. In K. Heremans (Ed.), High pressure research in the biosciences and biotechnology (pp. 225-228). Belgium: Leuven University Press.*

11. *Patterson, M. F, 2005, Microbiology of pressure-treated foods. Journal of Applied Microbiology, 98, 1400-1409.*

13. *Manas, P., and Mackey, B. M, 2004, Morphological and physiological changes induced by high hydrostatic pressure in exponential- and stationary-phase cells of Escherichia coli: relationship with cell death. Applied and Environmental Microbiology, 70(3), 1545-1554.*

B. Effects of HHP on Microbiota [1]

The utility of HHP technology is due to the destruction suffered by the microbial population, which allows a substantial increase in shelf-life and improves food safety [2]. HHP applied at ambient temperature destroys vegetative cells and inactivates certain enzymes [3], with a minimal change in the organoleptic properties [4]. The effectiveness of the treatment depends primarily on the pressure applied and on the holding time. The resistance of micro-organisms is highly variable, depending mainly on the type of organism and the food matrix. The pressure can also impair sub lethal injury to the cell.

i. **Prokaryotes compared to eukaryotes:** High hydrostatic pressure usually has a higher destructive effect in organisms with a greater degree of organization and structural complexity. **Prokaryotes are usually more resistant, compared to eukaryotes [4a]**. The destruction of protozoa and parasites is achieved with relatively low pressures.

ii. **Molds and yeasts: They have intermediate resistance [5, 6]**. As for moulds, mycelia are particularly susceptible, but mould spores particularly resistant to high pressures have been reported (Neosartorya, Talaromyces) [7, 8].

Viability Loss of Vegetative Pathogens and Foodborne Viruses by HHP [1]

Sl. No.	*Pathogens and Viruses*	*Substrate*	*P (MPa)*	*Time (min)*	*Temp. (°C)*	*Inactivation*
1	**Food borne viruses (PFU)**					
	Hepatitis A virus	Oysters	400	1	10	>3 log
	Norovirus	Oysters	400	5	5	4 log
2	**Vegetative pathogens (CFU)**					
	Campylobacter jejuni	Pork slurry	300	10	25	6 log
	Salmonella senftenberg 775W	Strained baby food	340	10	23	<2 log
	Salmonella enteritidis	Broth	345	10	35	22 log
	Escherichia coli O157: H7	Poultry meat	600	15	20	3 log
	Staphylococcus aureus	Poultry meat	600	15	20	3 log
	S. aureus 765	Broth	345	10	35	4 log
	Listeria monocytogenes CA	Poultry meat	375	15	20	2 log
	L. monocytogenes	Broth	345	10	35	5 log
	Vibrio parahaemolyticus O3: K6	Oysters	300	3	10	5 log

iii. **Pathogens:** In studies with food borne vegetative **pathogens [9], large viability losses** (reductions in excess of 8 logs) are shown when relatively low pressures (300 MPa) are combined with intermediate temperatures

(50°C, 5 min). Large differences among pathogenic micro-organisms (*Listeria monocytogenes, Staphylococcus aureus, Escherichia coli, Salmonella typhimurium*) have been reported, with reductions in the range of 0.5-8.5 log units.

iv. **Spores:** They **show great resistance to inactivation** by high pressure. The genera Bacillus and Clostridium comprise significant species as food borne spore forming pathogens, such as *Clostridium botulinum, Clostridium perfringens* and *Bacillus cereus*. **For most forms of vegetative bacteria, significant reductions (usually higher than 4 log units) in the population are achieved when 400-600 MPa at room temperature are used. However, under these conditions, significant reductions in the load of spores are not achieved.**

v. **Viruses:** Viruses possess a **wide range of pressure resistance**, depending on their structural diversity. Enveloped viruses are usually more sensitive to the pressure than naked viruses. Reduced time-pressure combinations are effective in decreasing the infective power of hepatitis A virus [10, 11] and norovirus [12].

Viability Loss of Foodborne Pathogenic Spores by HHP [1]

Sl. No.	*Pathogenic Spores*	*P (MPa)*	*Time (min)*	*Temp. (0C)*	*Inactivation*
1	*B. cereus*	400	25	30	0.5 log
2	*B. cereus*	600	-	60	6-7log
3	*C. perfringens*	500	30	25, 45 and 65	Minimal or no reduction
4	*C. perfringens* type A	650	15	75	3.7 log
5	*C. botulinum* type E	827	5	50-55	Approx. 5 log
6	*C. botulinum* non-proteolytic type B	827	20	75	> 6 log
7	*C. botulinum* type A	827	20	75	2-3 log
8	*C. botulinum* proteolytic Type B	600	70	80	5 log
9	*C. botulinum* proteolytic Type B	800	4	80	2.3 log
10	*C. botulinum* proteolytic Type A	600	6	80	5 log

Sources

1. *E. Rendueles, M. K. Omer, O. Alvseike, C. Alonso-Calleja, R. Capita, M. Prieto, Microbiological food safety assessment of high hydrostatic pressure processing: Areview, journal homepage: www.elsevier.com/locate/lwt.*

2. *Considine, K. M., Kelly, A. L., Fitzgerald, G. F., Hill, C., and Sleator, R. D, 2008, High pressure processing effects on microbial food safety and food quality. FEMS Microbiology Letters, 281(1), 1-9.*

3. *Simpson, R. K., and Gilmour, A, 1997, The effect of high hydrostatic pressure on the activity of intracellular enzymes of Listeria monocytogenes. Letters in Applied Microbiology, 25(1), 48-53.*

4. *Farkas, D., and Hoover, D. G, 2000, High pressure processing. Journal of Food Science, Supplement e Kinetics of Microbial Inactivation for Alternative Food Processing Technologies, 65, 47-64.*

4a. *Yuste, J., Capellas, M., Fung, D. Y. C., and Mor-Mur, M, 2001, High pressure processing for food safety and preservation: a review. Journal of Rapid Methods and Automation in Microbiology, 9(1), 1-10.*

5. *Palou, E., Lopez-Malo, A., Barbosa-Canovas, G. V., Welti-Chanes, J., Davidson, P. M., and Swanson, B. G, 1998a, Effect of oscillatory high hydrostatic pressure treatments on Byssochlamys nivea ascospores suspended in fruit juice concentrates. Letters in Applied Microbiology, 27(6), 375-378.*

6. *Shimoda, M., Kago, H., Kojima, N., Miyake, M., Osajima, Y., and Hayakawa, I, 2002, Accelerated death kinetics of Asper-gillus niger spores under high-pressure carbonation. Applied and Environmental Microbiology, 68(8), 4162-4167.*

7. *Smelt, J. P, 1998, Recent advances in the microbiology of high pressure processing. Trends in Food Science and Technology, 9, 152-158.*

8. *Chapman, B., Winley, E., Fong, A. S. W., Hocking, A. D., Stewart, C. M., and Buckle, K. A, 2007, Asco spore inactivation and germination by high pressure processing is affected by ascospore age. Innovative Food Science and Emerging Technologies, 8(4), 531-534.*

9. *Alpas, H., Kalchayan and, N., Bozoglu, F., Sikes, A., Dunne, C. P., and Ray, B, (1999), Variation in resistance to hydrostatic pressure among strains of food-borne pathogens. Applied and Environmental Microbiology, 65(9), 4248-4251.*

10. *Calci, K. R., Meade, G. K., Tezloff, R. C., and Kingsley, D. H, 2005, High-pressure inactivation of hepatitis A virus within oysters. Applied and Environmental Microbiology, 71(1), 339-343.*

11. *Kingsley, D. H., Guan, D., Hoover, G. D., and Chen, H, 2006, Inactivation of hepatitis A virus by high pressure processing: the role of temperature and pressure oscillation. Journal of Food Protection, 69, 2454-2459.*

12. *Buckow, R., Isbarn, S., Knorr, D., Heinz, V., and Lehmacher, A, (2008), Predictive model for inactivation of Feline Calicivirus, a norovirus surrogate, by heat and high hydrostatic pressure. Applied and Environmental Microbiology, 74(4), 1030-1038.*

C. Effect of High Hydrostatic Pressure on the Sensory and Nutritional Quality of Foods [1]

Covalent bonds of molecules are unaffected by pressure at the level used in food processing. As a result, small molecules that contribute to the color, flavor or nutritional quality of a food are relatively unchanged after exposure to pressure [2].

1. Tomatoes treated at 600 MPa for 1 h at 25°C underwent minimal change in lycopene and carotenoid content [3]. Moreover, the antioxidant capacity

of pressure-treated carrots and tomatoes was highly comparable to their untreated counterparts.

2. The inactivation of multiple vitamins suspended in a model system was also compared [4]. Their overall conclusions were that almost all vitamins and nutrients were minimally affected by pressure.
3. Although it is generally agreed that HHP has minimal effect on the sensory quality of foods, certain researchers have shown that inorganic transition metal ions can be released from their respective compounds under the effects of pressure, and these have been shown to adversely affect the flavor and shelf-life of the treated product due to accelerated lipid oxidation [5].
4. In addition, enzymes are functional proteinaceous macromolecules that are stabilized by ionic as well as hydrophobic forces, and often mediate reactions that dictate the various quality attributes of the product. Since pressure can disrupt the structure and function of enzymes, the organoleptic characteristics of HHP-treated foods may change upon processing [2]. HHP can also impart negative structural changes to foods with an open structure, such as queso fresco cheese, affecting its crumbling properties [6].

Sources

1. *Hudaa Neetoo (Faculty of Agriculture, University of Mauritius, Réduit, Mauritius) and Haiqiang Chen (Department of Animal and Food Sciences, University of Delaware, Newark, Delaware, US), Alternative Food Processing Technologies.*
2. *Pandrangi S, Balasubramaniam VM, 2005, High pressure processing of salads and ready-meals. In: Sun DW (ed) Emerging Technologies for Food Processing. California: Elsevier Academic Press, pp. 33-45.*
3. *Butz P, Edenharder R, Garcia AF, Fister H, Merkel C, Tauscher B, 2002, Changes in functional properties of vegetables induced by high pressure treatment. Food Research International 35: 295-300.*
4. *Sancho F, Lambert Y, Demazeau G, Largeteau A, Bouvier JM, Narbonne JF, 1999, Effect of ultra-high hydrostatic pressure on hydrosoluble vitamins. Journal of Food Engineering 39: 247-253.*
5. *Cheah PB, Ledward DA, 1997, Catalytic mechanism of lipid oxidation following high pressure treatment in pork fat and meat. Journal of Food Science 62: 1135-1139.*
6. *Hnosko J, San Martin Gonzales MF, Clark S, 2012, High pressure processing inactivates Listeria innocua yet compromises Queso Fresco crumbling properties. Journal of Dairy Science 95: 4851-4862.*

Factors (Intrinsic and extrinsic) that affects Pressure Resistance [1]

A. Observations in Model Systems

It was observed that the physiological status of microbial populations subjected to HHP processing influence pressure resistance.

1. Bacteria are more susceptible during the **logarithmic phase** as compared to **stationary phase**, similar to what occurs with other technological processes such as heating [2, 3, 4].
2. Membrane composition, stationary phase proteins and stress proteins are reported to affect pressure resistance in *L. monocytogenes* [2].
3. In foods, microbial populations are diverse and in very different physiological states. Exposure to cold temperatures before processing increases the percentage of polyunsaturated FA in cell membranes, and therefore the resistance to HHP processing.

B. Observations in Real Food System

The microbial susceptibility to HHP inactivation is clearly influenced by the conditions of the environment where micro-organisms are present. Therefore, the results obtained in model systems using artificial substrates cannot be directly compared to "real foods", and should be validated [5].

1. **The chemical composition of the food**: It is important, since the presence of fats, proteins, minerals and sugars serves as a protector and increases microbial resistance to pressure [6.7, 8]. Nutrient-rich foods, such as milk or poultry meat, can protect micro-organisms [9].
2. **Temperature**: The temperature of the product and pressure fluid can affect microbial resistance, with larger inactivation rates obtained above or below the ambient temperature. Therefore, process efficacy is influenced by the processing temperature [10, 11].
   a. The decrease in resistance to pressure at low temperatures may be due to changes in membrane structure and fluidity through weakening of hydrophobic interactions and crystallisation of phospholipids [12].
   b. Moderate heating (40-60°C) can also enhance microbial inactivation by pressure, which in some cases make application of lower pressure an option [13].
3. **pH**: Susceptibility to pressure increases visibly as pH deviates from neutral values [10]. The cells of *Lactobacillus plantarum* grown at pH 5.0 were more resistant to pressures of 250 MPa than were cells grown at pH 7.0 [14]. There is theoretical evidence that, during pressurization, a temporary reduction of pH and an increase in the dissociated form of the organic acids could be present [15]. Carboxylic acids normally employed as food preservatives have a high degree of dissociation under conditions of high pressure, and it is the non-dissociated form which has an antimicrobial effect.
4. **Water Activity(aw)**: Lower values of water activity increase microbial resistance [6, 16]. **This phenomenon is observed both in synthetic models and food. The efficacy of HHP processing decreases with reduced aw, and it is visibly observed in foods with values below 0.9.**

5. **Pressurizing Equipments**: Variations in pressurization equipment can yield different microbial survival results during storage, and different pressure time profiles (including come-up time and depressurization) at different locations in the vessel may result in a heterogeneous distribution of microbial inactivation within the product [17].

6. **Other systems in combination with HHP:**

   a. **Combination of HPP with Antimicrobial compounds**- The combined use of antimicrobial compounds (bacteriocins, lysozyme, chitosan, lactoperoxidase system, essential oils) together with HHP has been tested in searching for synergistic effects [17, 18].

      There are many bacteriocin-producing lactic acid bacteria that can be found naturally in food or intentionally added as starter cultures in fermented foods. Nisin and lysozyme have been commonly used in combination with HHP [10].

      i. Nisin: It is a peptide with activity at the cytoplasmic membrane, forming pores that affect stability.

      ii. Lysozyme: It targets the peptidoglycan layer, hydrolyzing glycosidic links between saccharides (N-acetyl muramic acid and N-acetyl- D-glucosamine). Under normal conditions its activity is restricted to Gram-positive bacteria, whose cell wall is not protected by lipopolysaccharides normally present in the outer cell membrane of Gram-negative bacteria. It can also be active in Gram-negative bacteria when the barrier properties of the membrane are removed.

   b. **Combination of HPP with lactoperoxidase-**Another system used in combination with HHP is the lactoperoxidase [19].

Sources

1. *E. Rendueles, M. K. Omer, O. Alvseike, C. Alonso-Calleja, R. Capita, M. Prieto, Microbiological food safety assessment of high hydrostatic pressure processing: A review, journal homepage: www.elsevier.com/locate/lwt.*

2. *Hayman, M. M., Anantheswaran, R. C., and Knabel, S. J, 2007, The effects of growth temperature and growth phase on the inactivation of Listeria monocytogenes in whole milk subject to high pressure processing. International Journal of Food Microbiology, 10, 220-226.*

3. *Linton, M., McClemens, J., and Patterson, M. F, 2001, Inactivation of pathogenic Escherichia coli in skimmed milk using high hydrostatic pressure. Innovative Food Science and Emerging Technologies, 2, 99e104.*

4. *Manas, P., and Mackey, B. M, 2004, Morphological and physiological changes induced by high hydrostatic pressure in exponential and stationary-phase cells of Escherichia coli: relationship with cell death. Applied and Environmental Microbiology, 70(3), 1545-1554.*

5. *Claeys, W. L., Van Loey, I. A. M., and Hendrickx, M. E, 2003, Review: are intrinsic TTIs for thermally processed milk applicable for high-pressure processing assessment ?Innovative Food Science and Emerging Technologies, 4, 1-14.*

6. *Black, E. P., Setlow, P., Hocking, A. D., Stewart, C. M., Kelly, A. L., and Hoover, D. G, 2007, Response of spores to high-pressure processing. Comprehensive Reviews in Food Science and Food Safety, 6(4), 103-119.*
7. *Hauben, K. J. A., Bernaerts, K., and Michiels, C. W, 1998, Protective effect of calcium on inactivation of Escherichia coli by high hydrostatic pressure. Journal of Applied Microbiology, 85(4), 678-684.*
8. *Molina-Hoppner, A., Doster, W., Vogel, R. F., and Ganzle, M. G, 2004, Protective effect of sucrose and sodium chloride for Lactococcus lactis during sublethal and lethal high pressure treatments. Applied and Environmental Microbiology, 70(4), 2013-2020.*
9. *Patterson, M. F., Quinn, M., Simpson, R. K., and Gilmour, A, 1995, Sensitivity of vegetative pathogens to high hydrostatic pressure treatment in phosphate buffered saline and foods. Journal of Food Protection, 58, 524-529.*
10. *Alpas, H., Kalchayanand, N., Bozoglu, F., and Ray, B, 2000, Interactions of high hydrostatic pressure, pressurization temperature and pH on death and injury of pressure-resistant and pressure-sensitive strains of food borne pathogens. International Journal of Food Microbiology, 60(1), 33-42.*
11. *Bayindirli, A., Alpas, H., Bozoglu, F., and Hizal, M, 2006, Efficiency of high pressure treatment on inactivation of pathogenic micro-organisms and enzymes in apple, orange, apricot and sour cherry juices. Food Control, 17(1), 52-58.*
12. *Cheftel, J. C, 1995, Review: high-pressure, microbial inactivation and food preservation. Food Science and Technology International, 1, 75-90.*
13. *Carlez, A., Rosec, J. P., Richard, N., and Cheftel, J. C, 1993, High pressure inactivation of Citrobacter freundii, Pseudomonas fluorescens and Listeria innocua in inoculated minced beef muscle. Lebensmittel-Wissenschaft und Technologie, 26, 357-363.*
14. *Wouters, P. C., Glaasker, E., and Smelt, J. P, 1998, Effects of high pressure on inactivation kinetics and events related to proton efflux in Lactobacillus plantarum. Applied and Environmental Microbiology, 64(2), 509-514.*
15. *Palou, E., Lopez-Malo, A., Barbosa-Canovas, G. V., and Swanson, B. G, 2007, High-Pressure treatment in food preservation. In M. S. Rahman (Ed.), Handbook of food preservation (pp. 815-853). Boca Raton: CRC Press.*
16. *Patterson, M. F, 2005, Microbiology of pressure-treated foods. Journal of Applied Microbiology, 98, 1400-1409.*
17. *Hugas, M., Garriga, M., and Monfort, J. M, 2002, New mild technologies in meat processing: high pressure as a model technology. Meat Science, 62(3), 359-371.*
18. *Karatzas, K. A. G., and Bennik, M. H. J, 2002, Characterization of a Listeria monocytogenes Scott A Isolate with high tolerance towards high hydrostatic pressure. Applied and Environmental Microbiology, 68(7), 3183-3189.*
19. *Garcia-Graells, C., Valckx, C., and Michiels, C. W, 2000, Inactivation of Escherichia coli and Listeria innocua in milk by combined treatment with high hydrostatic pressure and the lactoperoxidase system. Applied and Environmental Microbiology, 66 (10), 4173-4179.*

Benefits, and Shortcomings of Alternative Non-thermal Processing Method HHP [1]

Process considerations, benefits and shortcomings of alternative non-thermal processing method-HHP are summarized in following table:

Sl. No.	*Process Considerations*	*Benefits*	*Shortcomings*	*Examples of Applications*
1	Processing time	Enhances product safety	Equipment is cost -prohibitive	Fruit products
2	Treatment temperature	Extends shelf life of product	Phenomenon of "tailing" during microbial inactivation	Yogurts
3	Pressure level	Desirable textural changes possible	Changes in sensory quality possible	Smoothies
4	Product acidity	Production of "novel" products	Not suitable for foods with air spaces	Condiments
5	Water activity	Minimal effect on flavor, nutrients and pigment compounds	Not suitable for dry foods	Salad dressings
6	Physiological age of target organisms	Minimal textural loss in high-moisture foods. Food poisoning organisms can be destroyed leading to safer products.	Refrigeration needed for low-acid foods	Meats and vegetables
7	Product composition	Can eliminate spores when combined with high temperature	Elevated temperatures and pressures required for spore inactivation	Sauces
8	Vessel size	In-container and bulk processing possible	-	High-value commodities such as seafood
9	Packaging material integrity	Potential for reduction or elimination of chemical preservatives	-	-
10	Processing aids	Positive consumer appeal, No evidence of toxicity of HHP alone	-	-

Effect of Non-thermal Method (High Hydrostatic Pressure) on Inactivation of Micro-organism [2]

Sl. No.	*Products*	*Target Organism*	*Range Log Reduction(*, **)*
1	Strawberries Fresh	—	—
	Puree	*E. coli* O157: H7 *Salmonella*	>5 >5
2	Salsa	*Salmonella*	>5***

Sl. No.	Products	Target Organism	Range Log Reduction(*, **)
3	Oysters	*Norovirus surrogate*	0.8-4.3
		Total spoilage microflora	1-2
	Shucked	*Vibrio parahaemolyticus***	3.4-7.1
	Whole-in shell	*Vibrio parahaemolyticus***	5.0 - 6.5

Note: *: Depends on pressure, time and temperature; **" *Vibrio parahaemolyticus* (Vp) is more pressure-resistant than *Vibrio vulificus.* Vp reduction further impacted with frozen storage following pressure treatment; ***: Depends on pressure, time and temperature as well as type added acid and final pH used in preparation (vinegar, lemon juice).

Opportunities and Drawbacks [3]

The opportunities and drawbacks of HHP are discussed below:

1. Opportunities

1. Fresh-like quality
2. Retention of nutrients
3. Application for acid food and Low acid food: combined with heat or antibacterial agents (spores inactivation).

2. Drawbacks

1. Discontinuous or Semi discontinuous or Semi continuous process
2. Expensive equipments
3. Impact on texture, structure, appearance, functionality
4. Spores are not sensitive
5. Resistant bacteria

Sources

1. *Hudaa Neetoo (Faculty of Agriculture, University of Mauritius, Réduit, Mauritius) and Haiqiang Chen (Department of Animal and Food Sciences, University of Delaware, Newark, Delaware, US), Alternative Food Processing Technologies.*
2. *Lori F Pivarnik and Randy Worobo, Non-thermal or Alternative Food Processing Methods to Enhance Microbial Safety and Quality, Frequently Asked Questions, (USDA-NIFA #2011-68003-30005).*
3. *Ranjan Saha, New and Emerging Non-Thermal Technologies in Food Preservation, April 2015.*

Regulations (Ensuring Microbiological Safety of Food) and Future Prospect of HHP [1]

i. Regulations [1]

High-pressure-treated foodstuffs could be considered **"novel foods"** and in accordance related issues are covered by **Regulation (EC) No 258/97** [2]. In case a food is considered as novel, applications for authorization issued by the food company are received by the corresponding Member State (MS) in which the food is marketed, which carry out the scientific assessment of the safety of the product [3]. The rest of the MS can also object to the authorization procedure. Not only microbiological considerations but also toxicological, nutritional, intended use and allergenic aspects should be also part of a risk evaluation process. Several products, such as high-pressure-pasteurized **orange juice (France), cooked ham (Spain), oysters (Great Britain), fruits (Germany)** have been evaluated at national level with a common agreement that the high pressure treatment has not caused significant changes in the composition or the structure of the products affecting their nutritional value, metabolism or the amounts of undesirable substances [4].

ii. Future Prospect [1]

1. As mentioned, most vegetative pathogens are susceptible to pressure, and standard treatments achieve significant reductions. At international level agreement has been reached on standardized conditions for heat treatments (pasteurization, sterilization) according to certain time-temperature combinations.

   Likewise, the development of HHP would be favoured if standardized treatment conditions were proposed by international bodies such as **Codex Alimentarius, ICMSF** (International Commission on Microbiological Specifications for Foods) or **ILSI** (International Life Sciences Institute). The so-called Process Criteria (combinations of pressure-time-temperature for different foods and pathogens) can be calculated by predictive microbiology procedures, and would result from global values known as **Food Safety Objectives (FSO),** set for food-pathogen combinations. The establishment of these parameters would facilitate the development of procedures of Quantitative Microbiological Risk Assessment (QMRA) [5] (Smelt *et al.*, 2002).

2. Food companies can incorporate the stage of HHP processing **in HACCP plans**, as a Critical Control Point, which can be monitored, optimized, validated, and for which critical limits and corrective actions may be set.

Sources

1. *E. Rendueles, M. K. Omer, O. Alvseike, C. Alonso-Calleja, R. Capita, M. Prieto, Microbiological food safety assessment of high hydrostatic pressure processing: A review, journal homepage: www.elsevier.com/locate/lwt.*

2. *Regulation (EC) No 258/97 of the European Parliament and of the Council of 27 January 1997 concerning novel foods and novel food ingredients.*

3. *Heinz, V., and Buckow, R, 2010, Food preservation by high pressure. Journal für Verbrau-cherschutz und Lebensmittelsicherheit, 5, 73-81.*

4. *Eisenbrand, G, 2005, Safety assessment of high pressure treated foods. Opinion of the Senate Commission on Food Safety (SKLM) of the German Research Foundation (DFG). Molecular Nutrition and Food Research, 49, 1168-1174.*

5. *Smelt, J. P., Hellemons, J. C., Wouters, P. C., and van Gerwen, S. J, 2002, Physiological and mathematical aspects in setting criteria for decontamination of foods by physical means. International Journal of Food Microbiology, 78(1-2), 57-77.*

Applications of High Hydrostatic Pressure in Food Processing [1, 1a]

High hydrostatic pressure processing is applied at a wide range of pressures and temperatures in order to process and preserve foods while retaining their organoleptic and nutritional qualities. The food safety aspects of HHP processing with some mention of other applications such as pressure assisted freezing, pressure-assisted thawing [2], pressure-assisted extraction of bioactive compounds [3] and pressure-induced modification of food compounds, are important. The food is largely protected from the damaging force of the pressure since pressure is uniformly distributed around and throughout the food.

Some of the requirements for the suitability of a high-pressure system for food applications are given below [3d]:

1. Short cycle time for inactivating micro-organisms
2. Safe to operate
3. Easy to clean
4. Accurate and reliable pressure control
5. Low capital
6. Operating costs.

HHP technology has been applied to shellfish, sliced meats, vegetables, cheese, meat, fish, juices and smoothies, ready-to-eat meals, sauces and salsas successfully. Consumer-driven requirements for better tasting, additive-free foods with a longer shelf life are resulting in a drive to develop high pressure processing. The main applications of high pressure processing are preservation and preparation. For preservation applications, enzymes and micro-organisms are inactivated under high pressure. This results in a preserved product while remaining fresh taste, flavour and colour. Examples are pasteurisation of fruit juices, fruit desserts and guacamole [2a].

In Japan, for example, several rice-based foods are available with novel textures induced by high-pressure treatment. Other examples are gelatinisation of starches and proteins, preparation of pectin gels (jams) and tenderisation of meat. Commercial products have now been appearing in the US and Europe [2a].

1. Microbial Inactivation

High pressure processing kills micro-organisms by exposing foods to very high pressures. In 1899, the earliest investigation using high pressures in food processing involved the application of pressures in the range of 5000 to 7000 kg/cm^2 to reduce microbial levels in milk and meats [2b]. It was shown that a five- or six-log-cycle reduction in bacterial count was possible in milk when it was subjected to a high pressure of 6 800 kg/cm^2. Increased shelf life of meats was also observed when meat samples were pressurized to 5 400 kg/cm^2. The normal pressure used to kill pathogens in food using HPP can be 2000- 2, 500 times higher than normally found in an automobile tire and as high as 6000 times higher than the pressure we are exposed to at sea level. The high pressure causes the micro-organisms membrane to be disrupted and causes them to die. The amount of pressure needed to kill the pathogen will differ with the micro-organism and the food product. This processing method can be performed on pre-packaged food, minimizing risks of re-contaminating food [3a].

The mechanism of microbial control and inactivation lies in a combination of processes such as the breakdown of non-covalent bonds in large macromolecules, biochemical effects, effects on the genetic mechanisms of cells, morphological changes and the disruption and permeabilization of the cell membrane [4]. Food micro-organisms, such as vegetative bacteria, human infectious viruses, fungi, protozoa and parasites, can be significantly reduced when subjected to high pressure.

HPP technology can be successfully applied to milk processing for microbial deactivation without causing significant changes in quality. Under higher pressure, microbial cellular membranes suffer irreversible damage due to changes in membrane protein and others, causing microbial inactivation [1e, 1f].

A range of baro tolerances exists among the different microbial groups and even among the different strains of the same species [4, 5, 6]. Several researchers have investigated the efficacy of HHP to inactivate a variety of food-borne pathogens in different commodities to enhance their microbiological safety. The use of Safe Pac™ high-pressure pasteurization to inactivate micro-organisms and extend the shelf-life of ready-to eat products including meats, soups, wet salads, sauces, fruit smoothies, shellfish, and seafood by 200-300 per cent is just one example of the commercial adoption of HHP [7]. The synergistic effect of HHP and antimicrobials has also been reported [8]. Processing temperature also has an impact on microbial inactivation ratios: treatment temperatures above room temperature, in the range of 45-50°C, generally bring about dramatic increases in the inactivation rate [9].

i. Commercial Applications

Using validated processing parameters of pressure, temperature and time combinations, high pressure processing is used commercially for certain food products, *i.e.*oysters, salsa, delimeats, guacamole, juices, targeting specific bacterial pathogens (*e.g. Vibrio parahaemolyticus, Vibrio vulnificus, Salmonella, Listeria, E. coli* O157: H7). There is currently no commercial application of HPP on fresh produce, and HPP has not been used to target viruses [3a].

ii. In Milk Processing [1a]

a. *Niche Products (Yogurt, dairy desserts and smoothies)*

High-pressure processing used to manufacture these products to control syneresis, produce a firmer texture, increase shelf life [1d].

b. *Cheese Manufacturing*

HHP is also being used for milk processing to make cheese [3b]. High-pressure processing accelerates the aging of cheese, improve the texture of cheese and enhance safety in the raw milk cheese production where regulations allow it. Combined with cold pasteurization, high-pressure treatment maintains the bioactivity of probiotics and of the molecules of milk, colostrum and whey [1d].

c. *Milk Samples*

Milk samples were pressure treated in a FoodLab 900 (Stansted Fluid Power Ltd., U. K.) type laboratory pressure vessel. Effects of pressure treatments of fresh bovine milk at 200 MPa, 400 MPa, 600 MPa, 800 MPa for 5 min at room temperature were compared to the effects of heat treatment. Samples were stored at 4°C for a week. Changes in the total viable cell counts were followed, protein denaturation (PAGE) was examined, colour measurements were made, and changes in odour components were analysed. In case of milk samples, with the increase of pressure level, an increase in the inactivation of mesophile aerob cell counts was detected. While 200 MPa pressure treatment didn't cause significant loss in the total aerobic mesophile count, 400 MPa, 600 MPa, 800 MPa treatments increased the microbial inactivation with several orders of magnitude. During one-week storage at 4°C the growth of the microflora was delayed, the shelf-life of samples treated at 400 MPa, 600 MPa and 800 MPa proved to be longer than heat treated samples [3c].

Protein denaturation, caused by high hydrostatic pressure treatment, was investigated by DSC thermogram. DSC did not bring clear results in case of milk samples, the changes in protein structure was studied by gel electrophoresis, where protein fractions were separated by native PAGE. The electrophoretic pattern of milk samples showed that increased pressure level caused increased denaturation of certain protein fraction of milk. Casein and α-lactalbumin content seemed to be unchanged with the increasing pressure. The two isomers of β-lactoglobulin seemed to be the most responsive to pressure. The intensity of the β-lactoglobulin bands was decreasing and the less mobile β- lactoglobulin-A denatured first. Milk colour was not affected by hydrostatic pressure processing at a large extent [3c].

d. *High-pressure Homogenization*

The ability of High-pressure homogenization (HPH) to disactivate the *Listeria monocytogenes* and *Salmonella enteritidis* pathogens in raw milk, and to promote the proliferation and viability of inoculated probiotic bacteria in HPH milk was evaluated. Results reveal that *S. enteritidis* is more sensitive to HPH (up to the 4th cycle at 100 MPa) than *L. monocytogenes* (reduction of 5 Log cfu/ml with 8 cycles at 100 MPa). The possibility of producing probiotic cheeses with functional and sensory properties using HPH is greater because the cellular viability of *L. acidophilus* and of *Lactobacillus casei* is maintained [1b].

e. *Ultra-High Pressure Homogenization*

The study on ultra-high pressure homogenization (UHPH) as an alternative to high-pasteurization with similar microbial and physico-chemical characteristics, was made. UHPH of milk increases the inactivation of plasmin in proportion to the pressure exerted. Proteolysis and lipolysis were more intense in milk samples homogenized at the ultra-high pressure of 200 MPa than in samples of milk high-pasteurized or samples UHPH at 300 MPa pressure. Ultra-high pressure homogenization at 300 MPa could be therefore treated as an alternative to pasteurization [1c].

iii. In Processing of RTE [1a]

A widespread application in food processing is for RTE foods which are sliced after processing and packaging, to control the contamination introduced in this stage. Micro-organisms to consider would be *L. monocytogenes* and other post-processing contaminants, whose prevalence would depend on the hygienic conditions of process, equipment and personnel.

Sources

1. *Hudaa Neetoo (Faculty of Agriculture, University of Mauritius, Réduit, Mauritius) and Haiqiang Chen (Department of Animal and Food Sciences, University of Delaware, Newark, Delaware, US), Alternative Food Processing Technologies.*

1a. *E. Rendueles, M. K. Omer, O. Alvseike, C. Alonso-Calleja, R. Capita, M. Prieto, Microbiological food safety assessment of high hydrostatic pressure processing: A review, journal homepage: www.elsevier.com/locate/lwt.*

1b. *Francesca Patrignani, Alma Mater Studiorum, High-Pressure Homogenization as an Alternative to Thermal Treatment to Improve the Microbial Safety, Shelf-Life and Functionality of Milk and Dairy Products, Department of Food Sciences, Bologna University, Italy, Session-4, IDF Dairy Science and Technology Week - Quebec 2008.*

1c. *Julieta Pereda, Proteolysis and Lipolysis of Ultra-High Pressure Homogenized Treated Milk During Refrigerated Storage Special Centre for Research in Food Technology, Autonomous University of Barcelona, Bellaterra, Spain, Session-4, IDF Dairy Science and Technology Week-Quebec 2008.*

1d. *Geoffrey W. Smithers, Dairy Product Quality and Non-Thermal Processing Technologies Food Science Australia, Session-4, IDF Dairy Science and Technology Week - Quebec 2008.*

1e. *Soumitra Banerjee and Shanker Lal Shrivastava, Recent Trends in Milk Processing, Approches in poultry, dairy and Veterinary sciences 2015, Crimson Publishers, ISSN: 2576-9162.*

1f. *Datta N, Deeth HC (1999) High pressure processing of milk and dairy products. Australian Journal of Dairy Technology 54(1): 41-48.*

2. *Cheftel JC, Thiebaud M, Dumay E, 2002, Pressure-assisted freezing and thawing of foods: a review of recent studies. High Pressure Research 22: 601-611.*

2a. *Ranjan Saha, New and Emerging Non-Thermal Technologies in Food Preservation, April 2015.*

2b. *Benito, A. G., Ventoura, M., Cassadei, T. Robinson and Mackey, B. 1999. Variation in resistance of natural isolates of Escherichia coli O157: H7 to high hydrostatic pressure, mild heat, and other stresses. Appl. Environ Microbiol. 65: 1564-1569.*

3. *Prasad KN, Yang B, Zhao M, Sun J, Wei X, Jiang Y, 2010, Effects of high pressure or ultrasonic treatment on extraction yield and antioxidant activity of pericarp tissues of longan fruit. Journal of Food Biochemistry 34: 838-855.*

3a. *Lori F Pivarnik and Randy Worobo, Non-thermal or Alternative Food Processing Methods to Enhance Microbial Safety and Quality, Frequently Asked Questions, (USDA-NIFA #2011-68003-30005).*

3b. *Trujillo, A. J., Capellas, M., Buffa, M., Royo, C., Gervilla, R., Felipe, X., et al, 2000, Application of high pressure treatment for cheese production. Food Research International, 33(3-4), 311-316.*

3c. *Kálmánné Tuboly Eszter, PhD thesis on Effects of the use of high hydrostatic pressure technology on the microbiology and other quality parameters of some selected foodstuffs, Faculty of Food Science, Corvinus University of Budapest.*

3d. *Singh R. Paul, Professor of Food Engineering Department of Biological and Agricultural Engineering, Department of Food Science and Technology University of California Davis, FAO-2001, Technical Elements of New and Emerging Non-Thermal Food Technologies.*

4. *Patterson MF, 2005, Microbiology of pressure-treated foods. Journal of Applied Microbiology 98: 1400-1409.*

5. *Vanlint D, Rutten N, Michiels CW, Aertsen A, 2012, Emergence and stability of high-pressure resistance in different food borne pathogens. Applied and Environmental Microbiology 78: 3234-3241.*

6. *Whitney BM, Williams RC, Eifert J, Marcy J, 2007, High pressure resistance variation of Escherichia coli O157: H7 strains and Salmonella serovars in tryptic soy broth, distilled water and fruit juice. Journal of Food Protection 70: 2078-2083.*

7. *Safe Pac Pasteurization, 2010, Safe Pac™ expands the food safety universe. Philadelphia, PA. www.safepac.biz, accessed 11 November 2013.*

8. *Jofré A, Garriga M, Aymerich T, 2008, Inhibition of Salmonella sp., Listeria monocytogenes and Staphylococcus aureus in cooked ham by combining antimicrobials, high hydrostatic pressure and refrigeration. Meat Science 78: 53-59.*

9. *Mertens B, 1995, New Methods of Food Preservation. New York: Blackie Academic and Professional, pp. 135,*

2. Development of Novel "Value-Added" Food Products [1]

High hydrostatic pressure has also been used in the development of novel "value-added" food products [2]. Some of the acquired quality attributes of pressurized products are due to modifications related to the volume reduction that occurs due to the physics of pressure.

a. Protein Unfolding

High pressure induces protein unfolding, a new spatial organization and modified interactions with water molecules, leading to possible changes in food protein functionality such as color, texture, gelation ability, and emulsifying properties. Some examples are given below:

1. Soy and meat protein extracts can become a gel under pressure, without any heat treatment; these pressure-induced gels have different rheological properties from those of thermal gels.
2. HHP can also induce the modification of dairy proteins such as whey protein concentrate for improved functional properties [3, 4].
3. HHP also induces partial unfolding of egg albumen and yolk proteins, producing more elastic gels than heat induced gels.

b. Gelation of Polysaccharides

HHP can also lead to the gelation of polysaccharides. The effect of high pressure on the structure and physico-chemical properties of different starches has been of great interest, undergoing complete gelatinization at 600 MPa [5] and the properties of starch pastes and gels obtained under high pressure differ from those obtained from heat-gelatinized products, opening up opportunities for new **applications [6].**

Sources

1. *E. Rendueles, M. K. Omer, O. Alvseike, C. Alonso-Calleja, R. Capita, M. Prieto, Microbiological food safety assessment of high hydrostatic pressure processing: A review, journal homepage: www.elsevier.com/locate/lwt.*
2. *Torres JA, Velazquez G, 2009, Hydrostatic pressure processing of foods. In: Jun S, Irudayaraj JM (eds) Food Processing Operations Modeling: Design and Analysis. Boca Raton, FL: CRC Press, pp. 173-205.*
3. *Lim SY, Swanson BG, Clark S, 2008, High hydrostatic pressure modification of whey protein concentrate for improved functional properties. Journal of Dairy Science 91: 1299–1307.*
4. *Padiernos CA, Lim SY, Swanson BG, Ross CF, Clark S, 2009, High hydrostatic pressure modification of whey protein concentrate for use in low-fat whipping cream improves foaming properties. Journal of Dairy Science 92: 3049-3056.*
5. *Katopo H, Jane JL, Song Y, 2002, Effect and mechanism of ultrahigh hydrostatic pressure on the structure and properties of starches. Carbohydrate Polymers 47: 233-244.*
6. *Yaldagard M, Mortazavi SA, Tabatabaie F, 2008, The principles of ultra high-pressure technology and its application in food processing/preservation: a review of microbiological and quality aspects. African Journal of Biotechnology* **7***: 2739-2767.*

3. Application in Freezing and Thawing [1]

The research in other areas includes high-pressure freezing applications such as pressure-shift freezing (PSF), high-pressure thawing, and high-pressure non-frozen storage. It is thought that the use of HHP to assist freezing and thawing of foods can result in improved quality and reduced drip loss by generation of small ice crystals [2]. The interesting application of high pressure is to store food materials at subfreezing temperatures under high pressure without actually freezing the food [3]. The freezing point of water decreases with increasing pressure. Thus, a food under pressure may be kept at subfreezing temperature in an unfrozen state, minimizing the deleterious effects of ice crystal formation on food quality.

Sources

1. *E. Rendueles, M. K. Omer, O. Alvseike, C. Alonso-Calleja, R. Capita, M. Prieto, Microbiological food safety assessment of high hydrostatic pressure processing: A review, journal homepage: www.elsevier.com/locate/lwt.*
2. *Sanz PD, Otero L, 2005, High pressure freezing. In: Sun DW (ed) Emerging Technologies for Food Freezing. California: Elsevier Academic Press, pp. 627-647.*
3. *Hamilton, W. A. and Sale, A. J.1967. Effects of high electric fields on micro-organisms. II. Mechanism of action of the lethal effect. In J. Biochim. Biophys. Acta. 148: 789-800, 18.*

4. In Treatment of Fruit, Fruit Products and Fruit Juices [1]

Consumers are interested in the nutritional and health-related characteristics of fruits and vegetables. HPP effects the phenolic compounds as well as quality [2] of fruit/fruit juices and quality

a. Phenolic Compounds [1]

These are healthy substances, therefore amount of phenolic compounds in fruits/fruit juices are important for customers. Flavonoids are the most common and widely distributed group of plant phenolics. Fruits and fruit juices are among the best sources of polyphenols in the human diet because of their high content in most fruits. Processing fruits can change their content of phenolic compounds and heat treatments usually decrease phenolic compounds in fruits. Minimising losses of these bioactive compounds during processing and storage are one of the priorities and challenge for manufacturers. The application of HHP has the potential to produce high quality fruits that display characteristics of fresh foods, are microbiologically safe and extended shelf life. The impact of HHP processing conditions on the stability of phenolic compounds in fruits and their products after process, during storage was studied and compares with thermal process.

Matrix of food and processing parameters such as pressure, temperature, time are important for retaining of phenolic compounds. In general, several authors have proven that HHP treatment on different fruit products slightly modified their content of phenolic compounds and researches generally indicate that HHP is better than thermal processes on retaining phenolic compounds.

Most of the studies showed that total phenolic compounds (TPC) and some phenolic compounds (hesperetin, naringenin and anthocyanins) increased or unchanged by HHP. In general, several studies have proven that HHP on different fruit products (orange juice, lemon juice, apple juice, strawberry puree, kiwifruit puree, orange/carrot juice, apple/broccoli juice) slightly modified their content of phenolic compounds [3, 4]. The details related to the effect of HHP on anthocyanin and total phenolics in fruits, fruit juices, nectars and smoothies mousses and purees are given below:

Sl. No.	*Fruits and other Products*	*Process and Storage Conditions*	*Anthocyanin Content*		*Total Phenolics*
			After Process	*Storage*	
1	Fruits				
	Strawberry [5]	200-800 MPa/18-22°C/15 min., storage at 4-30°C for 9 days.	—	Greater stability at 800 MPa for P3G and P3R at storage temperature of 4°C for 9 days storage.	—
	Raspberry [5]	200-800 MPa,18-22°C,15 min., storage at 4-30°C for 9 days.	—	Greater stability at 800 MPa for C3G and C3S at storage temperature of 4°C for 9 days storage.	—
2	Fruit juices				
	Pomegranate juice [6]	400 MPa/25°C/5 min, 21-72 days storage at 4°C.	63 per cent of antho-cyanins retained.	65 per cent is retained at 21 days storage, 4°C; 63 per cent is retained at 72days storage, 4°C.	—
	Strawberry juice (cloudy and clear) [7]	600 MPa/4 min., 6 month storage at 4- 25°C.	—	29.76 per cent and 7.02 per cent decrease for cloudy and clear samples respectively at 4°C. The decrease at 25°C almost doubled.	16.22 per cent and 13. 82 per cent decrease for cloudy and clear samples respectively at the end of 6 months 4 ° C. The decrease at 25°C almost doubled.
	Orange juice milk beverage [8]	100- 400 MPa for 120-540 s or TT (90-98°C/15-21 s).	—	—	SN and NSN increases for all HHP treatments except 400 Mpa, 540s SN and NSN increases for all TT.
3	Nectars				
	Apricot nectar [9]	300-500MPa/5-20 min. or HTST (110°C/8.6s)	—	—	HHP treatments increased total and Individual phenolics in apricot nectars, which were significantly lower than those in HTST-treated apricot nectars.

Sl. No.	*Fruits and other Products*	*Process and Storage Conditions*	*Anthocyanin Content*		*Total Phenolics*
			After Process	*Storage*	
	Mango nectar [10]	600MPa/1min or HTST (110°C/8.6s) 16 weeks storage at 4 and 25°C, steam blanching was implemented prior both HHP and HTST.	—	—	No SN changes in TPC after HHP or HT ST. After 16 weeks storage, TPC decre-ased by 19.11 and 27.94 per cent in HHP-treated samples, and by 17.03 and 25.23 per cent in HTST-treated samples at 4 and 25°C, respectively.
4	Smoothies				
	Fruit smoothie [11]	450-600 MPa/20°C/ 1-120 min. or TT (P70 > 10 min).	—	—	TPC of HPP smoothies (450 MPa;5 min) was slightly higher (~11 per cent) tcompared to TT samples.
	Smoothie (mix of papaya, melon juices, carrot puree and skimmed milk) [12]	450-600MPa/ 20°C/3 min or TT (80°C/ 3 min.), 45 days storage at 4°C.	—	—	TPC increased after HHP treatment. TPC decreased 15 per cent, 12 per cent, 11 per cent and 8 per cent for untre-ated, HHP-450, HHP-600 and TT samples respectively at the end of storage period.
5-	Mousses				
	Wild Strawberry/ Strawberry Mousses [6]	500MPa/50°C/ 10min, 21-72 days storage at 4°C.	89 per cent of antho-cyanins retained.		72 per cent is retained at 21 days' storage, 4°C; 63 per cent (65 per cent for strawberry) is retained at 72 days storage, 4°C.
6	Purees				
	Apple Puree [13]	400-600 MPa/5min/ 20°C or TT (75°C/10 min.), 7-21 days storage.	—	—	No change at 400 MPa, SN decrease at 600 MPa. Storage revealed that mild pasteurization preserved higher levels of phenolics than pressure-treated samples.
	Strawberry and blackberry purees [14]	400-600 MPa/15 min/10-30°C or TT (70°C/2min).	No change at HHP, SN decrease at TT for all Samples.	—	**Strawberry**- No change for 400-500 Mpa, SN increase for 600 MPa, SN decrease for TT **Blackberry**-No change for 400MPa and TT. SN increase for 500, 600MPa.

Sl. No.	Fruits and other Products	Process and Storage Conditions	Anthocyanin Content		Total Phenolics
			After Process	Storage	
	Strawberry Puree [15]	600 MPa/20°C/5min. or TT (88°C/ 2min.), 3 months storage at 4°C.	SN decrease for HHP and TT samples.	SN decrease for HHP and TT samples.	SN decrease for HHP and TT samples after process. Slightly hig-her loss of TPC were observed during storage of HHP samples compared to TT samples.

Sources

1. *Muzeyyen berkel kapikci, Neriman badatliolu, High hydrostatic pressure treatment of fruit, fruit products and fruit juices: a review on phenolic compounds, Journal of Food and health Sciences, 2(1), 27-39, 2016.*

2. *Boynton, B. B., Sims, C. A., Sargent, S., Balaban, M. O., and Marshall, M. R., 2002, Quality and stability of precut mangos and carambolas subjected to high-pressure processing, Sensory and Nutritive Qualities of Food, 67(1), 409-415.*

3. *Cao, X., Zhang, Y., Zhang, F., Wang, Y., Yi, J, and Liao, X., 2011, Effects of high hydrostatic pressure on enzymes, phenolic compounds, anthocyanins, polymeric colour and colour of strawberry pulps. Journal of the Food Science of Food and Agriculture. 91, 877-885.*

4. *Barba, F. J., Jager, H., Meneses, N., Esteve, M. J., Frigola, A., and Knorr, D. 2011. Evaluation of quality changes of blueberry juice during refrigerated storage after high-pressure and pulsed electric fields processing. Innovative Food Science and Emerging Technologies, 14, 18-24.*

5. *Zabetakis, I., Leclerc, D., and Kajda, P., 2000, The effect of high hydrostatic pressure on the strawberry anthocyanins. Journal of Agriculture and Food Chemistry, 48, 2749-2754.*

6. *Ferrari, G., Maresca, P., and Ciccarone, R., 2011, The effects of high hydrostatic pressure on the poly-phenols and anthocyanins in red fruit products. Procedia Food Science 1, 847- 853.*

7. *Cao, X., Bi, X., Huang, W., Wu, J., Hu, X., and Liao, X., 2012, Changes of quality of high hydrostatic pressure processed cloudy and clear strawberry juices during storage. Innovative Food Science and Emerging Technologies, 16, 181-190.*

8. *Barba, F. J., Cortes, C., Esteve, M. J., and Frigola, A., 2012, Study of antioxidant capacity and quality parameters in an orange juice-milk beverage after high-pressure processing treatment. Food and Bioprocess Techology, 5, 2222-2232.*

9. *Huang, W., Bi, X., Zhang, X., Liao, X., Hu, X., and Jihong, W., 2013, Comparative study of enzymes, phenolics, carotenoids and color of apricot nectars treated by high hydrostatic pressure and high temperature short time. Innovative Food Science and Emerging Technologies, 18, 74-82.*

10. Liu, F., Wang, Y., Li, R., Bi, X., and Liao, X., 2014, Effects of high hydrostatic pressure and high temperature short time on antioxidant activity, antioxidant compounds and color of mango nectars. Innovative Food Science and Emerging Technologies, 21, 35-43.

11. Keenan, D. F., Brunton, N. P., Gormley, T. R., Butler, F., Tiwari, B. K., and Patras, A., 2010, Effect of thermal and high hydrostatic pressure processing on antioxidant activity and colour of fruit smoothies. Innovative Food Science and Emerging Technologies, 11, 551-556.

12. Andrés, V., Villanueva, M. J., and Tenorio, M. D., 2016, The effect of high-pressure processing on colour, bioactive compounds and antioxidant activity in smoothies during refrigerated storage. Food Chemistry, 192, 328-335.

13. Landl, A., Abadias, M., Sárraga, C., Viñas, I., and Picouet, P. A., 2010, Effect of high pressure processing on the quality of acidified Granny Smith apple purée product. Innovative Food Science and Emerging Technologies, 11, 557-564.

14. Patras, A., Brunton, N. P., Pieve, S., and Butler F., 2009b, Impact of high pressure on total antioxidant activity, phenolic, ascorbic acid, anthocyanin content and colour of strawberry and blackberry purees. Innovative Food Science and Emerging Technologies, 10, 308-313.

15. Terefe, N. S., Matthies, K., Simons, L., and Versteeg, C., 2009, Combined high pressure-mild temperature processing for optimal retention of physical and nutritional quality of strawberries (Fragaria×ananassa). Innovative Food Science and Emerging Technologies, 10, 297-307.

b. Quality and Colour of Fruit Juices [1, 3]

HHP-processed fruits and fruit products may exhibit long-term changes in flavor, nutritional properties, and color as a result of the residual activity of endogenous enzymes [2]. To reduce enzyme activity and preserve food quality during processing and storage, different methods have been successfully applied in combination with HHP, *e.g.* low pH conditions, refrigerated storage, high temperature treatment and anti-browning agents [1]. HHP of fruits and vegetables has been revealed as a useful tool to extend their shelf life and quality as well as to preserve their nutritional and functional characteristics.

i. Quality of Apple Juice [3]

The quality of apple juice was checked after processing by HPP and thermal processing. Initially, the apple juices were subjected [3] to HPP at 600 MPa for 3 minutes, afterwards a fresh set was subjected to thermal processing for 5 minutes. The juice that were treated by the HPP process retained its color better than the thermally processed juice. However, the enzyme inactivation was only possible through thermal processing. Although, the sugar and acids were less affected in both processes. The aldehyde, alcohol, ketone, and organosulfur were increased during the thermal processing technology. Also, the color and flavor of the apple juices remained better with the HPP process rather than the thermal heating process [4].

ii. Quality of Carrot Juice [3]

A comparative study was conducted [5] on the quality of carrot juices processed by HPP (subjected to HPP at 550MPa for 6 minutes) and HTST(subjected to for 110°C for 8.6 seconds). In both the processes, it was found the juices were microbiologically safe after being stored for 20 days at 4°C. In the storage process, the sample of juice, which was treated by the HPP process demonstrated greater nutritional, rheological, and antioxidant properties, when compared to the sample of HTST treated juice. When the sensory attributes were taken into consideration, the juice treated by the HPP methodology had fresher properties in aroma, taste, and overall suitability. However, the organoleptic properties of the two juices reduced after storage. Therefore, it was determined the HPP methodology can be used as an alternative to produce fresh carrot juice.

Sources

1. *Muzeyyen berkel kaþikci, Neriman baðdatlioðlu, High hydrostatic pressure treatment of fruit, fruit products and fruit juices: a review on phenolic compounds, Journal of Food and health Sciences, 2(1), 27-39, 2016.*
2. *Boynton, B. B., Sims, C. A., Sargent, S., Balaban, M. O., and Marshall, M. R., 2002, Quality and stability of precut mangos and carambolas subjected to high-pressure processing, Sensory and Nutritive Qualities of Food, 67(1), 409-415.*
3. *Harlin Kaur Saroya, Innovative non-thermal food processing technologies used by the food industry in the United States, MSc thesis, 2017, Western Kentucky University, Kentucky.*
4. *Yi, J., Kebede, B. T., Hai Dang, D. N., Buve, C., Grauwet, T., Van Loey, A., Hu, X., Hendrickx, M., Quality change during high pressure processing and thermal processing of cloudy apple juice. Food Science and Technology, 75, 85-92, 2017.*
5. *Zhang, Y., Liu, X., Wang, Y., Zhao, F., Sun, Z., and Liao, X., Quality comparison of carrot juices processed by high-pressure processing and high-temperature short-time processing. Innovative Food Science and Emerging Technologies, 33, 135-144, 2016.*

4. Application in Bakery [1]

HPP methodology was applied to batters and their layer cakes. Microbial, physical, and structural changes occurred when the products were subjected to a pressure range of 300 MPa-600MPa. It was proved that pressure helped in decreasing the molds, yeast, and aerobic mesophilic bacteria and additionally, the batter become more consistent and elastic. However, when the batter was baked, the cakes were of a lesser volume, had a darker crust, and had harder crumbs [2].

Sources

1. *Harlin Kaur Saroya, Innovative non-thermal food processing technologies used by the food industry in the United States, MSc thesis, 2017, Western Kentucky University, Kentucky.*
2. *Barcenilla, B., Román, L., Martínez C., Martínez, M. M., and Gómez, M. (2016). Effect of high pressure processing on batters and cakes properties. Food Science and Emerging Technologies, 33, 94-99.*

Pressure-Assisted Thermal Sterilization [1]

Although the application of hydrostatic pressure at a level of 400–800 MPa inactivates most pathogenic and spoilage vegetative bacteria, pressure inactivation of bacterial spores has been more problematic because of the marked resistance of spores to pressure [2]. It is generally accepted that high inactivation ratios of Bacillus and Clostridium spores cannot be realized with the application of high pressure alone.

Although HHP is often described as a non-thermal processing method, compression of the hydrostatic medium brings about the generation of adiabatic heat [3]. By capitalizing on the combined effect of high pressures and initial elevated temperatures (hurdle approach), spores can be inactivated by a process called pressure-assisted thermal sterilization (PATS).

During PATS, initial chamber temperatures are usually in the range of 70–90°C and can transiently reach temperatures of 80–130°C due to internal compression heating while pressures can vary from 600 to 1200 MPa. PATS thus takes advantage of the instantaneous and spatially homogeneous distribution of the adiabatic heat to control the temperature and heat exposure [2]. This can in turn eliminate viable spores, while enhancing product quality [4, 5].

- **i.** According to the study [6], PATS-induced lethality of viable spores is the result of a combined effect of thermally induced protein denaturation, enzyme inactivation and pressure-induced aggregation of proteins.
- **ii.** It was showed [7] that the application of high pressure at 600 MPa for 30 min at an initial temperature of 75°C resulted in variable reductions of 0–5.5 log colony-forming units (CFU)/mL, depending on the strain of *C. botulinum*.
- **iii.** The spore strips of C. sporogenes embedded in egg patties was inactivated at pressures of 690 MPa and a final product temperature of 110°C [8].
- **iv.** A caveat with PATS is the phenomenon of tailing [9]. This is thought to occur in response to stabilization of spores by pressure, super dormancy or even the non-linearity of inactivation kinetics reminiscent of studies dealing with large populations of cells [2].
- **v.** Some researchers have speculated that the high temperature of PATS treatments can affect both covalent and non-covalent bonds and thus these processes require the evaluation of chemical changes of treated foods [10]. Losses of nutrients and formation of hazardous compounds such as acrylamide, heterocyclic amines (HCAs), nitrosamines, polycyclic aromatic hydrocarbons (PAHs), furans, and heterocyclic amines have been reported as concomitants of conventional heating methods. Processing methods that indirectly heat food to very high temperatures, such as HHP, PEF, and ultrasound, may similarly result in the formation of undesirable compounds and significant loss of desirable ones such as vitamins, essential amino acids, and unsaturated fatty acids [11, 12]. Hence, the kinetics of formation of these compounds in PATS-treated food matrices should be evaluated, as this phenomenon is mostly unknown [13].

vi. In some regions of the world, processed food safety controls are already in place. For example, novel food laws approved in May 1997 in Europe regulate foods or food ingredients in which a production process gives rise to significant modifications in the composition and/or structure of the food or food ingredient [14].

vii. The safety assessment of these novel foods is based on the application of toxicological tests requiring a case-by-case comparison of the novel and conventional equivalent processed food. If undesirable compounds are found to be present in the traditional or novel product, a detailed risk assessment of these compounds needs to be conducted. This assessment includes identification and characterization of the hazards, an assessment of the human exposure, and final characterization of the toxicological risks. The formation of possible toxicological end-products can place an important constraint to the commercialization of PATS technologies, particularly in countries following the European Union novel food law model. PATS is affected by these regulations because this technology was not used before May 15 1997 [13]. In contrast, a PATS process submitted in 2008 in the US for review by the Food and Drug Administration (FDA) required no such characterization of toxicity risk and was approved in 2009 [13].

Sources

1. *Hudaa Neetoo (Faculty of Agriculture, University of Mauritius, Réduit, Mauritius) and Haiqiang Chen (Department of Animal and Food Sciences, University of Delaware, Newark, Delaware, US), Alternative Food Processing Technologies.*

2. *Black EP, Hoover DG, 2011, Microbiological aspects of high pressure food processing. In: Zhang HQ, Barbosa-Canovas GV, Balasubramaniam VM, Dunne CP, Farkas DF, Yuan JTC (eds) Non-thermal Processing Technologies for Food. Chichester: Wiley Blackwell, pp.51-71.*

3. *Knoerzer K, Buckow R, Sanguansri P, Versteeg C, 2010, Adiabatic compression heating coefficients for high-pressure processing of water, propylene-glycol and mixtures - a combined experimental and numerical approach. Journal of Food Engineering 96: 229-238.*

4. *Matser AM, Krebbers B, van den Berg RW, Bartels PV, 2004, Advantages of high-pressure sterilization on quality of food products. Trends in Food Science and Technology 15: 79-85.*

5. *Wilson MJ, Baker R, 1997, High temperature/ultrahigh pressure stabilization of low-acid foods. PCT 189601500/1997.*

6. *Rodriguez AC, Larkin JW, Dunn J et al., 2004, Model of the inactivation of bacterial spores by moist heat and high pressure. Journal of Food Science 69: 367–374.*

7. *Reddy NR, Tezloff RC, Solomon HM, Larkin JW, 2006, Inactivation of Clostridium botulinum nonproteolytic type B spores by high pressure processing at moderate to elevated high temperatures. Innovative Food Science and Emerging Technologies* **7**: *169-175.*

8. *Kouchma T, Guo B, Patazca E, Parisi B, 2005, High pressure-high temperature sterilization from kinetic analysis to process verification. Journal of Food Process Engineering 28: 610-629.*

9. *Margosch D, Ehrmann MA, Buckow R, Heinz V, Vogel RF, Ganzle MG, 2006, High pressure-mediated survival of Clostridium botulinum and Bacillus amyloliquefaciens endospores at high temperature. Applied and Environmental Microbiology 72: 3476-3478.*

10. *Ramirez R, Saraiva J, Lamela CP, Torres JA, 2009, Reaction kinetics analysis of chemical changes in pressure-assisted thermal processing. Food Engineering Review 1: 16-30.*

11. *Escobedo-Avellaneda Z, Moure MP, Chotyakul N, Torres JA, Welti-Chanes J, Lamela CP, 2011, Benefits and limitations of food processing by high-pressure technologies: effects on functional compounds and abiotic contaminants. CyTA Journal of Food 9: 351-364.*

12. *Vazquez-Landaverde PA, Qian MC, Torres JA, 2007, Kinetic analysis of volatile formation in milk subjected to pressure assisted thermal treatments. Journal of Food Science* ***72****: 389-398.*

13. *Segovia Bravo K, Ramirez R, Durst R et al., Formation risk of toxic and other unwanted compounds in pressure-assisted thermally processed foods. Journal of Food Science* ***77****: 1-10, 2011.*

14. *Constable A, Jonas D, Cockburn A et al., 2007, History of safe use of novel foods. Toxicology Letters 172: 2-12.*

Chapter 7

Ultrasonification

In addition to traditional methods such as heat treatment, mechanical agitation, drying and freezing, a variety of new technologies are now available to enhance the processing efficiency, shelf-life, safety and functional properties of dairy products and other processed foods and food ingredients **[1]**. Among various "emerging technologies" such as ultra-high pressure processing, pulsed electric fields, supercritical fluid extraction, microfluidization and ultraviolet light treatment, **ultrasonication has been identified as a particularly promising technology for processing specific food materials**, including dairy products [1a, 2, 3, 4]. Nowadays, power ultrasound is considered to be an emerging and promising technology for industrial food processing.

Ultrasound refers to sound waves, mechanical vibrations, which propagate through solids, liquids, or gases with a frequency greater than the upper limit of human hearing (16 Hz to 18 kHz). Ultrasound waves are propagated in upper frequencies from the limit of human hearing and have lower wavelength and more attenuation coefficient. Ultrasonic processing is an emerging technology in food processing. When ultrasound passes through a liquid, bubble nuclei present in the liquid grow by bubble coalescence and rectified diffusion. When these bubbles reach a critical size range, they collapse under near adiabatic conditions generating extreme conditions within the bubbles and in the surrounding liquid that include intense shear forces, turbulence and microstreaming effects. These ultrasound-induced physical effects are finding increasing use in food and dairy processing [1a]. Ultrasound technology has been used as an alternative processing technology over thermal processing techniques. This technology can be used for pasteurization and preservation of food by inactivation of enzymes and micro-organisms at low temperatures.

Ultrasonic waves (0.1-20 MHz) have garnered a great deal of interest in recent years in food processing due to their ability to influence the physical, biochemical or microbiological properties of food. This is because Ultra-sound (US) can interact with the food differently depending on the US wave and product characteristics. In recent years food technologists have discovered that it is possible to employ a more powerful form of ultrasound (>5 W/cm^2) at a lower frequency (generally around 40 kHz) [1i]. US, although it has interesting potential as a novel preservation method, it still has a long way to go before it can be utilized commercially for this purpose. It does, however, have numerous non-preservation applications, some of which are already being used commercially [1]. **Ultrasonic is a one of the fast, versatile, emerging, and promising non-destructive green technology** used in the food industry from last few years. The ultrasound is being carried out in various areas of food technology namely crystallization, freezing, bleaching, degassing, extraction, drying, filtration, emulsification, sterilization, cutting, *etc.* Ultrasound is being applied as an effective preservation tool in many food-processing fields *viz.* vegetables and fruits, cereal products, honey, gels, proteins, enzymes, microbial inactivation, cereal technology, water treatment, diary technology, etc [1b]. Ultrasound technology has been used as an alternative processing technology over thermal processing techniques. This technology can be used for pasteurization and preservation of food by inactivation of enzymes and micro-organisms at low temperatures.

Ultrasonic processing is still in its infancy and requires a great deal of future research in order to develop the technology on an industrial scale.

Principle [1]

Ultrasound [US] refers to sound waves above the frequency of human hearing (~ > 18 kHz). Ultrasound is energy generated by sound waves of 20,000 or more vibrations per second. When high-intensity ultrasound is passed through a medium, strong vibration of the medium occurs. Ultrasound technology as a key area of research and development in the food industry [1b, 1c] is based on mechanical waves at a frequency above the threshold of human hearing (>16 kHz) and can be categorized into following two frequency ranges:

i. Low Energy

Low-energy (low power, low intensity) ultrasound has frequencies higher than 100 kHz at intensities below 1 Wcm^{-2}. The low intensity ultra sound methodology is known as **non-destructive analytical** method. It is used in gauging the structure, composition, flow rate of foods and **change physical and chemical properties of food** [1f] by inducing pressure, shear, and temperature difference in the medium through which they propagate and is capable of producing cavitations in order to inactivate micro-organisms in foods [1g].

ii. High Energy

High-energy (high power, high-intensity) ultrasound uses intensities higher than 1 Wcm^{-2} (or W/cm^2) at frequencies between 20 and 500 kHz [1d] or (10-1000 W cm^{-2}) [1h]. **High-frequency ultrasound as an analytical technique is used to obtain information on the physico-chemical properties of food such as acidity,**

firmness, sugar content, ripeness, *etc.* High-intensity ultra sound methodology uses high frequencies, it causes physical damage to the tissues. Hence, they are used for cutting food or cleaning the equipment.

The representative range/typical limit for the frequency that is commonly applied in ultrasonic technology lies between 20 kHz and 500 MHz [1e]. In the ultrasound food processing industry, the aforesaid two divisions are being used [1h].

Ultrasound in higher frequencies and low powers has a good sensitivity and no effects on mechanical or chemical properties of material but only causes vibrations in the molecules. High power ultrasound with lower frequencies has the acoustic energy packages which induces physical, mechanical, and chemical changes in the material. The major reason in this process is the cavitation phenomenon which provides high temperature and pressure by collapsing the micro-bubbles.

The mechanical and chemical effects generated by low-frequency - high-intensity ultrasound may be useful for inactivating pathogens in food products and enhancing extraction and emulsification processes [1a].

The physical and chemical effects of ultra sound processing in liquid and solid medium have been significantly used [5c].

i. Solids

If the medium contains solid particles, for example, plant materials such as seeds, severe structural damage of these particles can occur. Such mechanical vibration effects can be used in cleaning and extraction applications.

ii. Liquid

In addition to the mechanical vibration effect, ultrasound also generates acoustic streaming within liquids [5], which can increase effective mass transfer leading to enhanced cleaning and extraction efficiencies in some industrial processes. In liquid medium, there are also some other extreme physical forces generated, such as cavitation, shear, micro-jet, and shockwaves. These forces further lead to the origin of ultra sound applications for food processing: emulsion, tenderization, and filtration [5c]

US is considered both a **thermal and a non-thermal treatment** by virtue of its mechanism of action [5a]. When US waves propagate into liquid media, alternating compression and expansion cycles are produced. During the expansion cycle, high intensity ultrasonic waves make small bubbles grow in liquid. When they attain a volume at which they can no longer absorb enough energy, they implode violently. This phenomenon is known as **cavitation.** In most ultrasonic processing applications, the observed effects are due to **acoustic cavitation** [5b], which involves the growth and collapse of pre-existing microbubbles in liquids. During the ultrasonication process, regions of **compression (positive pressure)** and **rarefaction (negative pressure)** are created. **Cavitation or bubbles form and grow in the liquid during rarefaction and collapse during compression,** creating micromechanical shocks (shear forces) that disrupt the structural and functional components of cells, ultimately leading to cell lysis [6]. When these cavitation bubbles explode or implode, they generate zones of highly localized extreme pressures (>108 Pa or estimated at 50000 kPa) and temperatures (approximately 4000 to 5000 K) that disrupt the cell structure and inactivate microbes. **The mechanism of inactivation can thus be summarized as a combination of thermal and non-thermal effects culminating in damage to cell walls, thinning of membranes (micro streaming) and the production of free radicals that lead to DNA damage [7, 8].**

When the applied acoustic energy exceeds the cavitation threshold, the oscillation of gas bubbles becomes intense leading either to **transient or stable** (repetitive transient) **cavitation** [1b, 5b].

Transient Cavitation

In transient cavitation, observed primarily **with horn-type sonicators** operated at 20 kHz and high acoustic intensities, bubble growth occurs within a few acoustic cycles leading into their resonance size range [8a], where the bubbles rapidly collapse. The bubbles disintegrate into fragments due to the violent nature of the collapse. Non-stable, rapidly collapsing and disintegrating into smaller bubbles are known as internal (transient) cavitations bubbles.

Stable Cavitation

In stable cavitation however, bubble growth occurs over thousands of acoustic cycles either by rectified diffusion and/or bubble coalescence pathways [8b]. Stable cavitation is commonly observed at high (> 200 kHz) ultrasound frequencies. Once the bubbles reach a resonance size range [8a], they collapse like transient bubbles; however, the collapse is less violent and hence the bubbles may grow and collapse continuously. Non-linear, forming large bubble clouds with equilibrium size during pressure cycles are known as stable cavitations bubbles.

The collapse of cavitation bubbles (both transient and stable) also leads to the generation of high temperatures and pressures within the collapsing bubbles. This is due to the fact that the collapse occurs in a very short time (< 1 μs) leading to near-adiabatic heating of the bubble interior. Experiments have suggested that the temperatures generated within the cavitation bubbles are in the range 2000-5000 K [5b].

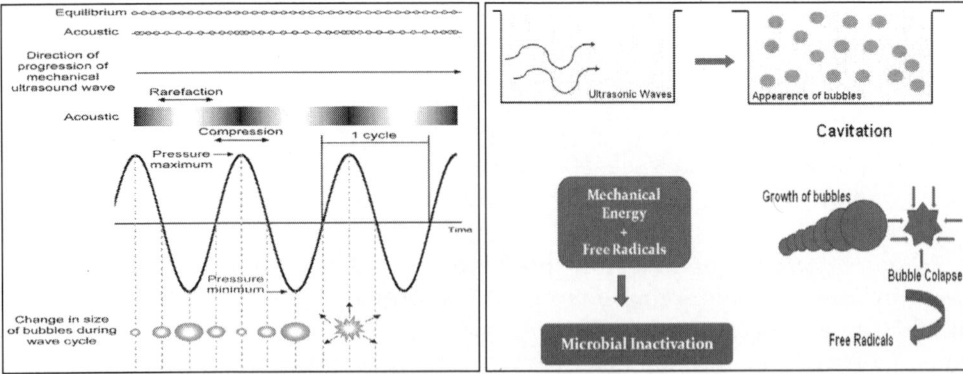

Cavitation Caused by Ultrsonification [1b].

The combined effects of mechanical vibration, acoustic streaming and cavitation can enhance mass transfer and hence reaction rates and extraction efficiencies. The non-cavitational physical effects (mechanical vibration and acoustic streaming) and cavitational physical effects (shear forces, shock waves, microjets, *etc.*) of ultrasound have been used for a number of applications that include emulsification, extraction, cleaning and welding [9, 10].

In aqueous solutions, the interior of the cavitation bubbles is filled with water vapor and gas molecules, such as N_2 and O_2 (in air saturated water) that diffuse into the bubbles during bubble oscillations in the growth phase. When subjected to the high temperature conditions generated during bubble collapse, the solvent and gas molecules undergo "endothermic" chemical reactions leading to the generation of a variety of highly reactive radicals. The details are given below:

1. The acoustic cavitation at low frequencies (~ 20 kHz) generates strong physical forces (shear, turbulence, *etc.*). The amount of reactive radicals generated is very low [5b].
2. Higher frequency acoustic cavitation (300-500 kHz) generates relatively weaker physical effects and significantly larger amounts of chemically active radicals.
3. Frequencies above 1 MHz are generally useful for non-destructive imaging purposes. However, these high frequencies have also been found to be most appropriate for the generation of fine aerosols from a free liquid surface, in a process referred to as ultrasonic nebulization [11].

Ultrasound is applied by the following three different methods:

1. Applying directly to the product
2. Coupling with the device
3. Submerging in an ultrasonic bath.

Sources

1. *Hudaa Neetoo (Faculty of Agriculture, University of Mauritius, Réduit, Mauritius) and Haiqiang Chen (Department of Animal and Food Sciences, University of Delaware, Newark, Delaware, US), Alternative Food Processing Technologies.*

1a. *Muthupandian ashok kumar, Raman bhaskaracharya, Sandra Kentish, Judy lee, Martin palmer, Bogdan zisu, The ultrasonic processing of dairy products -An overview, Dairy Sci. Technol. 90, 2010, 147-168.*

1b. *Ishrat Majid, Gulzar Ahmad Nayik, and Vikas Nanda Ultrasonication and food technology: A review, View further author informationArticle: 1071022, Published online: 03 Aug 2015.*

1c. *Ercan, S. S., and Soysal, Ç., 2011, Effect of ultrasound and temperature on tomato peroxidase. Ultrasonics Sonochemistry, 18, 689-695.*

1d. *Mason, T. J., Chemat, F., and Vinatoru, M., 2011, The extraction of natural products using ultrasound or microwaves. Current Organic Chemistry, 15, 237-247.*

1e. *Yusaf, T., and Al-Juboori, R. A., 2014, Alternative methods of micro-organism disruption for agricultural applications. Applied Energy, 114, 909-923.*

1f. *Soria, A. C., and Villamiel, M., 2010, Effect of ultrasound on the technological properties and bioactivity of food: A review. Trends in Food Science and Technology, 21, 323-331.*

1g. *Piyasena, P., Mohareb, E., and McKellar, R. C., 2003, Inactivation of microbes using ultrasound: A review. International Journal of Food Microbiology, 87, 207-216.*

1h. *Fellows, P., Food processing technologies: Princilpes and practices (3rd ed ed.), Boca Raton, FL: CRC Press, 2009.*

1i. *Ranjan Saha, New and Emerging Non-Thermal Technologies in Food Preservation, April 2015.*

2. *Butz P., Tauscher B., Emerging technologies: chemical aspects, Food Res. Int. 35, 2002, 279-284.*

3. *Mason T. J., Paniwnyk L., Lorimar J. P., The uses of ultrasound in food technology, Ultrason. Sonochem. 3, 1996, S253-S260.*

4. *Wan J., Mawson R., Ashokkumar M., Ronacher K., Coventry M. J., Roginski H., Versteeg K., Emerging processing technologies for functional foods, Aust. J. Dairy Technol. 60, 2005, 167-169.*

5. *Tho P., Manasseh R., Ooi A., Cavitation microstreaming patterns in single and multiple bubble systems, J. Fluid Mech. 576, 2007, 191-233.*

5a. *Feng H, Yang W, 2010, Power ultrasound. In: Hui YH (ed) and book of Food Science, Technology and Engineering. Boca Raton, FL: CRC Press, pp. 121-127.*

5b. *Ashok kumar M., Mason T., Sonochemistry, Kirk-Othmer Encyclopedia of Chemical Technology, John Wiley and Sons, 2007, pp. 1-34.*

5c. *Harlin Kaur Saroya, Innovative non-thermal food processing technologies used by the food industry in the United States, MSc thesis, 2017, Western Kentucky University, Kentucky.*

6. *Hoover DG, 2000, Ultrasound. Journal of Food Science 1 (suppl): 93-95.*

7. *Earnshaw RG, 1998, Ultrasound: a new opportunity for food preservation. In: Povey MJW, Mason T (eds) Ultrasound in Food Processing. London: Blackie Academic and Professional, pp.183-192.*

8. *Earnshaw RJ, Apple yard J, Hurst RM, 1995, Understanding physical inactivation processes: combined preservation opportunities using heat, ultrasound and pressure. International Journal of Food Microbiology 28: 197-219.*

8a. *Yasui K., Influence of ultrasonic frequency on multibubble sonoluminescence, J. Acoust. Soc. Am. 112, 2002, 1405-1413.*

8b. *Ashok kumar M., Grieser F., The effect of surface active solutes on bubbles in an acoustic field, Phys. Chem. Chem. Phys. 9, 2007, 5631-5643.*

9. *Balachandran S., Kentish S., Mawson R., Ashok kumar M., The use of ultrasound to enhance the supercritical extraction of ginger, Ultrason. Sonochem. 13, 2006, 471-479.*

10. *Mason T. J., Industrial sonochemistry: potential and practicality, Ultrasonics 30, 1992, 192-196.*

11. *Jimmy B., Kentish S., Grieser F., Ashok kumar M., Ultrasonic nebulisation in aqueous solutions and the role of interfacial adsorption dynamics in surfactant enrichment, Langmuir 24, 2008, 10133-10137.*

Ultrasound Equipment

Different types of apparatus exist for ultrasonic applications, depending on the scale of the process. Some examples of equipment that have been used include whistle reactors(Generator), ultrasonic baths, and probe systems(piezoelectric transducer**)** [1, 5].

a. A Whistle Reactor

It uses a mechanical ultrasonic source causing vibration of a stream of liquid moving past a metal blade. The frequency of vibrations is a function of the liquid flow rate and high enough flow rates can generate US, causing cavitation in the liquid. This set-up is used for food processing applications such as homogenization, emulsification or dispersion [2].

b. Ultrasonic Baths

These baths have a relatively simple and low-cost set-up, consisting of a metal bath with one or more transducers attached to the walls of the tank. The product to be treated is directly immersed into the bath [3].

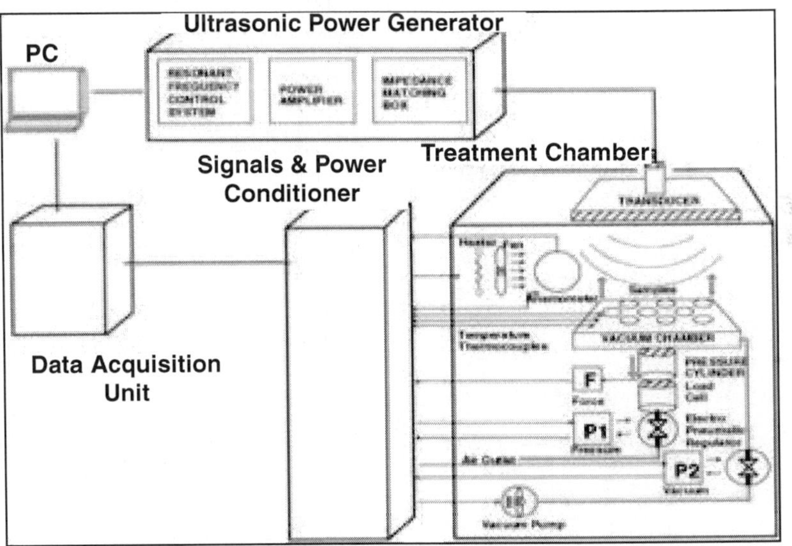

Ultrasonic in Food Processing [4].

c. Probe Systems

Another ultrasonic unit makes use of probe systems, comprising a metal horn linked to an ultrasonic transducer, which is made of a piezoelectric material, with the metal horn serving to amplify the vibration generated by the transducer [1]. The **piezoelectric transducer** diffuses ultrasound to the food through a horn and it is immersed in the liquid. This setup can either be placed in heating equipment for thermo-sonication (technique used for liquid food preservation) or in pressurized vessels food preservation [5]. The need to amplify sound waves is justified since the amplitude of the waves produced by the transducer is too low to deliver any appreciable effect. Probe systems can be directly placed in the fluid-like product with easily controlled power input [3]. **Technological advances to increase the efficiency and quality of this ultrasonic unit include use of "cup horn" flow cells as well as tube reactors.** While "cup horns" incorporate a cooling system to enable better temperature control [3], flow cells and tube reactors allow monitoring of the treatment time during which samples are exposed to US [3].

Sources

1. *Mason TJ, 1998, Power ultrasound in food processing - the way forward. In: Povey MJW, Mason TJ (eds) Ultrasound in Food Processing. London: Blackie and Academic and Professional, pp.105-126.*

2. *Torley PJ, Bhandari BR, 2007, Ultrasound in food processing and preservation. In: Rahman MS (ed) Handbook of Food Preservation. Boca Raton, FL: CRC Press, pp. 713-732.*

3. *Mason TJ, 1990, A survey of commercially available sources of ultrasound suitable for sono-chemistry. In: Mason TJ (ed) Sonochemistry: The Uses of Ultrasound in Chemistry. Cambridge: Royal Society of Chemistry, pp. 60-68.*

4. *Mathavi, V., Sujatha, G., Bhavani, S., and Karthika, D., New trends in food processing. International Journal of Advances in Engineering and Technology, 5(2), 176-187, 2013.*

5. *Harlin Kaur Saroya, Innovative non-thermal food processing technologies used by the food industry in the United States, MSc thesis, 2017, Western Kentucky University, Kentucky.*

Determination of Acoustic Power and Efficacy of Ultrasonic Treatments in Terms of Eliminating Microbes [1]

The most widely accepted method for determining the power from an acoustic horn into an aqueous solution is the calorimetric technique [2]. This method involves taking a known volume of water and applying ultrasound (for ca. 3 min) while moni-toring the change in temperature with time for various ultrasonic amplitudes. The ultrasonic power can be readily determined from the following equation:

$$P = m \times Cp \times \frac{\delta T}{\delta t}$$

where, P - the ultrasonic power, m - the mass of the solvent used, Csys - the specific heat capacity of the system, which is described by

$$Csys = CgMg + CwMw$$

where, Cg - the specific heat capacity of the container used (in this case glass) and Mg -the mass of the container the ultrasonic power can be deduced.

This simple equation has been widely used through-out the sonochemistry literature. The efficacy of ultrasonic treatments in terms of eliminating microbes was measured by their decimal reduction time (D). D value was calculated as the time (min) required to reduce the number of viable cells by one log cycle or to kill 90 per cent of population at a given temperature, time of ultrasonic treatment and sonic wave amplitude. D-values were calculated from the slope of the regression line plotted with the counts (CFU mL-1) of the straight portion of the survival curve. The D-value at 20 kHz was abbreviated as D_{US}.

$$\log \frac{N_1}{N_0} = -\frac{t}{D_{us}}$$

where, N_0 - number of total microbes before ultra-sound treatment, N_1 - number of microbes after ultrasound treatment at time t, D_{us} - decimal reduction time (min).

Sources

1. *Edita Juraga, Brankica Sobota Šalamon, Zoran Herceg, Anet Režek Jambrak, Application of high intensity ultrasound treatment on Enterobacteriae count in milk, Mljekarstvo, 61 (2), 125-134, 2011,.*

2. *Margulis, M. A., Maltsev, I. M. 2003: Calorimetric method for measurement of acoustic power absorbed in a volume of a liquid. Ultrasonic Sonochemistry 10, 343- 345.*

Benefits, and Shortcomings of Alternative Non-thermal Processing Method-US [1]

Process considerations, benefits and shortcomings of alternative non-thermal processing method- US are summarized in following table:

Sl. No.	*Process Considerations*	*Benefits*	*Shortcomings*	*Examples of Applications*
1	Amplitude of ultrasonic waves	Ultrasound effective against vegetative cells	Has little effect on its own	Any food that is heated
2	Exposure time	TS and MTS effective against vegetative cells and spores	Challenges with scaling up	
3	Microbial species	Reduced process times	Free radicals could damage product quality	
4	Volume of food	Amenable to batch and continuous processing	Can induce undesirable textural change	
5	Product composition required	Little adaptation required for existing processing plant	Can be damaging to eyes	
6	Treatment temperature	Possible modification of food structure and texture	Can cause burns and skin cancer	
		Energy efficient	Depth of penetration affected by solids and air in product	
		Several equipment options	Potential problems with scaling up of plant.	
		Effect on enzyme activity Can be combined with other unit operations.		

Sources

1. *Hudaa Neetoo (Faculty of Agriculture, University of Mauritius, Réduit, Mauritius) and Haiqiang Chen (Department of Animal and Food Sciences, University of Delaware, Newark, Delaware, US), Alternative Food Processing Technologies.*

Advantages and Disadvantages/Limitations of Ultrasonication [1]

a. Advantages and Disadvantages

Ultrasound applications offer numerous advantages in the food industry. Despite having lot of advantages, use of ultrasonication has also many disadvantages. Some of the advantages and disadvantages of US are tabulated below:

Sl. No.	*Advantages*	*Disadvantages/Limitations*
1	Ultrasound waves are non-toxic, safe, and environmentally friendly [2].	Ultrasound due to shear stress developed by swirls from the shock waves (mechanical effects) abuse inactivation of the released products [8].
2	Ultrasonication in combination with other non-thermal methods is considered an effective means of microbial inactivation [3].	Ultrasound application needs more input of energy which makes industrialists to think over while using this technique on commercial scale [9].
3	Ultrasonication involves lower running cost, ease of operation, and efficient power output.	Ultrasound induces physico-chemical effects which may be responsible for quality impairment of food products by development of off-flavors, alterations in physical properties, and degradation of components.
4	Ultrasonication does not need sophisticated machinery and wide range of technologies [4].	Ultrasonication leads to the formation of radicals as a result of critical temperature and pressure conditions that are responsible for changes in food compounds. The radicals (OH and H) produced in the medium deposit at the surface of cavitation bubble that stimulates the radical chain reactions which involve formation of degradation products and thus lead to considerable quality defects in product [10].
5	Use of ultrasound provides more yield and rate of extraction as compared to other conventional methods of extraction [5].	Frequency of ultrasound waves can impose resistance to mass transfer [11].
6	Ultrasonication involves minimum loss in flavor, superior consistency (viscosity, homogenization), and significant savings in energy expenditure [6].	Ultrasonic power is considered to be responsible for change in materials based on characteristics of medium. So, this power needs to be minimized in food industry in order to achieve maximum results [12].
7	Ultrasound has gained huge applications in the food industry such as processing, extraction, emulsification, preservation, homogenization, *etc.* [7].	

b. Application-wise Advantages of Ultrasound in Food Processing [13, 14]

The details related to the application wise advantages of US in food processing are tabulated below:

Sl. No.	*Applications*	*Advantages*
1	Filtration	Better filtration, Time saver.
2	Drying	Time saver, improves heat transfer and organoleptic quality.
3	Cooking	
4	Emulsification	Time saver
5	Oxidation	
6	De-foaming	Better hygiene, Time saver
7	Degassing	
8	Cutting	Accurate and repetitive cutting, Time saver
9	Freezing/Crystallization	Time saver, Fine crystals, Diffusion improvement, Rapid decrease in temperature

Sources

1. *Ishrat Majid Gulzar Ahmad Nayik and Vikas Nanda, Ultrasonication and food technology, A review, 1071022, published online,03Aug 2015.*
2. *Kentish, S., and Ashokkumar, M., 2011, The physical and chemical effects of ultrasound. In H. Feng, G. V. Barbosa-Canovas, and J. Weiss (Eds.), Ultrasound technologiesfor food and bioprocessing (pp. 1-12). London: Springer.10.1007/978-1-4419-7472-3.*
3. *Vercet, A., Sánchez, C., Burgos, J., Montañés, L., and Lopez Buesa, P. L., 2002, The effects of manothermosonication on tomato pectic enzymes and tomato paste rheological properties. Journal of Food Engineering, 53, 273-278.*
4. *Gallego-Juárez, J., Rodriguez, G., Acosta, V., and Riera, E., 2010, Power ultrasonic transducers with extensive radiators for industrial processing. Ultrasonics Sonochemistry, 17, 953-964.*
5. *Balachandran, S., Kentish, S. E., Mawson, R., and Ashokkumar, M., 2006, Ultrasonic enhancement of the upercritical extraction from ginger. Ultrasonics Sonochemistry, 13, 471-479.*
6. *Chouliara, E., Georgo-gianni, K. G., Kanello-poulou, N., and Kontominas, M. G., 2010, Effect of ultrasonication on microbiological, chemical and sensory properties of raw, thermized and pasteurized milk. International Dairy Journal, 20, 307-313.*
7. *Chemat, F., Zill-e-Huma, and Khan, M. K., 2011, Applications of ultrasound in food technology: Processing, preservation and extraction. Ultrasonics Sono-chemistry, 18, 813-835.*
8. *Lateef, A., Oloke, J. K., and Prapulla, S. G., 2007, The effect of ultrasonication on the release of fructosyl-transferase from Aureobasidium pullulans CFR 77, Enzyme and Microbial Technology, 40, 1067-1070.*

9. *Yusaf, T., and Al-Juboori, R. A., 2014, Alternative methods of micro-organism disruption for agricultural applications. Applied Energy, 114, 909-923.*

10. *Czechowska-Biskup, R., Rokita, B., Lotfy, S., Ulanski, P., and Rosiak, J. M., 2005, Degradation of chitosan and starch by 360-kHz ultrasound. Carbohydrate Polymers 60, 175-184.*

11. *Esclapez, M. D., García-Pérez, J. V., Mulet, A., and Cárcel, J. A., 2011, Ultrasound-assisted extraction of natural products. Food Engineering Reviews, 3, 108-120.*

12. *Feng, H., Barbosa-Canovas, G., and Weiss, J. (Eds.)., 2011, Ultrasound technologies for food and bioprocessing. New York, NY.*

13. *Harlin Kaur Saroya, Innovative non-thermal food processing technologies used by the food industry in the United States, MSc thesis, 2017, Western Kentucky University, Kentucky.*

14. *Chemat, F., Zill-e-Huma, and Khan, M. (2015). Applications of ultrasound in food technology: Processing, preservation and extraction, Ultrasonics Sonochemistry, 18(4), 813-835.*

Ultrasound with Heat and Pressure [1, 1a]

The application of US in conjunction with heat or pressure are termed as:

1. **Ultrasonication (US) [1a]**: It is the application of ultrasound at low temperature [2].
2. **Thermosonication (TS) [1]**: The conjunct use of US and heat(mild heat) is called thermosonication (TS). The effectiveness of TS varies as a function of the intensity, amplitude of sound waves, and treatment time. The product is subjected to ultrasound and moderate heat simultaneously. This method produces a greater effect on inactivation of micro-organismsthan heat alone [3, 4].
3. **Manosonication (MS) [1a]**: The conjunct use of US and pressure heat is called thermosonication (MS). Manosonication provides to inactivate enzymes and/or micro-organisms by combining ultrasound with moderate pressures (100 to 300 kPa) at low temperatures [5].
4. **Manothermosonication (MTS) [1]**: The combined application of pressure, heat and US is termed manothermosonication (MTS). In addition to the factors mentioned above, the effectiveness of MTS also depends on the magnitude of the applied pressure. It is thought that the application of pressure to the food material being ultrasonicated enables cavitation to be maintained at temperatures above the boiling point at atmospheric pressure [6]. It was found that manothermosonication inactivated both extracellular lipase and protease more effectively than heat treatment alone [7]. Micro-organisms (*Bacillus subtilis, Bacillus coagulans, Bacillus cereus, Bacillus sterothermophilus, Saccharomyces cerevisiae*) that have high thermo tolerance can be inactivated by manothermosonication.

Sources

1. *Hielscher Ultrasonic Technology, Ultrasonication and Its Manifold Applications in Food Processing, info @heilscher.com.*

1a. *Chetana B. Chaudhari, J. P. Prajapati and Suneeta V. Pinto, Ultrasound Technology for Dairy Industry, Department of Dairy Technology, SMC College of Dairy Science, AAU, Anand.*

2. *Zheng L. and Sun DW, 2006, Innovative applications of power ultrasound during food freezing processes -A review. Trends in Food Science and Technology, 17(1); 16-23.*

3. *Mason TJ, Paniwnyk L. and Lorimer JP, 1996, The uses of ultrasound in food technology. Ultrasonics Sonochemistry, 3, S253-60.*

4. *Villamiel M, Hamers-veld EH. and de Jong P., 1999, Effect of Ultrasound Processing on the Quality of Dairy Products. Milchwissenschaft., 54, 69-73.*

5. *Ercan SS. and Soysal C., 2013, Use of ultrasound in food preservation. Natural Science.5 (8 A2), 5-13.*

6. *Torley PJ, Bhandari BR, 2007, Ultrasound in food processing and preservation. In: Rahman MS (ed) Handbook of Food Preservation. Boca Raton, FL: CRC Press, pp. 713-732.*

7. *Vercet A., Lopez P., Burgos J., Inactivation of heat-resistant lipase and protease from Pseudomonas fluorescens by manothermosonication, J. Dairy Sci. 80, 1997, 29-36,*

Food Processing Applications of Ultrasound/Ultrasonification

Power ultrasound offers manifold possibilities for effective and reliable food processing applications. The most common applications in the food industry include mixing and homogenization, emulsification, dispersing, cell disruption and extraction of intra-cellular material, activation or deactivation of enzymes (which is dependent on the ultrasound intensity), preservation, stabilization, dissolving and crystallization, hydrogenation, meat tenderization, maturation, aging and oxidation, as well as degassing and spray drying [1]. The applications of US in food processing are tabulated below [7, 8]:

Sl. No.	*Applications*	*Conventional Methods*	*Ultrasound Principle*	*Food Products*
1	Filtration	Filters	Vibrations	Liquids
2	Drying	Atomization, Hot Gas stream Pulverization	Uniform heat transfer	Dehydrated Products
3	Cooking	Stove Fryer Water Bath		Meat, Vegetables
4	Emulsification	Mechanical treatment	Cavitation phenomenon	Sauces
5	Oxidation	Air contact		Alcohols
6	De-foaming	Thermal/Chemical/ Mechanical treatment		Fermented and carbonated products
7	Degassing	Mechanical treatment	Compression refraction phenomenon	Chocolates, Fermented Products

Sl. No.	*Applications*	*Conventional Methods*	*Ultrasound Principle*	*Food Products*
8	Cutting	Sharp tools	Cavitation phenomenon	Fragile Products (Cake, Cheeses)
9	Freezing/Crys-tallization	Freezer, Freezing by other products	Uniform heat transfer	Meat, Vegetables, Fruits, Milk Products.
10	Marinating	Brine	Increasing mass trans-fer	Cheese, Fish, Meat, Vegetables

Although ultrasound technology has a wide range of current and future applications in the food industry, including inactivation of micro-organisms and enzymes, presently, most developments for food applications are nonmicrobial.

Until recently the majority of applications of ultrasound in food technology involved **non-invasive analysis with particular reference to quality assessment.** Such applications use techniques that are similar to those developed in diagnostic medicine, or non-destructive testing, using high frequency (>1 MHz) low power (<1 W/cm^2) ultrasound. Examples of the use of such technologies are to be found in the location of foreign bodies in food the analysis of droplet size in emulsions of edible fats and oils and the determination of the extent of crystallization in dispersed emulsion droplets. In recent years food technologists have discovered that it is possible to employ a more powerful form of ultrasound (>5 W/cm^2) at a lower frequency (generally around 40 kHz) [1c].

1. Ultrasonic extraction of phenolic compounds from vacuolar structures by disrupting plant tissue.
2. Ultrasonic extraction of Betacyanin (red pigments, *e.g.* from beets) and Betaxanthin (yellow pigments).
3. Ultrasonic improvement of oil extraction from oil seeds.
4. Cell membrane permeabilization of fruits, such as grapes, plums, mango,.
5. Ultrasonic processing of
   a. Fruit juices, *e.g.* orange, grapefruit, mango, grape, plum.
   b. Purees.
   c. Sauces, *e.g.* tomato, asparagus, bell pepper, mushroom.
   d. Dairy products.
   e. Improve stability of dispersions, such as orange juice, *i.e.* reduce settling.
6. Microbial and enzyme inactivation (preservation) is another application of ultrasound in the food processing [1c].

The details related to the use of US in food processing, particularly its application for microbial inactivation, with some other applications, are given below:

A. Microbial Inactivation

Previous research has demonstrated the ability of US technology to pasteurize or even sterilize foods, alone or in combination with other preservation techniques [1a].

The shear forces and rapidly changing pressures created by US waves are effective in destroying vegetative microbial cells, especially when used in conjunction with other preservation treatments including heat, pH modification, and chlorination. There has been considerable interest in the potential use of high-power/high-intensity ultrasound, often in conjunction with mild heating and pressure for the inactivation of microbes and enzymes associated with spoilage, safety and quality deterioration in a range of liquid food systems [1b, 2, 2a]. The bactericidal effect of ultrasound is attributed to intracellular cavitation, that is, micro-mechanical shocks that disrupt cellular structural and functional components up to the point of cell lysis. The applications of US on microbial inactivation are summarized below:

1. Although US can inactivate vegetative cells, insignificant reduction of bacterial spores has been reported even with an extended US treatment [3].
2. However, the use of hurdle approaches such as TS and MTS was shown to have greater sporicidal effect, with MTS being more effective than TS [3].
3. Ultrasound can be used in the product sanitization process to improve the efficacy of the washing process [3].
4. It was demonstrated [4] that US combined with chlorinated water reduced the population of *S. typhimurium* on iceberg lettuce by 1.7 logs compared to 0.7-log reduction by US alone.
5. Ultrasound treatment was also found to be effective in reducing the population of S. typhimurium on chicken breast skin. Decontamination treatments in a chlorinated solution reduced *S. typhimurium* contamination by 0.2-0.9 logs while ultrasonication and chlorinated water delivered 2.4-3.9-log reductions [5].
6. The combined application of acetic acid and US treatments to clean eggshells has also been investigated with reported similar antimicrobial efficacy [6].

Sources

1. *Hielscher Ultrasonic Technology, Ultrasonication and Its Manifold Applications in Food Processing, info @heilscher.com.*

1a. *Earnshaw RJ, Apple yard J, Hurst RM, 1995, Understanding physical inactivation processes: combined preservation opportunities using heat, ultrasound and pressure. International Journal of Food Microbiology 28: 197-219.*

1b. *Knorr D., Zenker M., Heinz V., Lee D. U., Applications and potential of ultrasonics in food processing, Trends Food Sci. Technol.15, 2004, 261-266.*

1c. *Ranjan Saha, New and Emerging Non-Thermal Technologies in Food Preservation, April 2015.*

2. *Piyasena P., Mohareb E., McKellar R. C., Inactivation of microbes using ultrasound: areview, Int. J. Food Microbiol.87, 2003, 207-216.*

2a. *Muthupandian ashok kumar, Raman bhaskaracharya, Sandra Kentish, Judy lee, Martin palmer, Bogdan zisu, The ultrasonic processing of dairy products -An overview, Dairy Sci. Technol. 90, 2010, 147-168.*

3. *Guan D, Hoover DG, 2005, Novel non-thermal treatments. In: Sapers GM, Gorny JR Yousef AE (eds) Microbiology of Fruits and Vegetables. Boca Raton, FL: CRC Press, pp. 497-523.*

4. *Seymour IJ, 2002, Ultrasound decontamination of minimally processed fruits and vegetables. International Journal of Food Science and Technology 37: 547-550.*

5. *Lillard HS., 1993, Bactericidal effect of chlorine on attached salmonellae with and without sonication. Journal of Food Protection 56: 716-717.*

6. *Heath JL, Owes SL, Goble JW, 1980, Ultrasonic vibration as an aid in the acetic acid method of cleaning eggs. Poultry Science 59: 737-742.*

7. *Harlin Kaur Saroya, Innovative non-thermal food processing technologies used by the food industry in the United States, MSc thesis, 2017, Western Kentucky University, Kentucky.*

8. *Chemat, F., Zill-e-Huma, and Khan, M., 2015, Applications of ultrasound in food technology: Processing, preservation and extraction, Ultrasonics Sonochemistry, 18(4), 813-835.*

B. Other Applications

1. Applications of US in food processing operations such as drying, dewatering, filtration, membrane separation, salting, and osmotic dehydration have also been examined [1]. Combining US and air drying has been shown to enhance the rate of drying in various products, including carrot slices [2], onion slices [3], potato cylinders [4], wheat [5], corn [5], and rice [6]. Acoustic drying has an overriding advantage compared to conventional drying processes since heat sensitive foods can be dried faster and at a lower temperature (~50–60°C) than in conventional hot air driers (~100–115°C). There has also been considerable research into unit operations such as US-assisted membrane filtration and osmotic dehydration as well as processes such as cheese brining and meat curing.

   Ultrasonication singly or in combination with other preservation methods have been used to decrease the required processing temperature and time, or both, in pasteurization of liquid foods like milk, wine, and juices. It is used as a substitute or additional process to traditional thermal methods [6a]. Numerous other applications of ultrasound are reported in several foods including, cooking oils, bread, cereal products, and emulsified fat-based food products, food gels, aerated foods, and frozen foods [6b].

2. **Emulsification**: It is another application of power US. When a bubble collapses in the vicinity of the phase boundary of two immiscible liquids, the resulting shock wave can provide very efficient mixing of layers. Emulsions generated by US have a number of advantages, including their stability without the addition of surfactant, as well as narrow mean droplet size distribution [7].

3. Ultrasound technology has also been used in the making of US knives for "clean cutting" of sticky and brittle food products, including nut, raisins, and other hard fruits [1].
4. The use of ultrasonication to control enzymatic activity [1] as well as to extract a range of bioactive compounds has been widely investigated.
5. Several reviews have been published in the past on the use of US to extract plant origin metabolites and flavonoids from foods using a range of solvents [8] and bioactives from herbs [9, 10,]. The range of published extraction applications includes herbal extracts [10], almond oils [11], soy protein [12] and bioactives from plant materials, *e.g.* flavones [13] and polyphenolics [14].

Sources

1. *Feng H, Yang W, 2010, Power ultrasound. In: Hui YH (ed) Handbook of Food Science, Technology and Engineering. Boca Raton, FL: CRC Press, pp. 121-127.*

2. *Gallego-Juarez JA, Rodriguez-Corral G, Galvez-Moraleda JC, Yang TS, 1999, A new high intensity ultrasonic technology for food dehydration. Drying Technology 17: 597-608.*

3. *Da-Mota VM, Palau E, 1999, Acoustic drying of onion. Drying Technology 17: 855-867.*

4. *Bartolome LG, Hoff JE, Purdy KR, 1969, Effect of resonant acoustic vibrations on drying rates of potato cylinders. Food Technology 23: 321-324.*

5. *Huxsoll CC, Hall CW, 1970, Effects of sonic irradiation on drying rates of wheat and shelled corn. Transactions of the American Society of Agricultural Engineers 13: 21-24.*

6. *Muralidhara HS, Ensminger D, 1986, Acoustic drying of green rice. Drying Technology 4: 137-143.*

6a. *Valero, M., Recrosio, N., Saura, D., Muñoz, N., Martí, N., and Lizama, V., 2007, Effects of ultrasonic treatments in orange juice processing. Journal of Food Engineering, 80, 509-516.*

6b. *Gallego-Juárez, J., Rodriguez, G., Acosta, V., and Riera, E., 2010, Power ultrasonic transducers with extensive radiators for industrial processing. Ultrasonics Sonochemistry, 17, 953-964.*

7. *Mason TJ, Paniwnyk L, Lorimer JP, 1996, The Uses of Ultrasound in Food Technology. Ultrasonics Sonochemistry 3: 253-260.*

8. *Zhang R, Xu Y, Shi Y, 2003, The extracting technology of flavonoids compounds. Food and Machinery 1: 21-22.*

9. *Roldán-Gultiérrez JM, Ruiz-Jimenez J, de Castro MDL, 2008, Ultrasound- assisted dynamic extraction of valuable compounds from aromatic plants and flowers as compared with steam distillation and superheated liquid extraction. Talanta 75: 1369-1375.*

10. *Vinatoru M, 2001, An overview of the ultrasonically assisted extraction of bioactive principles from herbs. Ultrasonics Sonochemistry 8: 303-313.*

11. *Riera E, Golás Y, Blanco A, Gallego A, Blasco M, Mulet A, 2004, Mass transfer enhancement in supercritical fluids extraction by means of power ultrasound. Ultrasonics Sonochemistry 11: 241-244.*

12. *Moulton J, Wang C, 1982, A pilot plant study of continuous ultrasonic extraction of soybean protein. Journal of Food Science 47: 1127-1129.*

13. *Rostagno A, Palma M, Barroso C, 2003, Ultrasound-assisted extraction of soy isoflavones. Journal of Chromatography 1012: 119-128.*

14. *Xia T, Shi S, Wan X, 2006, Impact of ultrasonic-assisted extraction on the chemical and sensory quality of tea infusion. Journal of Food Engineering 74: 557-560.*

Application in Dairy Industry [1]

Ultrasound applications are classified into **high intensity and low intensity.** Analytical ultrasound involves monitoring and quality control of dairy products based on the physico-chemical variations during storage or processing. High intensity ultrasound has been employed for processing applications such as pasteurization, homogenization, fermentation, and extraction. Ultrasound has already made significant inroads into commercial dairy processing operations. Some of the applications of US in dairying are discussed below:

1. Ultrasonic Cutting Devices [1a]

This is most significant application. These machines utilize the high-frequency acoustic vibrations (20–35 kHz) to provide a smooth and clean cut for both hard and soft cheeses, minimizing wastage [2, 3].

2. Sealing Milk Containers and Packages

Ultrasound has also been comercialized as a method of sealing milk containers and packages of grated cheese in a process also referred to as ultrasonic welding, the temperature induced by the cavitation processes described above is sufficient to seal the plastic [4, 5].

3. Fermentation of Milk/Yoghurt [1, 15]

The use of low-intensity ultrasound increases mass transfer within the fermentation vessel and thus can increase enzyme productivity [1a]. It was reported [15a] that sonication increased the hydrolysis of lactose in milk in the presence of a starter culture (*Lactobacillus delbrueckii*) and hence the fermentation process.

Yogurt is a fermented milk product that can be produced by milk alone or by the addition of bacterial cultures. Bifidobacteria strains are common probiotics used for yoghurt fermentation. Ultrasonic cavitation applied to the bacterial cells can cause their destruction and simultaneously, the release of β-galactosidase (an hydrolase enzyme that is heavily used in the milk processing industry).

The ultrasonically assisted fermentation is accelerated due to a faster lactose hydrolysis resulting from the ultrasonically induced release of β-galactosidase from the bifidobacteria cells. The ultrasonic homogenization effects the break of the milk fat globules and a very fine-size distribution. **Ultrasonication can accerlerate the fermentation rate (reduction of total production time of up to 40 per cent** due to sonication increasing enzyme activity) **and improve the quality characteristics of yogurt, resulting in higher viscosity(attributed to increases in the water-retaining casein becoming available as the fat globule membrane surface area increased), stronger coagulum and superior texture [1, 15].** The rheological properties of the yogurt made from milk subjected to thermosonication was studied. The texture and firmness of the yogurt were significantly improved by the thermosonication procedure [16].

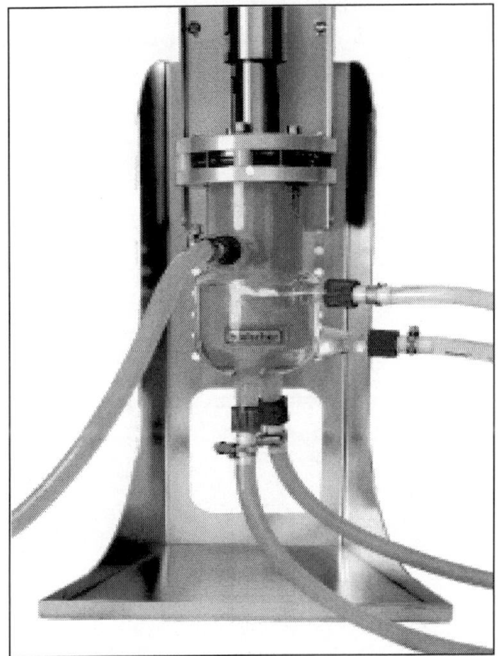

Ultrasonic Homogenizer.

4. Homogenization and Emulsion Preparation of Milk [1]

Emulsion Preparation

Ultrasonic emulsification is mainly driven by cavitation wherein the bubbles collapse at the interface of two immiscible continuous and dispersed phases [12a]. The preparation of very fine emulsions is of increasing interest. Low frequency ultrasound has been effectively used for the preparation of such food emulsions. A particular advantage of the ultrasonic approach is the ease of equipment cleaning relative to traditional homogenizers or the newer microfluidic devices. The emulsification efficiencies of **ultrasonic and micro-fluidization techniques** were compared. Both techniques were found to generate emulsion droplets in the size range 150-700 nm; however, the ultrasonic method was found to be better in terms of operation and cleaning [1a]. The analysis of the energy requirements also suggested that ultrasound may be a more cost-effective processing option than microfluidization [12]. It was suggested that ultrasonic homogenization is more energy intensive leading to higher costs [13].

Homogenization

Milk (*e.g.* cow, buffalo, goat or camel milk) is an emulsion or colloidal system that consists in butterfat globules within a water-based fluid that contains dissolved carbohydrates, proteins and minerals. As fat and water tend to separate into two phases, milk must be homogenized to obtain an even product. **Homogenization means the even distribution of the fat molecules in the milk liquid. Ultrasound is a well known method used for various applications in dairy processing. Ultrasonic treatment of milk results in homogenized fat globules, which are even and uniformly distributed.**

Sonication of fresh cow milk at 20 kHz resulted in a reduction in the size of fat globules. The ultrasonic homogenization of milk at 20 kHz was studied and compared the results with those of a conventional homogenizer [14]. The size distribution of the fat globules after the conventional homogenization process (at 200 bar and 55°C) was about 2-5 µm where as the size range of fat globules in ultrasonically homogenized milk samples was much smaller. The mean size and size distribution of the fat globules were dependent on the ultrasonic power and length of sonication. The observed smaller size of fat globules was attributed to the physical effects **generated during acoustic cavitation**. The highest homogenization efficiency and the smallest fat globule diameter were 3.22 and 0.725 µm at power level 100 (450 W) for 10 min, respectively [14a].

1. The homogenization by high-power ultrasound is also effective for (vegan/dairy free) milk substitues derived from plants such as coconut milk or soy milk.
2. The study shows that ultrasonic homogenization reduces the size of milk fat globules (MFG).
    i. Low amplitude (150W) had **not a satisfactory homogenization** effect, the MFG size and their distribution were similar to untreated milk.
    ii. Medium amplitude ultrasound (267.5, 375 W) had a **good homogenization effect**; MFG average diameter was 2 µm.
    iii. Higher amplitude (750W) ultrasound reduced the MFG size crucially, making them barely visible at the optical microscope (100x magnification); their average diameter size was 0.3 µm.

5. Ultrasonic Crystallization (Sonocrystallization) and Freezing (Ice cream Freezing [1, 1a]

Due to several factors, it is difficult to obtain uniform size of crystals in traditional methods of crystallization and freezing. Acoustic cavitation can promote nucleation in a phenomenon referred to as sonocrystallization [1a, 25]. Further, the microstreaming and general turbulence associated with this process accelerate the heat and mass transfer associated with the freezing process. Finally, the shear forces present can break crystals and hence result in a frozen product with much smaller crystal sizes. Ultrasound can thus accelerate the freezing process and lead to better product quality [1a, 25].

The studies would suggest that ultrasound could be particularly effective in the preparation of ice-cream. Sonication is known as an effective degasser and this would affect the final texture of the ice cream [1a]. The ice cream mix, which is required for ice cream production, consists of milk, milk powder, cream, butter or vegetable fat, sugar, dry mass, emulsifier, stabilizer as well as additives such as fruits, nuts, flavors and coloring. This special mixture has to be homogenized and pasteurized, then it is stirred slowly during the freeze process to prevent the forming of large ice crystal. Thereby, very small air bubbles are mixed in (so called aerating process) to froth the ice cream achieving a smoothly textured cold dessert. **This is the process step, where ultrasonication can be applied to enhance ice cream's quality.**

During the freezing process, crystals are formed from supercooled water. The morphology of the ice crystals plays an important role regarding the textural and physical properties of frozen and half-frozen food. As size and distribution of the ice crystals are of special importance for the quality of thawed tissue products, for ice cream, smaller ice crystals are preferred because large crystals results in an icy texture. **Nucleation is the most important factor to control the crystal size distribution during crystallization**. Thereby, the freeze rate is usually the parameter used for controlling size and size distribution of the ice crystals in ice cream. During the whipping and freezing, air is injected to achieve the smooth texture of ice cream. The so-called "over-run", the amount of air injected, is proportionated-specifically to the particular recipe - proportionally to the combined volume of solids and water. So, the over-run varies due to the different ice cream formulations and the processing streams. Standard ice cream shows an over-run of 100 per cent, which means that the final product consists of an equal volume of ice cream mix and air bubbles.

The use of Hielscher's **high power ultrasound devices** provides a better quality of ice cream by reducing the ice crystal size and avoiding the incrustation of a freezing surface. A better consistency and a more creamy mouth feeling is achieved due to the reduced ice cream crystal size and the enhanced air bubble distribution. Significantly shorter freezing times lead to a higher process capacity and a more energy-efficient production process. It was observed that increasing the ultrasound pulse time significantly decreased the freezing process time of ice cream, and improved sensory flavor, texture and mouth feel [5a].

6. Effect of US on Casein and Calcium [1]

The effect of ultrasonication on casein and calcium was investigated. They applied ultrasonic waves (20kHz) to samples of fresh skim milk, reconstituted micellar casein, and casein powder. They sonicated the samples until the milk fat globules were reduced to approx. 10nm. **The analysis of the sonicated milk shows that the size of the casein micelles is unchanged**. A small increase in soluble whey protein and a corresponding decrease in viscosity also occurred within the first few minutes of sonication. **The study was determined that casein micelles are stable**

during sonication and the soluble calcium concentration is not affected by the ultrasonic treatment.

7. High Intensity Ultrasound Treatment on Enterobacteriae Count in Milk [6]

The high intensity ultrasound was used to investigate the inactivation Enterobacteriae count in raw milk. Raw milk with 4 per cent of milk fat was treated with ultrasonic probe that was 12 mm in diameter and with 20 kHz frequency immerged in milk directly. For ultrasounds treatment, three parameters varied according to the statistical experimental design. Centre composite design was used to optimize and design experimental parameters: temperature (20, 40 and 60 °C), amplitude (120, 90 and 60 µm) and time (6, 9 and 12 minutes). All analyses were performed immediately after sonication and after 3 and 5 days of storage in refrigeration at 4 °C. The facts that substantially affect the inactivation of micro-organisms using ultrasound are the amplitude of the ultrasonic waves, the exposure/contact time with the micro-organisms, and the temperature of treatment. The achieved results indicate significant inactivation of micro-organisms under longer period of treatments with ultrasonic probe particularly in combination with higher temperature and amplitude. Output optimal value of Enterobacteriae count has been defined by Statgraphics where lowest Enterobacteriae count (1.06151 log CFU mL^{-1}) was as follows for specific ultrasound parameters: amplitude of 120 µm, treatment time for 12 min and temperature of 60 °C.

7. Ultrasonic Imaging [1a]

It is used for determining the elastic properties [7] (*e.g.* gelation of milk components [8], measurement of the velocity and the attenuation of the ultrasonic amplitude at different frequencies in the processed samples), the renneting properties [9] of casein solutions after heat treatment and for the development of different cheeses [1a] (*e.g.* cut-time in cheese making, the rheological properties of cheese, structural changes in cheese due to heating and cheese maturity).

i. Ultrasonic Degassing [1a]

The foaming of dairy solutions during processing can reduce the final product yield and may accelerate oxidative degradation. The 20 kHz pulsed ultrasound was used to degas reconstituted skim milk [10]. While gas bubbles were easily removed in < 5 min of sonication, the dissolved oxygen content could not be significantly reduced even after 20 min. It has also been observed that airborne ultrasound can be an effective approach to foam minimization [11].

8. Membrane Flux Enhancement and Milk Fouling and Detection of Fouling [1a, 1b]

Membrane Flux Enhancement and Milk Fouling [1a]

Fouling of ultrafiltration membranes is a major issue affecting the cost and efficiency of many dairy manufacturing operations. The ultrasound was used in both the production and cleaning cycles of whey ultrafiltration and found that sonication was effective in both cases. Ultrasound significantly improved membrane cleaning efficiencies [17, 18]. The ultrasound was effective in improving production flux values by between 40 per cent and 70 per cent. It is shown that the permeate flux was higher for low-frequency operation (50 kHz) compared to that at 1 MHz. On the basis of the several studies [17, 18 19, 20], the following specific effects are generated by ultrasonic processing:

i. Sonication can cause agglomeration of fine particles, thus potentially reducing pore blockage and cake compaction,
ii. The turbulence associated with ultrasound can be used to separate physical aggregates of proteins by disrupting the intermolecular forces,
iii. The asymmetric collapse of cavitating bubbles can scour surfaces and this leads to removal or control of the fouling cake layer.
iv. Acoustic streaming and/or cavitation causes turbulence which results in bulk water movement near the membrane surface, which reduces the effects of concentration polarization and thus increases the mass transfer coefficient at the membrane surface.

Detection of Fouling [1b]

When dairy products are processed in continuous high-temperature processing plants, the internal walls of the plant can become fouled by burnt on or chemically deposited material. The fouling layer will affect flow rate and also heat flow to or from the product. An ultrasonic sensor to detect and measure the thickness of these films in a dairy plant was developed [20a].

9. Inactivation of Micro-organism in Milk [1a]

It was observed [21] that the inactivation of *Listeria innocua* and mesophilic bacteria in raw whole milk is more efficient when thermosonication is used in place of purely thermal pasteurization. A combination of heat and ultrasound [22] was used to examine the survival rate of two strains of Bacillus subtilis in milk. It was observed that the simultaneous use of heat (70-95°C) and ultrasound (20kHz, 150W) was more effective in the deactivation process compared to individual treatment by heat or ultrasound alone. The viability of the spores decreased by between 70 per cent and 99.9 per cent after the thermosonication process. However, acceptable dairy applications of thermo-sonication have yet to be commercialized, as the treatment of milk systems with high-intensity ultrasound tends to generate undesirable off-flavors [23].

10. Spray Drying Using Ultrasonic Atomization [1a]

Atomizing devices provide a high surface-to-mass ratio during the spray drying of dairy products, enabling rapid heat transfer and high evaporation rates. The centrifugal (rotary) and pressure (nozzle) atomizers are the two most common atomizing devices being used in dairy. Now the ultrasonic atomizers offer a viable alternative [24]. **While ultra-sonic spray driers** containing nozzles are available for large-scale operations, **ultrasonic nebulizers** have not yet been developed for large-scale industrial use. Both devices have the technical advantage over classical spray-drying atomizers in that they produce a spray at a much lower velocity.

Ultrasound has also been used to determine the interaction of milk powder with solvent in order to evaluate the reconstitution of powders [24a] and is dependent on product porosity [24b].

11. Modification of Functional Properties of Dairy Ingredients [1a]

On the basis of the observations found in several studies, it could be concluded that the physical effects of ultrasound can be used effectively to reduce the viscosity of food systems by disrupting aggregates and reducing the interaction between neighboring structures. The modification of the physical properties of dairy systems by ultrasound can reasonably be attributed to the effect of shear forces alone. Some examples of effects of US on the functional properties of dairy ingredients are given below:

- **i.** The ultrasonic bath is used to disperse the aggregates of bacteria in refrigerated raw milk [26].
- **ii.** The sonication has disrupted casein micelles to generate free casein in the solution [27].
- **3.** The effect of high pressure processing, ultrasonic treatment and tribomechanical activation on the rheological and thermophysical properties of whey proteins (whey protein concentrate [WPC] and whey protein isolate [WPI]) were studied and observed that both ultrasound and tribomechanics increased the water solubility of the whey proteins by about 5-6 per cent [28].
- **4.** The sonication of medium heat Skim milk concentrate (SMC), age thickened by 24 hours storage, showed a significant reduction in viscosity (> 60 per cent reduction). The greatest reduction in viscosity was achieved in 1 min sonication, longer sonication times did not reduce the viscosity further. Experiments carried out with fresh SMC showed ~ 10 per cent reduction in viscosity.
- 5. The recent investigation [29] has shown that **ultrasonic processing can significantly alter the water solubility, particle size distribution, viscosity, gel strength and syneresis of reconstituted whey protein suspensions.**

12. Cheeses [1b]

The ultrasound speeds up the hardening of the curd and the final product showed a better firmness because of the activity on the chymosin, pepsin, and other related enzymes [13]. Chymosin has a proteolytic activity in liquid milk, coagulating the protein and conferring a solid-like coagulated gel. The activity of the chymosin increased with sonication and the nitrogen content of the extract decreased at the same time. During cheese making, the methods used to test curd firmness are destructive methods. The changes during enzymatic coagulation by ultrasound was studied [30] by using a 1-MHz transducer and the pulse-echo technique. During coagulation, no variation was observed in Ultrasonic velocity, but ultrasonic attenuation decreased when coagulation progressed. **This change in attenuation is mainly linked to changes in viscosity, which increases the viscous**

attenuation of ultrasound. Ultrasonics was used to classify defects (*e.g.* size and number of voids in kamaboko, identification of cracked cheeses, distance of the crack from the surface *etc.*) in cheese [31].

13. Lactose-Free Milk [1b]

Lactose-free milk(Milk without lactose) can be produced by fermentation of lactose-hydrolyzed milk or by the simultaneous addition of β-galactosidase and lactic acid bacteria. Ultrasound has the capability of raising the reaction activity of cells. The lactose hydrolysis was around 55 per cent, if the ultrasound was used in the processing of lactose-free milk where as, it was around 36 per cent, if traditional methods were used to produce lactose-free milk (fermentation) [32].

14. Detection of Water Addition in Milk [1b]

To detect the addition of water in milk, the variation of ultrasonic velocity in two types of milk, cow and buffalo, adulterated with different percentages of water was studied [33]. Ultrasonic velocity of cow and buffalo milk was found to be different due to the differences in composition and velocity decreased in line with the water addition and was dependent on temperature [33].

Sources

1. *Hielscher Ultrasonic Technology, Ultrasonication and Its Manifold Applications in Food Processing, info @heilscher.com.*

1a. *Muthupandian ashok kumar, Raman bhaskaracharya, Sandra Kentish, Judy lee, Martin palmer, Bogdan zisu, The ultrasonic processing of dairy products -An overview, Dairy Sci. Technol. 90, 2010, 147-168*

1b. *Chetana B. Chaudhari, J. P. Prajapati and Suneeta V. Pinto, Ultrasound Technology for Dairy Industry, Department of Dairy Technology, SMC College of Dairy Science, AAU, Anand.*

2. *Anonymous, Consonic Pty Ltd, http: //www. consonic.com/branson-ultrasonic-cutting. html, Accessed, 10 February 2009.*

3. *Lowe M. J. S., Ultrasonics, in: Braun S. G., Ewins D. J., Rao S. S. (Eds.), Encyclopedia of Vibration, vol. 3, Elsevier, Oxford, UK, 2002, pp. 1437-1441.*

4. *Herrmann T., Lynch B., Ultrasonic sealing of flexible pouches through contami-nated sealing surfaces, ANTEC Conference Proceedings 61, 2003, pp. 2629-3235.*

5. *Plastics Design Library, Staff Handbook of Plastics Joining, William Andrew Publishing/Plastics Design Library, Norwich, USA, 1997.*

5a. *Mortazavi A. and Tabatabaie F., 2008, Study of ice cream freezing process after treatment with ultrasound. World Applied Sciences Journal, 4(2), 188-190.*

6. *Edita Juraga, Brankica Sobota Šalamon, Zoran Herceg, Anet Režek Jambrak, Application of high intensity ultrasound treatment on Enterobacteriae count in milk, Mljekarstvo, 61 (2), 125-134, 2011.*

7. *Lee H. O., Luan H. C., Daut D. G., Use of an ultrasonic technique to evaluate the rheological properties of cheese and dough, J. FoodEng. 16, 1992, 127-150.*

8. *Corredig M., Alexander M., Dalgleish D. G., The application of ultrasonic spectroscopy to the study of the gelation of milk components, Food Res. Int. 37, 2004, 557-565.*

9. *Wang Q., Bulca S., Kulozik U., A comparison of low-intensity ultrasound and oscillating rheology to assess the renneting properties of casein solutions after UHT heat pre-treatment, Int. Dairy J.17, 2007, 50-58.*

10. *Villamiel M., Verdurmen R., de Jong P., Degassing of milk by high-intensity ultrasound, Milchwissenschaft 55, 2000, 123- 125.*

11. *Riera E., Gallego-Juarez J. A., Mason T. J., Airborne ultrasound for the precipitation of smokes and powders and the destruction of foams, Ultrason. Sonochem. 13, 2006, 107-116.*

12. *Kentish S. E., Wooster T., Ashokkumar M., Balachandran S., Mawson R., Simons L., The use of ultrasonics for nanoemulsion preparation, Innov. Food Sci. Emerg. Technol.9, 2008, 170-175.*

12a. *Mason TJ., 1998, Power ultrasound in food processing - The way forward. In: Ultrasound in Food Processing Povey M. J and Mason T. J. (Ed.), 105-26. London, UK: Thomson Science.*

13. *Villamiel M., van Hamersveld E. H., de Jong P., Review: effect of ultrasound processing on the quality of dairy products, Milchwissenschaft 54, 1999, 69-73.*

14. *Ertugay M. F., Sengul M., Sengul M., Effect of ultrasound treatment on milk homogenization and particle size distribution of fat, Turk. J. Vet. Anim. Sci. 28, 2004, 303-308.*

14a. *Villamiel M. and de Jong P. 2000. Influence of high-intensity ultrasound and heat treatment in continuous flow on fat, proteins, and native enzymes of milk. Journal of Agriculture and Food Chemistry, 48, 472-78.*

15. *Wu H., Hulbert G. J., Mount J. R., Effects of ultrasound on milk homogenization and fermentation with yogurt starter, Innov. Food Sci. Emerg. Technol. 1, 2008, 211-218*

15a. *Sakakibara M., Wang D., Ikeda K., Suzuki K., Effect of ultrasonic irradiation on productionof fermented milk with Lactobacillus delbrueckii, Ultrason. Sonochem. 1, 1994, S107-S110.*

16. *Vercet A., Oria R., Marquina P., Crelier S., Buesa P. L, Rheological properties of yoghurt made with milk submitted to manothermo-sonication, J. Agric. Food Chem. 50, 2002, 6165-6171.*

17. *Muthukumaran S., Kentish S. E., Ashok kumar M., Stevens G. W., Application of ultrasound in membrane separation processes: a review, Rev. Chem. Eng. 22, 2006, 155-194.*

18. *Muthukumaran S., Kentish S. E., Ashok kumar M., Vivekanand V., Mawson R., Power ultrasound offers an environmentally friendly approach to cleaning dairy UF membranes, Aust. J. Dairy Technol.59, 2004, 193.*

19. *Kobayashi T., Chai X., Fuji N., Ultrasound enhanced cross-flow membrane filtration, Sep. Purif. Technol.17, 1999, 31-40.*

20. *Kobayashi T., Kobayashi T., Fuji N., Effect of ultrasound on enhanced permeability during membrane water treatment, Jpn. J. Appl. Phys. 39, 2000, 2980-2981.*

20a. *Withers P., 1994, Ultrasonic Sensor for the Detection of Fouling in UHT Processing Plants, Food Control, 5(2): 67-72.*

21. *Bermúdez-Aguirre D., Barbosa-Cánovas G. V., Study of butter fat content in milk on the inactivation of Listeria innocua ATCC 51742 by thermo-sonication, Innov. Food Sci. Emerg. Technol. 9, (2008), 176-185.*

22. *Garcia M. A., Burgos J., Sanz B., Ordonez J. A., Effect of heat and ultrasonic waves on the survival of two strains of Bacillus subtilis, J. Appl. Bacteriol. 67, 1989, 619-628.*

23. *Riener J., Noci F., Cronin D. A., Morgan D. J., Lyng J. G., Characterisation of volatile compounds generated in milk by high intensity ultrasound, Int. Dairy J. 19, 2009, 269-272.*

24. *Luz P. P., Pires A. M., Serra O. A., A low cost ultrasonic spray dryer to produce spherical microparticles from polymeric matrices, Quimica Nova 30, 2007, 1744-1746.*

24a. *Richard, B., Toubal, M., Le Page, J. F., Nassar, G., Radziszewski, E., Nongaillard, B., Delaplace, G., 2012, Ultrasound tests in a stirred vessel to evaluate the reconstitution ability of dairy powders. Innovative Food Science and Emerging Technologies, 16, 233-242.*

24b. *García-Perez, J. V., Ozuna, C., Ortuno, C., Carcel, J. A., and Mulet, A., 2011, Modeling ultrasonically assisted convective drying of eggplant. Drying Technology, 29, 1499-1509.*

25. *Zheng L., Sun D., Innovative applications of power ultrasound during food freezing processes - a review, Trends Food Sci. Technol.17, 2006, 16-23.*

26. *Stone D. L., Fryer T. F., Disruption of bacterial clumps in refrigerated raw milk using an ultrasonic cleaning unit, N. Z. J. Dairy Sci. Technol. 19, 1984, 221-228.*

27. *Taylor M. J., Richardson T., Antioxidant activity of skim milk: effect of sonication, J. Dairy Sci. 63, 1980, 1938-1942.*

28. *Kresic G., Lelas V., Jambrak A. R., Herceg Z., Brncic S. R., Influence of novel food processing technologies on the rheological and thermophysical properties of whey proteins, J. Food Eng. 87, 2008, 64-73.*

29. *Ashok kumar M., Kentish S. E., Lee J., Zisu B., Palmer M., Augustin M. A., Processing of dairy ingredients by ultrasonication, PCT Int. Appl. WO 200907 9691 A1, 2009.*

30. *Ay C. and Gunasekaran S., 1994, Ultrasonic Attenuation Measurements for Estimating Milk Coagulation Time, Transactions of de ASAE, 37(3), 857-862.*

31. *Orlandini I. and Annibaldi S. (1983), New Techniques in Evaluation of the Structure of Parmesan Cheese: Ultrasonic and X-Rays, Sci. Latiero-Caseria, 34: 20-30.*

32. *Wang D. and Sakakibara M., 1997, Lactose hydrolysis and β-galactosidase activity in sonicated fermentation with lactobacillus strains. Ultrasonics Sonochemistry, 4, 255-61.*

33. *Bhatti SS, Bhatti R. and Singh S., 1986, Ultrasonic Testing of Milk, Acoustica 62: 96-99.*

34. *Welti J. and Gustavo V., 2002, Application of low intensity ultrasonics in the dairy industry, In: Engineering and Food for 21st century, Mulet. A., JA. Carcel, J. Benedito (Eds), CRC press, pp: 46.1-48, 8.*

Ultrasonication in Meat Technology [1]

Several applications of ultrasonic treatment are reported in meat technology. Some of these applications are given below:

1. Reduction of meat toughness due to large proportion of connective tissue [2].
2. Examining the composition of fish, poultry, raw, and fermented meat products by supporting genetic enhancement programs in case of livestock [3].
3. In the tenderization of meat products.

Sources

1. *Ishrat Majid Gulzar Ahmad Nayik and Vikas Nanda, Ultrasonication and food technology, A review, 1071022, published online,03Aug 2015.*

2. *Jayasooriya, S. D., Torley, P. J., D'Arcy, B. R., and Bhandari, B. R., 2007, Effect of high power ultrasound and ageing on the physical properties of bovine Semitendinosus and Longissimus muscles. Meat Science, 75, 628-639.*

3. *Gallego-Juárez, J., Rodriguez, G., Acosta, V., and Riera, E. (2010). Power ultrasonic transducers with extensive radiators for industrial processing. Ultrasonics Sonochemistry, 17, 953-964.*

Ultrasonication in Fruit and Vegetable Processing [1]

Ultrasonication is used to maintain both pre- and postharvest quality attributes in fresh fruits and vegetables [2] and is considered a substitute for washing of fruit and vegetable in food industry [3]. Ultrasonication has proved to be one such technique which helps in maintaining and improving the nutritional value of fruit juices [4, 5].

1. It is reported to retain fresh quality, nutritional value, and microbiological safety in following juices: guava juice [6], orange juice [7] and tomato juice [8].
2. US treatment can also be used to recover the nutrient loss occurred during blanching, resulting in achieving the collaborative benefit of both the techniques [9].
3. Ultrasonication cleaners (20-400 kHz) have been efficiently used to produce fruits and vegetables free of contamination [10]. It has been applied on strawberry fruits at 40 kHz and considerable reduction in decay and infection was observed along with quality maintenance [11].

Sources

1. *Ishrat Majid Gulzar Ahmad Nayik and Vikas Nanda, Ultrasonication and food technology, A review, 1071022, published online,03Aug 2015.*

2. *Gallego-Juárez, J., Rodriguez, G., Acosta, V., and Riera, E, 2010, Power ultrasonic transducers with extensive radiators for industrial processing. Ultrasonics Sonochemistry, 17, 953-964.*

3. *Alexandre, E. M. C., Brandao, T. R. S., and Silva, C. L. M., 2013, Impact of non-thermal technologies and sanitizer solutions on microbial load reduction and quality factor retention of frozen red bell peppers. Innovative Food Science and Emerging Technologies, 17, 199-205.*

4. *Bhat, R., Ameran, S. B., Voon, H. C., Karim, A. A., and Tze, L. M., 2011, Quality attributes of starfruit (Averrhoa carambola L.) juice treated with ultraviolet radiation. Food Chemistry, 127, 641-644.*

5. *Abid, M., Jabbar, S., Wu, T., Hashim, M. M., Hu, B., Lei, S., Zhang, X., 2013, Effect of ultrasound on different quality parameters of apple juice. Ultrasonics Sono-chemistry, 20, 1182-1187.*

6. *Cheng, Soh, Liew, and Teh, 2007 Cheng, L. H., Soh, C. Y., Liew, S. C. and Teh, F. F., 2007, Effects of sonication and carbonation on guava juice quality. Food Chemistry, 104, 1396-01.*

7. *Valero, M., Recrosio, N., Saura, D., Muñoz, N., Martí, N., and Lizama, V., 2007, Effects of ultrasonic treatments in orange juice processing. Journal of Food Engineering, 80, 509-516.*

8. *Wu, J., Gamage, T. V., Vilkhu, K. S., Simons, L. K., and Mawson, R., 2008, Effect of thermo-sonication on quality improvement of tomato juice. Innovative Food Science and Emerging Technologies, 9, 186-195.*

9. *Jabbar, S., Abid, M., Hu, B., Wu, T., Hashim, M. A., Lei, S., Zeng, X., 2014, Quality of carrot juice as influenced by blanching and sonication treatments. LWT - Food Science and Technology, 55, 16-21.*

10. *Lin, I., and Erel, D., 1992, Dynamic ultrasonic cleaning and disinfecting device and method. US Patent No. 5113881A. Washington, DC: U. S. Patent and Trademark Office.*

11. *Cao, S., Hu, Z., Pang, B., Wang, H., Xie, H., and Wu, F., 2010, Effect of ultrasound treatment on fruit decay and quality maintenance in strawberry after harvest. Food Control, 21, 529-532.*

Ultrasonication in Oil Technology [1]

Some of the applications of US are given below:

1. Ultrasonication stimulates the mixing and required reaction for conversion of soyabean oil to biodiesel, and can achieve optimum yield using 9: 1 oil to methanol ratio [2].
2. Ultrasonic irradiation is also used to increase the rate of transesterification [3].

Sources

1. *Ishrat Majid Gulzar Ahmad Nayik and Vikas Nanda, Ultrasonication and food technology, A review, 1071022, published online,03Aug 2015.*
2. *Santos, F. F. P., Rodrigues, S., and Fernandes, F. A. N., 2009, Optimization of the production of biodiesel from soybean oil by ultrasound assisted methanolysis. Fuel Processing Technology, 90, 312-316.*
3. *Deshmane, V. G., Gogate, P. R., and Pandit, A. B., 2009, Ultrasound-assisted synthesis of biodiesel from palm fatty acid distillate. Industrial and Engineering Chemistry Research, 48, 7923-7927.*

Ultrasonication in Honey [1]

Ultrasound applications in honey include the followings:

1. Use of velocity of ultrasonic wave propagation as a means to differentiate between different types of honey.
2. Determination of adulteration in honey and
3. Evaluation of the type of protein, aggregation state, and size [2].

Sources

1. *Ishrat Majid Gulzar Ahmad Nayik and Vikas Nanda, Ultrasonication and food technology, A review, 1071022, published online,03Aug 2015.*
2. *Gallego-Juárez, J., Rodriguez, G., Acosta, V., and Riera, E., 2010, Power ultrasonic transducers with extensive radiators for industrial processing. Ultrasonics Sonochemistry, 17, 953-964.*

Ultrasonication in Water Treatment [1]

Ultrasound treatment in combination with other water treatment methods (chlorination, ozonation) is considered efficient and economically feasible technique as in ultrasound equipment, energy requirement is huge [2]. Some of the examples of US in water treatment, are given below:

1. Ultrasonication is reported to remove all impurities such as worms, sludge, mold, fungi, bacteria, and agrochemicals [3].
2. Ultrasonication does not use chemicals for mineralization and destruction of recalcitrant organic compounds in water [4].
3. In anaerobic digestion process, ultrasonication is used to increase the process efficiency, leading to more methane production and significant decrease in digestion time. Anaerobic digestion process uses ultrasound treatment either as high or as low strength depending on the irradiation location.
   a. **High-strength ultrasonication (HS-ultrasonication)** is irradiated as a pretreatment to feed stock.
   b. **Low-strength ultrasonication (LS ultra-sonication)** is irradiated in the aerobic digestion process to the micro-organisms involved [5].

Sources

1. *Ishrat Majid Gulzar Ahmad Nayik and Vikas Nanda, Ultrasonication and food technology, A review, 1071022, published online,03Aug 2015.*

2. *Nithila, S. D., Anandkumar, B., Vanithakumari, S. C., George, R. P., Mudali, U. K., and Dayal, R. K., 2014, Studies to control biofilm formation by coupling ultra-sonication of natural waters and anodization of titanium. Ultrasonics Sonochemistry, 21, 189-199.*

3. *Cao, S., Hu, Z., Pang, B., Wang, H., Xie, H., and Wu, F., 2010, Effect of ultrasound treatment on fruit decay and quality maintenance in strawberry after harvest. Food Control, 21, 529-532.*

4. *Gogate, P. R., 2007, Application of cavitational reactors for water disinfection: Current status and path forward. Journal of Environmental Management, 85, 801-815.*

5. *Cho, S. K., Hwang, Y. H., Kim, D. H., Jeong, I. S., Shin, H. S., and Oh, S. E., 2013, Low strength ultrasonication positively affects the methanogenic granules toward higher AD performance. Part I: Physico-chemical characteristics. Bioresource Technology, 136, 66-72.*

Effect of Ultrasound on the Sensory and Nutritional Quality of Foods

Several researchers have observed an enhancement in the nutritional value of ultrasonicated foods while others have shown a reduction after sonication and the loss was exacerbated during subsequent storage. It was demonstrated [1] that a certain type of blueberry experienced greater than 60 per cent loss of anthocyanin and phenolics following ultrasound-assisted osmotic dehydration than when treated by osmotic dehydration only. Significant differences in total anthocyanins were found between all three osmotic concentration treatments (osmotically concentrated berries for 12h(O_{12}), osmotically concentrated berries for 3 h (O_3) and osmotically concentrated berries for 3 h with US (O_3 + U). Hence, the time of concentration and high-frequency US negatively influenced anthocyanin content in osmo-concentrated berries. Treatments O_3, O_3 +U and O_{12} underwent a decrease in anthocyanin content by 20 per cent, 42 per cent, and 59 per cent, respectively, when compared to the untreated sample. It was speculated that this could be due to surface cell rupture caused by cavitation, contributing to the release of anthocyanin and phenolics in berry samples. It was reported [2] that the level of L-ascorbic acid in distilled water was degraded by US, and the authors ascribed this to the generation of H• and OH• radicals. Degradation of astaxanthin by US as well as oxygen-labile nutrients such as ascorbic acid was reported in MTS-treated orange juice [3]. In contrast, the level of thiamin and riboflavin in milk was not adversely affected by MTS [3].

Ultrasound-treated milk also showed a higher degree of homogenization, brighter color, and better stability after processing [4]. Other researchers reported that US-treated fruit juices retained better color and flavor with minimal adverse impact on other organoleptic properties. The effect of US applications on the activity of enzymes affecting product quality has also been reported. It was examined [5] the effect of MTS on pectic materials and reported that they were unable to detect any residual activity of pectin methyl esterase (PME) in MTS-treated samples while polygalacturonase (PG) remained fully active. The authors attributed the decrease in PME activity to thermal and mechanical effects of the sonic energy cavitation. The ultrasonic treatment of tomato products was shown to inactivate enzymes and improve its rheological properties. Along the same lines, MTS also inactivated β-subunits of tomato endopolygalacturonase. These units have been shown to be problematic in tomato processing because they could potentially protect other degradative enzymes from thermal denaturation.

Texture of foods can also be variably affected by US, depending on the microstructure of the food. To some extent, the texture of US-treated foods is partly reliant on structural and functional changes of proteins including enzymes during sonication. For liquid food, US was shown to reduce particle sizes. It was demonstrated [5] that yogurt produced from US-treated milk had a stronger and firmer texture than untreated milk. It was demonstrated [6, 6a, 6b] improved meat tenderness following US-assisted cooking. Although there are limited studies investigating the effects of US on the flavor profile of foods, flavor improvement in Mahon cheese [7, 7a], generation of off flavor notes in edible oil [8] and loss of desirable aroma compounds in apple juice [9] have been reported.

Sources

1. *Stojanovic J, Silva JL, 2007, Influence of osmotic concentration, continuous high frequency ultrasound and dehydration on antioxidants, color and chemical properties of rabbiteye blueberries. Food Chemistry 101: 898-906.*

2. *Portenlänger G, HeusingerH, 1992, Chemical reactions induced by ultrasound and γ-rays in aqueous solutions of L-ascorbic acid. Carbohydrate Research 232: 291-301.*

3. *Vercet A, Burgos J, Lopez-Bues P, 2001, Mano- thermosonication of foods and food-resembling system: effect on nutrient content and nonenzymatic browning. Journal of Agricultural and Food Chemistry 49: 483-489.*

4. *Banerjee R, Chen H, Wu J, 1996, Milk protein-based edible film mechanical strength changes due to ultrasound process. Journal of Food Science 61: 824-828.*

5. *Vercet A, Sánchez C, Burgos J, Montañés L, Buesa PL, 2002, The effects of mano-thermosonication on tomato pectic enzymes and tomato paste rheological properties. Journal of Food Engineering 53: 273-278.*

6. *Pohlman FW, Dickeman ME, Zayas JF, Unruh JA, 1996, Effects of ultrasound and convection cooking to different end point temperatures on cooking characteristics, shear force and sensory properties, composition, and microscopic morphology of beef Longissimus and Pectoralis muscle. Journal of Animal Science 75: 386-401.*

6a. *Pohlman FW, Dickeman ME, Kropf DH, 1997a, Effects of high intensity ultrasound treatment, storage, time and cooking method on shear, sensory, instrumental color and cooking properties of packaged and unpackaged beef Pectoralis muscle. Meat Science 46: 89-100.*

6b. *Pohlman FW, Dickeman, ME, Zayas JF, 1997b, The effect of low-intensity ultrasound treatment on shear properties, color stability and shelf-life of vacuum-packed beef Semitendinosus and bicep femoris muscle. Meat Science 45: 329-337.*

7. *Sánchez ES, Simal S, Femenia A, Benedito J, Rosselló C, -2001a, Effect of acoustic brining on lipolysis and on sensory characteristics of Mahon cheese. Journal of Food Science 66: 892-896.*

7a. *Sánchez ES, Simal S, Femenia A, Liull P, Rosselló C, 2001b, Proteolysis of Mahon cheese as affected by acoustic-assisted brining. European Food Research and Technology 212: 147-152.*

8. *Chemat F, Grondin I, Shum Cheong Sing A, Smadja J, 2004, Deterioration of edible oils during food processing by ultrasound. Ultrasonics Sonochemistry 11: 13-15.*

9. *Feng H, Lee H, 2011, Effect of power ultrasound on food quality. In: Feng H (ed) Ultrasound Technologies for Food and Bio processing. Food Engineering Series. New York: Springer Science and Business Media, pp. 559-582.*

Chapter 8

Irradiation

Radiation processing [1] can be used for disinfestations of pests and disease-causing organisms from a range of products including fruits and vegetables. Radiation is one of the latest methods in food preservation and this technique makes the food safer to eat by destroying bacteria which is very much similar to the process of pasteurization. The irradiation of food is a process used in the food preservation industry to increase the shelf life, and improve the microbial safety of food. This is done without inducing any chemical pesticides by exposing food to ionizing radiation [1b]. In comparison with heat or chemical treatment, irradiation is more effective and appropriate technology to destroy food borne pathogens [1a]. Irradiation of food is one of the most effective ways of food preservation to inactivate micro-organisms and destroy insect pests [1a]. Irradiation is the treatment of food with x-rays or gamma radiation to kill bacteria and mold. It may be combined with vacuum packing to seal out microbes. Food irradiation is effective against a wide variety of pathogens including bacteria, fungi, viruses and parasites. As with all preservation techniques, irradiation should be used to prolong the life of appropriate food which is of high quality prior to treatment [1e].

Radiation will be moving fast to the status of a **'Wonder Technology'** to satisfy the sanitary and phyto-sanitary requirements of the importing countries. Food once irradiated, can be prone to re-contamination unless appropriately packed Therefore, if radiation treatment is intended to control microbiological spoilage or insect infestation, prepackaging becomes an integral part of the process [1a]. Irradiation is a technology that has been more widely investigated than any other novel preservation method.

The term "irradiation" is really a misnomer, since the food not exposed to electromagnetic radiation or beta rays (electrons produced by a radioactive source) [1e]. Food irradiation is permitted by over 60 countries, with about 500,000 metric tons of food annually processed worldwide [6, 7]. The regulations that dictate how food is to be irradiated, as well as the food allowed to be irradiated, vary greatly from country to country. In Austria, Germany and many other countries of the European Union only dried herbs, spices, and seasonings can be processed with irradiation and only at a specific dose, while in Brazil all foods are allowed at any dose [8, 9, 10, 11]. In the US, electron beam-irradiated beef burgers and ground meat "chubs" have been successfully introduced to the market. Over 5000 US retail stores in 48 states now carry products that have been pasteurized using electron beam irradiation. Though the market for irradiation is still in a nascent stage, the global market is expected to exceed $145 million by 2015 [1].

Types of Radiations

There are various radiations available for irradiation of food. However, only gamma radiation and accelerated electrons are used in the food processing applications. Other technologies, such as, alpha radiation or beta radiation may cause radioactivity. The accelerated electrons can also be converted into X-rays to utilize during food processing [1a, 1b, 2].

The ionizing ability helps in differentiating gamma rays, X-rays, and electrons. After the ionizing process, electrically charged ions and free neutral radicals are the byproducts. Together, they emit electrons that further cause the destruction of micro-organisms in food. The food, which contains high moisture content are treated by ionization. This process contributes in the breaking of chemical bonds by expelling the electrons from the water molecules [3].

Gamma irradiation is approved in the US and worldwide for a variety of food products. This process does not make food radioactive - energies from irradiation are far too low to induce radioactivity. Worldwide organizations have concluded that irradiation, used within regulatory parameters with upper levels clearly defined, is safe to use to control targeted pathogens in food products [3a].

Irradiated" and "Radioactive foods

There is the difference between the terms "Irradiated" and "Radioactive" food.

1. **Radiation processed foods:** These products are those that have been exposed to radiation to bring about the desired effect in food.
2. **Radioactive foods:** These products, on the other hand, are those that become contaminated with radionuclides. This type of contamination never occurs during food irradiation.

Cobalt 60, properly shielded, is primarily used as the gamma irradiation source. This irradiation has good food penetration. Doses used must be established for each specific food product taking into account product composition, texture, density and impacts on quality [3a].

1. Thus far, irradiation has been approved in low doses for a variety of food products that include lettuce, spinach, mushrooms and fresh fish.
2. Slightly higher doses of irradiation have been approved for use on dried spices, and herbs.

3. However, FDA approval, so far, has only targeted bacterial pathogens.
4. Researchers are currently studying the use of gamma irradiation on produce and shellfish to help kill viruses.
5. Food is safer without compromising texture, taste and nutrition and can be used effectively on frozen products.

Standards/Regulation and Labeling of Irradiated Foods

The Codex Alimentarius represents the global standard for irradiation of food, in particular under the WTO-agreement. Member states are free to convert those standards into national regulations at their discretion, therefore regulations about irradiation differ from country to country. The United Nations Food and Agricultural Organization (FAO) has passed a motion to commit member states to implement irradiation technology for their national phytosanitary programs; the General assembly of the International Atomic Energy Agency (IAEA) has urged wider use of the irradiation technology [6]. Radiation processed food cannot be recognized by sight, smell, taste or touch. Codex Alimentarius Commission has endorsed a green irradiation logo. The RADURA-logo is optional. Several countries use a graphical version that differs from the Codex-version. The suggested rules for labeling is published at CODEX-STAN - 1 (2005) [6a], and includes the usage of the Radura symbol for all products that contain irradiated foods. If whole foods are treated with ionizing radiation, then the product must be labeled with the **radura symbol** (as given in following table**)** or a written phrase **"treated with radiation"**. If irradiated ingredients are added to foods that have not been irradiated, no special labeling is required on retail packages [3a]. The country wise details related to labeling of irradiated foods are given below:

1. India

As per the PFA (Fifth Amendment) Rules, 1994 and now **Food Safety and Standarda Act -2006** (FSSA-2006), all packages of irradiated foods to be marketed in **India** will be labeled with this logo, along with the words "Processed by irradiation method" and the date of irradiation, licence number of the facility and the purpose of irradiation. Consumers will have a free choice to buy irradiated or non-irradiated commodity.

Pattern for Labeling of Irradiated Foods

Radura Symbol

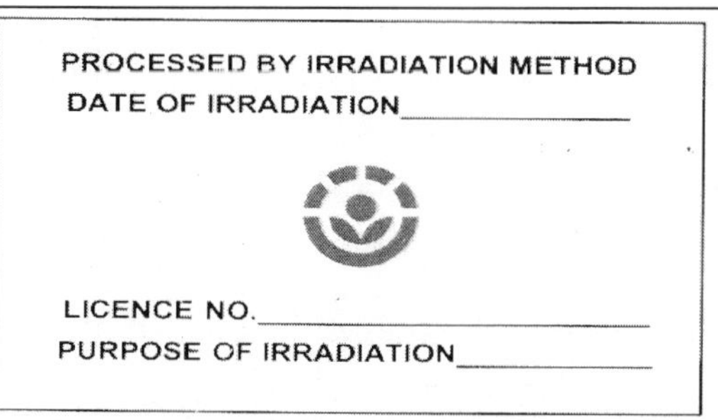

Labeling as per FSSA-2006.

2 The European Union [6]

It follows the Codex's provision to label irradiated ingredients down to the last molecule of irradiated food. The European Community does not provide for the use of the Radura logo and relies exclusively on labeling by the appropriate phrases in the respective languages of the Member States. The European Union enforces its irradiation labeling laws by requiring its member countries to perform tests on a cross section of food items in the market-place and to report to the European Commission [6]. European law dictates that all member countries must allow the sale of irradiated dried aromatic herbs, spices and vegetable seasonings. However, these Directives allow Member States to maintain previous clearances food categories the EC's Scientific Committee on Food (SCF) had previously approved (the approval body is now the European Food Safety Authority) [6].

3. The United States [6]

It defines irradiated foods as foods in which the irradiation causes a material change in the food, or a material change in the consequences that may result from the use of the food. Therefore, food that is processed as an ingredient by a restaurant or food processor is exempt from the labeling requirement in the US. This definition is not consistent with the Codex Alimentarius. All irradiated foods must bear a slightly modified [6a] Radura symbol at the point of sale and use the term "irradiated" or a derivative there of, in conjunction with explicit language describing the change in the food or its conditions of use [6b].

1. In the United States, each new food is approved separately with a guideline specifying a maximum dosage; in case of quarantine applications the minimum dose is regulated [6].
2. Packaging materials containing the food processed by irradiation must also undergo approval [6].
3. Food irradiation in the United States is primarily regulated by the FDA since it is considered a food additive. The United States Department of Agriculture (USDA) amends these rules for use with meat, poultry, and fresh fruit [6].
4. USDA has approved the use of low-level irradiation as an alternative treatment to pesticides for fruits and vegetables that are considered hosts to a number of insect pests, including fruit flies and seed weevils [6].

Mechanism for Reduction/Killing of Pathgogens by Irradiation [1]

Ionizing radiation is an energy, which can freeze the electrons, and separate them from their atomic bonds without directly contacting the food particles. This methodology helps by destroying pathogenic or spoilage bacteria. This contributes to the reduction of the risk of foodborne illness [1b].

Ionizing radiation can be used to decontaminate food. During irradiation, molecules absorb energy to form highly labile and reactive ions or free radicals that result in the breakage of a small percentage of chemical bonds [1c]. High-energy electrons fracture molecules (typically water molecules) into highly reactive species such as radicals, which can disrupt other cellular macromolecules such as DNA, proteins or structures such as cell membranes. It was mentioned [1d] that cell genetic material is particularly susceptible to radical damage during active replication or transcription.

Ionizing radiation can damage the nucleic acids and ultimately kill microbes by the following two ways:

1. By Direct Hit

The radiation collide with chemical bonds in the microbial DNA molecules, thereby breaking them and rendering the microbe dead.

2. By Indirect Hit

In the case of an indirect hit, damage to the nucleic acids occurs when the radiation ionizes an adjacent molecule, which in turn reacts with the genetic material. Because water is the largest component of most foods and microbes, it is often the adjacent molecule that ends up producing a lethal product [1c].

Ionizing radiation causes water molecules to lose an electron, producing H_2O^+ and an electron. These products react with other water molecules to produce a number of compounds, including hydrogen and hydroxyl radicals (OHÿ), molecular hydrogen, oxygen, and hydrogen peroxide (H_2O_2) [1c]. Hydroxyl radicals and hydrogen peroxide are very reactive and are known to interfere with the bonds between nucleic acids within a single strand or between opposite strands. Though biological systems do have a capacity to repair both single-stranded and double-stranded breaks of the DNA backbone [4], the **damage occurring from ionizing radiation is random [5] and extensive. Therefore, bacterial repair of radiation damage is a near impossibility.**

The relative sensitivity of different micro-organisms to ionizing radiation is based on their respective D10 values (which is the dose required to reduce the population by 90 per cent). Lower D10 values indicate greater sensitivity of the organism in question.

Sources

1. *Smith and Pillai, Food Technology Magazine, November 2004. 58 [11], 22-48-55.*

1a. *B. Kalyani and K. Manjula, Food Irradiation-Technology and Application, Int. J. Curr. Microbiol. App. Sci, 3(4): 549-555, 2014.*

1b. *Harlin Kaur Saroya, Innovative non-thermal food processing technologies used by the food industry in the United States, MSc thesis, 2017, Western Kentucky University, Kentucky.*

1c. *Grecz, N., Rowley, D. B., and Matsuyama, A.1983. The action of radiation on bacteria and viruses. In "Preservation of Foods by Ionizing Radiation, "Vol.2. CRC Press, Boca Raton, Fla..*

1d. *Arena, V.1971."Ionizing Radiation and Life." Mosby, St. Louis, Mo. Bagorogoza, K. and Bowers, M. 2001. The effect of irradiation and modified atmosphere packaging on the quality of intact chill-stored turkey breast. J. Food Sci. 66: 367-372.*

1e. *Ranjan Saha, New and Emerging Non-Thermal Technologies in Food Preservation, April 2015.*

2. *Fellows, P., Food processing technologies: Principles and practices (3rd ed ed.), Boca Raton, FL: CRC Press, 2009.*

3. *Knoerzer, K., Innovative food processing technologies: Advances in multiphysica simulation. Chichester, U. K: Wiley-Blackwell., 2011.*

3a. *Bhabha atomic research center- FAQ.*

4. *Bartek, J. and Lukas, J. 2003. Chk1 and Chk2 kinases in checkpoint control and cancer. Cancer Cell 3: 421-429.*

5. *Razskazovskiy, Y., Debije, M. G., Howerton, S. B., Williams, L. D., and Bernhard, W. A.2003. Strand breaks in X-irradiated crystalline DNA: Alternating CG oligomers. Rad. Resistance Res. 160: 334-339.*

6. *Wikipedia.*

6a. *General Standard for the Labelling of Prepackaged Foods, CODEX STAN 1-1985" (PDF). Retrieved March 19, 2014.*

6b. *CFR - Code of Federal Regulations Title 21". Access data.fda.gov. Retrieved March 19, 2014.*

7. *"Irradiation testing for correct labelling you can trust". Eurofins Scientific. January 2015. Retrieved February 9, 2015.*

8. *Food irradiation, Position of ADA, J Am Diet Assoc. 2000; 100: 246-253.*

9. *C. M. Deeley, M. Gao, R. Hunter, D. A. E. Ehlermann, The development of food irradiation in the Asia Pacific, the Americas and Europe; tutorial presented to the International Meeting on Radiation Processing, Kuala Lumpur, 2006.*

10. *Kume, T.et al., Status of food irradiation in the world, Radiat. Phys. Chem. 78, 2009, 222-226.*

11. *Farkas, J. et al., History and future of food irradiation, Trends Food Sci. Technol. 22, 2011, 121-126.*

Factors Affecting the Efficacy of Irradiation [1]

Several factors affect the efficacy of irradiation to achieve satisfactory microbial inactivation. The details are given below:

1. The dose of irradiation [2, 3].
2. The intrinsic or product-related factors, such as prevailing oxygen availability, also affect the bactericidal effect of irradiation in the order of aerobic > hypoxic > anoxic conditions [4].
3. Reduction of water activity protects bacterial cells against inactivation during irradiation.
4. Freezing has also been shown to increase microbial resistance to irradiation due to reduced availability of reactive water molecules [5].

5. In addition, other characteristics inherent in the product, such as the state of the food, product composition and matrix, and food thickness and particle size, have a bearing on product decontamination.
6. Although radiation can target the various microbial forms of life, micro-organisms can vary considerably in their sensitivity to ionizing radiation in the order of parasites > bacteria >viruses [6].

Sources

1. *Hudaa Neetoo (Faculty of Agriculture, University of Mauritius, Réduit, Mauritius) and Haiqiang Chen (Department of Animal and Food Sciences, University of Delaware, Newark, Delaware, US), Alternative Food Processing Technologies.*
2. *Ohlsson T, Bengtsson N, 2002, Minimal processing of foods with non-thermal methods. In: Ohlsson T, Bengtsson N (eds) Minimal Processing Technologies in the Food Industry. Cambridge: Woodhead Publishing, pp. 34-60.*
3. *Snyder OP, Poland DM, 1995, Food Irradiation Today. St Paul, MN: Hospitality Institute of Technology and Management Publications.*
4. *Lacroix M, Le TC, Ouattara B et al., 2002, Use of gamma irradiation to produce films from whey, casein, and soy proteins: structure and functional properties. Radiation Physics and Chemistry 63: 827-832.*
5. *Thayer DW, 1994, Wholesome-ness of irradiated foods. Food Technology 48: 132-135.*
6. *Niemira B, 2003, Irradiation of fresh and minimally processed fruits, vegetables and juices. In: Novak JS, Sapers GM, Juneja VK (eds) Microbial Safety of Minimally Processed Foods. Boca Raton, FL: CRC Press, pp. 279-299.*

Important Factors that Control the Resistance of Microbial Cells to Ionizing Radiation [1]

Microbial cells, whether pathogenic or comprising the normal microflora of foods, exhibit differences in their responses to ionizing radiation. The key factors that control the resistance of microbial cells to ionizing radiation are given below:

1. Size of the organism (the smaller the target organism, the more resistant it is to ionizing radiation).
2. Type of organism (*i.e.*, cell-wall characteristics and gram positive or gram negative in nature).
3. Number and relative "age" of the cells in the food sample
4. Absence or presence of oxygen.

In addition to the above mentioned factors, the physical and chemical composition of the food also affects microbial responses to irradiation. For example, as the temperature of ground turkey is decreased from 30°C to –30°C, the D10 value increases from 0.16 kGy to 0.29 kGy. D10 values change as the water in the product freezes, thereby decreasing the rate of migration of the ionization products, including free radicals, and requiring greater energy input to cause the collisions necessary to destroy the microbes [2].

Sources

1. *Smith and Pillai, Food Technology Magazine, November 2004. 58 [11]: 22, 48-55.*
2. *Thayer, D. W. 2004. Irradiation of food-Helping to ensure food safety. N. Engl. J. Med. 350(18): 1811-1812.*

Irradiation Sources, Processes, Equipment and Analysis of Irradiation [1, 1a]

Food irradiation uses radiant energy- **electron beams (**beta rays generated by electron beam or E-beam)**, gamma rays or x-rays** to rid food of harmful micro-organisms, insects, fungi and other pests, and to retard spoilage [1a]. None of these irradiation sources has sufficient energy to be capable of inducing radioactivity; however, they do have enough energy to remove electrons from atoms to form ions or free radicals. The freed electrons collide with chemical bonds in the microbial DNA molecules, thereby breaking them and rendering the microbe dead.

1. Irradiation Sources [1a]

According to the Codex Alimentarius General Standard (Food and agriculture organization and World health organization, 1984), the following three principal types of radiation source can be used in food irradiation :

a. **Gamma radiation from radionuclides such as 60Co or 137Cs**: The rays used in food processing are obtained from large 60Co radionuclide sources. This type of radiation is essentially monoenergetic (60Co emits simultaneously two photons per disintegration with energies of 1.17 and 1.33 MeV)
b. **Machine sources of electron beams with energies up to 10 MeV**: Electrons emitted by accelerators have fairly narrow spectral energy limits (usually less than ±10 per cent of the nominal energy). The energy of the electrons reaching the product is further controlled by the bending magnets of the beam handling system, if applicable. The range of an electron in a medium is finite (unlike that for photons) and is closely related to its energy [1b].
c. **Machine sources of bremsstrahlung (X rays) with electron energies up to 5 MeV**: Bremsstrahlung irradiator design principles are the same as those for electron irradiators. An extended source of X rays is achieved by distributing the primary electron beam over a target of sufficient size. In contrast to the radionuclide sources, which emit nearly monoenergetic photons, bremsstrahlung (X ray) sources emit photons with a broad energy spectrum [1c, 1d].

Because of their greater penetrating capability, Gamma rays and X-rays may be used for processing of relatively thick or dense products.

2. Irradiation Process [1a]

There are several processes that are collectively referred to as Food Irradiation. Each process has an optimal dose of ionizing energy (radiation) dependent on the desired effect. All three forms of ionizing energy have the same effect. Some of the major processes are as follow:

a. **Sanitation [1a]**: Irradiation is widely used to reduce organisms for spices, herbs and other dried vegetable substances. On certain fruits and tubers, irradiation delays ripening and/or sprouting. Some examples are given below:
   i. Irradiating berries to reduce mold.
   ii. Irradiating fresh fruits to extend their market reach.
   iii. Irradiating potatoes, onions and garlic to impair cell division and hence allow them to go through the off season without sprouting.
b. **Disinfestation [1a]:** Irradiation is used to stop reproduction of both storage and quarantine insect pests. Some examples are given below:
   i. Irradiating foreign produced mangoes to eliminate the seed weevil (Quarantined pest, for import to the US).
   ii. Irradiating papaya is to eliminate fruit flies(Quarantined pests, for import from Hawaii or foreign countries into the US mainland). All three forms of irradiation are referred to as a **cold process**.

   Food irradiation is sometimes referred to as **"cold pasteurization"** [4] or **"electronic pasteurization"** [5] because ionizing the food does not heat the food to high temperatures during the process, and the effect is similar to heat pasteurization. The term "cold pasteurization" is controversial because the term may be used to disguise the fact the food has been irradiated and pasteurization and irradiation are fundamentally different processes. Although all of the radiation energy is converted to heat during treatment, the process typically increases the product temperature by about 1 degree Celsius.
c. **Pasteurization (Pathogen Reduction)**-Irradiation is used to effectively eliminate disease causing organisms including bacteria and parasites. Irradiating ground beef and Irradiating live oysters make it safe from **E. coli O157: H7**and from **vibrio** respectively.
d. **Sterilization** Irradiation is used at a very high dose to eliminate all organisms so that refrigeration is not required (shelf stable). Certain foods are sterilized for NASA astronauts.

3. Dosimetry [1a, 1f]

The radiation absorbed dose is the amount energy absorbed per unit weight of the target material. Dose is used because, when the same substance is given the same dose, similar changes are observed in the target material [1a]. **The radiation absorbed dose is measured in units of rads (1rad= 100 erg/g) or grays (1Gy=100rads). A gray is a unit of energy equivalent to 1 joule per kilogram. This unit of measure is based on the metric system. Thus, 1 kilogray (kGy) is equal to 1,000 grays (Gy) [1f].** The SI unit for dose is grays (Gy or J/kg). Dosimeters are used to measure dose, and are small components that, when exposed to ionizing radiation, change measurable physical attributes to a degree that can be correlated to the dose received. Measuring dose (dosimetry) involves exposing one or more dosimeters along with the target material [6, 7].

4. Irradiation Equipment [1]

In the food irradiation process, three types of radiations (gamma radiation, X-rays, and electron) are used and hence there are three types of equipment where the food is exposed to ionizing radiation. This procedure is carried out in a special processing room, or in a chamber for a specific duration of time [2a, 2b].

i. Equipment for Gamma Radiation

Irradiation equipment typically consists of a high-energy isotope source to produce γ-rays. γ-Rays originating from cobalt-60 (^{60}Co) are often used commercially [1]. This process is also known as cobalt 60, and is the most common source of ionizing energy. This setup is placed in two stainless steel tubes called source pencils. They are placed one inside the other. This setup is placed on a rack, and when the system is not being used, the entire rack is immersed in water underground. The rack is removed from the water only when the processing has to be done. The packaged food is passed through the belt and goes into the room where the package is subjected to source pencils. The energy in the form of gamma ray's passes through and treats the food [2a, 2b].

ii. Equipment for X-rays

Less commonly, machine sources are also available that produce high-energy electrons and X-rays [2]. In this methodology, the electrons are focused by an electron beam accelerator on a metal plate where some of the energy is taken in and the rest is converted to X-rays. The X-rays can penetrate the food boxes up to 15 inches thick or more. Therefore, when the food is being processed, the majority of the radiation goes through without being absorbed by the food [2c].

iii. Equipment for Accelerated Electrons

This process is also referred to as an E-beam accelerator. The working of this setup is similar to that of a television tube. Rather than being extensively discrete, and striking a phosphorescent screen at a very low energy, the electrons are focused and fast-tracked at the speed of light. This effect fabricates quick reaction on the molecules. The electrons in this process accelerate energies to 5, 7.5 or 10Mev, along with beam power of 10 KW [2a, 2b]. Machine sources are electron accelerators, which consist of a heated cathode to supply electrons and an evacuated tube, in which electrons are accelerated by a high-voltage electrostatic field [1].

These three sources of ionizing radiation transfer their energies to materials by ejecting atomic electrons, which can then ionize other atoms through a self amplifying cascade of collisions. The source of choice for a particular application will depend on a multitude of factors, including the thickness and density of the food, the absorbed dose, and the overall economics of the process [2]. While γ- and X rays have high penetration power and the potential to treat foods in bulk, high-energy electrons have relatively low penetration power and are only used for thin foods.

Automatic Pallet Irradiator [2d].

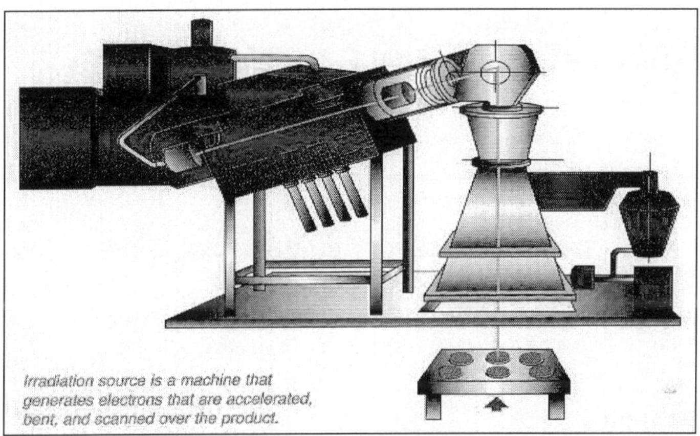

Accelerated Electron Generator [2d].

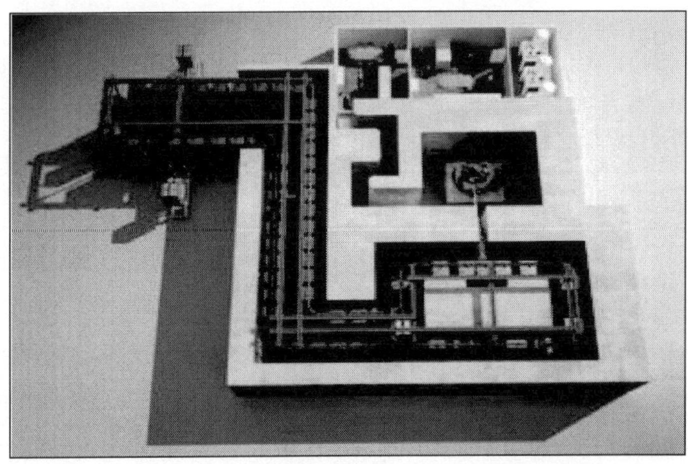

A Three-dimensional Drawing of an X-ray with Dual Pallet Carriers [2c].

Safety Measures for Irradiation

Irradiation (within the accepted energy limits, as 10 MeV for electrons, 5 MeV for X-rays [US 7.5 MeV]) and gamma rays from Cobalt-60) can not make food radioactive [1f]. Radiation does not leave the food items radioactive for following two reasons [1a]:

1. The gamma rays from cobalt-60 used in food radiation are not energetic enough to make it radioactive.
2. As the food never comes into contact with the source directly [1a], it is not possible for the food to become contaminated with radioactive material. It is impossible for food irradiators to induce radiation in a product. Irradiators emit electrons or photons and the radiation is intrinsically radiated at precisely known strengths (wavelengths for photons, and speeds for electrons). These radiated particles at these strengths can never be strong enough to modify the nucleus of the targeted atom in the food, regardless of how many particles hit the target material, and radioactivity can not be induced without modifying the nucleus [1f].

   Thick concrete walls and lead shields are often used to prevent leakage of radiation beyond the processing area. For the sake of personnel safety, strict procedures are in place to make sure that access to the irradiation plant is carefully monitored [3].

5. Radiation Processing of Food [1e, 1f]

Most irradiated food is processed by gamma irradiation., however the usage of electron beam and X-ray is becoming more popular as well.

Electron Irradiation [1f]

It uses a stream of electrons accelerated by an electric voltage or radiofrequency wave to a velocity close to the speed of light. Electrons have mass and negative electric charge, and therefore electron beams at the energies allowed for food irradiation do not penetrate into the product beyond several centimeters, depending on product density. Therefore, electron beams are preferred for thin products.

The advantages and disadvantages of E beam are tabulated below [1g]:

Sl. No.	*Advantages*	*Disadvantages*
1	E beams can be turned on only as needed.	Shallow depth of penetration. The disadvantage of the e-beam is its short penetration depth (about an inch), preventing its application to many foods and limiting the amount of food that can be processed in bulk.
2	E beams do not require replenishment of the source as does cobalt-60.	E-beams must be converted to x-rays to penetrate large items such as carcasses.
3	There is no radioactive waste.	High electric power consumption.
4	-	E-beams must be converted to x-rays to penetrate large items such as carcasses.
5	-	Complexity, and potentially high maintenance.

E-beam irradiation, though it uses the same term as gamma ray irradiation, is a completely different kind of treatment. High-energy electron beams are produced in an electron gun, a larger version of the cathode ray gun found in devices such as televisions and monitors. The electrons can be directed by a magnetic field to a target food. Nevertheless, the process has a similar effect to that of gamma ray irradiation. E-beam irradiation requires shielding as well, but nothing like the concrete bunkers used in gamma ray irradiation [1g].

Gamma Irradiation [1f]

It involves packets of light (photons) that originate from radioactive decay of atomic nuclei(Generally Cobalt-60). Gamma rays are used for products that are too thick for treatment with electron beams. Every radioactive element gives rise to Gamma rays with specific energies characteristic of the source element, these rays have zero mass and no electrical charge. The ones used for food irradiation are able to penetrate through food products. Gamma irradiation is most widely used technology because the deeper penetration of the gamma rays enables administering treatment to entire industrial pallets or totes (reducing the need for material handling) and it is significantly less expensive than using an X-ray source.

X-ray Irradiation [1f]

It is similar to irradiation by gamma rays but the rays produced have a distribution of energetic packets of light, still the maximum energy of the produced X-ray photon is limited by the energy put into the X-ray generator. X-rays are generated by colliding accelerated electrons with a dense target material (bremsstrahlung-conversion), and the X-rays so produced have a spectrum characteristic of the dense target material. Heavy metals, such as tantalum and tungsten(because of high atomic numbers and high melting temperatures) are used. Tantalum is usually preferred versus tungsten for industrial, large-area, high-power targets. Like electron beams, X-rays do not necessitate the use of radioactive materials and can be switched off when not needed. X-rays ability to penetrate the target is similar to gamma irradiation. X-ray machines produce better dose uniformity than Gamma rays, and their penetration depth is similar to slightly greater than that of gamma rays.

Radiation Process

It involves the following steps:

1. It is carried out inside an irradiation chamber shielded by 1.5-1.8 m thick concrete walls.
2. Food either pre-packed or in-bulk placed in suitable containers is sent into the irradiation chamber with the help of an automatic conveyor and dosimeters are placed on the pallet (at various locations) of food to serve as a check and ensure that the correct dose was achieved.
3. With gamma irradiation, special precautions are taken because gamma rays are continuously emitted by the radioactive material. In most designs, to nullify the effects of radiation, the radioisotope is lowered into a water-

filled storage pool, which absorbs the radiation but does not become radioactive. This allows pallets of the products to be added and removed from the irradiation chamber and other maintenance to be done [1f]. The conveyor goes through a concrete wall labyrinth, which prevents radiation from reaching the work area and operator room. When the facility is not in use the radiation source is stored under 6 m deep water. The water shield does not allow radiation to escape in to the irradiation chamber [1e]. Sometimes movable shields are used to reduce radiation levels in areas of the irradiation chamber instead of submerging the source. For x-ray and electron irradiation these precautions are not necessary as the radiation is generated by electricity, the source of the radiation can be switched off [1f]

4. For treating food, the source is brought to the irradiation position above the water level after activation of all safety devices.
5. The goods in aluminium carriers or tote boxes are mechanically positioned around the source rack and are turned round their own axis, so that contents are irradiated on both the sides.
6. The absorbed dose is determined by the residence time of the carrier or tote box in irradiation position and absorbed dose is checked by placing dosimeters at various positions in a tote box or carrier.

6. Detection of Proper Food Irradiation and Safety of Radiation Processed Food [1e]

a. Detection of Proper Food Irradiation

The specific tests developed for detection of irradiation for certain class of foods. No single method has yet been developed that reliably detects irradiation of all types of foods or radiation dose levels applied. The following techniques can be used for detecting radiation processed spices and meat containing bone tissue:

1. Thermoluminescence measurement.
2. Electron spin resonance spectroscopy.

b. Evaluation of the Safety of Radiation Processed Food

The safety of radiation processed food has been ascertained using an internationally accepted whole-someness testing protocol. The investigations include checking for induced radioactivity, monitoring changes in microflora, presence of any of mutant forms of microbes; assessing nutritional adequacy of irradiated foods by studying proximate analysis, vitamin losses, study of amino acids, protein efficiency ratio and in vitro digestibility; studying the chemical constituents of irradiated foods to rule out the possibility of any toxic compounds; and finally the multi-generation feeding studies lasting for 3-4 generations in at least two mammalian species.

Sources

1. *Hudaa Neetoo (Faculty of Agriculture, University of Mauritius, Réduit, Mauritius) and Haiqiang Chen(Department of Animal and Food Sciences, University of Delaware, Newark, Delaware, US), Alternative Food Processing Technologies.*

1a. *B. Kalyani and K. Manjula, Food Irradiation - Technology and Application, International Journal of current microbiology and applied Sciences ISSN: 2319-7706, Volume 3 Number 4, 2014, pp.549-555.*

1b. *Hayashi, T., Decontamination of dry food ingredients and seeds with soft electrons (low-energy electrons), Food Sci. Technol. Int. Tokyo 4, 1998, 114-120.*

1c. *Aikawa, Y, 2000, A new facility for X-ray irradiation and its application, Radiat. Phys. Chem. 57, 609 612.*

1d. *Cleland, M. R., Pageau, G. M., Comparison of X-ray and gamma-ray sources for industrial radiation processes, Nucl. Instrum. Methods, Sect. B 24/25, 1987, 967 972.*

1e. *Foods per cent 20 Irradiation/FAQ per cent 20Irradiation per cent 20 per cent 20 per cent 20 Bhabha per cent 20 Atomic.html.*

1f. *Wikipedia.*

1g. *Ranjan Saha, April 2015, New and Emerging Non-Thermal Technologies in Food Preservation.*

2. *Arvanitoyannis IS, Tserkezou P, 2010, Legislation on food irradiation: European Union, United States, Canada and Australia. In: Arvanitoyannis IS (ed)Irradiation of Food Commodities: Techniques, Applications, Detection, Legislation, Safety and Consumer Opinion. London: Elsevier, pp. 3-21.*

2a. *Harlin Kaur Saroya, Innovative non-thermal food processing technologies used by the food industry in the United States, MSc thesis, 2017, Western Kentucky University, Kentucky.*

2b. *Fellows, P., Food processing technologies: Princilpes and practices (3rd ed), Boca Raton, FL: CRC Press, 2009.*

2c. *Cleland, M. R, and Stichelbaut, F., Radiation processing with high-energy X-rays. Radiation Physics and Chemistry, 84, 91-99, 2013.*

2d. *Padua, G., Food Irradiation. Food Engineering Symposium, 1-9, 2009.*

3. *Fellows PJ, 2000, Food Processing Technology - Principles and Practice. Cambridge: CRC Press LLC/Wood head Publishing, pp. 196-208.*

4. *Cold Pasteurization of Food By Irradiation by Tim Roberts, Extension Specialist, Food Safety, Virginia Tech; Publication Number 458-300, posted August 1998.*

5. *The Truth about Irradiated Meat, Consumer Reports 34-37, August 2003.*

6. *Anon., Dosimetry for Food Irradiation, IAEA, Vienna, 2002, Technical Reports Series No. 409" (PDF). Retrieved March 19, 2014.*

7. *K. Mehta, Radiation Processing Dosimetry-A practical manual, 2006, GEX Corporation, Centennial, US.*

Impact of Irradiation on the Sensory and Nutritional Quality of Foods [1]

Ionizing radiation can change food quality but in general very high levels of radiation treatment (many thousands of gray) **are necessary to adversely change nutritional content, as well as the sensory qualities** (taste, appearance, and texture). Irradiation to the doses used commercially to treat food have very little negative impact on the sensory qualities and nutrient content in foods [1c].

Generally speaking, irradiation results in little adverse chemical change in food.

The macronutrient quality in food does not suffer due to irradiation [6 and WHO, 1999] and minerals have also been shown to remain stable [7]. **Extensive scientific studies have shown that irradiation has very little effect on the main nutrients such as proteins, carbohydrates, fats and minerals.**

Macronutrients [1, 1a]

Proteins and carbohydrates are relatively stable to radiation doses of up to 10 kGy [2]. However, irradiation applied at particularly high levels can potentially cleave certain amino acids, thereby changing the aroma and flavor of foods. Irradiation can also cause lipids to undergo auto-oxidation reactions, producing hydro-peroxides responsible for the development of off flavors in food. However, selection of appropriate treatment conditions can minimize these objectionable changes in food quality [3].

Vitamin Losses [1, 1a]

Nutrients such as vitamins are variably sensitive to irradiation. The extent of vitamin loss depends on the type of vitamins, the dose received, packaging conditions and the type and physical state of the food in question [1]. The general order of sensitivity of water-soluble vitamins to irradiation is: thiamin > ascorbic acid > pyridoxine > riboflavin > folic acid > cobalamin > nicotinic acid [4]. Water-soluble vitamins are more sensitive to irradiation than fat-soluble vitamins. Fat-soluble vitamins rank in the following order of decreasing sensitivity: vitamin E > vitamin A > vitamin K > vitamin D [5].

Vitamins A, C, and E are more sensitive and are thereby reduced at higher doses of irradiation, even though these losses are often similar to those occurring with thermal processing. Vitamin E is the most sensitive of the fat-soluble vitamins with significant losses (50 per cent) occurring when irradiated in the presence of oxygen [1a]. When oxygen was excluded or vacuum packaging was used, the losses were less than 10 per cent [8]. Significant losses were shown to occur in cream cheese (vitamin A) when air was not excluded [7] and in fruits and vegetables (vitamin C) treated with high doses. However, these findings are irrelevant because high-dose radiation is not used for such products. Significant losses in Thiamine (B1) can occur in irradiated meat products [6, 9]. However, the extent of such losses is dependent on processing conditions (temperature and dose) and can be minimized using packaging techniques [10]. Very little change in vitamin content is observed in food exposed to doses up to 1 kGy.

A comparative study of the extent of vitamin loss in irradiated and heat-sterilized chicken indicated similar inactivation kinetics. This therefore suggests that irradiation does not produce any special nutritional problems for treated foods [2, 4] and that the extent of nutritional degradation varies commensurate with various processing factors, similar to other preservation technologies [1, 2].

The Joint Expert Committee of the Food and Agriculture Organization (FAO), World Health Organization (WHO), and International Atomic Energy Agency (IAEA), in 1980 concluded that irradiation does not induce special nutritional problems in food [1b].

Impact on Sensory Quality [1a]

Foods such as milk, certain cheeses, eggs, and some fruits and vegetables are not likely candidates for irradiation because of the potential for undesirable offodors, flavors, and texture changes (WHO, 1999). The bulk of sensory work has focused on muscle foods, because most of the emphasis for this technology has been on these foods [11]. The researchers have shown that irradiation causes no significant **differences in the flavor, texture, or color of beef** irradiated at less than 3 kGy [12]. It was studied [13] that the **tropical fruits and found papayas, rambutans,** and Kau oranges were acceptable when treated with a quarantine level of 0.75 kGy (minimum dose required is 0.25 kGy). This study also found Chompoo and Biew Kiew fruit to be more acceptable when treated with 0.40 kGy than with the currently used hot-water immersion.

Sources

1. *Hudaa Neetoo (Faculty of Agriculture, University of Mauritius, Réduit, Mauritius) and Haiqiang Chen (Department of Animal and Food Sciences, University of Delaware, Newark, Delaware, US), Alternative Food Processing Technologies.*

1a. *Smith and Pillai, Food Technology Magazine, November 2004. 58 [11]: 22, 48-55.*

1b. *Foods per cent 20 Irradiation/FAQ per cent 20 Irradiation per cent 20 per cent 20 per cent 20 Bhabha per cent 20Atomic.html.*

1c. *Wikipedia.*

2. *Moreira RG, 2010, Food irradiation using electron-beam accelerators. In: Hui YH (ed) Handbook of Food Science, Technology and Engineering. Boca Raton, FL: CRC Press, 124: 1-8.*

3. *Olson DG, 1998, Irradiation of food. Food Technology 52: 56-61.*

4. *Dock LL, Floros JD, 2000, Thermal and non-thermal preservation methods. In: Schmidl M, Labuza TP (eds) Essentials of Functional Foods. Gaithersburg, MA: Aspen Publishers, pp. 49-88.*

5. *Kilcast D, 1994, Effect of irradiation on vitamins. Food Chemistry 49: 157-164.*

6. *Thayer, D. W. 1990. Food irradiation: Benefits and concerns. J. Food Qual. 13: 147-169.*

7. *Diehl, J. F. 1979. Vitamin A in bestrahiten Lebensmittel (Vitamin A in irradiated foodstuffs). Zeitschrift fur Lebensmittel-Untersuchung und-Forschung168: 29-31.*

8. *Josephson E. S., Thomas M. H., and Calhoun, W. K. 1975. Effects of treatments of foods with ionizing radiation. In "Nutritional Evaluation of Food Processing," ed. R. S. Harris and E. Karmas, pp. 393-411. AVI, Westport, Conn..*

9. *Fox, J. B., Lakritz, L., Hampson, J., Richardson, R., Ward, K., and Thayer, D. W. 1995. Gamma irradiation effects on thiamin and riboflavin in beef, lamb, pork, and turkey. J. Food Sci. 60: 596-598, 603.*

10. *Fox, J. B., Lakritz, L., and Thayer, D. W. 1997. Thiamin, riboflavin, and a-tocopherol retention in processed and stored irradiated pork. J. Food Sci. 62: 1022-1025.*

11. *Molins, R. A. 2001. Irradiation of meats and poultry. In "Food Irradiation: Principles and Applications," ed. R. A. Molins, p. 469. John Wiley and Sons, Hoboken, N. J..*

12. *Murano, P. S., Murano, E. A., and Olson, D. G. 1998. Irradiated ground beef: Sensory and quality changes during storage under various packaging conditions. J. Food Sci. 63: 548-551.*

13. *Follett, P. A and Sanxter, S. S. 2002. Longan quality after hot-water immersion and X-ray irradiation quarantine treatments. Hort. Sci. 37: 571-574.*

Benefits, and Shortcomings of Alternative Non-thermal Processing Method-Ionising Radiation [1]

Irradiation is a cold process and can be used to pasteurize and sterilize foods without causing changes in freshness and texture of food unlike heat. Unlike chemical fumigants, irradiation does not leave any harmful toxic residues in food and is more effective. It is efficient and can be used to treat prepacked commodities. In addition to food safety through destruction of pathogens, irradiation provides following other benefits [1a]:

1. Increasing shelf life of meats [1b] and fruits and vegetables [1c].
2. Improving quality of fruits and vegetables [1c].
3. Providing a suitable alternative to chemical treatments (*e.g.*, methyl bromide and ethylene oxide), especially for decontamination of fruits and vegetables [1c].
4. Providing economic savings due to reduced incidence of illness.

Process considerations, benefits and shortcomings of alternative non-thermal processing method-Ionising Radiation, are summarized in following table [1]:

Sl. No.	*Process Considerations*	*Benefits*	*Shortcomings*	*Examples of Applications*
1	Absorbed dose	Long history of use	High capital cost	Fresh produce
2	Water activity	High penetration power	Localized risks from radiation	Herbs and spices
3	Freezing	Suitable for sterilization (food and packages)	Hazardous operation	Packaging materials
4	Prevailing oxygen	Suitable for post package processing	Poor consumer accep-tance	Meat and fish

Sl. No.	*Process Considerations*	*Benefits*	*Shortcomings*	*Examples of Applications*
5	Microbial load	Suitable for non micro-biological applications (*e.g.* sprout inhibition)	Changes of flavor due to oxidation	—
6	Microbial species	Packaged and frozen foods can be treated	Loss of nutritional value	—
7	Product compo-sition	Low operating costs	Development of radiation resistant mutants	—
8	State of food	Can be scaled up	Microbial toxins could be present	—
9	Food thickness	Low and medium dose has minimal effect on product quality	Outgrowth of pathogens	—
10	Particle size	Suitable for low and high-moisture foods	—	—
11	Combination with other hurdles	Diverse applications	—	—

Effect of Non-thermal Method (Gamma Irradiation) on Inactivation of Micro-organism [2]

Sl. No.	*Products*	*Target Organism*	*Range Log Reduction*
1	Strawberries Fresh	Norovirus surrogate	2-3
	Puree	—	—
2	Oysters	Virus surrogate	1.5-2
	Shucked	—	—
	Whole-In shell	—	—

Sources

1. *Hudaa Neetoo (Faculty of Agriculture, University of Mauritius, Réduit, Mauritius) and Haiqiang Chen (Department of Animal and Food Sciences, University of Delaware, Newark, Delaware, US), Alternative Food Processing Technologies.*

1a. *Murano, P. S., Murano, E. A., and Olson, D. G.1998. Irradiated ground beef: Sensory and quality changes during storage under various packaging conditions. J. Food Sci. 63: 548 –551.*

1b. *Thayer, D. W. and Rajkowski, K. T. 1999. Developments in irradiation of fruits and vegetables. Food Technol. 53(11): 62-65.*

2. *Lori F Pivarnik and Randy Worobo, Non-thermal or Alternative Food Processing Methods to Enhance Microbial Safety and Quality, Frequently Asked Questions, (USDA-NIFA #2011-68003-30005).*

Food Irradiation and Food Safety [1]

The presence of microbial pathogens (*E. coli* O157: H7, *Salmonella, Campylobacter* spp., *Clostridium perfringens, Staphylococcus aureus,* and *Listeria monocytogenes*) on human foods is a serious global problem. Even in highly industrialized and developed countries, pathogen-contaminated foods and the resulting health and economic impacts are significant. The Hazard Analysis Critical Control Point (HACCP) system has been shown to greatly reduce the prevalence of pathogens, however, improved processing technologies, such as irradiation in combination with HACCP, can further advance postharvest food safety.

Issues and Concerns against Food Irradiation

Several extensive reviews of toxicological and other data by regulatory and health organizations (Health Canada-2003, FDA-1986, Codex Alimentarius Commission -CAC, 1983, European Commission's Scientific Committee on Food -EC, 2003) have determined that food irradiated below 10 kGy is safe. The CAC (2003) revised slightly its General Standard for Irradiated Foods, stating that the maximum absorbed dose delivered to a food should not exceed 10 kGy, except when necessary to achieve a legitimate technological purpose. A joint study group of the U. N.'s Food and Agriculture Organization (FAO), International Atomic Energy Agency (IAEA), and World Health Organization (WHO) concluded that food irradiated to any dose appropriate to achieve the intended technological objective is both safe to consume and nutritionally adequate. The group also concluded that no upper limit on absorbed dose was necessary because use of irradiation would be limited to doses that do not detrimentally impact the sensory attributes, thus creating a practical cut-off at about 50-75 kGy (WHO, 1999).

The Codex Alimentarius removed any upper dose limit for food irradiation as well as clearances for specific foods in 2003 and declaring that all are safe to irradiate. Pakistan and Brazil have adopted the Codex without any reservation or restriction and other countries, including New Zealand, Australia, Thailand, India, and Mexico, have permitted the irradiation of fresh fruits for fruit fly quarantine purposes, amongst others. Standards that describe calibration and operation for radiation dosimetry, as well as procedures to relate the measured dose to the effects achieved and to report and document such results, are maintained by the **American Society for Testing and Materials (ASTM international)** and are also available as ISO/ASTM standards [1b].

Inspite of the above facts, some groups of thoughts raised the following issues/ concerns on food irradiation [1]:

1. Presence of Radiolytic Products

Numerous volatile compounds have been isolated from irradiated products. The vast majority (more than 70 per cent) of the radiolytic volatile compounds found in irradiated foods are hydrocarbons, such as alkanes, alkenes, ketones, and aldehydes, that are commonly found in unprocessed and thermally processed foods [2, 3, 4] and are considered safe for human consumption.

The presence of following two groups of compounds in irradiated food have generated concern:

i. Benzene (and its derivatives)

The Federation of American Societies for Experimental Biology noted that small amounts of benzene could be detected in both irradiated (15 ppb) beef (56 kGy) and non-irradiated (3 ppb) beef [5] and concluded that such small amounts of benzene do not constitute a significant risk. Health Canada estimated that approximately 3 ppb of benzene would be formed in beef irradiated at the typical dose ranges (1.5-4.5 kGy). This level of benzene was noted to be much lower than naturally occurring levels in haddock (200 ppb) and eggs (average 62 ppb) [6]. Thus, the risk of benzene exposure from irradiated foods is considered negligible.

ii. Alkylcyclobutanones (ACBs)

ACBs were first identified in irradiated fats [7]. When pure triglycerides containing C6, C8, C10, C12, C14, C16, and C18 fatty acids were subjected to irradiation (60 kGy in vacuum), 2-substituted ACBs (2-ACBs) were formed. Thus, when the four major fatty acids present in most foods (palmitic, stearic, oleic, and linoleic acid) are irradiated, they are converted to their corresponding ACBs. As yet, these ACBs have not been found in raw or heat-processed foods and are considered unique radiolytic products [8, 9].

2. Mutagenic/Genotoxic

Regarding **Mutagenic/Genotoxic** effect of ACBs, on the basis of conclusion of several studies, the EC's Scientific Committee on Food released a revision of its 1986 opinion on food irradiation that addressed the ACB toxicity concerns (EC, 2003). It was the consensus of the committee that the genotoxicity of ACBs had not been established because there was no mutagenic effect in the Ames test or in standard cell lines.

3. Chromosomal Abnormalities (Abnormal chromosomes (Polyploidy) [1a]

National Institute of Nutrition (NIN), Hyderabad reported that the feeding of radiation processed food caused development of abnormal chromosomes (Polyploidy- occurrence of cells having more than twice the normal number of chromosomes) in malnourished children and laboratory animals but later on, number of researchers and scientific committees in India (GOI- 1976), Australia, Canada, Denmark, France, United Kingdom, United States, and a Joint FAO/ IAEA/WHO Expert Committee on the Wholesomeness of Irradiated Food have also evaluated the NIN data and rejected their claims. In the early 1980s, eight feeding studies using several radiation processed food items, including wheat, were conducted in China with human volunteers, More than 400 volunteers consumed radiation processed food under controlled conditions for 7-15 weeks. Results showed no significant differences between the control and the test groups. Incidence of polyploidy was within the normal range in participants irrespective of consuming irradiated or non-irradiated diet.

4. Increased Microbiological Hazards [1a]

Different studies have been shown that irradiation can neither increase the infectivity of pathogenic micro-organisms nor help them to grow better in radiation processed foods. Like in other processing methods, adequate care in packaging, handling and storage are necessary for ensuring safety of radiation processed foods.

5. Controlling of Irradiation Process to Ensure the Proper Irradiation of Foods [1a]

i. At National Level

Laws and regulations of **Atomic Energy Regulatory Board** govern operations of irradiators used to process non-food products, such as medical supplies. Three such irradiators are operating in India and about 160 around the world. The plants which must be approved by the Government before construction, are subject to regular inspection, safety audits, and other reviews to ensure that they are safely and properly operated. Similar controls would be applicable for radiation processing facilities.

ii. At the International Level

The **Codex Alimentarius Commission, of the United Nations Food and Agriculture Organization (FAO) and the World Health Organization (WHO)** have laid down standards for good manufacturing practices (GMP) and good irradiation practices (GIP) for a number of foods. They cover all aspects of treatment, handling, storage and distribution of food. The guidelines emphasize that, as with all food preservation techniques, effective quality control systems need to be installed and adequately monitored at critical control points at the irradiation facility.

6. Chemical Changes [1b]

Compounds known as free radicals form when food is irradiated. Most of these are oxidizers (*i.e.*, accept electrons) and some react very strongly. According to the free-radical theory of aging excessive amounts of these free radicals can lead to cell injury and cell death, which may contribute to many diseases. However, this generally relates to the free radicals generated in the body, not the free radicals consumed by the individual, as much of these are destroyed in the digestive process.

7. Nuclear Safety and Security [1b]

Interlocks and safeguards are mandated to minimize this risk. In a radiation processing facility, radiation specific concerns are supervised by special authorities, while **Ordinary** occupational safety regulations are handled much like other businesses. The regulators enforce a safety culture that mandates that all incidents that occur are documented and thoroughly analyzed to determine the cause and improvement potential. Such incidents are studied by personnel at multiple facilities, and improvements are mandated to retrofit existing facilities and future design.

The safety of irradiation facilities is regulated by the **United Nations International Atomic Energy Agency** and monitored by the different national Nuclear Regulatory Commissions. In the US the Nuclear Regulatory Commission

(NRC) regulates the safety of the processing facility, and the United States Department of Transportation (DOT) regulates the safe transport of the radioactive sources.

Concerns of Anti-Irradiation Groups [1]

These groups expressed the following concerns:

1. Misuse to Avoid Plant Sanitation

A common concern stated by these groups is that food irradiation would be used as an alternative to proper food-processing plant sanitation and cleanliness practices. A similar argument was used to dissuade implementation of milk pasteurization in the early 1900s. Today, milk pasteurization is a commonly used practice [10]. Food irradiation is intended as the final step of a comprehensive HACCP program.

2. Environmental Concern

i. There are lingering concerns among opponents to food irradiation regarding the environmental safety of irradiation facilities. Regulation of irradiation facilities is dependent on the source used. Gamma facilities have specific characteristics to protect workers (regulated by the Occupational Safety and Health Administration) and the surrounding environment from the radioactive isotopes and for storing the isotope material under water when not in use, which are regulated by the U. S. Nuclear Regulatory Commission. Cobalt-60, the isotope used in such facilities, requires 16-21 years to decay to approximately 6-12 per cent of its initial activity level, at which time it is shipped back to the manufacturer in hardened steel shipping canisters to be regenerated and reused. Unlike gamma facilities, E-beam and X-ray do not employ radioactive sources and thereby avoid such issues.

   On the basis of the more than 50 years experience, several health authorities (*e.g.* FDA *etc.*) stated that food irradiation can be used without posing a human health hazard and that furthermore, its use will improve the microbial safety of the food supply. This technology has been proven beneficial for not only controlling pathogens, but also increasing shelf life and maintaining food quality. Irradiation to ensure food safety is to be implemented as part of an overall HACCP plan and is not meant to replace existing control measures.

ii. **Meltdown of the irradiator and release of radioactivity [1a]:** There is no posiibility for a melt down to occur in a gamma irradiator. A radiation facility can not explode as Cobalt-60 is not a fissionable material and no neutrons are produced unlike in a nuclear reactor. Also no environmental contamination due to leakage of radioactivity can occur because the radioisotope is doubly encapsulated in stainless steel tubes.

iii. **Radioactive waste disposal problem [1a]:** The food irradiators do not have waste disposal problem. Cobalt-60 and Caesium-137 which are used as the source of radiation energy decay over many years to nonradioactive

Nickel and Barium, respectively. When the radioactivity falls to a low level the source pencils are returned to the supplier who has the option of reactivating them in a nuclear reactor or of storing them.

Sources

1. *Smith and Pillai, Food Technology Magazine, November 2004.58 [11]: 22, 48-55.*

1a. *Foods per cent 20Irradiation/FAQ per cent 20 Irradiation per cent 20 per cent 20 per cent 20 Bhabha per cent 20Atomic.html.*

1b. *Wikipedia.*

2. *Hannisdal, A.1993. Analysis of lipid-derived volatiles (alkanes and alkenes) in irradiated foods (salmon and chicken). In "Proc. Workshop on Recent Advantage on Detection of Irradiated Food, BCR Information, Chemical Analysis, Report EUR 14135 EN" ed. M. Leonardi, J. J. Raffi, J. J. Belliardo, pp.295-302, Commision of the European Communities, Brussels.*

3. *Morehouse, K. M., Ku, Y., Abrecht, H. L., and Yang, G. C., 1991, Gas chromatographic and electron spin resonance investigations of gamma irradiated frog legs. Radiat. Phys. Chem. 38: 62-68.*

4. *Nawar, W. W., Zhu, R., and Yoo, Y. J.1990. Radiolytic products of lipids as markers for the detection of irradiated foods. In "Food Irradiation and the Chemist, "ed. D. E. Johnston and M. H. Stevenson, pp. 13-24. Special Publication 86, Royal Society of Chemistry, Cambridge, UK.*

5. *Chinn, H. I. 1979. Evaluation of the health aspects of certain compounds found in irradiated beef. Chapter I: Further toxicological considerations of volatile products, pp 1-29, Life Sciences Research Office, Federation of American Societies for Experimental Biology, Bethesda, Md..*

6. *McNeal, T. P., Nyman, P. J., Diachenko, G. W., and Hollifield, H. C., 1993, Survey of benzene in foods by using head space concentration techniques and capillary gas chromatography. J. AOAC Intl. 76(6): 1213-9.*

7. *Letellier, P. R. and Nawar, W. W. 1972. 2-Alkylcyclobutanones from the radiolysis of triglycerides. Lipids 7(1): 75-6.*

8. *Crone, A. V. J., Hamilton, J. T. G., and Stevenson, M. H.1992a, Detection of 2-dodecylcyclo-butanones in radiation sterilized chicken meat stored for several years. Intl. J. Food Sci. Technol. 27: 691-696.*

9. *Stevenson, M. H. 1994. Identification of irradiated foods. Food Technol. 48(5): 141-144 and Satin, M. 1996, "Food Irradiation: A Guidebook." 2nd ed, CRC Press, Boca Raton, Fla. Sommers, C. H. and Schiestl, R. H. 2004.*

10. *Dodecyl-cyclobutanone does not induce mutations in the Salmonella mutagenicity test or intrachromosomal recombinations in Saccharomyces cerevisiae. J. Food Prot. 67: 1293-1298.*

Applications of Irradiation in Food Processing [1, 1b, 1d]

Irradiation will extend the shelf life of food because spoilage organisms are also reduced in number, and as with heat treatment, irradiation may also inactivate enzymes that would otherwise assist food spoilage. Irradiation is generally used for the following main purposes [1d]:

1. To reduce or eliminate the risk of food borne illnesses, prevent or slow down spoilage, arrest maturation or sprouting and as a treatment against pests.
2. to create safe foods for people at high risk of infection.
3. To create shelf-stable products.
4. To reduce postharvest losses.
5. To prevent the spread of invasive pest species through trade in fresh vegetables and fruits.

The details related to the application of irradiation in aforesaid area are given below:

A. Food Irradiation [1a]

For purposes of legislation, doses are divided into **low** (up to 1 kGy), **medium** (1 kGy to 10 kGy), and **high-dose** applications (above 10 kGy). High-dose applications are above those currently permitted by the regulators around the world [8]. Though these doses are approved for non commercial applications, such as sterilizing frozen meat or NASA astronauts (doses of 44 kGy) [9] and food for hospital patients. The details of three levels of food irradiation are given below [1a]:

1. Applications at Low Dose Levels (10 Gy- 1kGy)

Irradiation to a dose of between 1kGy and 10kGy is an effective method of reducing the microbial load of food. *Salmonella, listeria* or other harmful micro-organisms may contaminate foods such as dried herbs and spices, chicken meat, eggs, shrimps, cheese made from raw milk, frogs legs.

a. The inactivation of some pathogenic parasites of public health significance such as tapeworm and trichina in meat can be achieved at doses in the range 0.3 1 kGy.
b. Low dose irradiation (less than 1 kGy) is also used to inhibit sprouting (*e.g.* onion, garlic, potato). Inhibition of sprouting of Potatoes, onions, garlic, root ginger, yam at 0.05 - 0.15 kGy. Irradiation is effective in inhibiting the sprouting of potatoes. In Japan, doses of about 0.15 kGy have been used to control sprouting on potatoes, onions, and garlic [1b].
c. Insect disinfestations and parasite disinfection in Cereals and pulses, fresh and dried fruits, dried fish and meat, fresh pork, *etc.* at 0.15 - 0.5 kGy. Grains and tropical fruits may become infested with insects or larvae [1c]. Low dose (<1 kGy) has been shown to be effective for disinfestation of produce and also extends the shelf life by delaying ripening [1c].
d. Delay of physiological processes (Ripening) in Fresh fruits(*e.g.* of mango, papaya and other fruits) and vegetables at 0.25 - 1.0 kGy [1]. Irradiation arrests the ripening and maturation of fresh produce by inhibiting

hormone production and interrupting the biochemical processes of cell division and growth [1b].

e. It was showed [4] that low-dose γ-irradiation effectively reduced S. aureus in ready-to-eat ham and cheese sandwiches and proved to be more effective than refrigeration alone.

f **Effect of gamma radiation on millet grains-** A study related to the effect of gamma radiation on millet grains, was conducted [2a, 2b]. The grains were subjected to gamma radiation at different doses(0.25kGy,0.5kGy,0.75kGy, 1.0kGy and 2.0kGy). After the experiment, it was observed that a radiation level above 0.5kGy on the grain lowered the fungal incidence and free fatty acids content. At the same time, there was a reduction in anti-nutrients, such as, tannin and phytic acid. As a result, there was an increase in the vitro protein digestibility and protein solubility of the grain. This experiment proved gamma radiation was safe for shelf life extension of millet grains.

2. Applications at Medium Dose Levels (1-10 kGy) [1a]

Radiation enhances the keeping quality of certain foods through a substantial reduction in the number of spoilage causing micro-organisms [1a].

a. *Microbial inactivation [1, 1a]*

Various studies have demonstrated the ability of irradiation to destroy food-borne pathogens and spoilage microflora in various animal-derived and plant-derived products.

Irradiation can be used to extend the shelf life of foods by the following ways:

Radurization: Fresh meat, seafood, vegetables and fruits may be exposed to treatments with doses ranging from about 1 to 10 kGy for by destroying yeasts, molds and viable spoilage micro-organisms. This process of extending the shelf life is sometimes called radurization.

Radicidation: The process of improving the hygienic quality of food by inactivation of foodborne pathogenic bacteria (viable non-spore forming food-borne pathogens) and parasites is sometimes called radicidation. This medium dose application (using a dose of 1-8 kGy) is very similar to heat pasteurization, and is hence also **called radiopasteurization**.

Some examples of applications of radiation at medium dose level, are given below [1a]:

i. Extension of shelf-life of fresh fish, strawberries, mushrooms *etc.* at 1.0 - 3.0 kGy.

ii. Elimination of spoilage and pathogenic micro-organisms in Fresh and frozen seafood, raw or frozen poultry and meat, etc at 1.0 - 7.0 kGy.

iii. It was investigated [3] the effects of various doses of irradiation on the quality of fresh-cut iceberg lettuce and determined that 1-2 kGy was the optimum dose to ensure satisfactory inactivation of spoilage bacteria while minimizing sensory changes.

3. Applications at High Dose Levels (10-100kGy) [1a]

Radiation sterilization in the dose range 25-70 kGy extends the shelf life of precooked or enzyme inactivated food products in hermetically sealed containers almost indefinitely. This effect is achieved by the reduction of the number and/or activity of all organisms of food spoilage or public health significance, including their spores. This process is analogous to thermal canning in achieving shelf stability (long term storage without refrigeration) and is sometimes called **radappertization** (sterilization of products by killing both vegetative bacteria and spores using dose levels of 10-50 kGy). Some examples are given below:

- **i.** Industrial sterilization (in combination with mild heat) of Meat, poultry, seafood, prepared foods, sterilized hospital diets at 30 - 50 kGy,
- **ii.** Decontamination of certain food additives in Spices, enzyme preparations, natural gum, *etc.* and ingredients at10 - 50 kGy,
- **iii.** Other researchers have applied higher doses of up to 10 kGy to frozen poultry or shellfish (-18 °C) to destroy *Campylobacter* spp., *Escherichia coli* O157: H7 and *Vibrio* spp. [5],
- **iv.** For sterilization purposes, commodities such as herbs and spices are irradiated at a dose of 8-10 kGy to inactivate heat-resistant spore-forming bacteria [2],
- **v.** **Effect of e-beam on proteins [2a, 2c]**: A study was conducted [2c] to examine the effect of e-beam processing on the structural and functional properties of the following proteins:
  - **a.** **Albumin**: protein, which is soluble in water
  - **b.** **Globulin**: protein found within the blood
  - **c.** **Glutenin**: major protein within wheat flour
  - **d.** **Wheat germ protein** isolate.

  The end results indicated that e-beam irradiation with optimal conditions was a very effective process to enhance water absorption, oil absorption, and foaming properties of proteins.

  - **a** The best results were obtained below 60kGy because it provided the optimal condition for oil absorption activity in albumin, globulin, and glutenin.
  - **b.** 30kGy was best for water absorption, as well as glutenin and wheat gram protein isolate.
- **vi.** **Effcet of e-beam on Beef products** [2a, 2d]: A study related to the effect of e-beam processing of two beef products (steaks and hamburgers) on the shelf life extension, safety, and sensory attributes was conducted [2d]. Salmonella was inactivated by using different doses of irradiation, by first order kinetics. The optimization of the irradiation doses was done by making the most of sensory scores of samples and reducing the instrumental color changes. These optimum doses assured maximum worth of the product and proved that e-beam processing greatly affects the safety and sensory attributes of both poultry products.

vii. **Internal structure of pomegranate by X-ray [2a, 2e]**: A study was conducted [2e] using the X-ray computed tomography, to identify and measure the internal structure of pomegranate, The evaluation and volume estimation of the internal structure of the fruit was done by setting up a commercial X-ray system with a radiation source of 245 kW and electron current of 300μA. This helped in the generation of two-dimensional radioscopic images which were then reconstructed into three dimensional images. This technology helped in accurately foreseeing the physical attributes of the fruit, such as, length, diameter, and peel thickness.

The summary of aforesaid applications (Applications By Overall Average Dose) is tabulated below [1d]:

Sl. No.	*Level of Doses*	*Dose*	*Applications*
1	Low dose (up to 1 kGy)	0.03–0.15 kGy	Inhibit sprouting of bulbs and tubers.
		0.03–0.15 kGy	Delay fruit ripening.
		0.07–1.00 kGy	Stop insect/parasite infestations to help clear quarantine.
2	Medium dose (1 to 10 kGy)	1.50–3.00 kGy	Delay spoilage of meat.
		3.00–7.00 kGy	Reduce risk of pathogens in meat.
		10.00 kGy	Increase sanitation of spices.
3	High dose (above 10 kGy)	25.00–70.00 kGy	Sterilization of packaged meat.
		—	Increase juice yield.
		—	Improve re-hydration.

Combined Applications of Ionizing Radiation and Modified Atmosphere Packaging (MAP)

In addition to the application of irradiation alone, combined applications of ionizing radiation and modified atmosphere packaging (MAP) have been investigated and it was showed [6]that the combination of irradiation (1kGy) and MAP(10-20 per cent O_2, balance N_2) significantly reduced the microbial populations of fresh pork, although to the detriment of organoleptic quality. Similarly, it was showed [7] that the application of low-dose irradiation and MAP brought about a 4-5-fold extension in the shelf life of fresh catfish fillets. A combination of irradiation and MAP has been shown to deliver a synergistic effect in delaying ripening, lowering the dose required [2].

Sources

1. *Hudaa Neetoo (Faculty of Agriculture, University of Mauritius, Réduit, Mauritius) and Haiqiang Chen (Department of Animal and Food Sciences, University of Delaware, Newark, Delaware, US), Alternative Food Processing Technologies.*

1a. *B. Kalyani and K. Manjula, Food Irradiation - Technology and Application, International Journal of current microbiology and applied Sciences ISSN: 2319-7706, Volume 3 Number 4, 2014, pp. 549-555.*

1b. *Thomas P, (1999), Control of postharvest loss of grain, fruits and vegetables by radiation processing. In: Loaharanu P, Thomas P (eds) Irradiation for Food Safety and Quality. Lancaster: Technomic Publishing, pp. 93-102.*

1c. *Gupta SC, 1999, Irradiation as an alternative treatment to methyl bromide for insect control. In: Loaharanu P, Thomas P (eds) Irradiation for Food Safety and Quality. Lancaster: Technomic Publishing, pp.39-49.*

1d. *Wikipedia.*

2. *Fellows PJ, 2000, Food Processing Technology: Principles and Practice. Cambridge: CRC Press LLC/Wood head Publishing, pp. 196-208.*

2a. *Harlin Kaur Saroya, Innovative non-thermal food processing technologies used by the food industry in the United States, MSc thesis, 2017, Western Kentucky University, Kentucky.*

2b. *Mahmoud, N. S., Awad, S. H., Madani, R. M., Osman, F. A., Elmamoun, K., and Hassan, A. B., Effect of γ radiation processing on fungal growth and quality characteristcs of millet grains. Food Science and Nutrition, 4(3), 342-347, 2015.*

2c. *Wang, L., Ding, Y., Zhang, X., Li, Y., Wang, R., Luo, X., Li, Y., Li, J., Chen, Z., Effect of electron beam on the functional properties and structure of defatted wheat germ proteins. Journal of Food Engineering, 202, 9-17, 2017.*

2d. *Cárcel, J., Benedito, J., Cambero, M., Cabeza, M., and Ordóñez, J., Modeling and optimization of the E-beam treatment of chicken steaks and hamburgers, considering food safety, shelf-life, and sensory quality. Food and Bioproducts Processing, 96, 133-144, 2015.*

2e. *Arendse, E, Fawole, O., Magwaza, L., and Opara, U., Non-destructive characterization and volume estimation of pomegranate fruit external and internal morphological fractions using X-ray computed tomography. Journal of Food Engineering, 186, 42-49, 2016.*

3. *Fan X, Sokorai KJB, 2005, Assessment of radiation sensitivity of fresh-cut vegetables using electrolyte leakage measurement Postharvest Biology and Technology 36: 191-197.*

4. *Lamb JL, Gogley JM, Thompson MJ, Solis DR, Sen S, 2002, Effect of low-dose gamma irradiation on Staphylococcus aureus and product packaging in ready-to-eat ham and cheese sandwiches, Journal of Food Protection 65: 1800-1805.*

5. *Crawford LM, Ruff EH, 1996, A review of the safety of cold pasteurization through irradiation. Food Control* 7: *87-97.*

6. *Lambert Y, Demazeau G, Largeteau A, Bouvier JM, 2000, Packaging for high pressure treatments in the food industry. Packaging Technology and Science 13: 63-71.*

7. *Przybylski LA, Finerty MW, Grodner RM, Gerdes DL, 1989, Extension of shelf-life of iced fresh channel catfish fillets using modified atmosphere packaging and low dose irradiation, Journal of Food Science 54: 269-73.*

8. *Irradiated Food Authorization Database (IFA), Archived from the original on March 19, 2014, Retrieved March 19, 2014.*

9. *U. S. Food and Drug Administration, Center for Food Safety and Applied Nutrition. Office of Premarket Approval, Food Irradiation: The treatment of foods with ionizing radiation, Kim M. Morehouse, PhD Published in Food Testing and Analysis, June/July 1998 edition, Vol. 4, No. 3, Pages 9, 32, 35.*

B. Sterilisation and Requirement of Packing Materials [1]

Food once irradiated, can be prone to recontamination unless appropriately packed. Therefore, if radiation treatment is intended to control microbiological spoilage or insect infestation, prepackaging becomes an integral part of the process. The packaging materials for use in aseptic processing lines in the food, pharmaceutical and cosmetic industry are now being sterilized by ionizing radiation. Since packaging materials are also exposed to radiation during the treatment, these materials must also satisfy additional requirements such as resistance to radiation with respect to its functional properties. In addition, it should not transmit toxic substances into food nor impart any off odor to the products.

Depending on the type of radiation treatments, packaging materials used for irradiated food are broadly classified into the following two categories:

1. Processes requiring doses less than 10 kGy, such as extension of shelf-life of food.
2. Processes requiring doses from 10-60 kGy, for storing such items as meat and poultry for long periods without refrigeration.

The plastics offer unique advantages over the conventionally used rigid containers in terms of flexibility, low cost, light weight, and low weight to volume ratio. Multi-laminate packaging structure of polymers like nylon, EVOH, PVC, cellophane, PE and Polyester are used as a prominent barrier material in packaging of irradiated food [2].

Sources

1. *B. Kalyani and K. Manjula, Food Irradiation-Technology and Application, International Journal of current microbiology and applied Sciences ISSN: 2319-7706, Volume 3 Number 4, 2014, pp.549-555.*
2. *Agarwal, S. R. and Sreenivasan, A., 1972, Packaging of irradiated flesh food: A review. J. Food Technol., 8, 27-37.*

C. Other Applications

1. Certain types of fresh produce, such as strawberries and tomatoes, can be irradiated to extend the shelf life by about 2-3 times when stored at 10°C by slowing down the ripening process [1].
2. Irradiation is also used for disinfestation purposes. The recommended treatment for decontamination of dried spices, herbs, vegetable seasonings, and dry ingredients is irradiation at a dose of 3-30 kGy (ASTM, 1998).

   Irradiation of Spices: Most spices get heavily contaminated with microbes including pathogenic bacteria during sun drying. Contamination arises from deposition of excreta of insects, birds, rodents and other animals and from wind blown dust containing these microbes. Naturally present mold spores can grow under high humidity storage conditions and lead to formation of aflatoxins and other mycotoxins. It can also lead to caking and spoilage. Insect infestation is another major problem in stored spices often resulting in losses up to 20-25 per cent. Because of the aforesaid reasons, irradiation of spices is one of the important processes to make the spices safe for human consumption.

3. Improving technological properties of food at 2.0 - 7.0 kGy *e.g.* Grapes (increasing juice yield), dehydrated vegetables (reduced cooking time) [2].

Sources

1. *Thomas P, 1999, Control of postharvest loss of grain, fruits and vegetables by radiation processing. In: Loaharanu P, Thomas P, Irradiation for Food Safety and Quality. Lancaster: Technomic Publishing, pp. 93-102.*

2. *B. Kalyani and K. Manjula, Food Irradiation - Technology and Application, International Journal of current microbiology and applied Sciences ISSN: 2319-7706, Volume 3 Number 4, 2014, pp. 549-555.*

Chapter 9

Ultraviolet Light Treatment

Among the physical methods of food preservation, ultraviolet radiation **[1]** has been considered as an alternative to thermal treatment for controlling the microbial growth in liquids products mainly water. However, in recent years this technology has been used for disinfection of food products [1a]. Ultraviolet processing involves the use of radiation from the ultraviolet region of the electromagnetic spectrum (100 to 400 nm) for disinfection of solid surfaces and liquid [1a]. Compared to thermal processing, this promising technology can provide consumers with minimally processed, microbiologically safe and fresh-like products with minor effects on the nutritional and sensory properties of the product. On the other hand, this technology must not replace hygiene, good manufacturing or agricultural practice [1b].

UV light application can also be introduced as an alternative to the use of chemicals in food industry. Besides, the use of UV light does not generate chemical residues. Additionally, it offers some technological advantages especially in developing countries in a small-scale production due to its low maintenance cost, low installation cost and low operational cost with minimal energy use. The use of UV light must not only be considered for microbial inactivation but also for the development of novel dairy products. The UV-treated pasteurized cow's milk was authorized as a novel food in market by European Commission [1b].

Principle of UV Light Treatment [1, 1b]

The UV spectrum is customarily divided into the following three regions:

1. UV-A with a wavelength of 320-400 nm
2. UV-B with a wavelength of 280-320 nm
3. UV-C with a wavelength of 200–280 nm [1a, 2].

UV-C possesses germicidal properties and at a dose rate of 1000 J/m^2 or more, bacteria, yeasts, and viruses undergo as much as 4-log reductions. Because of the very low penetration depth of UV, it is used for surface treatments only. To accomplish microbial inactivation, the radiant exposure from UV should be more than 400 J/m^2 in all parts of the product [1, 1a]. Pulsed UV light is a modified and improved version of the UV-C light. Pulsed UV light is an application using devices containing ultraviolet lamps that emit ultraviolet light at high power at regular intervals [1b]. **The mechanism of inactivation and cell death is the absorption of UV by DNA and RNA [2]. At high levels, UV light causes damage to a micro-organism's DNA. It is this characteristic of UV light that is used to kill pathogens that are contained in food but does not impart any health concerns to the food [2a].** As UV light penetrates the microbial cell, it brings about a number of alterations to the cellular components [2].

In principle, the photochemical reactions of biomolecules of micro-organism primarily result in germicidal effect leading to inhibition of microbial growth or to inactivation of the cell. Germicidal effect of UV light on micro-organisms occurs because of cross-linking between the bases of adjacent pyrimidine dimers in the same DNA strand [1b]. UV light reportedly acts on the DNA by forming the cross-linkages between neighbouring pyrimidine nucleotide bases (thymine and cytosine) in the same DNA, which leads to blockage of transcription and replication, subsequently dis-functioning of cellular functions leading to cell death. It is also reported that the amount of crosslinking in DNA is proportional to the amount of UV exposure [1a]. Microbial inactivation is mainly the result of deleterious interactions between UV and nucleic acids. This interaction yields following photoproducts:

1. Photoproducts of which pyrimidine dimmers are the most important [3]. These are formed between two adjacent pyrimidine bases such as thymine and cytosine, on the same DNA strand.
2. Another type of "photoproduct, " called a "pyrimidine adduct, " is also formed between adjacent bases, but at reduced rates of formation compared to dimers. At sufficiently high UV doses, DNA-protein cross-links are formed while at higher doses still, DNA strand breakages may be induced [2].

Factors affecting Afficacy of UV Light [1, 1b]

The fate of micro-organisms following exposure to UV depends on a multitude of factors. The following various influential processing parameters/factors affect the efficacy of UV light similar to ionizing radiation or lethality of the process:

1. The range of wavelengths used to irradiate the cells.
2. Treatment time
3. Treatment intensity
4. Target species-composition, physical and surface properties
5. The treatment temperature
6. Concentration of hydrogen ions
7. Microbial load
8. Relative humidity.

The germicidal effects of UV radiation primarily depend on the UV dose (J/m^2) which refers to the **UV irradiance** or UV intensity flux and is defined as the function of the intensity and time of exposure. The UV intensity (W/m^2) is the total radiation from the specified area. In most cases, as the exposure time and intensity of UV light increase and the distance from light source to target decreases, inhibition rate of cells increases. The UV light sensitivity of the target micro-organism is an important parameter for the selection of the UV light dose. Physical, compositional and surface properties such as thickness, viscosity, density, optical properties, color differences, dirtiness, roughness *etc.* can change the process efficiency. UV light has a restricted penetrability. Viscosity and density determine the effectiveness of the transfer and flow model of the liquid in the system, while optical properties affect the UV light transmittance [1b]. Light transmission of food and packaging material in UV application to the surface of unpacked and packaged food is a critical factor for decontamination. Higher absorbsion of light is obtained in dark foods, causing decrease of available energy for microbial inactivation [1b].

Spores of bacteria are rather resistant to UV light compared to vegetative cells.

Some of the common terminologies used in UV processing are as follows [1a]:

i. Irradiance

Irradiance refers to the power passing through the surface. The irradiance is measured using radiometers, either thermal or photonic, UV sensors or actinometers. However, biodosimetry is considered as more reliable, wherein sample is inoculated with surrogate micro-organisms and log reductions are counted after the radiation exposure.

ii. Fluence and Fluence Rate

Fluence rate refers to the power passing through the sphere. The 'UV fluence rate' is the more appropriate term for UV disinfection as exposure of micro-organisms to UV occurs from any direction especially when multiple lamps are used.

Sources

1. *Hudaa Neetoo (Faculty of Agriculture, University of Mauritius, Réduit, Mauritius) and Haiqiang Chen (Department of Animal and Food Sciences, University of Delaware, Newark, Delaware, US), Alternative Food Processing Technologies.*

1a. *Nishant R Swami Hulle, Krishna Teja C and P Srinivasa Rao, Processing- Applications of UV Light, 2015, Food Marketing and Technology, magazine published by L. B. Associates Pvt Ltd.*

1b. *Nurcan Koca, Müge Urgu and Turkuaz Ecem Saatli (June 20th 2018). Ultraviolet Light Applications in Dairy Processing, Technological Approaches for Novel Applications in Dairy Processing, Nurcan Koca, IntechOpen, DOI: 10.5772/intechopen.74291.*

2. *Sharma G, 2010, Ultraviolet light. In: Hui YH (ed) Handbook of Food Science, Technology and Engineering. Boca Raton, FL: CRC Press, 122: 1-12.*

2a. *Lori F Pivarnik and Randy Worobo, Non-thermal or Alternative Food Processing Methods to Enhance Microbial Safety and Quality, Frequently Asked Questions, (USDA-NIFA #2011-68003-30005)*

3. *Harm H, Rupert CS, 1976, Analysis of photoenzymatic repair of UV lesions in DNA by single light flashes. XI. Light-induced activation of the yeast photoreactivating enzyme. Mutation Research 34: 75-92.*

Ultraviolet Equipment [1]

UV Light Sources

The following two different modes of UV treatments can be applied to food products [1a]:

i. Continuous Mode

The low and medium pressure mercury lamps and excimer lamp technologies can be used for food processing applications. Conventional low pressure mercury UV lamps used for surface decontamination emits UV light at 254 nm. UV-excimer lamps have the advantage of functioning at different wavelengths, which are produced by combining with different gases.

ii. Pulsed Mode

The pulsed UV light is reported to be effective in reducing microbial level; however, their low penetration and temperature rise in the product has been a concern.

Mercury Lamps [1]

Mercury lamps have been the sources of radiation in most UV-based disinfection systems. The low and medium pressure mercury UV lamp sources are reliable sources for disinfection applications which are beneficial for their performance, and low cost. They are based on the vapor pressure of mercury while the lamps are operating. Ultraviolet light is most commonly produced in low vapor pressure mercury lamps [2]. UV lamps are typically made of special quartz glass that allows 70-90 per cent of UV rays to pass through. Mercury vapor lamps contain an inert gas carrier and a small amount of mercury within the sealed glass tube. When an electric arc is created, it causes the mercury to vaporize and ionize, thereby emitting UV radiation. These form the basic components of a UV-emitting apparatus [3].

UV-C Lamps [1]

For optimal efficacy, the UV-C emitted should be able to access all parts of the treated food. UV-C lamps are placed to ensure that the light shines towards the center as well as the sides of the disinfection unit.

i. Thin Film UV-C Disinfection Systems [1]

Some researchers have developed "thin film UV-C disinfection systems" which deliver the (liquid) food via a nozzle in the form of a thin liquid film [4].

ii. A Convoluted Tube surrounded by UV-C Mercury Lamp [1]

Another approach involves a liquid flowing through a convoluted tube surrounded by UV-C mercury lamps. While novel thin film devices tend to be used for disinfecting liquids, conveyor belts carrying solid foods past UV sources have also been set up with lamps mounted onto the walls of the enclosures, creating "UV tunnels" [2].

A Commercially Available Bench-Top UV-C Test System at the University of Delaware.

Industrial UV Systems [1a, 5]

Different types of continuous system have been developed for processing of liquid products such as milk based beverages and fruit juices. The pointwise details are given below:

1. Many designs are presently under developmental stage.
2. **Reactors:** Reactors are devices used for UV light application to liquids. The first reactor design is a thin film UV reactor. Thin-film reactors are characterized by laminar flow with a parabolic velocity profile. Another reactor having laminar flow is laminar Taylor-Couette UV reactor. The second design approach is turbulent flow reactors [1b].

   The following different types of reactors include annular, thin film, dynamic mixers and coiled tube devices are being used:

   a. **Annular reactor**: The studies have been conducted on annular type reactors for treatment of beverages. The gap between two concentric cylinders and length of the annular reactor vary depending on the composition of the flowing liquid.

   b. **Thin film reactors**: In thin film reactors, the fluid is passed on fixed surface and the gap is usually kept less than 1 mm.

c. **Static mixers:** The static mixers continuously changes the flow pattern in the system thereby more exposure of microbial cells to the ultraviolet light.

d. **Dynamic Mixers:** In dynamic mixers, the reactor has stationary outer cylinder and a rotating inner cylinder, which brings fluid from the dark outer cylinder to high-intensity inner wall region.

3. There are several reports available on lab scale studies, the validation of the process on large scale is important to retain the process effectiveness.
4. With the current developments in food process engineering, it is now possible to use UV light for minimal processing of solid and liquid products for reducing the microbial population without significantly affecting the nutritional quality of the product.
5. There is still a need of efficient systems to apply UV technology for its commercial applications with validation of protocols for food applications.

UV Cabinets and Tunnels [1b]

Cabinet

UV cabinets are devices developed for UV light applications on the surface of solids. The number and position of lamps in the UV cabinet are the most critical factors for the disinfection of entire food surfaces. The UV processing units for solid food was well schematized by Manzocco and Nicoli [1c].

If one side of the solid food is exposed to UV light, the food is placed on a support. For the exposure to the top and bottom sides, the food can be placed on a film or turned upside down during treatment. If all the surfaces of the solid food are exposed to the UV light at the same time, it is needed to increase the number of lamps and place the food on a film. If there is no food support, the product flows near the lamps coated with waterproof quartz tubes in a vessel containing water.

In dairy industry, only upper surface of the yoghurt in package is enough to be treated by UV light while all surfaces are exposed for many cheeses.

Tunnels

It is also possible to design a tunnel with a dynamic system moving with the food. In this type of tunnel system, the food material is conveyed through UV tunnel and taken from the other end. The width and height of the tunnel are designed according to expectations of user. UV application time is adjusted by conveyor speed. Such tunnels provide convenience for industrial use [1b].

Sources

1. *Hudaa Neetoo (Faculty of Agriculture, University of Mauritius, Réduit, Mauritius) and Haiqiang Chen (Department of Animal and Food Sciences, University of Delaware, Newark, Delaware, US), Alternative Food Processing Technologies.*

1a. *Nishant R Swami Hulle, Krishna Teja C and P Srinivasa Rao, Processing- Applications of UV Light, 2015, Food Marketing and Technology, magazine published by L. B. Associates Pvt Ltd.*

1b. *Nurcan Koca, Müge Urgu and Turkuaz Ecem Saatli (June 20th 2018). Ultraviolet Light Applications in Dairy Processing, Technological Approaches for Novel Applications in Dairy Processing, Nurcan Koca, IntechOpen, DOI: 10.5772/intechopen.74291.*

1c. *Manzocco L, Nicoli MC. Surface processing: Existing and potential applications of ultraviolet light. Critical Reviews in Food Science and Nutrition. 2015; 55: 469-484.*

2. *Sharma G, (2010), Ultraviolet light. In: Hui YH (ed) Handbook of Food Science, Technology and Engineering. Boca Raton, FL: CRC Press, 122: 1-12.*

3. *Guerrero-Beltran JA, Barbosa-Canovas GV, (2011), Ultraviolet-C light processing of liquid food products. In: Zhang HQ, Barbosa-Canovas GV, Bala-subramaniam VM, Dunne CP, Farkas DF, Yuan JTC (eds) Non-thermal Processing Technologies for Food. Chichester: Wiley-Blackwell, pp. 262–270.*

4. *Sharma G, Peppiatt C, Biguzzi M, (1996), A novel thin film photoreactor. Journal of Chemical Technology and Biotechnology 65: 56-64.*

5. *Keyser, M., Miller, I. A., Cilliers, F. P., Nel, W., and Gouws, P. A. (2008). Ultraviolet radiation as a non-thermal treatment for the inactivation of micro-organisms in fruit juice. Innovative Food Science and Emerging Technologies, 9 (3), 348-354.*

Legislations on UV Light Application in Food Processing [1, 2, 3]

The following different protocols for validation of UV decontamination system are available:

1. The NSF International standard
2. The US EPA UV Disinfection Guidance Manual
3. NWRI guidelines.

USA [1]

The Food and Drug Administration, Department of Health and Human Services (US FDA) approved the use of UV radiation for controlling surface micro-organisms of food or food product, sterilization of water used in production and reduction of human pathogens and other micro-organisms in juice products under specific conditions defined by Code 21CFR179.39.

European Union [1]

In European Union, UV light is accepted as irradiation [4]. The use of irradiation is limited but authorized in many European countries. According to European Commission, treating food with ionizing radiation may be authorized if there is reasonable technological need, it poses no health hazard and benefits consumers, and if it does not replace hygiene, health or good manufacturing or agricultural practice.

Sources

1. *Nurcan Koca, Müge Urgu and Turkuaz Ecem Saatli (June 20th 2018). Ultraviolet Light Applications in Dairy Processing, Technological Approaches for Novel Applications in Dairy Processing, Nurcan Koca, IntechOpen, DOI: 10.5772/intechopen.74291.*

2. *Nishant R Swami Hulle, Krishna Teja C and P Srinivasa Rao, Processing- Applications of UV Light, 2015, Food Marketing and Technology, magazine published by L. B. Associates Pvt Ltd.*

3. *Keyser, M., Miller, I. A., Cilliers, F. P., Nel, W., and Gouws, P. A. (2008). Ultraviolet radiation as a non-thermal treatment for the inactivation of micro-organisms in fruit juice. Innovative Food Science and Emerging Technologies, 9 (3), 348-354.*

4. *Koutchma T, Forney LJ, Moraru CI. Ultraviolet Light in Food Technology: Principles and Applications. Boca Raton, FL, USA: Taylor and Francis Group; 2009. 278 p.*

Applications of Ultraviolet Light in Food/Milk Processing

The UV Light is mainly being used for surface treatment and liquid processing [1a].

1. Surface Treatment

UV light is used in disinfection of surface microflora of equipment which comes in contact with food during processing, utensils, packaging materials and conveyor systems. However, considering the limited penetration of UV light, its application for decontamination of food surfaces like meat products is limited. The packaging materials should be stable to the UV treatment and should not undergo depolymerisation reaction and change in elastic modulus.

2. Liquid Foods Processing

The processing of liquids using ultraviolet light on large scale began in 1906 for disinfection water. In North America more than 500 UV plants are reported to supply drinking water, whereas, in Europe more than 2000 plants use this technology for disinfection of water. Application of UV light in liquid food products has shown positive results as evident from the reduction in number of pathogenic micro-organisms. However, the efficacy of UV treatment in liquid foods is dependent on its constituents like pigments, organic solutes and suspended particles which limit the transmission of UV light through the liquid food [1a].

A. Applications of Ultraviolet Light in Food Processing [1]

The application of UV-C light technology for food products has been mainly confined to liquid foods and water.

1. Scientists demonstrated [2] the superficiality of UV-C light when shone into liquid food materials, showing that UV-C light was only able to travel through a distance of ~1 mm underneath the surface of the product (juices) and that 90 per cent of the light energy was absorbed.
2. A series of chambers to treat apple cider, achieving 3.8-log reduction in *E. coli* O157: H7 was used [3]. UV processing is being used in the juice and cider industries for pasteurization without heat targeting **E. coli O157: H7** and Cryptosporidium parvum [3a].
3. UV-C light to orange juice flowing as a thin film to maximize the exposed surface area/volume ratio was applied [4] and observed a two-fold

increase in shelf life with no adverse effects on the color or flavor of the product.

4. Various juices including grape, grapefruit and cranberry juices inoculated with S. cerevisiae with UV-C light for a treatment time of 30 min were processed [5] and reported a maximum log reduction of 0.5, 2.4, and 2.4 log CFU/mL, respectively.

   The US FDA has approved UV-light application for pasteurisation treatment of fruit juice products in 2000 and the basic criteria set for UV-application is 5 log reductions in number of pathogens present in fruit juices [1a].

5. The penetration effect of UV-C radiation depends on the type of product, its UV-C absorptivity, soluble solids in the liquid, and suspended matter in the liquid. The larger the amount of soluble solids and the darker the color of the medium, the lower the intensity or penetration of UV-C light will be in the liquid.
6. It was reported [5] that since grape juice was deep violet, only 0.53-log microbial reductions were obtained after 30 min of exposure to UV light. However, a higher reduction of 1.34-log was observed for *S. cerevisiae* inactivation in apple juice after 30 min of UV light exposure since apple juice is light brown in color and transparent in nature.
7. Even though the color and transparency of cranberry juice (red-brown and clear) and grape fruit juice (pink color and turbid) are entirely different, their response to UV light treatment was similar, suggesting that factors other than color and turbidity, such as the intrinsic composition, acidity and pH, could influence the efficacy of UV light.
8. In addition to juices, the effect of UV in extending the shelf life of fruits and vegetables has been reported. The use of UV light to extend the shelf life of Walla Walla onions at ambient storage temperature was studied [6].
9. The ability of UV-C radiation to preserve the postharvest state of ripe mangoes with minimal damage to their quality, was demonstrated [7].
10. UV-C treatments have also been used to control pathogenic as well as spoilage microbes on various meat, fish, and seafood products [8].
11. UV has not been used on other produce or shellfish for pathogen reduction, but is commonly employed for water treatment systems, shellfish wet storage, and municipal water [3a].
12. The UV application is also reported to reduce toxins like patulin mycotoxin in cider and allergenicity compound in peanut (Ara h2) and soybean (glycinin, β-conglycinin) based products [1a].
13. **Pulsed UV light application in honey [1a]**- In one of the studies on pulsed UV light application on inactivation of *Clostridium sporogenes* spores in clover honey, increase in reduction of spores was observed with increase in number of pulses; however complete inactivation was not achieved.

As honey is highly viscous it absorbs UV light strongly, resulting into incomplete inactivation of microbes in the sample.

B. Applications Ultraviolet Light in Processing of Milk [9]

1. UV irradiation of milk was first used in the mid-1900s for the purpose of vitamin D enrichment. It is reported that the treatment of the pasteurized milk with UV radiation results in an increase in the vitamin D3 (cholecalciferol) concentrations by conversion of 7-dehydrocholesterol to vitamin D3 [11a].
2. Raw milk has UV radiation absorption coefficient of 290 cm^{-1} at 253.7 nm wave length of UV radiation. Due to the photosensitive nature of milk, UV light radiation tends to have negative impact on milk [10].
3. The breaking of ***M. avium subsp. paratuberculosis*** cell clumps present in milk while passing the milk through UV chamber, was reported [12]. Short wavelength UV light may be utilized for Vitamin D synthesis.
4. **Pulsed UV light application in milk:** Researchers [11]reported pulsed UV light's potential in inactivation of pathogenic micro-organisms present in milk showing complete inactivation of *Staphylococcus aureus*, a milk pathogen.
5. Surface application of UV light after production can offer an attractive alternative method to eliminate or control the growth of post-processing contamination in milk products [11a].
6. Other promising uses of UV light are the disinfection of air and water used in dairy plant, and surface decontamination of food contact surfaces and packaging materials [11a].

C. Inactivation of Indigenous Enzymes [1a]

The effect of UV light on indigenous enzymes present in fruits juices such as pectinmethylesterase (PME), polyphenoloxidase (PPO) and peroxidase (POD) have been studied and partial inactivation of these enzymes is reported, hence the enzymatic changes in colour of the products cannot be completely controlled with the help of UV treatment.

Sources

1. *Hudaa Neetoo (Faculty of Agriculture, University of Mauritius, Réduit, Mauritius) and Haiqiang Chen (Department of Animal and Food Sciences, University of Delaware, Newark, Delaware, US), Alternative Food Processing Technologies.*

1a. *Nishant R Swami Hulle, Krishna Teja C and P Srinivasa Rao, Processing- Applications of UV Light, 2015, Food Marketing and Technology, magazine published by L. B. Associates Pvt Ltd.*

2. *Sizer CE, Balasubramaniam VM, 1999, New intervention processes for minimally processed juices. Food Technology 53: 64-67.*

3. *Wright JR, Sumner SS, Hackney CR, Pierson MD, Zoecklein BW, 2000, Efficacy of ultraviolet light for reducing Escherichia coli O157: H7 in unpasteurized apple cider. Journal of Food Protection 63: 563-567.*

3a. *Lori F Pivarnik and Randy worobo, Non-thermal or Alternative Food Processing Methods to Enhance Microbial Safety and Quality, Frequently Asked Questions, (USDA-NIFA #2011-68003-30005).*

4. *Farid MM, Chen XC, Dost Z, 2001, Ultraviolet sterilization of orange juice. In: Welti-Chanes J, Barbosa-Canovas GV, Aguilera JM (eds) Proceedings of the Eighth International Congress on Engineering and Food. Lancaster: Technomic Publishing, pp. 1567-1572.*

5. *Guerrero-Beltran J, Barbosa-Canovas GV, Welti-Chanes J, 2009, Ultraviolet-C light processing of grape, cranberry and grapefruit juices to inactivate Saccharomyces cerevisiae. Journal of Food Process Engineering 32: 916-932.*

6. *Lu IY, Stevens C, Yakubu P, Loretan PA, 1987, Gamma, electron beam and ultraviolet radiation on control of storage rots and quality of Walla Walla onions. Journal of Food Processing and Preservation 12: 53-59.*

7. *Gonzalez-Aguilar GA, Wang CY, Buta JG, Krizek DT, 2001, Use of UV-C irradiation to prevent decay and maintain postharvest quality of ripe "tommy atkins" mangoes. International Journal of Food Science and Technology 36: 767–773.*

8. *Rahman MS, (2007), Light energy in food preservation. In: Rahman MS (ed) Handbook of Food Preservation. Boca Raton, FL: CRC Press, pp. 751-757.*

9. *Soumitra Banerjee and Shanker Lal Shrivastava, Recent Trends in Milk Processing, Approches in poultry, dairy and Veterinary sciences 2015, Crimson Publishers, ISSN: 2576-9162.*

10. *Koutchma T, 2009, Advances in ultraviolet light technology for non-thermal processing of liquid foods. Food and Bioprocess Technology 2(2): 138-155.*

11. *Krishnamurthy K, Demirci A, Irudayaraj JM, 2007, Inactivation of Staphylococcus aureus in milk using flow-through pulsed UV-light treatment system. Journal of food Science 72(7): M233-M239.*

11a. *Nurcan Koca, Müge Urgu and Turkuaz Ecem Saatli (June 20th 2018). Ultraviolet Light Applications in Dairy Processing, Technological Approaches for Novel Applications in Dairy Processing, Nurcan Koca, IntechOpen, DOI: 10.5772/intechopen.74291.*

12. *Altic LC, Rowe M, Grant I 2007, UV light inactivation of Mycobacterium avium subsp. Paratuberculosis in milk as assessed by FASTP laque TB phage assay and culture. Applied and Environmental Microbiology 73(11): 3728-3733.*

Benefits, and Shortcomings of Alternative Non-thermal Processing Method-Ultraviolet Light [1]

Process considerations, benefits and shortcomings of alternative non-thermal processing method-Ultraviolet Light are summarized in table given on next page:

Sl. No.	Process Considerations	Benefits	Shortcomings	Examples of Applications
1	Transmissivity of product	Short processing time	Shadowing effect possible with complex surfaces	Bread
2	Geometric configuration of reactor	Minimal collateral effects on food	Has low penetration power	Cakes
3	Power	Low energy input	Ineffective against spores	Pizza
4	Wavelength	Suitable for high and low moisture food	Possible adverse sensory effects at high dosages	Fresh produce
5	Physical arrangement of source	Amenable for post package processing	Possible adverse chemical effects	Meats
6	Product shape/size	Medium cost	Reduced efficacy with high microbial load	Seafood
7	Product flow profile	—	Possible resistance in some microbes	Cheesed
8	Radiation pathlength	—	Reliability of equipment to be established	Food packages

Sources

1. *Hudaa Neetoo (Faculty of Agriculture, University of Mauritius, Réduit, Mauritius) and Haiqiang Chen (Department of Animal and Food Sciences, University of Delaware, Newark, Delaware, US), Alternative Food Processing Technologies.*

Effect of Ultraviolet Light on Inactivation of Micro-organism [1]

The effects of UV Light on inactivation of micro-organism are tabulated below:

Sl. No.	Products	Target Organism	Range Log Reduction (*)
1	Blueberries	*E. coli* O157: H7	1.3-5.2
		Salmonella	1.0
2	Strawberries	*E. coli* O157: H7	1.3-1.5
	Fresh	*Salmonella*	1.0
	Puree	—	—
3	Green Onions	*E. coli* O157: *H7*	1.3-1.5
		Salmonella	1.0

Note: *: Depends on wet or dry, inoculation site and type (calyx vs. skin vs. dip vs. spot).

Sources

1. *Lori F Pivarnik and Randy Worobo, Non-thermal or Alternative Food Processing Methods to Enhance Microbial Safety and Quality, Frequently Asked Questions, (USDA-NIFA #2011-68003-30005).*

Effect of Ultraviolet on the Sensory and Nutritional Quality of Foods [1]

Application of UV can have a variable impact on the organoleptic characteristics of food [2, 3, 4].

1. Deleterious effects manifested as surface discoloration, accelerated senescence or sprouting have also been shown to occur [5].
2. UV light has the potential to form lipid radicals, superoxide radicals, and hydrogen peroxide [6].
3. Fat-soluble vitamins and pigments can be affected by peroxides produced during extended UV light treatment [7]. Superoxide radicals can further induce carbohydrate cross-linking, protein cross-linking, protein fragmentation, peroxidation of unsaturated fatty acids [7] and loss of membrane fluidity function that could potentially have a deleterious effect on the textural quality of certain foods.
4. UV light can also degrade vitamins by photodegradation, especially vitamin A, B_2, and C. In high-moisture foods, water molecules absorb UV photons and produce OH- and H+ radicals, which in turn interact with other food components [7].
5. Furthermore, UV radiation may also denature proteins, enzymes, and aromatic amino acids, leading to changes in the composition of the food material. In addition, exposure to UV light can change the flavor profile in certain products.
6. It was reported [8] that during UV light exposure, the oxygen radicals formed induce the production of ozone, leading to the development of off-flavor notes in food products such as sour cream, whipped cream, butter, milk, mayonnaise, and dried vegetable soups. Therefore, UV light not only causes several undesirable chemical reactions, but can bring about deterioration in product quality.
7. Some sensory and nutritional changes in milk were reported. [9, 10].

Sources

1. *Hudaa Neetoo (Faculty of Agriculture, University of Mauritius, Réduit, Mauritius) and Haiqiang Chen (Department of Animal and Food Sciences, University of Delaware, Newark, Delaware, US), Alternative Food Processing Technologies.*
2. *Beaulieu JC, 2007, Effect of UV irradiation on cut cantaloupe: terpenoids and esters. Journal of Food Science 72: 272-81.*
3. *Matak KE, Sumner SS, Duncan SE et al., 2007, Effects of ultraviolet irradiation on chemical and sensory properties of goat milk. Journal of Dairy Science 90: 3178-86.*
4. *Rossitto PV, Cullor JS, Crook J, Parko J, Sechi P, Cenci-Goga BT, 2012, Effects of UV irradiation in a continuous turbulent flow UV reactor on microbiological and sensory characteristics of cow's milk. Journal of Food Protection 75: 2197-2207.*

5. *Sharma G, 2010, Ultraviolet light. In: Hui YH (ed) Handbook of Food Science, Technology and Engineering. Boca Raton, FL: CRC Press, 122: 1-12.*

6. *Kolakowska A (2003) Lipid oxidation in food systems. In: Sikorski ZE, Kolakowska A (eds) Chemical and Functional Properties of Food Lipids. New York: CRC Press, pp. 133-168..*

7. *Krishnamurthy K, Irudayaraj J, Demirci A, Yang W, (2009), UV pasteurization of food materials. In: Jun S, Irudayaraj JM (eds)Food Processing Operations Modeling: Design and Analysis. Boca Raton, FL: CRC Press, pp. 281-299.*

8. *Ohlsson T, Bengtsson N, (2002), Minimal processing of foods with non-thermal methods. In: Ohlsson T, Bengtsson N (eds) Minimal Processing Technologies in the Food Industry. Cambridge: Woodhead Publishing, pp. 34-60.*

9. *Soumitra Banerjee and Shanker Lal Shrivastava, Recent Trends in Milk Processing, Approches in poultry, dairy and Veterinary sciences 2015, Crimson Publishers, ISSN: 2576-9162.*

10. *Matak KE, Sumner SS, Duncan SE, Hovingh E, Worobo RW, et al. (2007) Effects of ultraviolet irradiation on chemical and sensory properties of goat milk. Journal of Dairy Science 90(7): 3178-3186.*

Chapter 10

Pulsed Light (PL) Treatment

The technique of pulsed light food processing was developed as a non-thermal food processing technique, that involves discharge of high voltage electric pulses (upto 70 Kilovolt/cm) into the food product placed between two electrodes for few seconds [1, 1a]. Pulsed light (PL) technology [1] is an innovative method for the decontamination and sterilization of foods using very high-power and very short-duration pulses of light emitted by inert gas flash lamps [2]. The pulsed light processing can be described as a sterilization or decontamination technique used mainly to inactivate surface micro-organisms on foods, packaging material and equipments [2a]. This process has shorter treatment times, with concomitantly higher throughput, rendering it very efficient. Exposure to PL can be regarded as a form of UV light, except that it is applied in pulses (rather than statically) and at higher intensity. It is one of the emerging technologies which are used for the replacement of traditional thermal pasteurization among non-thermal processes [2b]. It is a decontamination technique which aims at reducing the pests, spoilage micro-organisms and pathogens from food without much effect on its quality [2c].

Since these light pulses are of short duration and high energy, PL can achieve effective microbial inactivation without any major adverse effect on product properties [2]. This (PL) is also known as white light, and is used to ensure microbial decontamination on the surface of either food or packaging materials. The inactivation efficiency of pulsed light is directly proportional to intensity and number of pulses delivered [4].

Pulsed light (PL) is a non-thermal technology that uses short, intense pulses of white light which includes ultraviolet, infrared and visible light [1]. Ultraviolet light, broad spectrum white light and near infrared light can be used for pulsed light processing [1b]. Ultraviolet-C treatment for preserving food was discovered in 1930s [1c]. Pulsed light is an improved form of ultraviolet-C that is being given to foods. It is done with the help of Xenon lamps that can produce flashes several seconds. Pulsed light food processing technology uses intense and short-duration pulses of broad spectrum. Treatment of foods with **PL has been approved by the FDA (21 CFR179.41)** [3]. PL is the same as the light seen outside, but it is much more intense. **It is like a camera flash that is used to take pictures but far more intense. When this light is flashed on a food, it kills micro-organisms but has minimal impact on the food**. Short flashes of this intense light are used to prevent the temperature of the food from increasing [3].

This procedure, developed under the trade name PureBright (PurePulse Technologies, Inc., San Diego, California, USA), uses a light spectrum containing wavelengths ranging from ultraviolet to near-infrared. The light spectrum generated with the use of this equipment is similar to that of sunlight reaching the earth's surface [3a].

Sources

1. *Hudaa Neetoo (Faculty of Agriculture, University of Mauritius, Réduit, Mauritius) and Haiqiang Chen (Department of Animal and Food Sciences, University of Delaware, Newark, Delaware, US), Alternative Food Processing Technologies.*

1a. *Angersbach, A., Heinz, V. and Knorr, D., 2000, Effects of pulsed electric fields on cell membranes in real food systems. Innovative Food Science and Emerging Technologies 1(2): 135-149.*

1b. *Green, S., Basaran, N. and Swanson, B. 2005. Food preservation techniques. In Zenthen, P. and Bogh Sorenson, L. (Eds). washington, United States of America: Woodhead Publishing House, p. 365 CRC press.*

1c. *Artes, F. and Allende, A. 2005. Processing lines and alternative preservation techniques to prolong the shelf life of minimally processed leafy vegetables. European. Journal of Horticultural Science 70: 231- 245.*

2. *Palmieri L, Cacacea D, 2005, High intensity pulsed light technology. In: Sun DW (ed) Emerging Technologies for Food Processing. California: Elsevier Academic Press, pp. 279-306.*

2a. *Abida, J., Rayees, B. and Masoodi, F. A., Pulsed light technology: a novel method for food preservation, International Food Research Journal 21(3): 839-848, 2014.*

2b. *Heinz, V., Alvarez, I., Angersbach, A. and Knorr, D., 2002, Preservation of liquid foods by high intensity pulsed electric fields – basic concepts for process design. Trends in Food Science and Technology 12: 103–111.*

2c. *Bank, H. L., John, J., Schmehl, M. and Dracht, R. K., 1990, Bactericidal effectiveness of Modulated Ultravoilet Light. Applied and Environmental Microbiology 60: 3888-3889.*

3. *Lori F Pivarnik and Randy Worobo, Non-thermal or Alternative Food Processing Methods to Enhance Microbial Safety and Quality, Frequently Asked Questions, (USDA-NIFA #2011-68003-30005).*

3a. *Singh R. Paul, Professor of Food Engineering Department of Biological and Agricultural Engineering, Department of Food Science and Technology University of California*

Davis, FAO-2001, Technical Elements of New and Emerging Non-Thermal Food Technologies,.

4. *Harlin Kaur Saroya, Innovative non-thermal food processing technologies used by the food industry in the United States, MSc thesis, 2017, Western Kentucky University, Kentucky.*

Terminologies and Units [1]

This PL technology is always characterized by the use of following units:

1. **Fluence rate:** It is the energy received from the lamp by the sample per unit area per second. Its unit is Watt/meter2 (W/m^2).
2. **Fluence/Dose:** It is the energy received from the lamp by the sample per unit area during the treatment. Its unit is Joule/meter2 (J/m^2).
3 **Pulse width:** It is the time interval (fractions of seconds) during which energy is delivered.
4. **Exposure time:** It is the time period in seconds during which treatment is given.
5. **Peak power:** It is measured as pulse energy divided by the pulse duration. Its unit is Watt (W).
6. **Pulse-repetition-rate (PRR):** It is the number of pulses per second (Hertz [Hz]) or commonly expressed as pps (pulses per second).

Sources

1. *Abida, J., Rayees, B. and Masoodi, F. A., Pulsed light technology: a novel method for food preservation, International Food Research Journal 21(3): 839-848, 2014.*

Mechanism [1]

The lethality of Pulsed Light may be attributed to its rich broad spectrum ultraviolet content, its short duration, high peak power and the ability to regulate the pulse duration and frequency output of flash lamps [1a]. Several mechanisms have been proposed for the bactericidal action of PL. The UV portion of white light emitted at high power results in microbial inactivation through a photochemical, photothermal, and photophysical route [2]. The lethal effect is thus a combination of action of light pulses on certain cell constituents such as proteins and nucleic acids (photochemical effect), a transient temperature increase caused by heat dissipation of light pulses penetrating the product (photothermal effect) and morphological damage to cells caused by the constant disturbance originating from the high-energy pulses (photophysical effect).

Lethal Effect Due to Photochemical Mechanism [1a]

The lethal effect of pulsed light was explained by most of the authors on the basis of photochemical mechanism *e.g.*, the inactivation achieved by was associated with less than 1°C rise in temperature concluded that the lethality can be attributed to the photochemical action of the shorter ultraviolet wavelengths.

Lethal Effect due to Photo-thermal Mechanism [1a]

The lethal effect of Pulsed light can also be due to photothermal effect. Scientist proposed that with a fluence exceeding 0.5 Joule/cm^2, the disinfection is achieved through a rupture of bacteria during their temporary overheating caused by absorption of all ultraviolet light from a flash lamp.

Sources

1. *Hudaa Neetoo (Faculty of Agriculture, University of Mauritius, Réduit, Mauritius) and Haiqiang Chen (Department of Animal and Food Sciences, University of Delaware, Newark, Delaware, US), Alternative Food Processing Technologies.*

1a. *Abida, J., Rayees, B. and Masoodi, F. A., Pulsed light technology: a novel method for food preservation, International Food Research Journal 21(3): 839-848, 2014.*

2. *Krishnamurthy K, Tewari J, Irudayaraj J, Demirci A, 2010, Microscopic and spectroscopic evaluation of inactivation of Staphylococcus aureus by pulsed UV light and infrared heating. Food Bioprocess-ing Technology 3: 93-104.*

Pulsed Light Equipment and Working [1]

Pulsed light has a similar spectrum as sunlight, which has wavelengths ranging from ultraviolet 170 nm to infrared 26 nm. It peaks between 400-500 nm. As pulsed light is produced in short pulses that range from 1μs-0.1 s, they are approximately 20,000 times the intensity of sunlight at sea level [1a]. The Pulsed light equipment has the following parts:

i. Generator

The generation of pulsed light is done by charging the capacitors. To generate PL, capacitors release the power as a high voltage, high-current pulse of electricity to lamps filled with inert gas [2]. Inert gases such as xenon or krypton are used to ensure a high conversion efficiency of electrical to light energy. The electricity goes through the gas into the lamp allowing it to emit a very strong pulse of light. These rapid pulses of lights (generated at a speed of tens per second) are directed to the surface or packaging of the food. The current ionizes the gas, resulting in a broad-spectrum white light flash, spanning wavelengths of 200 nm (UV) to 1 mm (infrared). White light emitted comprises approximately 25 per cent UV, 45 per cent visible light, and 30 per cent infrared [3]. Light pulsed at a rate of 1-20 pulses/sec is directed over the material to be treated [4].

ii. Ammeter

Pulsed light systems are equipped with ammeters to measure the lamp current for each flash, which in turn determines the light intensity and spectrum.

iii. Detector

Silicon photodiode detectors are also used to measure the energy received from the lamp by the sample per unit area during the treatment ("**fluence**"). Current and fluence are tightly controlled, such that deviations from the set points cause the system to shut down automatically to avoid under processing. The flashing frequency can also be adjusted for the processing operation in question. The unit may also comprise one or more inert gas lamps that flash in unison or in tandem [5].

In the process, the energy that is exposed on the food or on the packaging materials is measured as fluence and is quoted in units of J cm-2. A number of synchronized mechanisms, for example, chemical modification to proteins, membranes, and nucleic acids lead to the inactivation of micro-organisms [1a].

iv. Automatic Stop

The system was programed to stop automatically if any of the above arrangements did not function. In general, the processing of food using this technology was done at 1-20 pulses with energy densities ranging between 0.1-50 Jcm^{-2}. The fuence and spectral density can be regulated for different applications. The entire process can be optimized for various food by adjusting the number of lamps frequency of flashes [6].

Working

The power unit is used to generate high voltage, following which resulting high-current pulses are employed in the lamp. AC power is first converted to high voltage DC power, which is then used to charge a capacitor. After the capacitor is charged to certain voltage, a high voltage switch discharges the capacitor into a lamp. The system is properly contained in order to protect personnel from high voltage. Cooling water is used to minimize heating of the treated product. The treatment unit contains one or more inert gas lamps. When a high current pulse is applied to the lamp, gas contained within the lamp emits an intense pulse of light. The frequency of flashing, number of lamps and flashing configuration depend on the treatment application [7].

The PureBright system involves illuminating the desired treatment area with 0.1 to 3 J/cm^2 per flash, with total accumulated fluences of 0.1 to 12 J/cm^2. Flashes are, in general, applied at a rate of 0.5 to 10 Hertz, for duration of several hundred microseconds [7].

Sources

1. *Hudaa Neetoo (Faculty of Agriculture, University of Mauritius, Réduit, Mauritius) and Haiqiang Chen (Department of Animal and Food Sciences, University of Delaware, Newark, Delaware, US), Alternative Food Processing Technologies.*

1a. *Harlin Kaur Saroya, Innovative non-thermal food processing technologies used by the food industry in the United States, MSc thesis, 2017, Western Kentucky University, Kentucky.*

2. *Dunn J, Clark W, Ott T, 1995, Pulsed light treatment of food and packaging. Food Technology 49: 95-98.*

3. *Dunn J, (1996), Pulsed light and pulsed electric field for foods and eggs. Poultry Science 75: 1133-1136.*

4. *Demirci A, Krishnamurthy K, 2011, Pulsed ultraviolet light. In: Zhang HQ, Barbosa-Canovas GV, Balasubramaniam VM, Dunne CP, Farkas DF, Yuan JTC (eds) Non-thermal Processing Technologies for Food. Chichester: Wiley-Blackwell, pp. 249-261.*

5. *Barbosa-Canovas GV, Pothakamury UR, Palou E, Swanson BG, 1998, Light-pulse sterilization of foods and packaging. In: Barbosa-Canovas GV, Pothakamury UR, Palou E, Swanson BG (eds) Non-thermal Preservation of Foods. New York: Marcel Dekker, pp.139-159.*

6. *Fellows, P. (2009). Food processing technologies: Principles and practices (3rd ed ed.). Boca Raton, FL: CRC Press.*

7. *Singh R. Paul, Professor of Food Engineering Department of Biological and Agricultural Engineering, Department of Food Science and Technology University of California Davis, FAO-2001, Technical Elements of New and Emerging Non-Thermal Food Technologies.*

Benefits, and Shortcomings of Alternative Non-thermal Processing Method–Pulsed UV Light [1]

Process considerations, benefits and shortcomings of alternative non-thermal processing method- Pulsed UV Light are summarized in following table:

Sl. No.	*Process Considerations*	*Benefits*	*Short-comings*	*Examples of Applications*
1	Transmissivity of product	Short processing time	Shadowing effect possible with complex surfaces	Bread
2	Geometric configuration of reactor	Minimal collateral effects on food	Has low penetration power	Cakes
3	Power	Low energy input	Ineffective against spores	Pizza
4	Wavelength	Suitable for high and low moisture food	Possible adverse sensory effects at high dosages	Fresh produce
5	Physical arrangement of source	Amenable for Post package processing	Possible adverse chemical effects	Meats
6	Product shape/size	Medium cost	Reduced efficacy with high microbial load	Seafood
7	Product flow profile	—	Possible resistance in somemicrobes	Cheeses
8	Radiation path length	—	Reliability of equipment to be established	Food packages

Effect of Non-thermal Method (Pulsed Light) on Inactivation of Micro-organism [2]

Sl.No.	*Products*	*Target Organism*	*Range Log Reduction (*, **)*
1	Blueberries	*E. coli* O157: H7	2.8-5.8
		Salmonella	3.1-5.9
2	Raspberries	*E. coli* O157: H7	3.9
		Salmonella	4.5
3	Green	*E. coli* O157: H7	0.9-4.9
	Onions	*Salmonella*	0.8-4.6

Note: *: Depends on treatment time, and inoculation cite (calyx vs. skin vs. stems vs. leaves);
**: Depends on wet or dry, inoculation site and type (calyx vs. skin vs. dip vs. spot).

Advantages and Disadvantages of Pulsed Light Treatment [3]

Advantages

The intensity of light, that lasts for only a second, is 20,000 times brighter than sunlight, but there is no thermal effect, so quality and nutrient content are retained [4]. The xenon-flash lamps used in pulsed light treatment are more eco-friendly than the mercury vapour lamps used in ultraviolet (UV) treatment [5]. Pulsed white light is not strictly a non-thermal, but the thermal action, due to its very short duration, it doesn't show much adverse effect on the nutrients [6].

Disadvantages

A possible problem of this preservation method is that folds or fissures in the food may protect microbes from being exposed to the pulsed light [4]. There might be some strains of micro-organisms which are resistant to the pulsed light treatment, for example *Listeria monocytogenes* [7]. This technique for decontamination of micro-organisms is useful mostly in case of liquid foods and surface of solid foods and hence limiting its application.

Sources

1. *Hudaa Neetoo (Faculty of Agriculture, University of Mauritius, Réduit, Mauritius) and Haiqiang Chen (Department of Animal and Food Sciences, University of Delaware, Newark, Delaware, US), Alternative Food Processing Technologies.*
2. *Lori F Pivarnik and Randy Worobo, Non-thermal or Alternative Food Processing Methods to Enhance Microbial Safety and Quality, Frequently Asked Questions, (USDA-NIFA #2011-68003-30005).*
3. *Abida, J., Rayees, B. and Masoodi, F. A., Pulsed light technology: a novel method for food preservation, International Food Research Journal 21(3): 839-848, 2014.*
4. *Brown, A. C., 2008, Understanding Food: Principles and Preparation. In Brown, A. C. (3rd Eds). Belmont, CA: Thompson/Wadsworth publishing, p. 47.*
5. *Gomez-Lopez, V. M., Ragaerta, P., Debevere, J. and Devlieghere, F., 2007, Pulsed light for food decont-amination: A review. Trends in Food Science and Technology 18: 464-473.*
6. *Ohlsson, T. and Bengtsson, N., 2002, Minimal processing technologies in the food industry. In Ohlsson, T. and Bengtsson, N. (Eds). Cambridge, England: Woodhead Publishing, p. 112. CRC Presss.*
7. *Caminiti, I. M., Noci, F., Munoz, A., Whyte, P., Morgan, D. J. and Cronin, D. A., 2011a, Impact of selected combinations of non-thermal processing technologies on the quality of an apple and cranberry juice blend. Journal of Food Chemistry 124: 124 (144) 13871389.*

Food Processing Applications of Pulsed Light [1, 1a]

The influence of surface topography on bacterial inactivation by PL has been shown to be variable. The following food processing applications are discussed:

A. Reduction of Micro-organism from Surfaces

a. Reduction of Micro-organism from Equipment (Steel) and Packing Materials

1. The study showed greater than 4-log reduction of *Listeria innocua* on rough stainless steel compared to smooth counterparts [2]. The application of pulsed light can reduce up to 9 logs of vegetative micro-organisms and more than 7 logs of bacterial spores on smooth, nonporous surfaces such as those of packaging materials. When the surfaces are more complex and porous, such as in case of food materials, microbial reduction is of the order of only 2 to 3 log cycles [2a].
2. On the other hand, study has reported [3] higher microbial reductions of pathogens on simple surfaces than complex surface aspects. Indeed, efficient light absorption depends on the distance through which light is passing, the thickness of the product, and the thickness of the package.
3. **Decontamination of packaging materials [1a]**-Paper-polyethylene was artificially inoculated with spores such as *Cladosporium herbarum, Aspergillus niger, Aspergillus repens* and *Aspergillus cinnamomeus* and then exposed to pulsed light with fluence ranging from 0.244 to 0.977 Joule/ cm^2. The highest level of inactivation of 2.7 log reduction was achieved. The colour of the spores affected their resistance to pulsed light. Different spores required different fluences for their inactivation [1b].
4. **PL traetment on food processing equipment [1a]**-Pulsed ultraviolet (UV) light treatment was studied for its applicability in decontamination of the stainless steel surface contacting meat from *Listeria monocytogenes* and *Escherichia coli* O157: H7.
   - ☆ A four lamp batch scale apparatus which generated 3 Joule/cm^2 with an input voltage of 3000 Volts was used.
   - ☆ The study was performed on stainless steel slicing knife. The type of meat product in contact with the treatment surface and the time between contamination and intense pulse treatment decide the effectiveness of the microbial inactivation.
   - ☆ When the knife surface was in contact with meat product containing lower fat and protein content and the time between contamination and treatment was 60 seconds, highest effectiveness of inactivation of 6.5 log colony forming units (CFU)/side of knife was achieved.

b. Reduction of Micro-organism from Products [1]

1. A study [11] was conducted to analyze the impact of pulsed light on the decontamination, quality, and bacterial attachment of fresh raspberries. The qualities of the raspberries were evaluated during the 10 days of storage at 4°C and the duration of pulsed light was 5s, 15s and 30s. On the 10th day, all the raspberries, which were treated by pulsed light maintained significantly lower total bacterial count, and total yeast, and mold count [11].

2. There are companies commercially using pulsed light for the UV germicidal component to disinfect food contact surfaces but not for produce or seafood applications [12].
3. Researchers [13] performed a research study in which the effects of pulsed light on selected wheat gluten properties were investigated. The gluten photo-reactivity was highly effected by hydration, which leads it to be higher in 1 per cent gluten suspension as compared to gluten powder. The pulsed light can be regarded as a non-thermal decontamination technology, as well as, a well-organized approach to steer the structure and functionality of protein rich ingredients.
4. **PL treatment to eggs for surface decontamination [1a]**-PL was used to decontaminate eggs from *Salmonella enteritidis*.
   - ✰ Eggs and egg-based products were frequently associated with salmonellosis outbreaks caused by *Salmonella enteritidis* [1c]. Eggs were treated with pulsed light of flashes of 2.1 Joule/cm^2 and 10.5 Joule/cm^2.
   - ✰ Exposure to 2.1 Joule/cm^2 leads to death of Salmonella cells (5 log colony-forming units (CFU) per egg shell) on the egg surface with slight increase in temperature.
   - ✰ Increase to 10.5 Joule/cm^2 did not cause penetration of Salmonella cells to the egg contents from the shell. No adverse effect on quality of egg albumin was observed. No effect on sensory and functional properties [1d].
5. The effect of product thickness on PL-induced microbial inactivation was demonstrated [4] and showed a reduction of 73.9 per cent of *C. sporogenes* spores in clover honey in samples of 2mm thickness compared to 14.2 per cent in samples that were four-fold thicker.

B. Pulsed Light Treatment to Chicken

1. Similarly, the efficacy of PL to inactivate Campylobacter on chicken breast was demonstrated [5] and also showed an inverse correlation between the microbial reduction and package thickness.
2. The study demonstrated [6] the optimal efficacy of pulsed UV light to decontaminate boneless chicken breasts for 15 sec when placed at a distance of 5 cm from the quartz window in the pulsed UV light chamber. It was reported that the treatment resulted in an approximately 2-log reduction in the population of *Salmonella typhimurium*, with minimal impact on the quality and color of the sample tested.
3. **PL treatment to chicken to decontaminate from food pathogens [1a]**: High-power pulsed light of 1,000 pulses, treatment duration 200 seconds and total ultraviolet light dose 5.4 Joule/cm^2 was found to reduce viability of *Salmonella typhimurium* and *Listeria monocytogenes* inoculated on the surface of chicken by 2-2.4 log10 (N/N0) colony-forming units (CFU)/ml [6a].

C. PL Treatment to Beer [1]

1. Moreover, the applicability of PL to pasteurize bottled beer has also been demonstrated previously [7].

D. Extension of Shelf Life of Food Products [1]

The use of PL to enhance the microbiological quality and extend the shelf life of foods has also been examined. For instance:

1. It was showed [8]that the application of PL led to a decrease in coliform and psychrotrophic bacterial counts in catfish fillets.
2. Similarly, it was showed [9] that PL can be used to inactivate mold spores and consequently extend the shelf life of foods, such as sliced white bread and cakes, from a few days to over 2 weeks.
3. The shelf life of cakes, pizza, and bagels was also extended to 11 days during storage at ambient temperature following PL treatment [10].
4. The use of UV irradiation in conjunction with MAP also extended the shelf life of beef steak by 10 days [11a].
5. Pulsed light treatment has been reported to be effective in extending the shelf life of foods such as bread, shrimp and meats. Pulsed light is effective in treating water, owing to the transparency of water and the fact that it permits the penetration of light [2a].

E. Continuous Flow Pulsed Light System for Bacterial Inactivation in Fruit Juices and Milk [1a]

i. Juices

Apple juice and orange juice were inoculated with gram positive (*Listeria innocua* 11288) and gram negative (*Escherichia coli* DH5-α) bacteria. These were then subjected to continuous pulsed light system. Xenon-flash lamp emitting light in the range of 100-110 nanometre (nm) and with the flashes at constant frequency of 3 Heartz and lasting for 360 microseconds (μs) was used. With treatment of 4 Joule/cm^2, the microbial load reductions in apple and orange juices for *Escherichia coli* were 4.00 and 2.90 Log-cycle respectively and for *Listeria innocua* were 2.98 and 0.93 Log-cycles respectively [11b].

ii. Milk

Continuous flow pulsed light technique was also used for inactivation of *Staphylococcus aureus* in milk and has a potential in treatment of milk pathogens [11c] carried out measurements of the temperature increase during the pulsed light treatment of milk in a continuous flow system. Milk temperature increased up to 38°C, depending on the residence time as well as the distance of the product from the light source. This temperature increase caused a fouling effect as well as possible changes in milk quality.

F. Decontamination of Food Powders using Pulsed Ultraviolet (UV) Light [1a]

Food powders were decontaminated using pulsed ultraviolet (UV) light. 58 Joule/cm^2 of pulsed light was required for *Saccharomyces cerevisiae* decontamination and reducing the microbial load by 7 log. It was found that the thermal effect rather than the Ultraviolet (UV) effect of pulsed light worked for decontamination of coloured powders [11d].

Sources

1. *Hudaa Neetoo (Faculty of Agriculture, University of Mauritius, Réduit, Mauritius) and Haiqiang Chen (Department of Animal and Food Sciences, University of Delaware, Newark, Delaware, US), Alternative Food Processing Technologies.*

1a. *Abida, J., Rayees, B. and Masoodi, F. A., Pulsed light technology: a novel method for food preservation, International Food Research Journal 21(3): 839-848, 2014.*

1b. *Turtoi, M. and Nicolau, A., 2007, Intense light pulse treatment as alternative method for mould spores destruction on paper–polyethylene packaging material. Journal of Food Engineering 83: 47-53.*

1c. *Braden, C. R., 2006, Salmonella enterica serotype Enteritidis and eggs: A national Epidemic in the United States. Clinical Infectious Diseases: an official publication of the Infectious Diseases Society of America 43: 512-517.*

1d. *Lasagabster, A., Arboleya, J. C. and Martinez, D. M., 2011, Pulsed light technology for surface decontamination of eggs, Impact on Salmonella inactivation and egg quality. Innovative Food Science and Emerging Technologies 12 (2): 124-128.*

2. *Uesugi AR, Woodling SE, Moraru CI, 2007, Inactivation kinetics and factors of variability in the pulsed light treatment of Listeria innocua cells. Journal of Food Protection 70: 2518-2525.*

2a. *Singh R. Paul, Professor of Food Engineering Department of Biological and Agricultural Engineering, Department of Food Science and Technology University of California Davis, FAO-2001, Technical Elements of New and Emerging Non-Thermal Food Technologies.*

3. *Dunn J, Clark W, Ott T, 1995, Pulsed light treatment of food and packaging. Food Technology 49: 95-98.*

4. *Hillegas SL, Demirci A, 2003, Inactivation of Clostridium sporogenes in clover honey by pulsed UV light treatment. Agricultural Engineering International 5: 1-7.*

5. *Haughton PN, Lyng JG, Morgan DJ, Cronin DA, Fanning S, Whyte P, 2011, Efficacy of high-intensity pulsed light for the microbiological decontamination of chicken, associated packaging, and contact surfaces. Food borne Pathogens and Disease 8: 109-117.*

6. *Keklik NM, Demirci A, Puri VM, 2010, Decontamination of unpackaged and vaccum-packaged boneless chicken breast with pulsed ultraviolet light. Poultry Science 89: 570-581.*

6a. *Paskeviciute, E., Buchovec, I. and Luksiene, Z., 2011, High Power Pulsed light for decontamination of chicken from food pathogens: A study on antimicrobial efficiency and organoleptic properties. Journal of Food Safety 31: 61-68.*

7. *Palmieri L, Cacacea D, 2005, High intensity pulsed light technology. In: Sun DW (ed) Emerging Technologies for Food Processing. California: Elsevier Academic Press, pp. 279-306.*

8. *Figueroa-Garcia JE, Silva JL, Kim T, Boeger J, Cover R, 2002, Use of pulsed-light to treat raw channel catfish fillets. Journal of the Mississippi Academy of Science 47: 114-120.*

9. *Rice J, 1994 Sterilizing with light and electrical impulses: technological alternative to hydrogen peroxide, heat and irradiation. Food Processing* **7**: *66.*

10. *Ohlsson T, Bengtsson N, 2002, Minimal processing of foods with non-thermal methods. In: Ohlsson T, Bengtsson N (eds) Minimal Processing Technologies in the Food Industry. Cambridge: Woodhead Publishing, pp. 34-60.*

11. *Xu, W., and Wu, C. (2016). The impact of pulsed light on decontamination, quality, and bacterial attachment of fresh raspberries. Food Microbiology, 57, 135-143.*

11a. *Djenane D, Sanchez A, Beltran J, Roncales P, 2003, Extension of the shelf life of beef steaks packaged in a MAP by treatment with rosemary and display under UV-free lighting. Meat Science 64: 417-426.*

11b. *Pataro, G., Munoz, A., Palgan, I., Noci, F., Ferrari, G. and Lyng, J. G., 2011, Bacterial inactivation in fruit juices using a continuous flow Pulsed Light (PL) system. Food Research International 44: 1642-1648.*

11c. *Krishnamurthy, K., Demirci, A. and Irudayaraj, J. M., 2007, Inactivation of Staphylococcus aureus in milk using flow-through Pulsed ultraviolet light treatment System. Journal of Food Science 72: 233-239.*

11d. *Fine, F. and Gervais, P. 2004. Efficiency of pulsed ultravoilet light for microbial decontamination of food powders. Journal of Food Protection 67: 787-792.*

12. *Lori F Pivarnik and Randy Worobo, Non-thermal or Alternative Food Processing Methods to Enhance Microbial Safety and Quality, Frequently Asked Questions, (USDA-NIFA #2011-68003-30005).*

13. *Panozzo, A., Manzocco, L., Lippe, G., and Nicoli, M. C. (2016). Effect of pulsed light on structure and immune-reactivity of gluten. Food Chemistry, 194, 366-372.*

Pulsed Light Field Technology in Combination with other Non-thermal Processing Technologies [1]

PL technology in combination with other non-thermal processing technologies was experimented on a blend of apple and cranberry juice and the efficacy of the combination of technologies was determined on the basis of quality attributes such as odour and flavour.

- ✰ The non-thermal technologies likes ultra-violet light (5.3 Joule/cm^2), high intensity pulsed light (3.3 Joule/cm^2), pulsed electric field processing

(34 kilovolt/cm, 18 Hertz, 93 microsecond) and manothermosonication (5 bar, 43°C, 750 Watt, 20 kilohertz) were studied. A blend of apple and cranberry juice in the ratio of 90: 10 (volume/volume) was taken and stored at -20°C pre- and post processing. A light based technology (ultraviolet or high intensity light pulses) in combination with pulsed electric field or manothermosonication was applied. The followings were concluded:

- Ultraviolet and pulsed electric field combination or high intensity light pulses and pulsed electric field combination was found to maintain product quality better than any combination with manothermosonication under the applied conditions which lead to adverse effects on product quality [2].
- High intensity light pulses in combination with pulsed electric field were used to inactivate *Escherichia coli* in apple juice. The optimum combination was obtained and sensory analysis was performed as well for quality effects. The optimum combination did not affect the quality [3].
- High intensity light pulses in combination with thermosonication were experimented for inactivation of *Escherichia coli* in orange juice could be developed as hurdle technology.

Sources

1. *Abida, J., Rayees, B. and Masoodi, F. A., Pulsed light technology: a novel method for food preservation, International Food Research Journal 21(3): 839-848, 2014.*
2. *Caminiti, I. M., Noci, F., Munoz, A., Whyte, P., Morgan, D. J. and Cronin, D. A., 2011a, Impact of selected combinations of non-thermal processing technologies on the quality of an apple and cranberry juice blend. Journal of Food Chemistry 124: 124 (144) 13871389.*
3. *Caminiti, I. M., Palgan, I., Noci, F., Munoz, A., Whyte, P., Cronin, D. A., Morgan, D. J. and Lyng, J. G., 2011b, The effect of pulsed electric fields (PEF) in combination with high intensity light pulses (HILP) on Escherischia coli and quality attributes in apple juice. Innovative Food Science and Emerging Technologies 12: 118-123.*

Mitigation of Allergen using Pulsed Ultraviolet Light [1]

Peanut allergy is a severe Immunoglobin E mediated reactions with food. Peanut allergy can be prevented by complete avoidance. But pulsed ultraviolet light treatment of peanut extracts and peanut butter showed to deactivate Ara h 2, the most potent allergic protein present in peanut. Protein band intensity for Ara h1, Ara h2, Ara h3 reduced at energy levels ranging from 111.6-223.2 Joule/cm^2 [2]. Pulsed ultraviolet light treatment of soy extracts has found to decrease the levels of soy allergens (glycinin, β-conglycinin). But clinical data are needed for development of products from such treated soy extracts [3].

Sources

1. *Abida, J., Rayees, B. and Masoodi, F. A., Pulsed light technology: a novel method for food preservation, International Food Research Journal 21(3): 839-848, 2014.*

2. *Yang, W. W., Mwakatage, N. R., Goodrich-Schneider, R., Krishnamurthy, K. and Rababah, T. M., 2011, Mitigation of major peanut allergens by pulsed ultraviolet light. Food and Bioprocess Technology 87: 1-11.*

3. *Yang, W. W., Chung, S. Y., Ajayi, O., Krishnamurthy, K., Konan, K. and Goodrich-Schneider, R., 2010, Use of pulsed ultraviolet light to reduce allergenic potency of soybean extracts. International Journal of Food Engineering 6 (3): 1-12.*

Effect of Pulsed Light on the Sensory and Nutritional Quality of Foods [1]

Pulsed light treatment of food applied at high doses can negatively affect the sensory and nutritional quality of foods.

1. It is evidenced by the manifestation of surface discoloration, loss of vitamins and other pigment compounds, as well as textural changes [2].
2. A sensory evaluation of PL-treated vegetables was conducted [3] and demonstrated that the overall effect depended on the type of product under consideration.
   a. White cabbage treated by PL acquired a transient "plastic" off-odor; however, the taste scores, where 1 = fresh, 3 = acceptable, 5 = spoiled, were 2.1 to 2.7 and overall visual quality (OVQ) scores, where 1= excellent, 5 = fair, 9 = extremely poor, were 1.0 to 4.9. These scores were not significantly different from untreated counterparts with taste and OVQ scores of 1.6 to 2.5.
   b. Treated iceberg lettuce received higher scores in terms of odor (2.3-3.7), where 1=no off-odor and 5 = severe off-odor, taste (1.8-2.6) and OVQ (1.5-6.9) compared to their untreated controls. In particular, the extent of leaf edge browning in treated iceberg lettuce (2.9-3.3), where 1=none and 5 =severe, was lower than the untreated product (3.6-3.7) after 3 and 5 days of chilled storage at 7°C. This clearly points to the usefulness of pulsed UV in enhancing the organoleptic characteristics [2].

Sources

1. *Hudaa Neetoo (Faculty of Agriculture, University of Mauritius, Réduit, Mauritius) and Haiqiang Chen (Department of Animal and Food Sciences, University of Delaware, Newark, Delaware, US), Alternative Food Processing Technologies.*

2. *Demirci A, Krishnamurthy K, 2011, Pulsed ultraviolet light. In: Zhang HQ, Barbosa-Canovas GV, Balasubramaniam VM, Dunne CP, Farkas DF, Yuan JTC (eds) Non-thermal Processing Technologies for Food. Chichester: Wiley-Blackwell, pp. 249-261.*

3. *Gomez-Lopez VM, Devileghere F, Bonduelle V, Debevere J, 2005, Intense light pulses decontamination of minimally processed vegetables and their shelf-life. International Journal of Food Microbiologyogy 103: 79-89.*

Chapter 11

Cold Plasma Technology

Quality, both in terms of objective product characteristics and subjective consumer perception, is an essential factor for the success of any food product [1, 2, 3,]. Thermal processing of food has been in use for more than two centuries and is still the major food processing technique used in the food industries [4]. The use of severe heat leads to undesirable effects such as change in color, texture, loss of nutrients *etc.*, motivating researchers to explore non-thermal alternatives for food processing. Cold Plasma (CP) is one of the non-thermal technologies which has shown significant potential in food processing [1].

In general, plasma is categorized in thermal or non-thermal. The thermal plasma needs high pressure and temperature with dense electrons, whereas, non-thermal plasma is generated at atmospheric temperatures that range between 30°C-60°C with low energy [1a]. A **second surface technology** is **cold plasma**. Plasma is a high-energy gas that is created when an electrical current is passed through a gas. A well-known application of this principle is fluorescent lighting [1b]. CP technology has proven very effective as an alternative tool for food decontamination and shelf-life extension. CP's inactivation of pathogenic and spoilage micro-organisms could result in minimally processed, safe food products with extended shelf-life [2]. The impact of CP on food quality is very crucial for its acceptance as an alternative food processing technology. Due to the non-thermal nature, CP treatments have shown no or minimal impacts on the physical, chemical, nutritional and sensory attributes of various products. CP is very versatile technology with great potential to benefit areas of the food industry and plasma parameters need to be optimised for the application (*e.g.* Food vs. Contact surfaces), It is closer to being a commercial surface decontamination technology and have great potential for food decontamination and shelf life extension but more work is needed to understand the quality impacts [5]. The negative impacts and limitations posed by CP technology for food products were also reported.

The limited studies on interactions of CP species with food components at the molecular level offers future research opportunities. It also highlights the need for optimization studies to mitigate the negative impacts on visual, chemical, nutritional and functional properties of food products. The design versatility, non-thermal, economical and environmentally friendly nature of CP offers unique advantages over traditional processing technologies. However, CP processing is still in its nascent form and needs further research to reach its potential [2].

Until recently, plasmas could only be created at relatively high temperatures in a vacuum and the use of plasma in the agro-sector and the packaging industry, was therefore impractical. In the last few years technological breakthroughs have made it possible to produce cold plasma under atmospheric conditions. Thus, it is now possible to apply the plasma technology in the food industry [1b]. Earlier cold plasma was used for sterilization of sensitive materials and now it is being used in food industries as a novel technology. Plasma sterilization provides high efficacy, preservation and does not introduce toxicity to the medium [1]. The cold plasma techniques are preservation treatments that are effective at ambient temperatures, there by minimum thermal effects on nutritional and sensory quality parameters of food with no chemical residues. Low temperature, atmospheric pressure plasmas have been shown to possess very effective germicidal characteristics. Their relatively simple and inexpensive designs, as well as their non-toxic nature, give them the potential to replace conventional sterilization methods in the near future [1b].

For years cold plasma processing has been viewed as useful for microbial inactivation. However, this process is not effective for in vitro model food systems for inactivation of microbes or enzymes which are present in intact tissues, as it is a surface phenomenon. Cold plasma technology is also used to inactivate endogenous enzymes which are responsible for browning reactions particularly polyphenoloxidase and peroxidases. Plasma technology is considered as modern non conventional technique which is used for the preparation of modified starches, altering its physical and chemical properties. The microbial destruction on different food substrates like fruits, meat products, cheese *etc.* was reported as overall application of cold plasma. CP is also used to alter the germination rate of seeds. It is an eco-friendly process which is used in the preservation of food and other potential applications as an alternative to common techniques [1].

Sources

1. *Rohit Thirumdas, Chaitanya Sarangapani and Uday S. Annapure, Cold Plasma- A novel Non-Thermal Technology for Food Processing, Foods biophysics, 10(1), November 2014.*

1a. *Harlin Kaur Saroya, Innovative non-thermal food processing technologies used by the food industry in the United States, MSc thesis, 2017, Western Kentucky University, Kentucky.*

1b . *Ranjan Saha, New and Emerging Non-Thermal Technologies in Food Preservation, April 2015.*

2. *Shashi K. Pankaj, Zifan Wan and Kevin M. Keener, Effects of Cold Plasma on Food Quality: A Review, 2018, MDPI, Basel, Switzerland, Foods, 7, 4, 2018.*

3. *Grunert, K. G. Food quality and safety: Consumer perception and demand. Eur. Rev. Agric. Econ. 2005, 32, 369-391.*

4. *Pankaj, S. K. Thermal processing of food. In Advances in Food Biotechnology; JohnWiley and Sons Ltd., Hoboken, NJ, USA, 2015; pp. 681-692.*

5. *Danny Bayliss, New Technologies Specialist, Cold plasma: Applications and future perspectives within the food and drink industry, Danny.bayliss@ campdenbri.co.uk.*

States of Matter and Plasma [1]

Matter on earth exists mostly in three distinct phases (gas, liquid and solid) but when universe is considered as fourth state of matter which abundantly exists. So, Plasma is hence referred to as the fourth state of matter, next to solids, liquids and gases. The term **'Plasma'** was first employed to define this fourth state of matter which is partially or wholly ionized state of gas and discovered plasma oscillations in ionized gas [2]. The change of phase from solid to liquid and further to gas occurs as we increase the energy input likewise increasing the energy input beyond a certain level in gas state causes ionization of molecules which yields the plasma state [3]. It was reported [4] that plasma can be obtained either in low temperature, non-equilibrium glow discharge or high temperature, equilibrium thermal plasma.

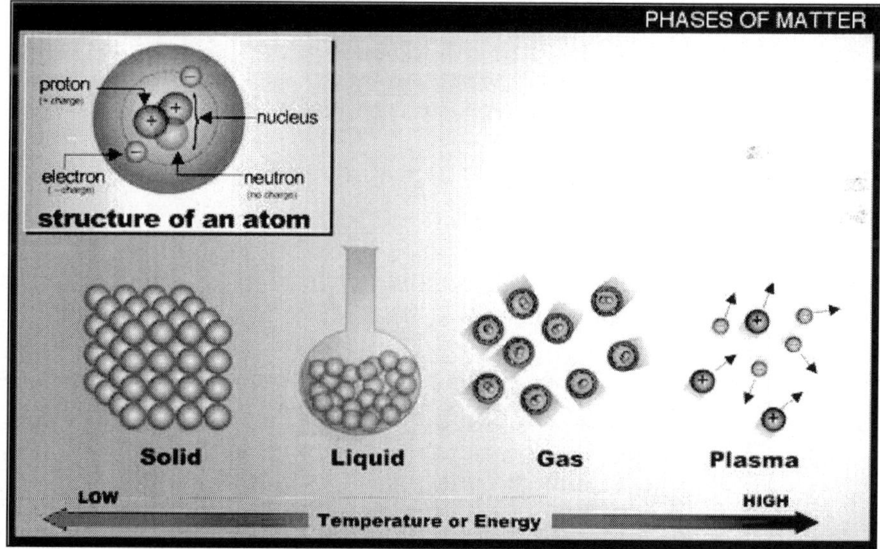

Cold Plasma (5).

Cold Plasma system provide a source of highly reactive and biologically relevant chemical species close to room temperature using a gas and electricity. Cold plasma is a dry treatment so would also be an ideal disinfection tool for processors of low water activity products [5].

Sources

1. *Rohit Thirumdas, Chaitanya Sarangapani and Uday S. Annapure, Cold Plasma: A novel Non-Thermal Technology for Food Processing, Foods biophysics, 10(1), November 2014.*
2. *I. Langmuir, Proceedings of the National Academy of Sciences of the U. S. A 14 627, 1928.*
3. *Q. Z. Luo, N. Angelo, R. L. Merlino, Dept of phys and Astron Retrieved 20, 1998.*
4. *R. P. D'Agostino, P. Favia, C. Oehr, M. R. Wertheimerv, Plasma Process. Polym. 2, 7, 2005.*
5. *Danny Bayliss, New Technologies Specialist, Cold plasma: Applications and future perspectives within the food and drink industry, Danny.bayliss@campdenbri.co.uk.*

Plasma Physics and Sources [1]

Plasma is a quasi-neutral ionized gas state composed of ions, free electrons, atoms and molecules in their fundamental or excited states with a net neutral charge [2]. Plasma can be classified in different groups depending upon the thermal equilibrium and pressure conditions.

A. Classification Based on the Thermal Equilibrium

Plasma can be equilibrium and plasma can be classified as thermal and low-temperature plasma.

1. **Thermal Plasma**-In this plasma, all species exist in a thermodynamic equilibrium (*e.g.*, arc plasma, electron temperature heavier species temperature 10,000 K).
2. **Low-temperature plasma**- In this class, the temperatures of all species are the same in localized areas in the plasma. This plasma can be further subdivided into
   a. **Thermal plasma** (quasi-equilibrium plasma), which are in a local thermal equilibrium state,
   b. **Non-thermal plasma** (non-equilibrium plasma), where species are in thermal non-equilibrium (*e.g.*, glow discharges; electron temperature ≈ 10,000–100,000 K, heavier species temperature ≈ 300–1000 K) [3]. The non-thermal plasma, where electrons and heavier species are in thermal non-equilibrium, is referred to as **cold plasma**.

B. Classification Based on the Pressure Conditions

Depending on the pressure conditions, plasma can also be classified in following three groups:

1. High-pressure.
2. Atmospheric pressure: In atmospheric pressure plasma, plasma is generated at the normal atmospheric pressure, eliminating the need for cost intensive reaction chambers to maintain pressure.
3. Low-pressure plasma.

Physical Quality

CP is generally considered as a tool for surface treatments. In fact, CP has been used by the polymer and packaging industries for decades for surface modification and functionalization of polymers [4]. However, during processing of food, food products might be placed in strong electric field and are subjected to numerous reactive gas species that could affect physical quality attributes such as color and texture.

Sources

1. *Shashi K. Pankaj, ZifanWan and Kevin M. Keener, Effects of Cold Plasma on Food Quality: A Review, 2018, MDPI, Basel, Switzerland, Foods, 7, 4, 2018.*
2. *Pankaj, S. K.; Misra, N. N.; Cullen, P. J. Kinetics of tomato peroxidase inactivation by atmospheric pressure cold plasma based on dielectric barrier discharge. Innov. Food Sci. Emerg. Technol, 2013, 19, 153–157.*
3. *Misra, N. N.; Pankaj, S. K.; Segat, A.; Ishikawa, K. Cold plasma interactions with enzymes in foods and model systems. Trends Food Sci. Technol, 2016, 55, 39-47.*
4. *Pankaj, S. K.; Keener, K. M. Cold plasma: Background, applications and current trends. Curr. Opin. Food Sci, 2017, 16, 49-52.*

Plasma Chemistry: Process and Categories [1]

Plasma chemistry is a complex science involving numerous species in a myriad of chemical reactions occurring in different time scales [1b]. For example, air plasma involves over 75 different chemical species in almost 500 chemical reactions, making it more complex to understand their interaction with food components [1a]. Different plasma parameters significantly effect the plasma chemistry. In plasma processing ionization is always considered as first important element followed by other factors like reaction rate, rate constants, the mean free path, the electron energy distribution [2]. Plasma chemical process can be divided into following two categories based on reactions:

1. **Homogeneous gas-phase reaction** - for example generation of N_3 from N_2.
2. **Heterogeneous reaction** where plasma comes in contact with the solid or liquid medium. The heterogeneous reaction is further classified into the following three sub categories:
   a. In **the first sub category,** material is removed from the surface termed as etching or ablation,
   b. **In the second sub category,** material is added on the solid surface in the form of thin film observed during plasma polymerization by a process called plasma enhanced chemical vapour deposition.
   c. **In third sub category** there is no material added or removed but the substrate surface is modified physically and chemically during exposure to plasma [3].

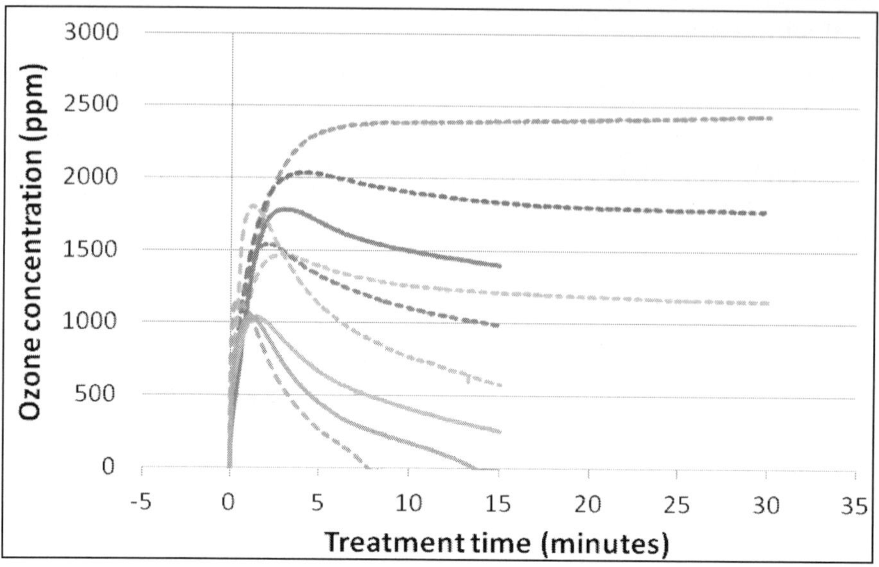

Plasma Generation [1, 1a]

Cold atmospheric plasma can be developed by applying electromagnetic waves on gas at lower pressure; thereby, leading to a thermo-dynamic non-equilibrium nature. **In the earlier days, plasma treatments were performed in vacuum chambers.** However, with the advancements in technologies, researchers have discovered new techniques, such as, **atmospheric pressure plasma**. This technology helps in reducing cost and increases the treatment speed [1c]. Different types of possible mechanisms that the gas used for plasma generation may act on the substrate during the plasma processing. Plasma can be produced by subjecting a gas to an electric field (between two electrodes), either of constant (direct current field) or alternating amplitude (usually high frequency field) to ionize the gases. Plasma state can be attained by the application of energy in several forms including, thermal, optical (UV light), electric or magnetic fields, X-ray electromagnetic radiation magnetic fields and radioactive (gamma radiation) or microwave frequencies, which increase the kinetic energy of the electrons resulting in increased number of collisions in the gas forming plasma products like electrons, ions, radicals and radiation of varying wavelengths including that in the UV ranges. However, electric or electromagnetic fields are widely used for CP generation [4]. The various approaches used for plasma generation includes the **corona discharge, dielectric barrier discharges (DBD), radiofrequency plasma (RFP) and the gliding arc discharge** [4a]. Cold plasmas, including low pressure DC and RF discharges (silent discharges), discharges from fluorescent (*e.g.*, neon) illuminating tubes, DBDs may be found both at low pressure or atmospheric pressure [5].

The versatility of CP generation sources offers unique designs which are compatible with current food industry equipment. With regards to food processing, **dielectric barrier discharge and jet plasma are most commonly used** [1, 1a].

i. Dielectric Barrier Discharge (DBD)

The dielectric barrier discharges (DBDs), historically also called **'silent discharges'**. They also operate at approximately atmospheric pressure (typically 0.1-1 atm). An A. C. voltage with amplitude of 1-100 kV and a frequency of a few Hz to MHz is applied to the discharge, and a dielectric layer (made of glass, quartz, ceramic material or polymers) is again placed between the electrodes. When a potential difference is applied between cathode and anode, a continuous current will flow through the discharge; giving rise to direct current (D. C.) glow discharge. These devices consist of two metal electrodes, where at least one of these electrodes is covered with a dielectric barrier. Dielectric barriers act as a stabilizing material, avoiding any arc transition, and help in creating a large number of micro-discharges for homogeneous treatments. Capacitively coupled (CC) radio-frequency (RF) discharges are produced when alternating voltage is applied between the two electrodes, so that each electrode will act alternately as the cathode and anode. The frequencies generally used for these alternating voltages are typically in the radiofrequency (RF) range (1 kHz–103 MHz; with a most common value of 13.56 MHz). Non-thermal gas discharges at atmospheric pressure are of interest for the food industries.

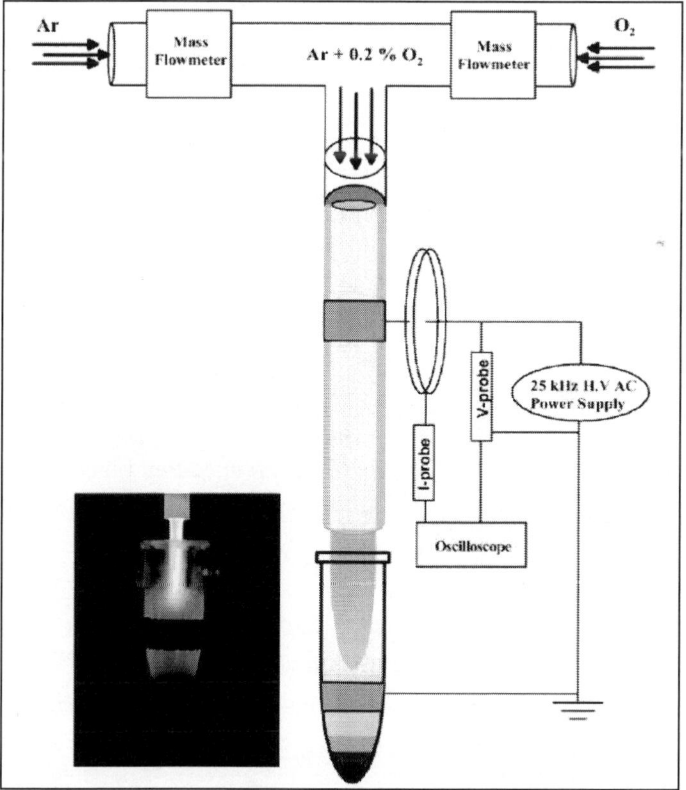

Atmospheric Pressure Argon Plasma Experimental Setup (Right) and a Picture of Plasma Equipment Treating Wastewater Sample (Left) [1e].

ii. Plasma Jet Devices

These devices consist of two concentric electrodes, where the inner electrode is typically connected to a radiofrequency power at high frequency causing ionization of the working gas, which exits the nozzle and gives a 'jet-like' appearance.

iii. Gliding Arc Discharge [1c]

For the generation of the cold plasma, also referred to as gliding arc, two components (**power supply and plasma emitter**) are required. The equipment requires an AC power of 60Hz and an operating output rate of 60mA at 15kV. The equipment operates in the open with electrodes of 3mm thickness, which are attached at the top and bottom along with stainless steel lugs. The rods are secured at a definite distance of 3mm from the plasma creating point and are bent at an angle of 45°. The minimum area between the gas inlet and plasma creating point is 8mm [1d].

Sources

1. *Rohit Thirumdas, Chaitanya Sarangapani and Uday S. Annapure, Cold Plasma: A novel Non-Thermal Technology for Food Processing, Foods biophysics, 10(1), November 2014.*

1a. *Shashi K. Pankaj, ZifanWan and Kevin M. Keener, Effects of Cold Plasma on Food Quality: A Review, 2018, MDPI, Basel, Switzerland, Foods, 7, 4, 2018.*

1b. *Pankaj, S. K.; Keener, K. M. Cold plasma applications in food packaging. In Reference Module in Food Science; Elsevier: Amsterdam, The Netherlands, 2017.*

1c. *Harlin Kaur Saroya, Innovative non-thermal food processing technologies used by the food industry in the United States, MSc thesis, 2017, Western Kentucky University, Kentucky.*

1d. *Dey, A., Rasan, P., Choudhury, A., Singh, J., Maisnam, D., and Rasane, P., Cold plasma processing: A review. Journal of Chemical and Pharmaceutical Science, 9(4), 2980-2984, 2016.*

1e. *Mohamed, A., Al-Shariff, S., Ouf, S., and Benghanem, M., Atmospheric pressure plasma jet for bacterial decontamination and property improvement of fruit and vegetable processing wastewater. Journal of Physics D: Applied Physic, 49(19), 2016.*

2. *A. Fridman, Plasma chemistry, Cambridge Univ Press, New York, 2008.*

3. *R. P. D'Agostino, P. Favia, C. Oehr, M. R. Wertheimerv, Plasma Process. Polym. 2, 7, 2005.*

4. *Pankaj, S. K., Keener, K. M. Cold plasma applications in food packaging. In Reference Module in Food Science; Elsevier: Amsterdam, The Netherlands, 2017.*

4a. *H. Conrads, H. Schmidt, Plasma Sources Sci. Technol. 9, 441, 2000.*

5. *O. Kylian, J. Benedikt, L. Sirghi et al., Plasma Process Polym. 6, 255, 2009.*

Plasma System

The following designs help in understanding the working of plasma system:

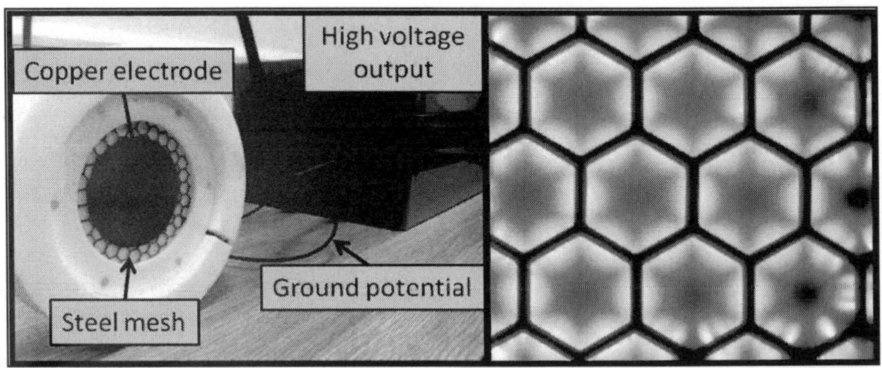

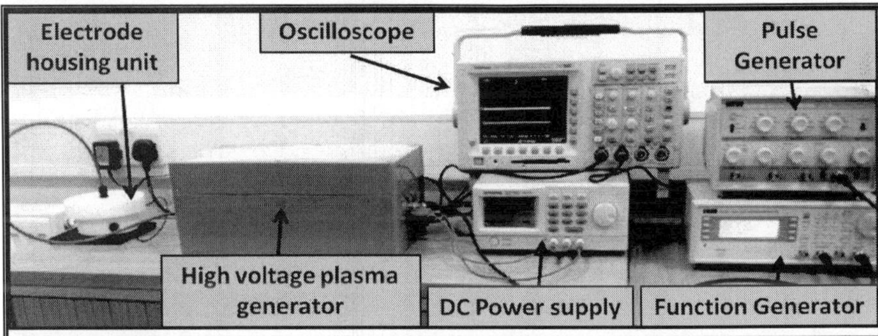

Plasma System (Designed by Dr James Walsh from the University of Liverpool).

Advantages/Benefits of Cold Plasma System [1]

This system has following benefits:

1. Operates in open air (no need for expensive gases)
2. Works at Low power (~60 W average input power)
3. Scalable to larger surface areas and geometries.

Sources

1. *Danny Bayliss, New Technologies Specialist, Cold plasma: Applications and future perspectives within the food and drink industry, Danny. Bayliss @ campdenbri.co.uk.*

Limitations/Disadvantages and Toxicology of Plasma Treatment [1]

There are following limitations of plasma processing:

1. Increase in oxidation of lipids
2. Reduction in colour
3. Decrease in firmness of fruits
4. Increase in acidity.

The details of some limitation/disadvantages are given below:

1. In decontamination of the surface of walnuts and peanuts using cold plasma technology, the main important problem encountered is increase in peroxide value by 20 per cent in walnuts at higher power and time of treatment. Similar kind of results was observed for strawberries treated with cold atmospheric plasma [2].
2. The discolouration and wilting effect of spinach leaves was observed after treating in atmospheric non equilibrium plasma for 5 min [3].
3. Another main disadvantage of this technology, it is not possible to use for inactivation of endogenous enzymes which are present intact in the whole fruits because plasma effect is a surface phenomenon.
4. The application of direct plasma causes decrease in firmness of fruits [2].
5. It is reported [4] that with the application of plasma there is decrease in pH of milk after 10 min treatment.

Till now no investigation has been carried out on the formation of any toxic compounds after the application of cold plasma in food products.

Sources

1. *Rohit Thirumdas, Chaitanya Sarangapani and Uday S. Annapure, Cold Plasma: A novel Non-Thermal Technology for Food Processing, Foods biophysics, 10(1), November 2014.*
2. *N. N. Misra, S. Patil, T. Moiseev, P. Bourke, J. P. Mosnier, K. M. Keener, P. J. Cullen, J. Food Eng. 125, 131, 2014.*

3. *P. A. Klockow and K. M. Keener (ASABE) Meeting Presentation 1, 2008.*

4. *H. J. Kim, H. I. Yong, S. Park, K. Kim, W. Choe, C. Jo, Food Cont.47, 451, 2015.*

Effect of Cold Plasma/Functionality of Plasma [1, 1a]

Cold Plasma has effect on organoleptic properties (colour, Texture), Microbial cells and chemical properties of different foods. The details related to the aforesaid effects are given below:

A. Physical/Organoleptic Properties (Colour, Texture) of Foods

1. Color

Color of food products is an important attribute which has a direct effect on consumer perception. The color of food products are mostly due to presence of pigments and chemical reactions. Any undesirable change in the color of food products due to processing technique will be a big obstacle for its acceptability.

The changes in the color could be due to the partial degradation of pigments such as chlorophyll and anthocyanin, as reported in some studies [1, 8]. Overall, these results demonstrate that CP processing has a minimal effect on the color of food products at lower treatment times. The product type (whole or cut, solid or liquid), plasma treatment parameters (input voltage, time, power, working gas) and storage conditions are some of the critical factors affecting the color. Varying effects of CP treatments on the color of fresh fruits and vegetables have been reported depending on the severity of treatment conditions. Some exmples(fruits, foods) are given below:

a. Impact on Colour of Fruits

1. There is no significant loss of color after CP treatments of strawberry, apples, kiwifruit, cherry tomatoes, lettuce and carrots [2, 3, 4].
2. Some researchers reported minor changes after the CP treatments [5, 6].
3. In some cases, such as blueberry, researchers [7, 8] reported loss of color at higher treatment times.
4. Similarly, total color difference after CP treatment of fruit juices were also found minimal and not perceptible by naked eyes [9].

b. Impact on Colour of Foods

CP processing was also reported to lead to certain desirable effects on the color of a few food products.

1. Researchers [10] have reported an increase in the brightness and whiteness index of brown rice after plasma treatment.
2. In another study, researchers [11] have used CP in the manufacturing of pork jerky without adding sodium nitrite. They used specific plasma processing parameters to achieve similar redness/color in the pork jerky without using any chemical nitrite additive.

3. **Effect of CP on anthocyanins and color in pomegranate juice [7a]-** It was investigated [12] and the results of the treatment were observed at following three working conditions:
   a. Treatment Time,
   b. Treated Juice Volume
   c. Gas Flow.

The highest anthocyanin stability was achieved at 3 min, 5cm^3 volume and 0.75 dm^3/min gas flow. The use of this technology also helped in the increment of anthocyanin content from 21 per cent to 35 per cent and color change.

2. Texture

Many of the reported studies suggest the retention of texture of food products after CP processing. In case of fresh fruits and vegetables, no significant difference was observed after CP treatment of strawberry, apple, melons and cherry tomatoes-[4, 5].

1. However, a decrease in firmness was reported after CP treatment of blueberries [7, 8]. The softening of the blueberries was attributed to the mechanical damage due to the high air-flow rates of the plasma jet and the temperature rise during the treatment. In another study on CP treatments of strawberry in modified atmosphere packaging, the firmness retention was found to be better in a high-oxygen environment than a nitrogen-rich environment [5]. This study demonstrates that plasma gas is an important factor influencing the firmness of treated products.
2. CP treatment of grains and legumes resulted in a decrease in hardness and chewiness [7]. These groups also reported a decrease in soaking/cooking time for the plasma-treated products, which was deemed desirable for the industries.

3. Physical Properties of Defatted Soybean Meal (DSM)-based Edible Film [7a]

The effect of cold plasma (using different plasma forming gases) on the physical properties of defatted soybean meal (DSM)-based edible film (DSM film) was investigated [7b]. The effects of cold plasma treated DSM film in packing smoked salmon, for storage stability, was studied and it was determined the storage stability increases. The DSM film was created by molding film-forming solution, which was prepared by high pressure homogenization. Next, the ink adhesion test was performed that helped to understand, O_2^-, N_2–, He–, and Ar-CPT (cold plasma treatment) increased the stretchability of DSM film, N_2– and Ar-CPTs helped in increasing the lightness, and O_2^- and air-CPTs improved the film's printability. The CPT increased the tensile and moisture properties of the DSM film.

Sources

1. *Shashi K. Pankaj, ZifanWan and Kevin M. Keener, Effects of Cold Plasma on Food Quality: A Review, 2018, MDPI, Basel, Switzerland, Foods, 7, 4, 2018.*

1a. *Rohit Thirumdas, Chaitanya Sarangapani and Uday S. Annapure, Cold Plasma: A novel Non-Thermal Technology for Food Processing, Foods biophysics, 10(1), November 2014.*

2. *Bermúdez-Aguirre, D., Wemlinger, E., Pedrow, P., Barbosa-Cánovas, G., Garcia-Perez, M. Effect of atmospheric pressure cold plasma (APCP) on the inactivation of Escherichia coli in fresh produce. Food Control, 2013, 34, 149-157.*

3. *Ramazzina, I., Berardinelli, A., Rizzi, F., Tappi, S., Ragni, L., Sacchetti, G.; Rocculi, P. Effect of cold plasma treatment on physico-chemical parameters and antioxidant activity of minimally processed kiwifruit. Postharvest Biol. Technol.2015, 107, 55-65.*

4. *Ziuzina, D., Misra, N., Cullen, P.; Keener, K., Mosnier, J., Vilaro, I., Gaston, E.; Bourke, P. Demonstrating the potential of industrial scale in-package atmospheric cold plasma for decontamination of cherry tomatoes. Plasma Med, 2016, 6, 397-412.*

5. *Misra, N., Moiseev, T., Patil, S., Pankaj, S., Bourke, P., Mosnier, J., Keener, K., Cullen, P. Cold plasma in modified atmospheres for postharvest treatment of strawberries. Food Bioprocess Technol, 2014, 7, 3045-3054.*

6. *Wang, R., Nian, W., Wu, H., Feng, H., Zhang, K., Zhang, J., Zhu, W., Becker, K., Fang, J. Atmospheric pressure cold plasma treatment of contaminated fresh fruit and vegetable slices: Inactivation and physiochemical properties evaluation. Eur. Phys. J. D, 2012, 66, 1-7.*

7. *Sarangapani, C., O'Toole, G., Cullen, P., Bourke, P. Atmospheric cold plasma dissipation efficiency of agrochemicals on blueberries. Innov. Food Sci. Emerg. Technol, 2017, 44, 235-241.*

7a. *Harlin Kaur Saroya, Innovative non-thermal food processing technologies used by the food industry in the United States, MSc thesis, 2017, Western Kentucky University, Kentucky.*

7b. *Oh, Y. A, Roh, S. H, and Min, S., Cold plasma treatments for improvement of the applicability of defatted soybean meal-based edible film in food packaging. Food Hydrocolloids, 58, 150-159, 2016.*

8. *Lacombe, A.; Niemira, B. A., Gurtler, J. B., Fan, X., Sites, J., Boyd, G., Chen, H. Atmospheric cold plasma inactivation of aerobic micro-organisms on blueberries and effects on quality attributes. Food Microbial, 2015, 46, 479-484.*

9. *Kovacevic, D. B., Putnik, P., Dragovic-Uzelac, V., Pedisic, S., Jambrak, A. R., Herceg, Z. Effects of cold atmospheric gas phase plasma on anthocyanins and color in pomegranate juice. Food Chem, 2016, 190, 317-323.*

10. *Thirumdas, R., Saragapani, C., Ajinkya, M., Deshmukh, R., Annapure, U. Influence of low pressure cold plasma on cooking and textural properties of brown rice. Innov. Food Sci. Emerg. Technol, 2016, 37, 53-60.*

11. *Yong, H. I., Lee, S. H., Kim, S. Y., Park, S., Park, J., Choe, W., Jo, C. Color development, physio-chemical properties, and microbiological safety of pork jerky processed with atmospheric pressure plasma. Innov. Food Sci. Emerg. Technol, 2017.*

12. *Bursac-Kovacevi, Putnik, P., Dragovic-Uzelac, V., Pedisiæ, S., Rezek-Jambrak, A., and Herceg, Z., Effects of cold atmospheric gas phase plasma on anthocyanins and color in pomegranate juice. Food Chemistry, 190, 317-323, 2016.*

B. Effect on Chemical Properties of Fruits/Foods [1]

The Plasma reactive species are considered to be the major factor for all the observed changes in the chemical quality attributes of the treated products. The plasma reactive species are largely dependent on the gas used for plasma generation, making this one of the most critical factors for chemical changes. The details related to the effect of CP on chemical changes in food products are given below:

a. Effect on Macronutrients of Food [1]

CP has following significant effects on the macro nutrients (Proteins, Fat/ Lipids, Carbohydrates *etc.*) of food:

i. Fat/Lipids

Fat/Lipid is important nutrient of food. Lipid oxidation is a major concern for foods(specially muscle foods), which could lead to undesirable changes in the color, taste, odor and shelf life. Lipid oxidation, a complex process involves free radical chain mechanisms forming fatty acyl peroxides or other oxidation products [2]. Thiobarbituric acid reactive substance (TBARS) and Peroxide value (PV) are commonly used to measure fat oxidation. Since CP is often considered as an advanced oxidation process, it is essential to analyze its influence on the lipids present in the muscle foods. The available studies on the effects of CP on lipids in different food products are very limited. However, based on the reported studies, treatment time and plasma gas could be considered as critical factors affecting lipid oxidation [1]. The following references help in understanding the effect of CP on fat/Lipid:

- ✰ No significant effect on lipid oxidation were observed after CP treatment in fresh and frozen pork [3], beef jerky [4] and raw pork [5] but an increase in lipid oxidation in fresh pork and beef after treating it for an extended time period of 10 min., is reported [6].
- ✰ CP treatment led to a significant lipid oxidation in fresh mackerel fillets [7]. Researchers observed an increase in PV from 6.89 to 37.57 meq. active oxygen/kg lipids and dienes from 1.42 to 5.56 m mol of hydroperoxides/ kg lipid after plasma treatment at 80 kV for 5 min. A decrease in oleic acid (C18: 1, n-9) and eicosapentaenoic acid (C20: 5, n-3) after plasma treatments was also repoeted [7].
- ✰ The hydrogen plasma was used for the manufacturing of partially hydrogenated soybean oil without any trans-fatty acid [8]. CP technology has shown unique advantages over the current hydrogenation processes as it can be performed at room temperature, under atmospheric pressure without any catalyst.

ii. Protein

Some studies on effect of CP on proteins, are available. The different factors (The type of protein/enzyme, type of plasma, reactive gas, processing parameters, sample volume and enzyme media) play an important role on the protein denaturation and enzyme inactivation by CP. The mechanisms of protein denaturation by CP could be due to the interaction of plasma reactive species with amino acids [9] and secondary structure due to loss of α helix and β sheet [10].

- ☆ The effects of CP in muscle protein were studied in fresh mackerel and a decrease in immobilized water located in the protein-dense myofibrillar network was observed [7].
- ☆ Another study on wheat flour also suggested changes in protein structure due to oxidation of sulfhydryl groups and formation of disulphide bonds, affecting its structural and functional properties.

iii. Carbohydrates

Carbohydrates play an important role in defining and maintaining the quality of different food products. The summary of the effect of CP on carbohydrates is given below:

- ☆ CP treatment of cashew apple juice resulted in the degradation of all the reducing sugars, such as fructose and glucose and non-reducing sucrose [11]. An increase in sucrose content after long exposure to CP, was also reported which they attributed to the degradation of the oligosaccharides with a high degree of polymerization.
- ☆ A decrease in the fructose, increase in the sucrose and degradation of oligosaccharides with a high degree of polymerization was also reported after CP treatment of prebiotic orange juice [12].
- ☆ The effect of CP on polysaccharides has been mainly focused on starch in legume and grain products. An increase in the water uptake rate in black gram was reported [13], which was due to the surface etching and increase in water binding sites due to fragmentation of starch and protein by plasma reactive species. Overall, it could be concluded that CP treatment lead to de-polymerization and cross-linking of starch affecting its structural, functional and rheological properties.

iv. Vitamins

Some vitamins, such as riboflavin (B2), pyridoxine (B6) and biotin, are usually stable, others, such as Thiamin (B1) and vitamins A, C and E, are relatively labile. Most of the reported studies on CP treatment of food products have only focused on vitamin C (ascorbic acid) stability. Most of the studies on CP treatment of whole fruits and vegetables have reported no significant reduction in ascorbic acid content after plasma treatment.

- ☆ Up to 4 per cent reduction in ascorbic acid content was observed after plasma treatment of cut fruits and vegetables [14].

- ✰ The reduction in ascorbic acid was also observed after CP treatment of orange juice [15] and cashew apple juice [11].
- ✰ The degradation of ascorbic acid could be attributed to the reaction with ozone and other oxidizing plasma species during the processing. Sample type (whole/cut), processing time and plasma gas were critical factor for ascorbic acid degradation.

b. Effect on pH/Acidity and Antioxidant Activity [1]

CP has following effects on the pH/Acidity and antioxidant activity:

i. pH and Acidity [1]

pH and acidity are important and a closely regulated quality attribute in most of the processed food products. Any drastic change in these properties could lead to an undesirable impact on the taste, texture and shelf life of the food. In the case of fresh fruits and vegetables, there are significant variations due to differences in cultivation practices, varietal differences, environmental parameters *etc.*

- ✰ Several studies revealed that CP treatment effects the pH of food products [12, 16]. The pH and acidity changes after plasma treatment were mostly attributed to the interaction of plasma reactive gases with the moisture present in the food products.
- ✰ **In solid food product-** Plasma species reacts with the surface water, forming acidic compounds only on the surface.

  In liquid products- Effects were more pronounced.
- ✰ Many researchers also reported no pH effect of CP treatments in food products with buffering capacity [15].

ii. Antioxidant Activity [1]

The antioxidant activity is not a direct quality attribute used in the food industries but it is a close indicator of various polyphenols, flavonoids and flavanols present in the food products.

- ✰ The antioxidant effects of phenolic compounds could be due to their redox properties, which include possible mechanisms such as free-radical scavenging activity, transition metal-chelating activity and singlet-oxygen quenching capacity [17].
- ✰ The antioxidant activities in food are generally analyzed using 3-ethylbenzothiazoline-6-sulfonic acid (ABTS) radical scavenging activity, oxygen radical absorbance capacity (ORAC), 2, 2-diphenyl-1-picrylhydrazyl (DPPH) scavenging activity, and ferric reducing ability of plasma (FRAP) assay.
- ✰ No significant changes in the antioxidant capacity after CP treatment were reported in radish sprouts, kiwifruits, red chicory and onion powder [18, 19].

c. Effect on Functionality of Wheat Flour [1a, 1b]

In another study on CP treatment of wheat flour, researchers [1] reported an increase in the peak integral, elastic modulus, viscous modulus and dough strength. Cold atmospheric pressure was applied to wheat flour at a low level of cold plasma (powers of 0.19W/cm^2 and 0.43 W/cm^2) at different time periods. The process did alter the molecular weight distribution of wheat protein polymers. In the process of treatment, the cold plasma oxidized free fatty acids and phospholipids. Ultimately, all these qualities enhanced the functionality of wheat flour leading to the formation of strong dough [1b].

Sources

1. *Shashi K. Pankaj, ZifanWan and Kevin M. Keener, Effects of Cold Plasma on Food Quality: A Review, 2018, MDPI, Basel, Switzerland, Foods, 7, 4, 2018.*

1a. *Harlin Kaur Saroya, Innovative non thermal food processing technologies used by the food industry in the United States, MSc thesis, 2017, Western Kentucky University, Kentucky.*

1b. *Bahrami, N., Bayliss, D., Chope, G., Penson, S., Perehinec, T., and Fisk, I., Cold plasma: A new technology to modify wheat flour functionality. Food Chemistry, 202, 247-253, 2016.*

2. *D. Ladikos, V. Lougovois, Lipid oxidation in muscle foods: A review. Food Chem. 35, 295-314, 1990.*

3. *S. Choi, P. Puligundla, C. Mok, Corona discharge plasma jet for inactivation of Escherichia coli O157: H7 and Listeria monocytogenes on inoculated pork and its impact on meat quality attributes. Ann. Microbiol., 66, 685-694, 2016.*

4. *J. S. Kim, E. J. Lee, E. H. Choi, Y. J. Kim, Inactivation of Staphylococcus aureus on the beef jerky by radio-frequency atmospheric pressure plasma discharge treatment. Innov. Food Sci. Emerg. Technol., 22, 124-130, 2014.*

5. *N. Ulbin-Figlewicz, A. Jarmoluk, Effect of low-pressure plasma treatment on the color and oxidative stability of raw pork during refrigerated storage. Food Sci. Technol. Int. 22, 313-324, 2016.*

6. *D. D. Jayasena, H. J. Kim, H. I. Yong, S. Park, K. Kim, W. Jo. C. Choe, Flexible thin-layer dielectric barrier discharge plasma treatment of pork butt and beef loin: Effects on pathogen inactivation and meat-quality attributes. Food Microbiol., 46, 51-57, 2015.*

7. *I. Albertos, A. Martin-Diana, P. Cullen, B. Tiwari, S. Ojha, P. Bourke, C. Álvarez, D. Rico, Effects of dielectric barrier discharge (DBD) generated plasma on microbial reduction and quality parameters of fresh mackerel (Scomber scombrus) fillets. Innov. Food Sci. Emerg. Technol., 44, 117-122, 2017.*

8. *X. V. Yepez, K. M. Keener, High-voltage atmospheric cold plasma (HVACP) hydrogenation of soybean oil without trans-fatty acids. Innov. Food Sci. Emerg. Technol. 38, 169-174, 2016.*

9. *Y. Li, A. Kojtari, G. Friedman, A. D. Brooks, A. Ji. H. Fridman, Decomposition of L-valine under non-thermal dielectric barrier discharge plasma. J. Phys. Chem. B, 118, 1612-1620, 2014.*

10. A. Segat, N. N. Misra, P. J. Cullen, N. Innocente, Effect of atmospheric pressure cold plasma (ACP) on activity and structure of alkaline phosphatase. Food Bioprod. Process., 98, 181-188, 2016.

11. O. Rodríguez, W. F. Gomes, S. Rodrigues F. A. Fernandes, Effect of indirect cold plasma treatment on cashew apple juice (Anacardium occidentale L.). LWT-Food Sci. Technol. 84, 457-463, 2017.

12. F. D. L. Almeida, R. S. Cavalcante, P. J. Cullen, J. M. Frias, P. Bourke, F. A. Fernandes, S. Rodrigues, Effects of atmospheric cold plasma and ozone on prebiotic orange juice. Innov. Food Sci. Emerg. Technol., 32, 127-135, 2015.

13. C. Sarangapani, R. Y. Devi, R. Thirumdas, A. M. Trimukhe, R. R. Deshmukh, U. S. Annapure, Physico-chemical properties of low-pressure plasma treated black gram. LWT-Food Sci. Technol., 79, 102-110, 2017.

14. R. Wang, W. Nian, H. Wu, H Feng, K. Zhang, J. Zhang, W. Zhu, K, Becker, J. Fang, Atmospheric-pressure cold plasma treatment of contaminated fresh fruit and vegetable slices: Inactivation and physiochemical properties evaluation. Eur. Phys. J. D, 66, 1-7, 2012.

15. L. Xu, A. L. Garner, B. Tao, K. M. Keener, Microbial inactivation and quality changes in orange juice treated by high voltage atmospheric cold plasma. Food Bioprocess Technol., 10, 1-14, 2017.

16. K. H. Lee, H. J. Kim, K. S. Woo, C. Jo, J. K. Kim, S. H. Kim, H. Y. Park, S. K. Oh, W. H. Kim, Evaluation of cold plasma treatments for improved microbial and physico-chemical qualities of brown rice. LWT-Food Sci. Technol., 73, 442-447, 2016.

17. B. Shan, Y. Z. Cai, M. Sun, H. Corke, Antioxidant capacity of 26 spice extracts and characterization of their phenolic constituents. J. Agric. Food Chem. 53, 7749-7759, 2005.

18. Pasquali, F.; Stratakos, A. C.; Koidis, A.; Berardinelli, A.; Cevoli, C.; Ragni, L.; Mancusi, R.; Manfreda, G.; Trevisani, M. Atmospheric cold plasma process for vegetable leaf decontamination: A feasibility study on radicchio (red chicory, Cichorium intybus L.). Food Control, 60, 552-59, 2016.

19. Kim, J. E.; Oh, Y. J.; Won, M. Y.; Lee, K.-S.; Min, S. C. Microbial decontamination of onion powder using microwave-powered cold plasma treatments. Food Microbial., 62, 112-123, 2017.

C. Effect on Microbiological Cells and its Inactivation [1]

Cold Plasma effects the microbial cells and its inactivation.

i. Effect on Microbiological Cells

The plasma treatment effects the microbial cells is mainly due to the plasma ions and cell interactions. The reactive species in plasma have been widely associated with the direct oxidative effects on the outer surface of microbial cells. The plasma effect is highly dependent on the presence of water and highest effect in moist organism where as lowest effect in dry organism was observed [2]. The potential application of plasma in inactivation is based on the fact that plasma reactive species damage the deoxyribonucleic acid (DNA) in the chromosomes. The micro-organisms are more susceptible to singlet state oxygen leading to deformation of cells [3]. The lipid bi-layer of microbial cell is more susceptible to atomic oxygen as the reactivity of atomic oxygen is much higher than the molecular oxygen, which can degrade lipids, proteins and DNA of cells. The damage of the double bonds in lipid bi-layer cause impaired transportation of molecules in and out of cell. The bombardment of reactive oxygen species (ROS) on the surface of bacterial cell also disrupts the membrane lipids [4-6].

During application of plasma, micro-organisms are exposed to an intense bombardment by the radicals most likely provoking surface lesions that the living cell cannot repair sufficiently fast this process is termed **"etching"**. Plasma etching is based on the interaction of relative energetic ions and activated species with the molecules of the substrate.

Microbial Inactivation using Cold Plasma [1]

Sl. No.	*Micro-organism*	*Substrate*	*Plasma Source*	*Exposure Time*	*Log Reduction*
1	*S. enteritidis* (01)	Table Egg [7]	RBD prototype	90 min	4–5
2	*L. monocytogenes*	Sliced cheese [8]	APP	120 s	s > 8]
		Bacon [13]	APP	90 s	2–3
3	*E.coli*	Chicken skin [9]	Pulsed gas plasma discharge	24 s	up to 8
		Red apples [10]	CP Gliding arc	3 min	up to 3.7
		Almonds [11]	Dielectric discharge	30 s	1.8–5
		Apple juice [10]	Needle/plate system	-	7
		Mango and melon skin [12]	APJ	5 s	up to 3
4	*S. typhimurium*	Lettuce	-	15 min	up to 2.72
		Strawberry [14]	CAP	15 min	1.76
		Tomatoes [15]	DBD	300 s	3.8
5	*S. aureus*	Orange juice [16]	DBD	25 s	>5
6	*A. hydrophila*	Lettuce [17]	COP	5 min	5
7	*A. parasiticus, Penicillium*	Grains and cereals [18]	SF6 CP	15 min	3

Note: DBD: Dielctric barrier discharge; APJ: Atmospheric plasma jet; CP: cold plasma; COP: Cold oxygen plasma; APP: Atmospheric pressure plasma.

Sources

1. *Rohit Thirumdas, Chaitanya Sarangapani and Uday S. Annapure, Cold Plasma: A novel Non-Thermal Technology for Food Processing, Foods biophysics, 10(1), November 2014.*

2. *D. Dobrynin, G. Fridman, G. Friedman, A. Fridman doi: 10.1088/1367-2630/11/11/115020.*
3. *M. F. A. Aziz, E. A. Mahmoudb, G. M. Elaragi, J. Stored Prod. Res., 2014.*
4. *R. Mogul, A. A. Bolshakov, S. L. Chan, R. M. Stevens et al., Biotechnol. Prog. 19, 776, 2003.*
5. *Z. B. Guzel-Seydim, A. K. Greene, A. C. Seydim, Leben-Wissenund-Technol. 37, 453, 2004.*
6. *T. C. Montie, K. Kelly-Wintenberg, J. R. Roth, Plasma Sci. IEEETrans. 28, 41, 2002.*
7. *N. Rowan, S. Espie, J. Harrower, J. Anderson, L. Marsili, S. MacGregor, J. Food Protect.70, 2805, 2007.*
8. *H. J. Lee, H. Jung, W. Choe, J. S. Ham et al., Food Microbiol. 28, 1468, 2011.*
9. *E. Noriega, G. Shama, A. Laca, M. Diaz et al., Food Microbiol. 28, 1293, 2011.*
10. *J. Montenegro, R. Ruan, H. Ma, P. Chen, J. Food Sci. 67, 646, 2002.*
11. *S. B. Deng, R. Ruan, C. K. Mok, G. W. Huang et al., J. Food Sci. 72, 62, 2007, 12-S. Perni, G. Shama, M. G. Kong, J. Food Protect. 71, 1619, 2008.*
13. *B. Kim, H. Yun, S. Jung, Y. Jung et al., Food Microbiol.28, 9, 2011.*
14. *E. Fernandeza, B. Noriega, A. Thompson, Food Microbiol. 33, 24, 2012.*
15. *D. Ziuzina, S. Patil, P. J. Cullen, K. M. Keener, P. Bourke, Food Microbiol. 42, 109, 2014.*
16. *X. M. Shi, G. J. Zhang, X. L. Wu, Y. X. Li et al., IEEE Trans. Plasma Sci. 39, 1591, 2011.*
17. *I. K. Jahid, N. Han, S. Ha, Food Res. Int.55, 181, 2014.*
18. *M. Selcuk, L. Oksuz, P. Basaran, Bioresour. Technol. 99, 5104, 2008.*

ii. Effect on Inactivation of Micro-organism [1]

Cold plasma is a unique technology which is responsible for microbial destruction and surface modification of substrate as conventional preservatives techniques as some detrimental effects on nutritional quality [1]. An effective inactivation of a wide range of micro-organisms including spores [2-4] and viruses [5] by plasma treatment, is reported. It was reported [2] that cold plasma technology preferred as alternative source for surface sterilization and disinfection process which can act on both vegetative cells and spores with shorter periods of time. The chemical composition of plasma contains free radicals, highly reactive species and radiations are often generated in varying range from UV to visible. It is believed that the role of different constituent depends on the gas and operating pressure. The destruction of microbial DNA by UV irradiation, volatilization of compounds from spore, so-called **"etching"** of the spore surface by adsorption is because of reactive species like free radicals [6]. The efficiency of microbial inactivation depends upon the surface of treating produce, plasma device, gas composition, and mode of exposure.

CP can be successfully employed for microbial destruction on fresh products to increase shelf life. Some examples are given below:

1. It is reported [7] in recent investigation that decrease in total mesophilic count was 12-85 per cent, yeast and mould count by 44-95 per cent in cold plasma treated strawberries.
2. Raw milk was treated for destruction of *E. coli* using low temperature plasma [8]. Researchers observed 54 per cent reduction of *E. coli* after the treatment of 3 min.
3. Application of air plasma and SF6 on nuts (pea nuts, hazel nuts and pistachio) found that 50 per cent reduction in total aflatoxins using air plasma and 20 per cent reduction using SF6 [9].
4. It was observed [10] that treating with plasma for 15 min can achieve 2.72, 1.76 and 0.94 log reductions of *S. typhimurium* on lettuce, strawberry and potato respectively.
5. Produces like potatoes, strawberries took more time for complete destruction of microbes due to grooves and uneven surfaces [10, 11].
6. Researchers [12] applied the gas plasma for the decontamination of *Salmonella enteritidis* and *Listeria monocytogenes* from table eggs. The maximum reduction was observed to be 4 and 5 log reduction of *Salmonella enteritidis* and *Listeria monocytogenes* respectively [12]. A plasma treatment of 2 minutes, (this means each part on surface was exposed to 1s of plasma) inactivated approximately 90 per cent of the salmonella on carrot slices and 80 per cent on cucumber and pear slices [13].
7. It was reported [14] that cold plasma was suitable to inactivate milk from *E. coli*, but changes in biochemical composition was observed during the study, *i.e.* changes in aldehyde composition were reported.

Sources

1. *Rohit Thirumdas, Chaitanya Sarangapani and Uday S. Annapure, Cold Plasma: A novel Non-Thermal Technology for Food Processing, Foods biophysics, 10(1), November 2014.*

2. *J. Feichtinger, A. Schulz, M. Walker, U. Schumacher, Surf Coat Technol.174, 564, 2003.*

3. *K. Kelly-Wintenberg, A. Hodge, T. Montie, L. Deleanu et al., J. Vacuum Sci. Technol. 17, 1539, 1999.*

4. *K. Lee, K. Paek, W. T. Ju, Y. Lee, J. Microbiol.44, 269, 2006.*

5. *O. Terrier, B. Essere, M. Yver, M. Barthelemy et al., J. Clin. Virol.45, 119, 2009.*

6. *N. Philip, B. Saoudi, M. C. Crevier, M. Moisan, J. Barbeau, J. Pelletier, IEEE Trans. Plasma Sci. 30, 1429, 2002.*

7. *N. N. Misra, S. Patil, T. Moiseev, P. Bourke, J. P. Mosnier, K. M. Keener, P. J. Cullen, J. Food Eng. 125, 131, 2014.*

8. C. Gurol, F. Y. Ekinci, N. Aslan, M. Korachi, Int. J. Food Microbiol.15, 157, 2012.

9. P. Basaran, N. Basaran, L. Oksuz, Food Microbiol. 25, 626, 2008.

10. E. Fernandeza, B. Noriega, A. Thompson, Food Microbiol. 33, 24, 2012.

11. D. Ziuzina, S. Patil, P. J. Cullen, K. M. Keener, P. Bourke, Food Microbiol. 42, 109, 2014.

12. L. Vannini, C. L. Montanari, A. Berardinelli et al., World's poultry science association. Netherlands, 2009.

13. Harlin Kaur Saroya, Innovative non-thermal food processing technologies used by the food industry in the United States, MSc thesis, 2017, Western Kentucky University, Kentucky.

14. Korachi M, Ozen F, Aslan N, Vannini L, Guerzoni ME, et al. (2015) Biochemical changes to milk following treatment by a novel, cold atmospheric plasma system. International Dairy Journal 42: 64-69.

D. Effect on Endogenous Enzymes [1]

The plasma can also be applied to simpler biological compounds, like enzymes [2]. Fruits and vegetables mostly spoil due to the enzymatic browning. The endogenous enzymes particularly poly-phenoloxidase and peroxidase are the major causes for enzymatic browning as they oxidize phenols at the expense of H_2O_2 leading to off flavours [3].

1. Different methods used to prevent the enzymatic browning are heating, blanching, commercial sterilization [4, 5]. Enzymes are inactivated through oxidation reactions mediated by free radicals and atomic oxygen [6].
2. The other non conventional techniques used in inactivation of endogenous enzyme are the pulsed electric field [7, 8], irradiation [9], high pressure processing [10]. It was observed that there is decrease in enzymatic activity of trypsin (zero at 4 Jcm^{-2}) after the application of plasma [2]. It was reported that the plasma was able to change the 3D structure of proteins in trypsin enzymes due to cleavage of peptides bonds. In here in this investigation, it was found that the activity of polyphenoloxidase in cold plasma treated guava (*Psidium guajava*) pulp and whole fruit was reduced by 70 per cent and 10 per cent in 300 s at 2 kv respectively.
3. In recent investigations, the kinetics of inactivation of tomato peroxidase enzymes fitting in different kinetic modeling like first-order, Weibull and logistic models was studied [11]. They observed that there is a decrease in the enzymes activity at different voltages using atmospheric air dielectric barrier discharge plasma.
4. The hypothesis [12] showed there is increase in the activity of lipase and dehydrogenase enzymes in the hypocotyls of tomato root cells. They concluded that there is no detrimental effect of magnetized plasma on the tomato growth and yield but treatments should be optimized. The relative lipase activity was increased from 1 to 1.4 with the increase in treatment

time from 0 to 50 s using helium RF atmospheric pressure discharge [13]. They also reported that there is increase in activity of lipase with increase in plasma treatment time due to changes in the molecular structure of lipase protein which were confirmed by circular dichroism [CD] and fluorescence spectrum.

5. The changes in activity of antioxidant enzymes was studied in the body tissues of plasma treated Indian meal moth larvae [14] and it was reported that the activity of catalyse, lipid peroxide and glutathione S-transferase enzymes was observed significant increase and no change in glutathione peroxidase activity.

The effects of CP on various food enzymes are studied and findings are tabulated below [15]:

Sl. No.	Food	Enzymes	Plasma	Observations
1	Fresh-cut apples	Polyphenol Oxidase [16]	DBD, 150 W,15 + 15,30 + 30 min, Air,1.5 m/s.	Noticeable reduction in superficial browning but not proportional to treatment time,Variable effects on PPO activity, Effect were strictly cultivar dependent.
2	Fresh-cut melon	Peroxidase [17]	DBD, 15 kV, 12.5 kHz, 15 + 15, 30 + 30 min, Air	Residual activity were 91 per cent and 82 per cent after 15 + 15 and 30 + 30 min treatment, respectively.
		Pectin Methylesterase [17]	DBD, 15 kV, 12.5 kHz, 15 + 15, 30 + 30 min, Air.	15 + 15 min treatment was ineffective, Residual activity was 94 per cent after 30 + 30 min treatment.
3	Mushrooms (*Agaricus bisporus*)	Superoxide Dismutase [18]	Plasma jet, 18 kV, 10 kHz, 98 per cent Ar + 2 per cent O2, 5 L/min	SOD activity was higher in plasma treated mushroom during storage

Sources

1. *Rohit Thirumdas, Chaitanya Sarangapani and Uday S. Annapure, Cold Plasma: A novel Non-Thermal Technology for Food Processing, Foods biophysics, 10(1), November 2014.*
2. *D. Dobrynin, G. Fridman, G. Friedman, A. Fridman doi: 10.1088/1367-2630/11/11/115020.*
3. *B. Surowsky, A. Fischer, O. Schlueter, D. Knorr, Inn. Food Sci. Emerg. Technol. 19, 146, 2013.*
4. *G. E. Anthon, D. M. Barrett, J. Agric. Food Chem. 50, 4119, 2002.*
5. *P. Elez-Martinez, I. Aguilo-Aguayo, O. Martín-Belloso, J. Sci. Food Agric. 86, 71, 2006.*
6. *N. K. Rastogi, Food Rev. Int. 19, 229, 2003.*
7. *J. Giner, V. Gimeno, G. V. Barbosa-Canovas, O. Martin, Food Sci. Technol. Int. 7, 339, 2001.*

8. K. Zhong, J. Wu, Z. Wang, F. Chen et al., Food Chem. 100, 115, 2007.
9. L. Zhang, Z. Lu, F. Lu, X. Bie, Food Cont. 17, 225, 2006.
10. B. R. Thakur, P. E. Nelson, Food Rev. Int. 14, 427, 1998.
11. S. K. Pankaj, N. N. Misra, P. J. Cullen, Innov. Food Sci. Emerg. Technol., 19, 153, 2013.
12. Y. Meiqiang, H. Mingjing, M. Buzhou, M. Tengcai, Plasma Sci. Technol. 7, 3143, 2005.
13. H. P. Li, L. Y. Wang, G. Li et al., Plasma Process Polym. 8, 224, 2011.
14. M. F. A. Aziz, E. A. Mahmoudb, G. M. Elaragi, J. Stored Prod. Res., 2014.
15. Shashi K. Pankaj, ZifanWan and Kevin M. Keener, Effects of Cold Plasma on Food Quality: A Review, 2018, MDPI, Basel, Switzerland, Foods, 7, 4, 2018.
16. S. Tappi, L. Ragni, U. Tylewicz, S. Romani, I. Ramazzina, P. Rocculi, Browning response of fresh-cut apples of different cultivars to cold gas plasma treatment. Innov. Food Sci. Emerg. Technol. 2017.
17. S. Tappi, G. Gozzi, L. Vannini, A. Berardinelli, S. Romani, L. Ragni, P. Rocculi, P. Cold plasma treatment for fresh-cut melon stabilization. Innov. Food Sci. Emerg. Technol., 33, 225-233, 2016.
18. Y. Xu, Y. Tian, R. Ma, Q. Liu, J. Zhang, 2016, J. Effect of plasma activated water on the postharvest quality of button mushrooms, Agaricus bisporus. Food Chem., 197, 436-444.

E. Effect on Phenolic and Antioxidant Compounds [1]

Antioxidants are considered as first line of defence against free radicals. Antioxidants protect cells against the damaging effects of ROS, such as singlet oxygen, superoxide, peroxyl radicals, hydroxyl ions and peroxynitrite *etc.* [2]. The following few literatures are available on effect of plasma treatment on the phenolic compounds:

1. It is reported [3] that several plant species are tolerant to UV radiations accumulating flavonoid metabolites in epidermal cells. During plasma generation the UV radiations formed may be responsible for the formation of phenolic compounds which are extracted from the cells of upper epidermis of leaves.
2. Recent investigations [4] using low-pressure oxygen plasma reveals that a time and structure dependent degradation flavonoids can be observed, due to plasma immanent reactive species such as ozone or hydroxyl radicals.
3. **Cold oxygen plasma:** To evaluate possible protective effects of the plant's matrix against plasma-induced degradation of secondary plant metabolites researchers [5] worked on flavonols and phenolic acid of lamb's lettuce leaves which were exposed to the cold oxygen plasma. It was reported that there is increase in phenolic contents particularly protocatechuic acid and luteolin was observed doubled upon plasma

treatment. The diosmetin content was found to be increased more than others, whereas there is no change in chlorogenic acid content.

4. The use of cold plasma technique over the conventional sources like pasteurization for the treatment of sour cherry marasca juice showed a higher percentage of phenolic acid and anthocyanin content [6].

Sources

1. *Rohit Thirumdas, Chaitanya Sarangapani and Uday S. Annapure, Cold Plasma: A novel Non-Thermal Technology for Food Processing, Foods biophysics, 10(1), November 2014.*
2. *M. Percival, Advanced Nutrition Publications Inc., 1996.*
3. *J. B. Harborne, C. A. Williams, Phytochem 55, 481, 2000.*
4. *F. Grzegorzewski, . W. Rohn Kroh, M. Geyer, O. S. Schluter, Food Chem. 122, 1145, 2010.*
5. *F. Grzegorzewski, S. Rohn, A. Quade et al., Plasma Proc. Polym. 7, 466, 2010.*
6. *I. E. Garofuli, A. R. Jambrak, S. Milosevi et al., LWT- Food Sci. Technol., 2014.*

F. Effect on Starch Granules and its Modification [1]

Modified starches are prepared to alter its functional properties to increase its efficiency to use as food additives in several food preparations. Tensile strength and mechanical properties were improved in starch films made with the use of modified starches. The details of the some alternative methods being used in manufacturing of modified starches are given below:

1. **Cold plasma technology**: During preparation of modified starches, weak chemicals are used to convert smooth hydrophobic surfaces to rough hydrophilic surfaces as the natural polymers available are sensitive to chemicals particularly to strong acids for etching of surface. It is reported [2] that cold plasma technology is an alternative technique for dry etching for surface modification and surface cleaning of biopolymers. It is also reported that introduction of oxygen containing groups like hydroxyl, carboxyl, and carbonyl groups increases surfaces hydrophilicity.
2. Starch can be modified by using cold plasma. Most probably starch is modified by two main mechanisms, deploymerization and cross linking of starch granules. The depolymerisation of starch granules can occur at amylopectin side chains or break down of glycoside bonds which are close to side chains. The ultimate goal of cold plasma is to create more hydrophilic surfaces of the starch films.
3. **Glow plasma**: Researchers [3] reported that **Glow plasma** could induce the graft-polymerization of ethylene onto sweet potato and rice starches, while the homopolymerization of ethylene on the granules took place for cassava, potato, corn, and waxy corn starches.

4. The plasma treatment has a significant effect on the crystallinity of the solid starch granules [4]. It was also reported that there is a high degree effect on the supramolecular region and a higher decrease in the molecular weight of plasma treated starch granules. The cross linking agent joins two starch molecules by the removal of hydrogen bonds forming stronger covalent bonds.
5. Previous studies reported that modification of starches can also be prepared by exposing starch to plasma at different varying condition of pressure and time. It is also [4a] reported that the starch modification can be occur using plasma as per the mechanism shown in following equation. The two starch molecules are cross linked by the release of water molecule. The water molecule formed will de disassociated to hydrogen and hydroxyl ion with no residue of chemical compound.

   Starch–H + OH–Starch → Starch–O–starch + H_2O (1)
6. It is reported [3] that reaction of plasma treated starch with iodine reagent showed an increase in shift in wavelength in UV-VIS band this shows that there is an increase in availability of amylose for the reaction with iodine. There is decrease in pH of plasma treated starch solution due to cross linking of starch molecules resulting in formation of carboxylic acid [3]. Due to the oxidation, the surface of starch granules are covered by irregular deposits like subtle occurs by a process called etching [5].
7. The effect of plasma on crystalline structure of starch was explained [6], it was observed slight decrease in crystallinity after exposure to ethylene plasma. The changes in crystalline structure is due to the physical damage caused by the bombardment of high energetic particles formed during the plasma generation breaking down the chemical bonds [7].
8. **RF plasma**: Researchers [8] studied the changes in properties of potato starch film using **RF plasma** produce from methane gas to use as filler in polyethylene composite films. The plasma treatment can be used to reduce the capillary elevation by de-agglomeration of starch granules. In untreated samples agglomeration of starch granules was observed, but in treated samples starch granules were separated attributing to decrease in capillary elevation.
9. **Cold glow butene plasma**: Researchers [9] worked on the plasma polymerization of maize starch film to reduce its water absorption using **cold glow butene plasma**. It is reported [10] that due to rearrangement of intermolecular bonds of starch resulting in disintegration causing lower water absorption capacity.
10. **Atmospheric plasma**: Researchers [11] studied the effect of atmospheric plasma on the dough rheology of wheat flour and reported that there is increase in viscous and elastic moduli of strong wheat flour after the plasma treatment. With the application of cold plasma the secondary structure of proteins were affected by changing the number of beta sheets and alpha helix.

Sources

1. *Rohit Thirumdas, Chaitanya Sarangapani and Uday S. Annapure, Cold Plasma: A novel Non-Thermal Technology for Food Processing, Foods biophysics, 10(1), November 2014.*
2. *I. Pashkuleva, P. M. Lopez-Perez, R. L. Reis. (Woodhead Publishing Series in Biomaterials, 2008, p.165.*
3. *C. Y. Lii, C. D. Liao, L. Stobinski, P. Tomasik, Carbo. Polym. 49, 499, 2002a.*
4. *B. Zhang, S. Xiong, X. Li et al., Food Hydrocoll. 37, 69, 2014.*

4a. *J. J. Zou, C. J. Liu, B. Eliasson, Carbo. Polym. 55, 23, 2004.*

5. *F. Starzyk, C. Y. Lii, P. Tomasik, Polym. J. FoodNutr. Sci. 10, 27, 2001.*
6. *C. Y. Lii, C. D. Liao, L. Stobinski, P. Tomasik, Carbo. Polym. 49, 449, 2002.*
7. *S. M. Mirabedini, H. Arabi, A. Salem, S. Asiaban, Prog. Org. Coat.60, 105, 2007.*
8. *H. Szymanowskia, M. Kaczmareka, M. Gazicki-Lipmana et al., Surf. Coat. Technol. 200, 539, 2005m.*
9. *C. T. Andrade, R. A. Simao, R. M. S. M. Thire, C. A. Achete, Carbo. Polym. 61, 407, 2005.*
10. *M. Lares, E. Perez, Plant Food Hum. Nutr. 61, 109, 2006.*
11. *N. N. Misra, S. Kaur, B. K. Tiwari et al., Food Hydrocoll. 44, 115, 2015.*

G. Effect on Packing Materials [1]

Food packaging materials are responsible for protection of foods/dairy products from the outside environment during handling, transportation and distribution. Cold plasma is used in decontamination of packing material externally where chance of shadow effect is negligible as plasma flows all round the surface [2] and Plasma processing is well known to change or make surface modifications in the case of packing materials. CP serves purposes for surface treatment such as cleaning, coating, printing, painting, and adhesive bonding [3].

1. **Low temperature gas plasma**: This (LTGP) sterilization allows fast and safe sterilization of packaging materials such as plastic bottles, lids and films without adversely affecting the properties of the material or leaving any residues.
2. **Cold plasma**: It can be used for sterilization of heat sensitive packing materials like polythene ethylene and polycarbonate as the temperature is low.
3. Surfaces of polymers particularly for edible packaging films nature of surface should be more hydrophobic with lower surface energies [4]. Using plasma as the transport mechanism and the catalyst, one material can be deposited (in a very thin layer) onto the surface of another material; thereby transferring some of its qualities.

4. Researchers studied the oxygen transport properties of the SiOx coating on polyethylene terephthalate [PET], low and high-density polyethylene [LDPE, HDPE] and polypropylene [PP] films, obtained using cold plasma technology and compared experimental data with computer model and found diffusivity was less than normal material [5].
5. Plasma deposition of heat-sensitive materials such as vitamins, antioxidants and antimicrobials into the packaging material may be sought as potential alternatives in the emerging field of antimicrobial and active packaging. Applications of nanotechnology in packaging as become wide spread to improve the barrier properties of packing material and this can be achieved by cold plasma processing.
6. The surface topography of zein film using atomic force microscopy [AFM] was studied [6]. The roughness of plasma treated zein film increased with a root mean square of 100 nm from 20 nm in untreated films this is due to surface etching occurred during plasma treatment. The finding of changes in the properties of packing materials after the exposure to cold plasma is tabulated below [1]:

Sl. No.	*Packing Materials*	*Treatment Plasma, Applied Voltage*	*Results*
1	BOPP film [7]	RF Air plasma, 10–50 W	Decrease in contact angle, increase roughness ageing effect
2	HDPE film [8]	RF Ar: O_2 Plasma, 150 W	Decrease in crystallinity, contact angle, increase in roughness
3	PET films [9]	Glow discharge, 10 W	Decrease in contact angle, increase in roughness, crystallinity and degradation yield
4	LDPE [10]	RF Argon plasma, 25–100 W	Decrease in contact angle and ageing effect, increase in crystallinity
5	PET film [11]	Jet plasma DC discharge, 35 W	Increase in hydrophobic nature
6	PP film [12]	Diode plasma discharge, 8.3 W	Decrease in contact angle, increase in surface energy
7	Polypropylene [13]	Air Corona 30 kHz, 1.7 j/cm^2	Decrease in contact angle, increase in adhesion

Sources

1. *Rohit Thirumdas, Chaitanya Sarangapani and Uday S. Annapure, Cold Plasma: A novel Non-Thermal Technology for Food Processing, Foods biophysics, 10(1), November 2014.*
2. *M. Shakila, P. Sasikala, D. Aruna, V. Kavitha, R. Lavanya, Int. J. Emerg. Trends Eng. Dev. 4, 803, 2012.*
3. *S. K. Pankaja, C. Bueno-Ferrera, N. N. Misra et al., Trends Food Sci. Technol. 35, 5, 2014.*

4. A. Vesel, M. Mozetic, Vacuum 86, 634, 2012.

5. M. S. Hedenqvist, K. S. Johansson, Surf. Coat. Technol. 172, 7, 2003.

6. S. K. Pankaj, C. Bueno-Ferrer, N. N. Misra et al., J. Renew. Mater., 2014.

7. S. M. Mirabedini, H. Arabi, A. Salem, S. Asiaban, Prog. Org. Coat. 60, 105, 2007.

8. I. Banik, K. S. Kim, Y. I. Yun et al., J. Adhes. Sci. Technol. 16, 1155, 2002.

9. K. N. Pandiyaraj, V. Selvarajan, R. R. Deshmukh, M. Bousmina, Surf. Coat. Technol. 202, 4218, 2008.

10. M. Ataeefard, S. Moradian, M. Mirabedini, M. Ebrahimi, S. Asiaban, Prog. Org. Coat. 64, 482, 2009.

11. Y. Akishev, M. Grushin, N. Dyatko et al., J. Phys. 41, 235, 2008.

12. P. Slepicka, A. V. Kolsk, Z. Luxbacher et al., Beam Interact. Mater. Atoms 268, 2111, 2010.

13. D. Dixon, B. J. Meenan, J. Adhes. Sci. Technol. 26, 2325, 2012.

H. Wastewater Treatment [1]

The complete oxidation is required for the treatment of wastewater and transfer of contaminants is not complete in methods like photocatalysis, ultrasonication, UV/ozone [2]. Developing an innovative advanced oxidation process for treatment of wastewater is a big challenge. Physico-chemical effects of plasma generates the formation of oxidizing species: radicals (H^-, O^-, OH^-) may diffuse into the liquids and molecules (H_2O_2, O_3, *etc.*), shockwave, ultraviolet light and electrohydraulic cavitation may degrade the pollutant in wastewater or decomposes the pollutant into other compound [3]. In liquid and gas plasma can be created either directly in the liquid, or in the gas above the liquid, or, in case of hybrid rectors, both in liquid and in gas. The more efficient way which requires less energy for wastewater treatment can be done by diffusing gaseous phase species into liquid [4].

Extensive research for commercial design of non-thermal plasma wastewater treatment equipment should involve in theoretical aspects of diversified fields of Engineering, Electronics and Applied Sciences. Plasma oxidation mechanism of degradation of parent pollutants, different other properties to optimise the plasma plant process and also the interaction of liquid and gas on liquid gas interface should be examined [5].

Sources

1. *Rohit Thirumdas, Chaitanya Sarangapani and Uday S. Annapure, Cold Plasma: A novel Non-Thermal Technology for Food Processing, Foods biophysics, 10(1), November 2014.*

2. *P. M. K. Reddy, C. Subrahmanyam, Ind. Eng. Chem. Res. 51, 11097, 2012.*

3. *B. Jiang, J. Zheng, X. Lu et al., Chem. Eng. J.215, 969, 2013.*

4. *B. Jiang, J. Zheng, S. Qiu et al., Chem. Eng. J.236, 348, 2014.*

5. S. Ognier, D. Iya-sou, C. Fourmond, S. Cavadias, Plasma Chem. Plasma Process. 29, 261, 2009.

Applications of Cold Plasma [1]

Cold atmospheric plasma is a unique food processing technology, which is widely used for air decontamination of airborne bacteria, elimination of organic chemicals or bacteria by the liquid treatment, and for decontaminating the surface of foods, equipment packaging, and work surfaces [1]. The details related to the Potential plasma applications are given below [1a]:

1. Air decontamination of airborne bacteria (particularly important for turbulent air flows such as those in fridges),
2. Liquid treatment for removal of organic chemicals, bacteria and biofilms
3. Surface decontamination of Foods, process equipment, packaging and work surfaces.

Based on the properties of plasma, this technology is used in following various fields [2]:

1. Textile
2. Electronics
3. Life sciences
4. Packaging *etc.*

a. Application

The application of plasma technology as a surface cleaning tool has been commercially adopted for the removal of disinfection chemicals applied to medical devices manufactured from heat sensitive plastics [3].

b. Biomedical Sector

In this sector plasma technology is used for cold sterilization of instruments and prostheses as well as many thermo labile materials used in the biomedical technology sector for its particular advantages, including its moderate or negligible impact on substrate materials and use on nontoxic compounds [4].

c. Sterilisation of Packaging Materials

Cold plasma is used efficiently for sterilization and modifications of packaging polymers purpose. Details are given under effect of cold plasma.

d. Food Processing

1. There is a huge application of cold plasma technology in food processing. Conventionally, sterilization methods such as heat, chemical solutions are used for the surface disinfection of fruits, seeds, and spices *etc.*, which are often time-consuming and damaging or have toxic residues [5]. During processing of food, food products might be placed in strong electric field

and are subjected to numerous reactive gas species that could affect physical quality attributes such as color and texture. A summary of effects of CP on food products has been tabulated below [1b]:

Sl. No.	Juices, Foods, Fruits etc.		Plasma	Observations	
				Quality	Microbial
1	Fruit Juices	Orange Juice [5a]	DBD, Air/MA65 (65 per cent O_2, 30 per cent CO_2, 5 per cent N_2), 90 kV, 30–120 s.	No significant change in Brix or pH, Vit. C is reduced by 22 per cent in air, Maximum total color difference is less than 1.2.	Up to 5 log_{10} reduction of *Salmonella enterica*
		Cashew apple juice [5b]	PE-100, 80 kHz, N_2, 10–50 mL/min, 5–15 min, 30 kPa.	Decrease in vitamin C at higher flow rate, Increase in sucrose content while glucose and fructose contents decreased, Longer treatment promoted higher polyhphenol and total flavonoid content.	—
		Prebiotic orange juice [5c]	DBD 70 kV (50 Hz)	Degradation of oligo saccharides in the juice, Decrease in pH, Increase in L* value and slight reduction in chroma and hue angle, Decrease in total phenolic content and antioxidant capacity in some cases.	—
2	Fruits	Blueberry [5d]	Plasma jet, 47 kHz, 549W, air, 4–7 cubic feet/min, 7.5 cm,0–120 s	Significant reductions in firmness, color and anthocyanins at higher treatment times.	Upto 2 log_{10} reduction in total aerobic plate count.
		Strawberry [5e]	DBD, 60 kV, 50 Hz, air, 5 min, indirect exposure.	No significant change in color, firmness andrespiration rate.	2 log_{10} reduction in background Microflora (aerobic mesophilic bacteria, yeast and mould).
		Golden delicious apples [5f]	Gliding arc plasma 60 Hz.	No changes in color and texture.	~3.5 log_{10} reduction in *Salmonella and E. coli* O157: H7 reduction.
3	Grains	Wheat, bean, chick pea, soy bean, barley, oat, rye, lentil and corn. [5g]	Low pressure cold plasma, 1 kHz, 20 kV, 500 mTorr, 300 W, air and SF6, 5–20 min.	Slight change in moisture content of legume and wheat, No difference in water soaking, yield and cooking time of legumes, No change in wet gluten content, gluten index and sedimentation in wheat.	3 log10 reduction of *Aspergillus* spp. and *Penicillum* spp. after 15 min-treatment in SF6.

Sl. No.	*Juices, Foods, Fruits etc.*		*Plasma*	*Observations*	
				Quality	*Microbial*
		Brown Rice [5h]	Radiofrequency plasma, air	Decrease in cooking time, hardness, chewiness, contact angle, and moisture content, Higher degree of elatinization, Increase in water uptake, L value and whiteness index.	—
		Black gram [5i]	Radiofrequency plasma, 2 Pa, air (0.15 mbar), 13.56 MHz, 30-50W, 5-15 min	Surface etching and hydrophillization of surface, Decrease in hardness, cooking time, ash and moisture content	—
4	Meat and Meat products	Fresh pork [5j]	Microwave plasma, air, 5-10 min, 2.45 GHz, 1.2 kW, 20 slm.	Increased a value and decreased b values of pork meat, Difference in reflectance and fluorescence, Significant changes in pH.	Aerobic viable count remained between 10^2 and 10^3 cfu/g during the storage period of 20 days.
		Fresh pork and beef [5k]	Thin-layer DBD plasma 1–10 min, 100W, $N_2 + O_2$.	No significant effect on texture, L* and b* value, Decrease in a* values after 5 min exposure, Significant lipid oxidation after 10 min exposure, No change sensory parameters except taste, which was negatively influenced.	Up to 2.7 log_{10} cfu/g reduction of *Listeria monocytogenes, E. coli* O157: H7 and *Salmonella tryphimurium* in porkand beef.
		Beef jerky [5l]	RF plasma Ar 20,000 sccm, 200W,0–10 min.	No significant change in fatty acid composition, color and shear force.	1.8 log_{10} reduction in *Staphylococcus aureus* after 8min treatment.
5	Others	Radish sprouts [5m]	Microwave plasma, 2.45 GHz, 900W, 669 Pa, 1-20 min, N_2, 1 L/min,	No change in color, water activity, ascorbic acid concentration and antioxidant activity, Lower moisture content during storage	2.6 log_{10} reduction in *Salmonella typhimurium,* 0.8 log_{10} reduction in total Meso-hilic aerobes
		Onion powder [5n]	Microwave plasma, 170 and 250 m Wm^{-2}, 2.45 GHz, 400–900 W, 10–40 min,0.7 kPa, He, 1 L/min,	No effect on color, antioxidant activity and quercetin concentration.	2.1 log_{10} spores/cm^2, 1.6 log_{10} spores/cm^2 and 1.9 cfu/cm^2 reduction of *Bacillus cereus, A. brasiliensis* spore, and *E. coli* O157: H7, respectively.

2. The effect of cold plasma on bacterial spores is more than the conventional techniques like heat, chemicals and UV treatment, was reported [6]. Effect of cold plasma on microbial inactivation and structural modifications of packaging materials and the effect of cold plasma on endogenous enzymes, starch modifications and limitations for its potential application in food sector as novel technology were studied.
3. The applications of CP for food industries have been demonstrated for food decontamination [7], microbial inactivation(Details given under effect of cold plasma), enzyme inactivation (Details given under effect of cold plasma) [8], toxin removal [9], food packaging modifications (Details given under effect of cold plasma) [10], starch modifications (Details given under effect of cold plasma) and wastewater treatment (Details given under effect of cold plasma) [11]. Particularly for food processing, CP has been shown to be effective against major food-borne pathogenic micro-organisms such as *Escherichia coli* [12], *Salmonella typhimurium* [13], *Staphylococcus aureus* [14], and *Listeria monocytogenes* [15].

One of the important challenges associated with cold plasma technology is ensuring high microbial inactivation while maintaining sensory qualities that ensure there fresh appearance [1].

Sources

1. *Rohit Thirumdas, Chaitanya Sarangapani and Uday S. Annapure, Cold Plasma: A novel Non-Thermal Technology for Food Processing, Foods biophysics, 10(1), November 2014.*

1a. *Danny Bayliss, New Technologies Specialist, Cold plasma: Applications and future perspectives within the food and drink industry, Danny.bayliss@campdenbri.co.uk.*

1b. *Shashi K. Pankaj, ZifanWan and Kevin M. Keener, Effects of Cold Plasma on Food Quality: A Review, 2018, MDPI, Basel, Switzerland, Foods, 7, 4, 2018.*

2. *J. R. Roth, S. Nourgostar, T. A. Bonds, IEEE Trans. Plasma Sci. 35, 233, 2007.*

3. *M. Moisan, J. Barbeau, S. Moreau et al., Int. J. Pharmaceut. 2, 226, 2001.*

4. *I. A. Soloshenko, Plasma Phys. Rep. 26, 792, 2000.*

5. *P. Muranyi, J. Wunderlich, M. Heise, J. Appl. Microbiol. 103, 1535, 2007.*

5a. *A. L. Xu. L. Garner, B. Tao, K. M. Keener, Microbial inactivation and quality changes in orange juice treated by high voltage atmospheric cold plasma. Food Bioprocess Technol., 10, 1-14, 2017.*

5b. *O. Rodríguez, W. F. Gomes, S. Rodrigues, F. A. Fernandes, F. A. Effect of indirect cold plasma treatment on cashew apple juice (Anacardium occidentale L.), LWT-Food Sci. Technol., 84, 457-463, 2017.*

5c. *F. D. L. Almeida, R. S. Cavalcante, P. J. Cullen, J. M. Frias, P. Bourke, F. A. Fernandes, S. Rodrigues, Effects of atmospheric cold plasma and ozone on prebiotic orange juice. Innov. Food Sci. Emerg. Technol., 32, 127-135, 2015.*

5d. C. Sarangapani, G. O'Toole, P. Cullen, P. Bourke, Atmospheric cold plasma dissipation efficiency of agro-chemicals on blueberries. Innov. Food Sci. Emerg. Technol., 44, 235-241, 2017.

5e. N. Misra, S. Patil, T. Moiseev, P. Bourke, J. Mosnier, K. Keener, P. Cullen, In-package atmospheric pressure cold plasma treatment of strawberries. J. Food Eng., 125, 131-138, 2014.

5f. B. A. Niemira, J. Sites, Cold plasma inactivates Salmonella Stanley and Escherichia coli O157: H7 inoculated on golden delicious apples. J. Food Prot., 71, 1357-1365, 2008.

5g. M. Selcuk, L. Oksuz, P. Basaran, Decontamination of grains and legumes infected with Aspergillus spp. and Penicillum spp. by cold plasma treatment. Bioresour. Technol., 99, 5104-5109, 2008.

5h. R. Thirumdas, C. Saragapani, M. Ajinkya, R. Deshmukh, U. Annapure, Influence of low pressure cold plasma on cooking and textural properties of brown rice. Innov. Food Sci. Emerg. Technol., 37, 53-60, 2016.

5i. C. Sarangapani, R. Y. Devi, R. Thirumdas, A. M. Trimukhe, R. R. Deshmukh, U. S. Annapure, Physico-chemical properties of low-pressure plasma treated black gram. LWT-Food Sci. Technol., 79, 102-110, 2017.

5j. A. Fröhling, J. Durek, U. Schnabel, J. Ehlbeck, J. Bolling, O. Schlüter, Indirect plasma treatment of fresh pork: Decontamination efficiency and effects on quality attributes. Innov. Food Sci. Emerg. Technol., 16, 381-390, 2012.

5k. D. D. Jayasena, H. J. Kim, H. I. Yong, S. Park, K. Kim, W. Jo. C. Choe, Flexible thin-layer dielectric barrier discharge plasma treatment of pork butt and beef loin: Effects on pathogen inactivation and meat-quality attributes. Food Microbiol., 46, 51-57, 2015.

5l. J. S. Kim, E. J. Lee, EH. Choi, Y. J. Kim, Inactivation of Staphylococcus aureus on the beef jerky by radio-frequency atmospheric pressure plasma discharge treatment. Innov. Food Sci. Emerg. Technol., 22, 124-130, 2014.

5m. Y. J. Oh, A. Y. Song, S. C. Min, Inhibition of Salmonella typhimurium on radish sprouts using nitrogen-cold plasma. Int. J. Food Microbiol., 249, 66-71, 2017.

5n. J. E. Kim, Y. J. Oh, M. Y. Won, K. S. Lee, S. C. Min, Microbial decontamination of onion powder using microwave-powered cold plasma treatments. Food Microbial., 62, 112-123, 2017.

6. H. B. Van de Veen, H. Xie, E. Esveld et al., Food Microbiol, 2014.

7. N. N. Misra, B. K. Tiwari, K. S. M. S Raghavarao, P. J. Cullen, Non-thermal plasma inactivation of food-borne pathogens. Food Eng. Rev., 3, 159-170, 2011.

8. N. N. Misra, S. K Pankaj, A. Segat, K. Ishikawa, Cold plasma interactions with enzymes in foods and model systems. Trends Food Sci. Technol., 55, 39-47, 2016.

9. N. N., Misra, The contribution of non-thermal and advanced oxidation technologies towards dissipation of pesticide residues. Trends Food Sci. Technol., 45, 229-244, 2015.

10. *S. K. Pankaj, C. Bueno-Ferrer, N. N. Misra, V. Milosavljevi´c, C. P. O'Donnell, P. Bourke, K. M. Keener, P. J. Cullen, Applications of cold plasma technology in food packaging. Trends Food Sci. Technol., 35, 5-17, 2014.*

11. *C. Sarangapani, N. N. Misra, V. Milosavljevic, P. Bourke, F. O'Regan, P. J. Cullen, Pesticide degradation in water using atmospheric air cold plasma. J. Water Process Eng., 9, 225-232, 2016.*

12. *D. Bermúdez-Aguirre, E. Wemlinger, P. Pedrow, G. Barbosa-Cánovas, M. Garcia-Perez, Effect of atmospheric pressure cold plasma (APCP) on the inactivation of Escherichia coli in fresh produce, Food Control, 34, 149-157, 2013.*

13. *A. Fernández, E. Noriega, A. Thompson, Inactivation of Salmonella enterica serovar Typhimurium on fresh produce by cold atmospheric gas plasma technology. Food Microbiol., 33, 24-29, 2013.*

14. *J. S. Kim, E. J. Lee, E. H. Choi, Y. J. Kim, Inactivation of Staphylococcus aureus on the beef jerky by radio-frequency atmospheric pressure plasma discharge treatment. Innov. Food Sci. Emerg. Technol., 22, 124-130, 2014.*

15. *D. Ziuzina, S. Patil, P. J. Cullen, K. M. Keener, P. Bourke, Atmospheric cold plasma inactivation of Escherichia coli, Salmonella enterica serovar Typhimurium and Listeria monocytogenes inoculated on fresh produce. Food Microbiol., 42, 109-116, 2014.*

Section-B

Alternate Thermal Processing Technologies

Chapter 12

Microwave Heating Technology

Microwave heating [1] has vast applications in the field of food processing such as cooking, drying, pasteurization sterilization, thawing, tempering, baking and preservation of food materials [1a]. Unlike conventional heating, which relies on the slow march of heat from the surface of the material to the interior, heating with MW energy is in effect bulk heating in which the electromagnetic field interacts with the food as a whole. The microwave heating of foods has garnered scientific and consumer interest due to its volumetric (bulk) heating, rapid increase in temperature and relative ease of cleaning [2]. Unlike more traditional forms of thermal processing, such as pasteurization and retorting, which are characterized by a slow thermal diffusion process, the volumetric nature of heat generated by microwaves can substantially reduce total heating time, thereby minimizing the overall severity of the process and leading to a greater retention of the desirable quality attributes of the product [3]. As per the study [4], the time required to come to target process temperature is attained within one-quarter of the time typically reached by conventional heating processes. Microwave technology is also amenable to batch processing and offers the flexibility of being easily turned on or off. The federal communication commission (FCC) allocates 915 MH_z and 2450 MHz bands for microwave heating application [5].

Microwave pasteurization has the ability to achieve destruction of micro-organisms at temperatures lesser than that of conventional pasteurization due to significant enhancement or magnification of thermal effects. Applications of microwave drying include microwave assisted hot air drying, microwave vacuum drying and microwave freeze drying. Microwave drying combined with other conventional drying enhances the drying characteristics of sole effect of microwave drying.

Microwave heating has gained momentum in food processing due to its different abilities given on next page:

1. To achieve high heating rates,
2. Significant reduction in cooking time,
3. More uniform heating,
4. Safe handling,
5. Ease of operation
6. Low maintenance [6, 7].

Microwave heating might change flavor and nutritional qualities of food in a lesser extent as opposed to conventional heating during cooking or reheating process [8].

Sources

1. *Hudaa Neetoo (Faculty of Agriculture, University of Mauritius, Réduit, Mauritius) and Haiqiang Chen (Department of Animal and Food Sciences, University of Delaware, Newark, Delaware, US), Alternative Food Processing Technologies.*

1a. *S. Chandrasekaran, S. Ramanathan and Tanmay Basak, Microwave food processing- a review, Department of Chemical Engineering, Indian Institute of Technology Madras, Chennai.*

2. *Ahmed J, Ramaswamy HS, 2007, Microwave pasteurization and sterilization of foods. In: Rahman MS, (ed) Handbook of Food Preservation. Boca Raton, FL: CRC Press, pp. 691-712..*

3. *Sumnu G, Sahin S, 2005, Recent developments in microwave heating. In: Sun DW (ed) Emerging Technologies for Food Processing. California: Elsevier Academic Press, pp. 419-444.*

4. *Tewari G., 2007, Microwave and radio-frequency heating. In: Tewari G, Juneja V K (eds) Advances in Thermal and Non-thermal Food Preservation. Ames, IA: Blackwell Publishing, pp. 91-98.*

5. *Nalaka Nugapola, Microwave Food Processing, UVa Wellassa University of Srilanka.*

6. *Salazar-Gonzalez, C., San Martin-Gonzalez, M. F., Lopez-Malo, A., and Sosa-Morales, M. E., 2012, Recent studies related to microwave processing of fluid foods. Food Bioprocess and Technology, 5, 31-46.*

7. *Zhang, M., Tang, J., Mujumdar, A. S., and Wang, S., 2006, Trends in microwave-related drying of fruits and vegetables. Trends in Food Science and Technology, 17, 524-534.*

8. *Vadivambal, R., and Jayas, D. S., 2010, Non-uniform temperature distribution during microwave heating of food materials - a review. Food and Bioprocess Technology, 3, 161-171.*

Mechanism and Equipment [1]

Microwaves are a form of electromagnetic radiation characterized with a frequency between 300 MHz to 300 GHz. Domestic microwave appliances operate generally at a frequency of 2.45 GHz, while industrial microwave systems operate at frequencies of 915 MHz and 2.45 GHz [1a, 1b].

Mechanism

Microwaves might be absorbed owing to its dipole or ionic polarization and these are not absorbed by the material due to its electronic or atomic polarization. Dipole polarization is significant at frequencies above 1 GHz while ionic losses are predominant at frequencies below 1 GHz [1a, 1e].

Microwave Heating

It is caused by the ability of the materials to absorb microwave energy and convert it into heat [1a]. Microwave (MW) heating is generated by the absorption of microwave by a dielectric material, resulting in the microwaves giving up their energy to the material with a concomitant rise in temperature. There are two main mechanisms by which microwaves produce heat in dielectric materials- **dipole rotation and ionic polarization.**

i. Dipole Rotation

Food materials contain polar molecules such as water. These molecules generally have a random orientation. When an oscillating electric field is incident on the water molecules, the permanently polarized dipolar molecules try to realign in the direction of the electric field (the molecules orient themselves according to the polarity of the field). Due to high frequency electric field, this realignment occurs at million times per second and causes internal friction of molecules with the surrounding medium resulting in the volumetric heating of the material. The moisture or water content (as it is good absorber of microwaves) present in a food material is deciding factor for determining the dielectric properties of the food material.

ii. Ionic Polarization

Microwave heating might also occur due to the oscillatory migration of ions in the food which generates heat in presence of high frequency oscillating electric field [1c]. When an electric field is applied to food solutions containing ions, the ions move at an accelerated pace due to their inherent charge. The resulting collisions between the ions cause the conversion of kinetic energy of the moving ions into thermal energy [1].

Several factors *e.g.* dielectric properties of food materials, penetration depth, microwave oven design (oven size and geometry), microwave frequency, placement of food material inside the oven, moisture content, density, composition, load, shape and the size of food materials affect the microwave food processing [1a, 1d]. The most important factors that affect the microwave heating are **dielectric properties of food materials and penetration depth [1a].**

Dielectric Properties [1a]

The real part of dielectric property, termed as dielectric constant, signifies the ability to store electric energy and the imaginary part of dielectric property, termed as dielectric loss, signifies the ability and convert electric energy into heat,

$$\varepsilon^* = \varepsilon' - j\,\varepsilon''$$

where, ε′ and ε″ are dielectric constant and dielectric loss respectively and $j = \sqrt{1}$

The ratio of dielectric loss to dielectric constant is given by loss tangent and is expressed as,

$$Tan\delta = \frac{k''}{k'} = \frac{\varepsilon''}{\varepsilon'}$$

where, k′ and k″ are relative dielectric constant and relative dielectric loss respectively, which are given as $k' = \varepsilon'/\varepsilon_0$ and $k'' = \varepsilon''/\varepsilon_0$. Here, ε_0 is the permittivity of free space ($\varepsilon_0 = 8.854 \times 10^{-12}$ F/m).

The dielectric properties are mainly affected by the operating temperature and the microwave frequency used and based on the microwave absorption, materials are classified into the following categories [1a]:

1. Absorbers or high dielectric loss materials which are strong absorbers of microwave,
2. Transparent or low dielectric loss materials where microwave passes through the material with little attenuation and
3. Opaque or conductors which reflect the microwaves.

Hence, a knowledge of dielectric properties is necessary to differentiate the materials into the above three categories.

The following factors affect the dielectric properties of foods:

1. The dielectric properties of food materials are mainly determined by their chemical composition and to a less extent of physical structure.
2. The dielectric characteristics of the food materials may also vary with their particle size, structure and density of the material.
3. Dielectric properties are also affected by the apparent density of the air-particle mixture of a granular or particulate material [1d].
4. The dielectric properties of food materials such as bread, flour, fruits and vegetables depend mostly on their water content. The dielectric property of water varies depending on whether it is in free or bound state. In the presence of an electric field, the polar molecules of water in free state orient more freely than those of bound water. At microwave frequencies of 915 and 2450 MHz, the dielectric properties were found to be dependent on temperature for low moisture content samples. On the other hand, at higher moisture contents, the dielectric properties were found to be less significant on temperatures.
5. The variation of dielectric properties with temperature and microwave frequencies for solutions containing salt, sugar and carboxymethylcellulose (CMC) was studied. CMC does not have any significant effect on the dielectric properties. dielectric constant increases with the temperature and the sugar concentration. The sugar, salt and CMC can be used to mimic

the dielectric properties and rheological properties of the food product to be processed [1f].

Penetration Depth [1a]

The power penetration depth is defined as the distance at which the power density drops to a value of *1/e* from its value at the surface.

Modeling of Microwave Heating of Food Materials [1a]

It is based on **Maxwell's equations and Lambert's law equations** were established along with their applications [1a]. Modeling of microwave heating involves the use of electro magnetic equations and energy equations to predict the temperature distribution as well as microwave power absorption inside the food products. The electromagnetic equations (*e.g.* Lambert's law and Maxwell's field equations) as well as heat and mass transport equations, are generally used to describe microwave absorption and to predict temperature and moisture distribution during microwave heating of food materials.

Lambert's law- It is based on the exponential decay of microwave absorption within the product. As Lambert's law is limited to semi-infinite samples, this law leads to poor approximation for various practical situations.

Maxwell's equations – These equations provide exact solution for the propagation of microwave radiation within the samples. The Maxwell's equations which govern the propagation of microwave radiation in a dielectric medium are given by

$$\nabla . \boldsymbol{D} = \nabla .(\varepsilon {}^{*}\boldsymbol{E}) = \rho$$

$$\nabla . \boldsymbol{B} = \nabla (\mu . \boldsymbol{H}) = 0$$

$$\nabla x E = \frac{\partial B}{\partial \tau}$$

$$\nabla x H = J + \frac{\partial D}{\partial \tau}$$

where, ***H*** and ***E*** are the magnetic and electric field intensities respectively.

J and ∂ D/∂τ denote the current density and displacement current density, respectively.

D and ***B*** signify the electric flux density and magnetic flux density respectively; μ is the magnetic permeability.

ε* is the permittivity; ρ is the density of the food material and *t* is the time required for heating. Note that, ***D, B, J, H*** and ***E*** are vector quantities whereas ε*, μ, and ρ are scalar quantities.

Microwave heating of solid food with **rectangular and cylindrical** geometries were theoretically investigated [1g]. Lambert's law was used to determine the temperature distribution and microwave power absorption during microwave heating of solid foods and the equations were solved using two dimensional finite element software. For cylindrical samples, the temperature distribution predicted by the model was similar at all regions, except at the central region. For **slab sized** potatoes, the predicted temperature and moisture distribution were found to vary slightly with the measured values.

The temperature distribution in 2 per cent agar gel cylinders using pulsed and continuous microwave heating were predicted using Lambert's law and Maxwell's equations [1h]. The temperature predictions provided by the Maxwell's equations were more accurate than that of Lambert's law equation. This is due to the fact that Maxwell's equations consider standing wave effect inside the sample whereas Lambert's law does not account for standing wave effect, due to the assumption of semi-infinite sample length of materials.

Problems Associated with the Microwave Heating [1a]

The microwave heating is volumetric and hence is more uniform compared to many traditional heating methods but still non-uniform temperature distribution is the major problem associated with this. Because of uneven temperature distribution, micro-organisms are not fully eradicated during microwave pasteurization [2]. The significant parameters which can affect the non-uniformity are penetration depth, microwave flux and the duration of microwave heating [2a].

1. Microwave pasteurization of ready-to-eat meals shows that the multimode (2450 MHz, 1.5 kW) and single mode tunnel (896 MHz, 7 kW) microwave systems were shown to display different temperature distributions. In multimode microwave ovens, the use of turntable reduces the non-uniformity in the temperature of food products [1a].
2. The size and shape of the food materials affect the temperature distribution. Three different shapes such as, brick, cylinder and hexagonal prism with three different volumes were studied to determine non-uniform temperature distribution of potato samples and found the following observations [2b] :

Brick Shaped Samples

The hot spot occurred at the corner and the cold spot occurred at the geometric center.

Cylinder Shaped Products

The hot spot occurred at the center.

Hexagonal Prism Samples

The hot spot was found to be at the boundary regions. It was also reported that the hexagonal prism shaped products provided more uniform temperature distribution than cylinder or brick shaped products.

The following solutions were proposed to decrease non-uniformity in temperature distribution during microwave heating:

1. Combining conventional and microwave heating,

2. Controlling the food geometry,
3. Providing shielding using metallic bands at suitable spacing and orientation,
4. Providing suitable microwave oven design,
5. Manipulating the heating cycle
6. Heating with reduced microwave power for long duration [2].

Equipment

Microwave ovens come in a variety of designs although the underlying principles are the same. Each MW system typically consists of three basic parts: a MW source, a waveguide, and an applicator. The most widely used source of microwaves for either industrial or commercial applications is the magnetron tube [2c]. A magnetron consists of a vacuum tube with a central electron-emitting cathode of highly negative potential surrounded by a structured anode [3].

The magnetron requires several thousand volts of direct current and converts the power supplied into MW energy, emitting high-frequency radiant energy. The polarity of the emitted radiation changes between negative and positive at high frequencies. The power of the magnetron can range from 300 to 3000 W and for industrial equipment various magnetrons are used to increase the global power output. A waveguide channels the microwaves into the cavity that holds samples for heating. Domestic ovens are designed with reflecting cavity walls that produce several modes of microwaves, thereby maximizing the efficiency of the heating process [4]. Single-mode ovens are also available, which distribute the microwaves into the reactor in a precise way. Typically, MW food processing makes use of two frequencies: 2450 and 915 MHz. Of these two, the 2450 MHz frequency is used for home ovens, while both are used in industrial heating [3].

In the combination oven, halogen lamps were provided at the top (oven ceiling) and bottom (oven floor) and a rotary table was provided to improve heating uniformity of the samples. The results showed that the temperature were higher in microwave-infrared combination, intermediate in microwave-impingement combination and lower in JET oven [1a].

Advantages of Microwave Heating [5]

The followings are the advantages of microwave heating:

1. Rapid heating of foods reduces processing time.
2. Large potential for reduced.
3. Energy consumption.
4. Higher product quality, due to reduced processing time and volumetric heating (less shear forces)
5. Increased production capacity and flexibility.
6 Potential for reducing the acryl amide content in certain products.
7. Can be used in the development and production of innovative products.

Sources

1. *Hudaa Neetoo (Faculty of Agriculture, University of Mauritius, Réduit, Mauritius) and Haiqiang Chen (Department of Animal and Food Sciences, University of Delaware, Newark, Delaware, US), Alternative Food Processing Technologies.*

1a. *S. Chandrasekaran, S. Ramanathan and Tanmay Basak, Microwave food processing - a review, Department of Chemical Engineering, Indian Institute of Technology Madras, Chennai, https: //www.intechopen.com/microwave-heating/ microwave-applications-in-thermal.*

1b. *Datta, A. K., and Anantheswaran, R. C., 2000. Handbook of microwave technology for food Applications, New York: Marcel Dekker Inc.*

1c. *Datta, A. K. and Davidson, P. M., 2000, Microwave and radiofrequency processing. Journal of Food Science Science, 65, 32-41.*

1d. *Icier, F., and Baysal, T. (2004a), Dielectrical properties of food materials-1: Factors affecting and industrial uses. Critical Reviews in Food Science and Nutrition, 44, 465-471.*

1e. *Ryynanen, S., 1995, The electromagnetic properties of food materials: a review of the basic principles. Journal of Food Engineering, 26, 409-429.*

1f. *Coronel, P., Truong, V-D., Simunovic, J., Sandeep, K. P., and Cartwright, G. D., 2005, Aseptic processing of sweetpotato purees using a continuous flow microwave system. Journal of Food Science, 70, E531-E536.*

1g. *Lin, Y. E., Anantheswaran, R. C., and Puri, V. M., 1995, Finite element analysis of microwave heating of solid foods. Journal of Food Engineering, 25, 85-112.*

1h. *Yang, H. W., and Gunasekaran, S. (2004), Comparison of temperature distribution in model food cylinders based on Maxwell's equations and Lambert's law during pulsed microwave heating. Journal of Food Engineering, 64, 445-453.*

2. *Vadivambal, R., and Jayas, D. S., 2010, Non-uniform temperature distribution during microwave heating of food materials - a review. Food and Bioprocess Technology, 3, 161-171.*

2a. *Lobo, S., and Datta, A. K. (1998), Characterization of spatial non-uniformity in microwave reheating of high loss foods. Journal of Microwave Power and Electromagnetic Energy, 33, 158-166.*

2b. *Vilayannur, R. S., Puri, V. M., and Anantheswaran, R. C., 1998, Size and shape effect on nonuniformity of temperature and moisture distributions in microwave heated food materials: Part I simulation. Journal of Food Process Engineering, 21, 209-233.*

2c. *Metaxas AC, Meredith RJ, 1983, Industrial Microwave Heating. IEE Power Engineering Series. London: Peregrinus.*

3. *Regier M, Rother M, Schuchmann HP, 2010, Alternative heating technologies. In: Ortega-Rivas E (ed) Processing Effects on Safety and Quality of Foods. Boca Raton, FL: CRC Press, pp, 188-245.*

4. *Orsat V, Raghavan V, 2005, Microwave technology for food processing- an overview. In: Helmar S, Regier M (eds) The Microwave Processing of Foods. Cambridge: Woodhead, pp. 105-115.*

5. *Microwave Heating Applications, Industrial solutions, Technical Literature, SP, Box 5401, SE-402 29 Goteborg.*

Food Processing Applications of Microwaves [1, 1a]

There are six major classes of applications for MW processing for foods:

1 Pasteurization
2. Sterilization
3. Tempering
4. Dehydration
5. Blanching
6. Cooking [2].

Other uses have also been mentioned, including the use of MW in the following fields:

1. Baking
2. Coagulation
3. Coating
4. Gelatinization
5. Puffing
6. Roasting.

Pasteurization and Sterilization [1, 1a]

1. Microwaves Pasteurization

Several thermal and Non-thermal methods are being used for pasteurisation of food and dairy products. Microwave pasteurization is one of these methods. Since MW energy can heat many foods effectively and rapidly, considerable research has focused on the use of MW heating for pasteurization and sterilization applications. Microwave pasteurization was found to be more effective in the destruction of pathogens or in the inactivation of enzyme, due to significant enhancement or magnification of thermal effects. Microwaves pasteurization is a gentler heat treatment, occurring between 60°C and 82°C.

Mechanism [1a, 2a]

Destruction of microbes or enzymes by microwave or radiofrequency waves at sublethal temperatures was explained by one or more of the following theories:

1. **Selective Heating**: The selective heating theory suggests that the micro-organisms are selectively heated due to microwaves and reach temperature higher than that of surrounding fluid. This causes the micro-organisms to be destroyed more quickly.

2. **Electroporation**: According to the electroporation theory, the electrical potential across the cell membrane causes pores, which results in the leakage of cellular materials.
3. **Cell membrane rupture**: In the cell membrane rupture theory, cell membrane is ruptured due to the voltage applied across the cell membrane.
4. **Magnetic field coupling**: According to the magnetic field coupling theory, the internal components of the cell are disrupted due to the coupling of electromagnetic energy with critical molecules such as protein or DNA [2a].

Applications

a Liquid Foods

The application of microwave pasteurization has been largely applied to fluid foods such as pasteurization of fresh juices and sterilization of milk. Microwave pasteurization of fresh juice and microwave sterilization milk were reviewed [2b]. Microbial and enzyme inactivation of various fluid foods such as apple juice, apple cider, coconut water, grapewater juice, milk and sweet potato puree were studied. The suitability of MW pasteurization to inactivate *E. coli* in apple juice was demonstrated [3]. It was showed [4] that MW pasteurization of milk by continuous flow achieved a satisfactory level of microbial reduction without excessive damage to sensory quality. In addition, the use of MW heating combined with other methods such as ultraviolet (UV) light, hydrogen peroxide, and γ-irradiation has shown synergistic effects, appearing to be promising food decontamination strategies.

b. Solid Foods

Microwave pasteurization of solid food products has some advantages over conventional treatment in terms of duration, whereas other novel techniques may also need to be combined to obtain the optimal process sequence. The Microwave pasteurization of egg shell and pickled asparagus was studied. Microwave pasteurization was able to achieve 2-fold reduction in the number of Alicyclobacillus acidoterrestris spores in a cream of asparagus [1a].

2. Microwave Sterilization

MW sterilization operates in the temperature range of 110-130°C. The Microwave sterilization was found to save cost, thermal processing time and improve the quality(Fresh-like taste and Texture and improving visual appeal), increase retention of nutrients, better energy usages, and post packaging process. Glass, paper and ceramic were used for microwave packaging [4a]. Significant reduction in bacterial count (*Pseudomonas fragi* and *Escherichia coli*) was achieved, when UV, laser and microwave heat treatments were applied in sequence [4b]. The normal shelf life of microwave sterilized food is 2-3 years or longer [4d].

Recently, U. S. Food and Drug Administration has approved microwave sterilization process for mashed potatoes in trays and salmon fillet in sauce in pouches [4c].

Microwave Baking

It has also been extensively investigated [5, 6, 7]. MW technology has made an immense contribution to accelerated baking, leading to an enhanced throughput with negligible additional space required for MW power generators. Accelerated baking, through the combination of MW and conventional or infrared baking, enhances throughput whilst ensuring product quality, appropriate degrees of crust formation, and surface browning [8].

It is required to obtain browning and good texture at a fixed moisture level in bread baking process [8a]. Conventional baking using hot air provides suitable color and texture where as in microwave baking, sufficient brown color on the surface of breads and crust formation were not possible because in microwave heating, the air surrounding the food product is cold and water evaporating from food gets condensed on contact with cold air, which results in the lack of crispness of food product. Susceptors(microwave absorbing materials, which convert microwave to heat) were placed at the bottom of the sample to provide crust formation and surface browning of the food product [1a].

In contrast to conventional baking, MW heating quickly inactivates enzymes such as α-amylase, due to a fast and uniform temperature rise in the whole product, to prevent the starch from extensive digestion, and releasing sufficient carbon dioxide and steam to produce a uniform porous texture [8]. The moisture content of microwave-baked breads was found to be lowest among the samples. Microwave-baked samples were found to have highest hardness values, setback viscosities, total mass crystallinity values and retrogradation enthalpies among other heating modes [1a].

Microwave Cooking [1a]

It is also the major application of microwave and several studies on the effects of microwave on cooking parameters such as quality, taste and color retention for various food materials are conducted. Microwave cooking is affected by the presence of moisture and fat content in food materials. Microwave cooked products have the advantages of retaining more taste, color, quality and nutritional value compared to those cooked by other conventional methods.

1. Microwave cooking of rice provided shorter cooking time period than other cooking methods. The uniformity of moisture distribution is better for microwave cooking.
2. Microwaves cooking of bacon (streaky and back type) in domestic and industrial microwave oven was studied. Cooking was more uniform in industrial microwave than in domestic microwave oven, due to uniformity in magnetron position and the movement of material through the microwave system. The industrial microwave oven yielded a better and economical product by reducing the weight loss, cooking time as well as by providing uniform cooking [1a, 8b].
3. The effects of various fat levels (5, 10, 15 and 20 per cent) on microwave cooking of goat meat patties were studied. Microwave cooking time was

found to decrease with an increase in fat level, as the dielectric constant and loss factor decrease with fat content. The sample with high fat content might possess lower specific heat capacity which might lead to decrease in the heating rate. The product yield was found to be significantly lower for 20 per cent fat level due to high total cooking loss (15.2 per cent). The cooking loss during microwave heating can be reduced by adding salt or sodium content to the meat patties. The amount of fat content in food materials influences the microwave heating in terms of heating rate, uniformity of temperature distribution and fat retention [1a, 8c].

Microwave Drying [1, 1a]

Drying of food materials is a process in which moisture is removed from food materials without affecting their physical and chemical composition [1a]. Microwave drying has the advantages of achieving fast drying rates and improving the quality of some food products. The energy absorption level is controlled by the wet products which can be used for selective heating of interior parts of the sample containing moisture and without affecting the exterior parts. Microwave drying is considered very useful during falling rate period. The shrinkage of food materials is prevented in microwave drying. [1a]. Microwaves can also enhance drying for low-moisture food products [2]. It helps in preserving the food products and enhancing their storage stability The overriding advantage of MW drying and dehydration is that damage to nutrients and sensory quality, typically caused by prolonged drying times and elevated surface temperatures, is avoided [9]. Although the MW oven is a common household appliance, MW heating has found increasing applications in the food industry in various processing operations.

1. The microwave drying of spinach leaves, carrot slices and sliced mushroom were studied [1a].
2. One of the **disadvantages of microwave drying** is that excessive temperature along the corner or edges of food products results in scorching and production of off-flavors especially during final stages of drying [1a].

Microwave Assisted Dryings [1a]

The Microwave assisted dryings are used in food processing [1a]. Microwave combined with other drying methods such as air drying or infrared or vacuum drying or freeze drying gave better drying characteristics compared to their respective drying methods or microwave drying alone. Then following microwave assisted drying techniques are being used:

1. Microwave Assisted Air Drying [1a]

In the microwave assisted air drying, hot air drying is combined with microwave heating in order to enhance the drying rate. Microwave can be combined with hot air in the following three different stages of the drying process:

i. **At the initial stage**: At this stage, microwave is applied at the beginning of the dehydration process, in which the interior gets heated rapidly.

ii. At rapid drying period, a stable temperature profile is established in such a way the vapor is forced outside due to improved drying rate.

iii. **At the final stage**: At the reduced drying rate period or at the final stage of drying, the drying rate begins to fall where the moisture is present at the centre and with the help of microwave heating, vapor is forced outside in order to remove bound water [9c]. The microwave assisted air drying is found to be helpful at the final stages of drying food products especially for fruits and vegetables. Besides increasing the drying rate, microwave assisted air drying enhances the rehydration capacity of dried products and also overcomes shrinkage problems.

   a. Drying of banana is difficult as it falls under the falling rate period. But, hot air drying combined with microwave finish drying reduced the drying time by 64 per cent as compared to convective air drying [9d].

   b. During microwave assisted air-drying of apple and mushroom, a minimum air velocity of 1 m/s was required in order to prevent browning of the food samples. When compared to hot air drying, microwave assisted air drying reduced the drying time by a factor of two for apples and by a factor of four for mushrooms.

   c. The garlic cloves, thin layer carrot cubes, osmotically dehydrated strawberries, pre-cooked soybeans were successfully dried by microwave assisted air-drying method.

2. *Microwave Assisted Vacuum Drying [1a]*

In vacuum drying, high energy water molecules diffuse to the surface and evaporate due to low pressure and because of this, water vapor concentrates at the surface and the low pressure causes the boiling point of water to be reduced. Thus vacuum drying prevents oxidation due to the absence of air, and thereby maintains the color, texture and flavor of the dried products. In the absence of convection, either conduction or radiation or microwaves can be combined with vacuum drying to improve its thermal efficiency [9c].

a. Vacuum microwave drying of banana slices was examined at a microwave power supply of 150 W and found the quality of the product to be good and was comparable to that of freeze dried product. The dried product also provided excellent taste and flavor with no shrinkage or change in color [9e].

b. A study on microwave vacuum drying of carrot slices showed that the microwave vacuum dried products had higher α-carotene content and vitamin C content, softer texture, had higher rehydration potential and lesser color deterioration than that of air drying [9f].

c. The microwave vacuum drying is applied for heat sensitive materials such as banana, carrot, potato, *etc.* The loss of nutritional qualities (vitamins, α and β-carotenes *etc.*) of food products by microwave vacuum drying is minimum due to non exposure of heat and oxygen.

3. *Microwave Assisted Freeze Drying [1a]*

Freeze drying is considered as a gentle dehydration technique applied for heat sensitive foods, pharmaceutical and biological materials [9c]. In freeze drying, the temperature is lowered and, by applying vacuum or low pressure, the frozen water is directly transferred to the vapor phase without going through the liquid phase. Microwave freeze drying can be applied in following two different ways:

i. Freeze drying accompanied concurrently with the help of microwave - In the first type, the whole drying process takes place under vacuum environment and microwave field is applied to supply the heat of sublimation required for freeze drying.

ii. Microwave drying applied after freeze drying [9g]- In second type, the drying process was divided into two stages, a- freeze drying followed by b- microwave/vacuum microwave drying [9g]. Free drying combined with microwaves offers advantages like reduced processing time and better product quality [9c].

iii. The quality of freeze dried products is better than other conventional drying methods due to its low processing temperature and lack of oxygen in the process but freeze drying is an expensive and lengthy dehydration process, which lead to small throughput and high capital and energy costs [9c, 9g].

iv-. In microwave freeze drying of cabbage, the drying rate of microwave freeze drying was twice greater than that of vacuum freeze drying. Microwave freeze drying affected the drying rate of falling rate drying period more significantly than that of constant rate drying period. The drying rate of microwave freeze drying was found to increase with a decrease in the material thickness and the cavity pressure [1a].

v. High maturity samples displayed better quality parameters such as color, rehydration ratio and hardness than those of low and medium matured samples but high maturity samples lead to uneven heating due to more sugar content. This results in the poor appearance of the product caused by the expansion of sugar molecules due to uneven heating [1a, 9h]. The microwave assisted freeze drying of high maturity banana slices containing high sugar content at high drying rates, was studied and it was noted that sugar molecules have higher loss factor than ice crystals and hence they absorb more microwave energy and convert it to heat efficiently [9h].

Microwave Blanching

It is generally used for color retention and enzyme inactivation, which is carried out by immersing food materials in hot water, steam or boiling solutions containing acids or salts.

i. Microwave blanching of herbs such as marjoram and rosemary was carried out by soaking the herbs in a minimum quantity of water and exposed to microwaves [9a].

ii. Microwave blanching of peanuts and found that microwave blanching was better than traditional blanching techniques in terms of energy and time savings [9b].

iii. Microwave blanching was observed for maximum retention of color, ascorbic acid and chlorophyll contents than that of water and steam blanching.

iv. Blanching is also used to remove seed coat or testa which may reduce enzyme activity and moisture content and might interfere with further processing into specific products. Blanching also helps to remove damaged or discolored seeds, foreign material and dust.

MW processing has been successfully applied on a commercial scale for meat sausage cooking, bacon precooking. It is also used for the tempering of products and Pasteurizing diverse products as given in following table:

Sl. No.	*Processing*	*Products*
1	Tempering	Beef meats [10]
		Fish blocks [11]
		Frozen potato purée [12]
2	Pasteurizing diverse products	Raw meats [13]
		Beef frankfurters [14]
		In-shell eggs [15]
		Mashed potatoes [16]
		Orange juice [17]
		Apple cider [18]

Sources

1. *Hudaa Neetoo (Faculty of Agriculture, University of Mauritius, Réduit, Mauritius) and Haiqiang Chen (Department of Animal and Food Sciences, University of Delaware, Newark, Delaware, US), Alternative Food Processing Technologies.*

1a. *S. Chandrasekaran, S. Ramanathan and Tanmay Basak, Microwave food processing - a review, Department of Chemical Engineering, Indian Institute of Technology Madras, Chennai, https: //www.intechopen.com/microwave-heating/microwave-applications-in-thermal.*

2. *Fu, YC, 2010, Microwave heating in food processing. In: Hui YH(ed) Handbook of Food Science, Technology and Engineering, Boca Raton, FL: CRC Press, pp. 1-11.*

2a. *Kozempel, M. F., Annous, B. A., Cook, R. D., Scullen, O. J. and Whiting, R. C.1998, Inactivation of micro-organisms with microwaves at reduced temperatures. Journal of Food Protection, 61, 582-585.*

2b. *Salazar-Gonzalez, C., San Martin-Gonzalez, M. F., Lopez-Malo, A., and Sosa-Morales, M. E., 2012, Recent studies related to microwave processing of fluid foods. Food Bioprocess and Technology, 5, 31-46.*

3. *Canumir JA, Celis JE, de Bruijn J, Vidal LV, 2002, Pasteurisation of apple juice by using microwaves. Lebensmittel Wissenschaft and Technologie 35: 389-392.*

4. *Villamiel M, Lopez-Fandino R, Olano A, 1997, Microwave pasteurization of milk in a continuous flow unit: effects on the cheese-making properties of goat's milk. Milchwissenschaft 52: 29-32.*

4a. *Guillard, V., Mauricio-Iglesias, M., and Gontard, N., 2010, Effect of novel food processing methods on packaging: Structure, composition, and migration properties. Critical Reviews in Food Science and Nutrition, 50, 969-988.*

4b. *Maktabi, S., Watson, I., and Parton, R., 2011, Synergistic effect of UV, laser and microwave radiation or conventional heating on E. coli and on some spoilage and pathogenic bacteria. Innovative Food Science and Emerging Technologies, 12, 129-134.*

4c. *Brody, A. L., 2012, The coming wave of microwave sterilization and pasteurization. Food Technology Magazine, 66(3), 78-80.*

4d. *Nalaka Nugapola, Microwave Food Processing, UVa Wellassa University of Srilanka.*

5. *Icoz D, Sahin S, Sumnu G, 2004, Color and texture development during microwave and conventional baking of breads. International Journal of Food Properties 7: 201-213.*

6. *Sumnu G, Sahin S, 2005, Recent developments in microwave heating. In: Sun DW(ed) Emerging Technologies for Food Processing. California: Elsevier Academic Press, pp. 419-444.*

7. *Sumnu G, Keskin SO, Rakesh V, Datta AK, Sahin S, 2007, Transport and related properties of breads baked using various heating modes. Journal of Food Engineering 78: 1382-1387.*

8. *Ohlsson T, Bengtsson N, (2002), Minimal processing of foods with non-thermal methods. In: Ohlsson T, Bengtsson N (eds) Minimal Processing Technologies in the Food Industry. Cambridge: Woodhead Publishing, pp. 34-60.*

8a. *Icoz, D., Sumnu, G., and Sahin, S., 2004, Color and texture development during microwave and conventional baking of breads. International Journal of Food Properties, 7, 201-213.*

8b. *James, C., Barlow, K. E., James, S. J., and Swain, M. J., 2006, The influence of processing and product factors on the quality of microwave pre-cooked bacon. Journal of Food Engineering, 77, 835-843.*

8c. *Das, A. K., and Rajkumar, V., 2011, Effect of different fat level on microwave cooking properties of goat meat patties. Journal of Food Science and Technology, DOI 10.1007/s13197.*

9. *Brewer M, 2005, Microwave processing, nutritional and sensorial quality. In: Helmar S, Regier M (eds) The Microwave Processing of Foods. Cambridge: Woodhead, pp.76-94.*

9a. *Singh, M., Raghavan, B. and Abraham, K. O., Processing of marjoram (Marjona hortensis Moench) and rosemary (Rosmarinus officinalis L.). Effect of blanching methods on quality, 1996, Nahrung, 40, 264-266.*

9b. *Schirack, A. V., Drake, M. A., Sanders, T. H. and Sandeep, K. P., 2006a, Characterization of aroma-active compounds in microwave blanched peanuts. Journal of Food Science, 71, C513-C520.*

9c. *Zhang, M., Tang, J., Mujumdar, A. S., and Wang, S., 2006, Trends in microwave-related drying of fruits and vegetables, Trends in Food Science and Technology, 17, 524-534.*

9d. *Maskan, M., 2000, Microwave/air and microwave finish drying of banana. Journal of Food Engineering, 44, 71-78.*

9e. *Drouzas, A. E., and Schubert, H., 1996, Microwave application in vacuum drying of fruits, Journal of Food Engineering, 28, 203-209.*

9f. *Lin, T. M., Durance, T. D., and Scaman, C. H., 1998, Characterization of vacuum microwave, air and freeze dried carrot slices. Food Research International, 31, 111-117.*

9g. *Duan, X., Zhang, M., Mujumdar, A. S., and Wang, R., 2010a, Trends in microwave-assisted freeze drying of foods. Drying Technology, 28, 444-453.*

9h. *Jiang, H., Zhang, M., and Mujumdar, A. S., 2010a, Microwave freeze-drying characteristics of banana crisps, Drying Technology, 28, 1377-1384.*

10. *Farag KW, Cronin DA, Morgan DJ, Lyng JG, 2008, Dielectric and thermo-physical properties of different beef meat blends. over a temperature range of -18 to +10°C, Meat Science 79: 740-747.*

11. *Ramaswamy H, Tang J, 2008, Microwave and radiofrequency heating. Food Science and Technology International 14: 423-427.*

12. *Seyhun N, Sahin S, Ahmed J, Sumnu G, (2009), Comparison and modeling of microwave tempering and infrared assisted microwave tempering of frozen potato puree. Journal of Food Engineering 92: 339-344.*

13. *Huang L, Sites J, 2010, New automated microwave heating process for cooking and pasteurization of microwaveable foods containing raw meats. Journal of Food Science 75: 110-115.*

14. *Huang L, Sites J, 2007, Automatic control of a microwave heating process for in-package pasteurization of beef frankfurters. Journal of Food Engineering 80: 226-233.*

15. *Dev SRS, Gariepy Y, Raghavan GSV, 2008, Dielectric properties of egg components and microwave heating for in-shell pasteurization of eggs. Journal of Food Engineering 86: 207-214.*

16. *Guan D, Cheng M, Wang Y, Tang J, 2004, Dielectric properties of mashed potatoes relevant to microwave and radiofrequency pasteurization and sterilization processes. Journal of Food Science 69: 30-37.*

17. *Cinquanta L, Di Matteo M, Cuccurullo G, Albanese D, 2010, Effect on orange juice of batch pasteurization in an improved pilot-scale microwave oven. Journal of Food Science 75: 46-50.*

18. *Gentry TS, Roberts JS, 2005, Design and evaluation of a continuous flow microwave pasteurization system for apple cider. LWT Food Science and Technology 38: 227-238.*

Effect of Microwaves on the Sensory and Nutritional Quality of Foods [1]

Milk is a diverse source of vitamins and MW heating has been shown to have a variable effect on its nutrients.

1. It was reported [2] an insignificant loss in vitamin A, B1 (thiamine) and B2 (riboflavin) although losses of the vitamins E and C, by 17 per cent and 36 per cent respectively, were observed.
2. However, the heat stability of thiamine and riboflavin in microwave-heated and traditionally heated milk was compared [3] and reported virtually no difference.
3. The chemical and sensory changes in milk pasteurized by microwave and conventional systems during cold storage was compared [4]. Milk heated in the microwave oven or in the conventional system for 15 sec was not distinguished by a sensory panel using a triangle test procedure either after processing or during the storage period at 4.5 °C for up to 15 days.
4. The effects of microwave pasteurization on various quality parameters of orange juice such as cloud stability, color, carotenoid compounds, and vitamin C content were investigated [5]. It was observed that the carotenoid content, responsible for the sensory and nutritional quality in fresh juices, decreased by about 13 per cent after MW pasteurization at 70°C for 1 min. However, the decrease in vitamin C content was minimal with retention ranging from 96.1 per cent to 97 per cent.
5. The effect of MW heating on the various quality parameters of Granny Smith apple purée such as vitamin C stability, total polyphenol content, viscosity, color, and titratable acidity was examined [6]. It was demonstrated that MW treatment at 652W for a short duration of 35 sec followed by storage at 5 °C for 15 days resulted in an average loss of vitamin C of the order of 50 per cent although the viscosity and titratable acidity of the product were unaffected during the storage period.

Sources

1. *Hudaa Neetoo (Faculty of Agriculture, University of Mauritius, Réduit, Mauritius) and Haiqiang Chen (Department of Animal and Food Sciences, University of Delaware, Newark, Delaware, US), Alternative Food Processing Technologies.*
2. *Lopez-Fandino R, Villamiel M, Corzo N, Olano A, 1996, Assessment of the thermal treatment of milk during continuous microwave and conventional heating. Journal of Food Protection 59: 889-892.*
3. *Sierra I, Vidal-Valverde C, Olano A, 1999, The effects of continuous flow microwave treatment and conventional heating on the nutritional value of milk as shown by influence on vitamin B1 retention. European Food Research and Technology 209: 352-354.*

4. *Valero E, Martinez-Castro I, Sanz J, Villamiel M, 2000, Chemical and sensorial changes in milk pasteurised by microwave and conventional systems during cold storage. Food Chemistry 70: 77-81.*

5. *Cinquanta L, Di Matteo M, Cuccurullo G, Albanese D, 2010, Effect on orange juice of batch pasteurization in an improved pilot-scale microwave oven. Journal of Food Science 75: 46-50. Cole JC, 1993, The User's Guide to Cuisine Sous-Vide. Buckinghamshire: JC Group Associates, pp. 63.*

6. *Picouet PA, Catellari M, Viclas I, Landl A, Abadias M, 2009, Minimal processing of a Granny Smith apple purée by microwave heating. Innovative Food Science and Emerging Technologies 10: 545-550,*

Chapter 13

Radiofrequency Heating Technology

On the basis of direct and indirect heating, electro heating can be sub-divided into the followings **[1]**:

1. Direct electro heating where electrical current is applied directly to the food *e.g.* ohmic heating (OH).
2. Indirect electro heating where the electrical energy is firstly converted to electromagnetic radiation which subsequently generates heat within the product *e.g.* microwave (MW) or radiofrequency (RF) heating [1c].

Radiofrequency (RF)heating is another form of dielectric heating, which has potential for the rapid heating of solid and semi-solid foods. RF heating refers to the heating of dielectric materials with electromagnetic energy at frequencies between 1 and 300 MHz [1]. The RF heating technology is in research for many decades, the industrial application of it is still limited towards heterogeneous food stuff like meat, disinfection of agricultural commodities and post baking, where as there is a large potential source available in utilizing RF technology [1c]. **Radiofrequency** (**RF**) **heating** is a promising, an advanced and emerging technology for **food** applications because of the associated rapid and uniform **heat** distribution, large penetration depth and lower energy consumption Due to this reason, RF has entered the food industry to sterilize, pasteurize, disinfect food and agricultural commodity like fruits and dry nuts, post baking and in many other applications [1c].

Radiofrequency heating has been successfully applied for drying, baking and thawing of frozen meat and in meat processing. However, its use in continuous pasteurization and sterilization of foods is rather limited. During RF heating, heat is generated within the product due to molecular friction resulting from oscillating molecules and ions caused by the applied alternating electric field. RF heating is influenced principally by the dielectric properties of the product when other conditions are kept constant. The frequency level, temperature and properties of food, such as viscosity, water content and chemical composition affect the dielectric properties and thus the RF heating of foods. Therefore, these parameters should be taken into account when designing a radiofrequency heating system for foods [1a]. The demand for safe, hygienic, tastier, no fat and preservative free food has widened up RF application in the modern industries [1c]. RF heating in current scenario is the preferred method in modern industries for product cooking, baking, drying, pasteurization and other applications [1b].

Dielectric **heating**, also known as electronic **heating**, **RF** (**radiofrequency**) **heating**, and high-**frequency heating**, is the process in which a **radiofrequency** alternating electric field, or **radio** wave or microwave electromagnetic radiation heats a dielectric material. Radiofrequency heating is accomplished through a combination of dipole heating and electric resistance heating resulting from the movement of dissolved ions present in the food. Although identical to the other electromagnetic heating methods, radiofrequency with electromagnetic waves in 30-300MHz spectrum has the added advantage of uniform heating in homogeneous foods and most important of all, high penetration depth that could be used to pasteurize or sterilize liquid products. Rapid heating, as a primary advantage of RF heating allow faster line speeds and shorter line lengths for semi continuous as well as continuous unit operation in food industries [1c, 1d].

Principle of RF Heating [1c]

The mechanism of radiofrequency heating is that the molecules within a product placed in an RF environment re-orient themselves (27 million times/s at 27 MHz) continuously in response to the applied field. This response initiates volumetric heating within the entire product due to frictional interaction between the molecules, thus selectively heats only the product and not the air or surrounding it. Radiofrequency heating is accomplished through a combination of dipole heating and electric resistance heating resulting from the movement of dissolved ions present in the food. Radiofrequency with electromagnetic waves in 30-300MHz spectrum has the added advantage of uniform heating in homogeneous foods and most important of all, high penetration depth that could be used to pasteurize or sterilize liquid products.

In general food materials are poor electric insulators. Food materials have the ability to store and dissipate electric energy when subjected to an electro-magnetic field. The principle of absorption of this energy is described by Maxwell's equations. In practice, a relative complex permittivity $\varepsilon^* = \varepsilon/\varepsilon o$ is used where $\varepsilon 0 = 8.8542 \times 10^{-12}$ F/m is the permittivity of free space. This relative complex permittivity ε^*, is composed of two properties ε (Dielectric constant) and ε'' (Dielectric loss factor), are the component responsible for dielectric heating. This dimensionless number measures the ability of the material to interact with the electric field of RF.

- ✰ The dielectric constant is the measure of the food material's ability to store electromagnetic energy.
- ✰ The dielectric loss factor is the material's ability to dissipate electromagnetic energy.

Radiofrequency Equipment and Working [1, 1c]

Working [1c]

In a **radiofrequency heating** system the **RF** generator creates an alternating electric field between two electrodes. The material to be **heated is** conveyed between the electrodes causing polar water molecules in the material to continuously reorient to face opposite electrodes. Friction resulting from this molecular movement causes the materials to rapidly heat volumetrically.

The following figure depicts a radiofrequency heating system with material between the electrodes:

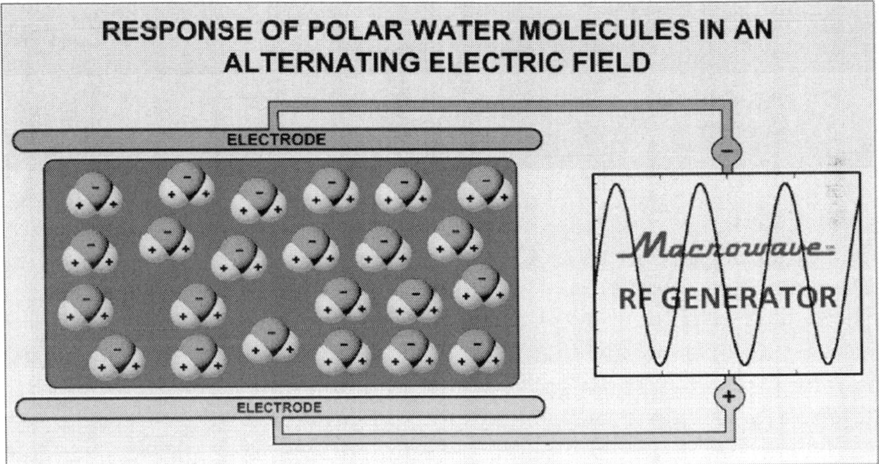

In rediofrequency heating, the energy absorption involves the following two primary mechanisms :

1. Dipolar relaxation and
2. Ionic conduction.

i. Dipolar Relaxation

Water in the food is often the primary component responsible for dielectric heating. Due to their dipolar nature, water molecules try to follow the electric field associated with electromagnetic radiation as it oscillates at very high frequencies, thus producing heat.

ii. Ionic Conduction

This mechanism of heating is through the oscillatory migration of ions in the food that generates heat. The amount of heat generated in the product is determined by the frequency, the square of the applied voltage, dimensions of the product and the dielectric loss factor of the material, which is essentially a measure of the ease with which the material can be heated by radio waves.

RF power is produced when electricity is applied to an RF generator whose signal is amplified and delivered to a parallel electrode system (RF cavity), in which the selected material is placed.

Equipment [1]

The equipment needed for RF heating consists of the following two fundamental components:

1. **Generator**: This is responsible for the generation of RF waves [2].
2. **Applicator**: This is responsible for the application of RF power to the food (the applicator, the main part of which is the electrodes) [2].

There are basically two groups of units used to produce and apply RF power to foods:

1. Conventional RF heating equipment-where the applicator is part of the RF generator circuit,
2. A more recent model of RF equipment, whose RF applicator is separate from the RF generator and linked only by a high-power coaxial cable [3].

The design of electrodes in RF heating equipment is one of the most crucial aspects. A number of different electrode configurations are available, depending on factors such as field strength and physical characteristics of the sample such as the moisture content, thickness, geometry, *etc.* There are following three main configurations for the electrodes [2]:

- ☆ Through field electrodes for thick samples,
- ☆ Fringe field electrodes for thin samples,
- ☆ Staggered through field electrodes for samples of intermediate thickness [2].

Electrodes can thus be designed to form unique electric field patterns and uniform heating patterns to suit foods of different geometry [2].

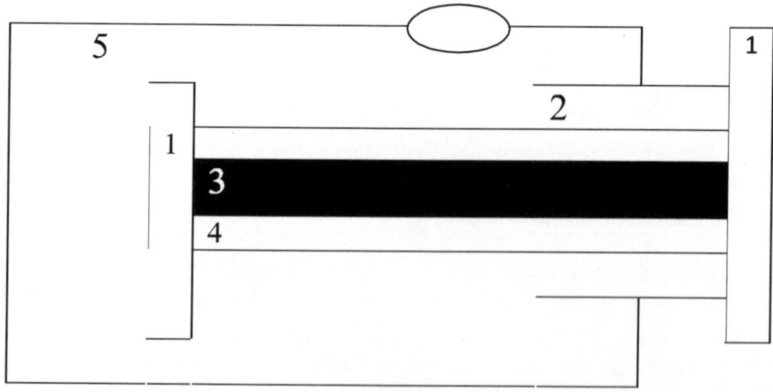

A Schematic of a RF Dielectric Heating Unit [4].
1: Flange; 2: Electrode; 3: Product; 4: Insulating Tube; 5: Radiofrequency Generator.

Sources

1. *Huḍaa Neetoo (Faculty of Agriculture, University of Mauritius, Réduit, Mauritius) and Haiqiang Chen (Department of Animal and Food Sciences, University of Delaware, Newark, Delaware, US), Alternative Food Processing Technologies.*

1a. *Piyasena P, Dussault C, Koutchma T, Ramaswamy HS, Awuah GB, Radiofrequency heating of foods: principles, applications and related properties-a review, Crit Rev Food Sci Nutr. 2003, 43(6), 587-606.*

1b. *Orsat, V., Raghavan, G, S., 2005, "Radio-Frequency Processing", Bio resource, 446-450.*

1c. *Sona Jojo and R. Mahendran, Radiofrequency heating and its application in food processing: A review, International Journal of Current Agricultural Research 2013, Vol. 1, No. 9, PP. 042-046.*

1d. *Rao. M. A., Syed S. H., Rizvi., Ashim. K. Datta., 2005, "Engineering properties of food". III edition, 105-108.*

2. *Vicente A, Castro IA, 2007, Novel thermal processing technologies. In: Tewari G, Juneja VK (eds) Advances in Thermal and Non-thermal Food Preservation. Ames, IA: Blackwell, pp. 99-131.*

3. *Rowley AT, 2001, Radiofrequency heating. In: Richardson PS (ed) Thermal Technologies in Food Processing. Cambridge: Woodhead Publishing, pp. 469-493.*

4. *Zhao Y, Flugstad B, Kolbe E, Park JW, Wells JH, 2000, Using capacitive (radiofrequency) dielectric heating in food processing and preservation - a review. Journal of Food Process Engineering 23: 25-55.*

Developmental Year-wise Steps for Potential Use of RF Technology for Food Processing [1]

The details are tabulated below:

Sl. No.	*Years*	*Details*
1	1946	The potential use of RF technology for food processing was recognized after World War II and described 'electric heat'.
2	Before 1960	RF energy employed for applications such as the cooking of processed meat products, heating of bread, dehydration and blanching of vegetables. However, there was no any commercial installations.
3	By 1960	The application of RF energy to foods was focused on the defrosting of frozen products, which resulted in several commercial production lines [2].
4	1974	It was also showed [3] that juices (peach, quince and orange) sealed in bottles and carried on a conveyer belt through an RF applicator had better bacteriological and organoleptic qualities than the juices treated by conventional thermal methods.
5	1980	The next generation of commercial applications for RF energy in the food industry was post-bake drying of cookies and snack foods [4].
6	In 1990s	The area of RF pasteurization was studied to improve energy efficiency and to solve technical problems such as run-away heating [5].

Sl. No.	Years	Details
7	2008 Onward	The most recent developmental step is: i. On drying, ii. Post-baking, pasteurization of juices especially apple juices and other high value produce, iii. Wastewater treatment in diary industries, iv. Disinfection in agricultural produce especially walnut, red and black pepper, poultry and fish industries [6, 7].

Sources

1. *Sona Jojo and R. Mahendran, Radiofrequency heating and its application in food processing: A review, International Journal of Current Agricultural Research, 2013, Vol. 1, No. 9, PP. 042-046.*
2. *A. C., Jason and H. R. Sanders, 1962, "Dielectric thawing of fish. I. Experiments with frozen herrings". Food Technology, 16 (6), 101-106.*
3. *M., MDemeczky, 1974, "Continuous pasteurisation of bottled fruit juices by high frequency energy". International Congress on Food Science and Technology, 4, 11-20.*
4. *J. Rice, 1993, "RF technology sharpens bakery's competitive edge", Food Processing 6, 18-24.*
5. *Y. Zhao, B. Flugstad, E. Kolbe, J. E. Park and J. H. Wells, J. H., 2000, "Using capacitive (radiofrequency) dielectric heating in food processing and preservation - a review". Journal of Food Process Engineering, 23, 25-55.*
6. *D. J., Geveke and C., Brunkhorst, 2008, "Radiofrequency electric fields inactivation of Escherichia coli in apple cider", Journal of Food Engineering, 85 (2), 215-221.*
7. *J., Wang, K., Lue-chapattanaporn, Y., Wang, J., Tang, 2012, "Radio-frequency heating of heterogeneous food - Meat lasagna", Journal of Food Engineering, 108, 183-193.*

Effect of Radiofrequency on the Sensory and Nutritional Quality of Foods [1]

The effects of RF heating on the food constituents, as well as the overall quality, have been rarely documented in the literature.

1. It was observed [2] that there is no significant differences in moisture, protein, fat, ash, and sodium chloride content in RF and steam-cooked turkey breast rolls.
2. The contents of water-soluble vitamins (thiamine and riboflavin) were the same in both cases and the texture profile analysis also gave comparable results.
3. However, the researchers did observe that the rate of lipid oxidation was significantly lower in RF-cooked meats compared to their steam-cooked counterparts. Taken together, it can be inferred that product quality of RF-processed foods is maintained if not enhanced, especially by virtue of the rapid volumetric heating characteristics of the technology.

Sources

1. *Hudaa Neetoo (Faculty of Agriculture, University of Mauritius, Réduit, Mauritius) and Haiqiang Chen (Department of Animal and Food Sciences, University of Delaware, Newark, Delaware, US), Alternative Food Processing Technologies.*
2. *Tang X, Cronin DA, Brunton NP, 2005, The effect of radiofrequency heating on chemical, physical and sensory aspects of quality in turkey breast rolls. Food Chemistry 93: 1-7.*

Merits and Demerits of RF Heating [1]

RF heating system has following merits and demerits :

a. Advantages

RF heating has following advantages:

1. Rapid and require less time to come up to the desired process time.
2. RF heating includes the ease of targeting the product, not the air surrounding it, tends to drive the moisture from inside out, and hence leveling the moisture, considerably decreasing the processing time, avoiding overheating and dehydration on the surface of the product.
3. RF heating system can be turned on or off instantly [2] and can be provided with temperature measurement systems (thermocouple).

b. Disadvantages

RF heating has following disadvantages:

1. RF heating is restricted by limited understanding of RF photons interact with materials and by high cost of its source electronics.
2. Subject to fluctuations of electrical costs is another drawback of Rf heating.
3. RF technology has major obstacle that the factors such as product dielectric properties, size, shape, position between the RF electrodes and the electrode configuration may affect temperature uniformity thus providing non uniform heating of food produce.
4. The RF heating system should be provided such that all generators and applicators must be properly shielded and specially designed to meet the product specific requirements.
5. The lack of in-depth information on area such as its impact on all aspects of product quality is another factor of concern(Meat product).

Sources

1. *Sona Jojo and R. Mahendran, Radiofrequency heating and its application in food processing: A review, International Journal of Current Agricultural Research, 2013, Vol. 1, No. 9, PP. 042-046.*
2. *V. Orsat, G. S. Raghavan, 2005, "Radio-Frequency Processing", Bio resource, 446-450.*

Applications of Radiofrequency in Food Processing [1]

In food processing, the applications of RF heating are less common than that of Microwave heating. The thermal sensitivity is in inversely proportional to biological complexity in living matter and therefore, organisms with a high degree of biological complexity have similarly high sensitivity to thermal energy [1a]. Radiofrequency has a long history of use in the food processing industry. Food applications described in the literature include blanching, thawing, drying, heating of bread/baking, meat processing, pasteurization, and sterilization [1]. RF drying has been described as a "self-leveling" phenomenon, as more energy is dissipated in wetter locations of the samples than drier ones, with the net effect of improvement in quality and consistency of the final products [2]. The applications of RF heating in different fields of food processing are discussed below:

1. Processing of Agriculture Products

RF postbaking and RF-assisted baking of biscuits, crackers and snack foods(Pasta *etc.*) is one of the most accepted and widely used applications of RF heating [1b]. RF heating of cookies and crackers can help even out the moisture distribution after baking, by targeting the remaining moisture pockets [1a]. Researchers [1a] proposed the use of RF technology in high value and highly perishable commodities like walnut for the purpose of drying as well as disinfection of the agricultural commodity.

1. **Drying of Grains**: Other applications include RF drying of grains such as cowpea grains, moisture removal, and "moisture leveling" in finished goods [2].
2. RF cooking of pumpable foods has also gained importance because heating is uniform and rapid [3].
3. Researchers studied the effect of RF on wheat flour [1a].

2. Meat and Seafoods

Another application of RF that has attracted much recent attention is defrosting of frozen meats [4a, 4b] and seafood [5]. Meat, fish, fruits and vegetables require higher preservation techniques [1a]. The research work using RF heating mechanism in meat and fish product processing have significant effect on nutrition, quality, quantity and storage of the produce [1a]. The RF heating at 27MHz and applying an alternating current of 6kW can be used to process pre-packaged heterogeneous foods (meat lasagna in large polymeric containers -295 x 253x 42 mm) to retain product quality better compared to conventional method of heating [5a]. A team from UCD Dublin, Ireland, who heated products using a 27.12 MHz RF applicator with a maximum power output of 600W included RF as the best and upcoming method to be considered in meat and fish processing industries [1a].

3. Disinfection of Agriculture Products [1a, 1c]

Researchers [1c] applied the RF heating to inactivate *Salmonella typhimurium* and *Escherichia coli* O157: H7 on black and red pepper spice and observed the results given on next page:

1. by RF heating for 50s, 2.80 to 4.29 log CFU/g reductions of S. Typhimurium and *E. coli* O157: H7 in black peppers.
2. RF heating of red peppers for 40 s reduced pathogens by 3.38 log CFU/g to more than 5 log CFU/g without affecting the color change, provided 27.12MHz RF heating system.
3. With regard to sterilization applications, researchers have demonstrated the efficacy of RF energy to inactivate heat-resistant spores to produce shelf-stable foods such as scrambled eggs.

4. Liquid Food Processing

RF technology has possibility of sterilizing and pasteurizing a food product at time temperature value much lower than those required for conventional heating techniques. Some examples of processing of liquid foods by RF are given below:

a. **Milk**: Researchers [1a, 6] used a 2 kW, 27.12 MHz, RF applicator in order to find best conditions to inactivate surrogates of both Listeria and *E. coli* cells in milk under continuous laminar flow conditions.

b. **Apple and Orange Juices**: The impact of RF treatment on microbial inactivation in orange and apple juices and in apple cider was assessed [1a, 7, 8].

c. **Soya Milk**: Scientists [1a, 9] reported the inactivation of Bacillus subtilis spores in soybean milk by radio-frequency flash heating at 28MHz by heating up to 115°C for 0.4sec. Fouling of the electrode surface by the protein(soybean milk) with high electric field was later rectified by covering the electrode surface with a Teflon film.

Sources

1. *Hudaa Neetoo (Faculty of Agriculture, University of Mauritius, Réduit, Mauritius) and Haiqiang Chen (Department of Animal and Food Sciences, University of Delaware, Newark, Delaware, US), Alternative Food Processing Technologies.*

1a. *Sona Jojo and R. Mahendran, Radiofrequency heating and its application in food processing: A review, International Journal of Current Agricultural Research, 2013, Vol.1, No. 9, PP. 042-046.*

1b. *Fu YC, 2010, Microwave heating in food processing. In: Hui YH (ed) Handbook of Food Science, Technology and Engineering. Boca Raton, FL: CRC Press, pp. 1-11.*

1c. *K. S., Youn, H. G. Sagong, H. S. Choi, S. Ryu and D. H. Kang, 2011. "Radio-frequency heating to inactivate Salmonella Typhimurium and Escherichia coliO157: H7 on black and red pepper spice". International Journal of Food Microbiology, 153, 171-175.*

2. *Jones PL, Rowley AT, 1997, Dielectric dryers. In: Baker CGJ (ed) Industrial Drying of Foods. London: Blackie Academic and Professional, pp.156-178.*

3. *Ohlsson T, 1999, Minimal processing of foods with electric heating. In: Oliviera FAR, Oliviera JC (eds) Processing Foods- Quality Optimization and Process Assessment. New York: CRC Press.*

4a. Farag KW, Cronin DA, Morgan DJ, Lyng JG, 2008, Dielectric and thermophysical properties of different beef meat blends over a temperature range of –18 to +10°C, Meat Science, 79, 740-747.

4b. Farag KW, Cronin DA, Lyng JG, Duggan E, Morgan DJ, 2009, Comparison of conventional and radiofrequency defrosting of lean beef meats: effects on water binding characteristics. Meat Science 83: 278-284.

5. Archer M, Edmonds M, George M, 2008, Seafood thawing systems. In: Archer M, Edmonds M, George M (eds) Seafood Thawing. Research and Development Department, Campden and Chorleywood Food Research Association, SR 598, pp. 21-33.

5a. Wang, J., Luechapattanaporn, K., Wang, Y., Tang, J., 2012, "Radio-frequency heating of heterogeneous food -Meat lasagna". Journal of Food Engineering, 108, 183-193.

6. G. B. Awuah, H. S. Ramaswamy, A. Economides and K., Mallikarjuanan, K., 2005, "Inactivation of Escherichia coli K-12 and Listeria innocua in milk using radiofrequency (RF) heating". Innovative Food Science and Emerging Technologies, 6 (4), 396-02.

7. D. J. Geveke and C. Brunkhorst, 2004, "Inactivation of Escherichia coli in apple juice by radiofrequency electric fields", Journal of Food Science, 69 (3), 134-138.

8. D. J. Geveke, C. Brunkhorst and X. Fan, 2007, Radio-frequency electric fields processing of orange juice. Innovative Food Science and Emerging Technologies, 8 (4), 549-554.

9. K. Uemura, C. Takahashi and I. Kobayashi, 2011, "Inactivation of Bacillus subtilis spores in soybean milk in radio-frequency flash heating". Journal of Food Engineering, 100, 622-626.

Chapter 14

Infrared Heating Technology

Infrared (IR) **[1]** uses electromagnetic radiation generated from a hot source (quartz lamp, quartz tube or metal rod) resulting from the vibrational and rotational energy of molecules. Thermal energy is thus generated following the absorption of radiating energy. By exposing an object to infrared (IR) radiation (wavelength of 0.78 to 1000 μm), the heat energy generated can be absorbed by food materials. Along with microwave, radio-frequency (RF), and induction, IR radiation transfers thermal energy in the form of electromagnetic (EM) waves and encompasses that portion of the EM spectrum that borders on visible light and microwaves [2]. IR radiation can be classified into the following 3 regions [4]:

1. Near infrared (NIR)- spectral ranges of 0.75 to 1.4.
2. Mid infrared (MIR)- spectral ranges of 1.4 to 3.
3. Far infrared (FIR)- spectral ranges of 3 to 1000 μm.

In general, FIR radiation is advantageous for food processing because most food components absorb radiative energy in the FIR region [5].

Infrared (IR) radiation heating has been considered as an alternative method for food and agricultural processing to improve product quality and safety, increase energy and processing efficiency and reduce water and chemical usage. The basic characteristics of infrared heating are as follows:

1. The high heat transfer capacity,
2. Heat penetration directly into the product,
3. Fast process control
4. No heating of surrounding air.

Infrared (IR) heating provides significant advantages over conventional heating [2, 3]. The details are given on next page.

1. Reduced heating time,
2. Uniform heating (homogeneous temperature distribution),
3. Reduced quality losses,
4. Absence of solute migration in food material,
5. Versatile, simple,
6. Compact equipment [2], and significant energy saving up to 50 per cent [2, 3],
7. Low operating and maintenance costs,
8. Safer and cleaner process and the higher drying rate [3],
9. High Therrmal Efficiency
10. Alternate source of energy.

Infrared (IR) heating provides following disadvantages [2]:

1. Low penetration power,
2. Prolonged exposure of biological materials may cause fracturing
3. Not sensitive to reflective properties of coatings.

Infrared heating can be applied to various food processing operations, namely, drying, baking, roasting, blanching, pasteurization, and sterilization. Combinations of IR heating with microwave heating and other common conductive and convective modes of heating have been gaining momentum because of increased energy throughput [2]. Unlike microwave heating, suitable levels of heating can be achieved at the surface and at the core of the body. The industrial application of this technology is relatively new and IR treatment of food has mainly been limited to experimental or pilot-scale processes [1].

Foods are complex mixtures of different biochemical molecules, biological polymers, inorganic salts and water. The infrared spectra of such foods (Complex mixtures) originate with the mechanical vibrations of molecules or particular molecular aggregates within a very complex phenomenon of reciprocal overlapping [2, 6].

- ☆ Amino acids, polypeptides and proteins reveal 2 strong absorption bands localized at 3 to 4 and 6 to 9 µm.
- ☆ Lipids show strong absorption phenomena over the entire infrared radiation spectrum with 3 stronger absorption bands situated at 3 to 4, 6, and 9 to 10 µm.
- ☆ Carbohydrates yield 2 strong absorption bands centered at 3 and 7 to 10 µm [7].

Sources

1. *Hudaa Neetoo (Faculty of Agriculture, University of Mauritius, Réduit, Mauritius) and Haiqiang Chen (Department of Animal and Food Sciences, University of Delaware, Newark, Delaware, US), Alternative Food Processing Technologies.*

2. *Krishnamurthy, K., Khurana, H., Sujin, J. and Irudayaraj, J., Infrared Heating in Food Processing: An Overview, Comprehensive Reviews in Food Science and Food Safety, 2008, 7(1), 2-13.*

3. *Vicente A, Castro IA, 2007 Novel thermal processing technologies. In: Tewari G, Juneja VK (eds) Advances in Thermal and Non-thermal Food Preservation. Ames, IA: Blackwell, pp. 99-131.*

4. *Sakai, N and Hanzawa, T., 1994, Applications and advances in far-infrared heating in Japan, Trends Food Sci Technol, 5: 357-62.*

5. *Sandu, C., 1986, Infrared radiative drying in food engineering: a process analysis, . Biotechnol Prog 2(3), 109-19.*

6. *Halford, R. S., 1957, The influence of molecular environment on infrared spectra, Ann New York Acad Sci, 69: 63-9.*

7. *Rosenthal, I., 1992, Electromagnetic radiations in food science, Berlin, Heidelberg: Springer-Verlag.*

Infrared Equipment [1, 3]

The IR heating component consists of the followings:

i. A Radiator

It basically radiates in all directions. The maximum energy flux of radiators can range from 50 kW/m^2 (long wave) to 4010 kW/m^2 (ultra short wave). Industrial IR radiators can be categorized as either gas-heated radiators or electrically heated radiators where ohmic heating of a metal within a gas atmosphere (long waves), a ceramic body (long waves) or a quartz tube (medium or short waves) produces the IR radiation [2]. The capital cost of gas heaters is higher, while the operating cost is cheaper than that of electric infrared systems. The operating efficiency of an electric IR heater ranges from 40 per cent to 70 per cent, while that of gas-fired IR heaters ranges from 30 per cent to 50 per cent [3]. Infrared radiation is transmitted through water at short wavelength, whereas at longer wavelengths it is absorbed at the surface [4]. Hence, drying of **thin layers seems** to be more efficient at the **FIR region, while drying of thicker bodies should give better results at the NIR region.**

ii. A Reflector

It is responsible for directing the heat radiation to the target location. Reflectors consist of polished metallic surfaces with a low absorption and high emission capacity to reflect the IR. Commercial reflectors can take the form of metallic/ gold reflectors, glit twin quartz tube reflectors, or flat metallic/ceramic cassette reflectors [2].

Sources

1. *Hudaa Neetoo (Faculty of Agriculture, University of Mauritius, Réduit, Mauritius) and Haiqiang Chen (Department of Animal and Food Sciences, University of Delaware, Newark, Delaware, US), Alternative Food Processing Technologies.*

2. *Regier M, Rother M, Schuchmann HP, 2010, Alternative heating technologies. In: Ortega-Rivas E (ed) Processing Effects on Safety and Quality of Foods. Boca Raton, FL: CRC Press, pp. 188-245.*

3. *Krishnamurthy, K., Khurana, H., Sujin, J. and Irudayaraj, J., Infrared Heating in Food Processing: An Overview, Comprehensive Reviews in Food Science and Food Safety, 2008, 7(1), 2-13.*

4. *Sakai, N and Hanzawa, T., 1994, Applications and advances in far-infrared heating in Japan, Trends Food Sci Technol, 5: 357-62.*

Basic Laws and Interaction of Infrared Radiation [1]

There are several basic laws that help to understand IR radiation interaction. The amount of the Infra red radiation that is incident on any surface (of food or other materials) has a spectral dependence because energy coming out of an emitter is composed of different wavelengths and the fraction of the radiation in each band, dependent upon the temperature and emissivity of the emitter. The wavelength at which the maximum radiation occurs is determined by the temperature of the IR heating elements. The following basic laws for blackbody radiation, are helpful in understanding the aforesaid relationship [2]:

Sl. No.	*Basic Laws*	*Aspect Explanation*
1	Planck's law	Gives spectral blackbody emissive power distribution $E_b\lambda(T, \lambda)E_b\lambda(T, \lambda) = 2\ \pi hc^2{}_0/n^2\ \lambda^5\ [e^{hco/n}\ \lambda^{kT}\ -1]$
2	Wien's displacement law	Gives the peak wavelength (λ_{max}), where spectral distribution of radiation emitted by a blackbody reaches maximum emissive power. $\lambda_{max = 2898/T}$
3	Stefan-Boltzmann's law	Gives the total power radiated ($E_b(T)$) at a specific temperature from an infrared source. $\boldsymbol{E_b(T) = n^2\sigma T^4}$
4	Modified Beer's law	Gives the transmitted spectral irradiance ($H\lambda$ W/m^{2}µm) in nonhomogeneous systems.
		$\boldsymbol{H\lambda = H\lambda_0 \exp(-\sigma^*\lambda u)}$, $\rho+\alpha+\tau = 1$ Reflectivity (ρ): ratio of reflected part of incoming radiation to the total incoming radiation, absorptivity (α): ratio of absorbed part of incoming radiation to the total incoming radiation, and transmissivity (τ): ratio of transmitted part of incoming radiation to the total incoming radiation.
	Notes	*k*: Boltzmann's constant (1.3806×10^{-23} J/K), *n*: refractive index of the medium (*n* for vacuum is 1 and, for most gases, n is very close to unity), λ: the wavelength (µm), *T*: source temperature (K), c_0: speed of light (km/s), *h*: Planck's constant (6.626×10^{-34} J·s), σ: Stefan–Boltzmann constant, λ_{max}: peak wavelength, $H\lambda_o$: incident spectral irradiance (W/m^{2}µm), · *u*: mass of absorbing medium per unit area (kg/m^2) and $\sigma^*\lambda$: spectral extinction coefficient (m^2/kg).

When radiant electromagnetic energy impinges upon a food surface, it may induce changes in the electronic, vibrational, and rotational states of atoms and molecules. As food is exposed to infrared radiation, it is absorbed, reflected, or scattered (a blackbody does not reflect or scatter). Absorption intensities at different wavelengths differ by food components. The type of mechanisms for energy absorption determined by the wavelength range of the incident radiative energy can be categorized as [3]:

1. Changes in the electronic state corresponding to the wavelength range 0.2 to 0.7 μm (ultraviolet and visible rays).
2. Changes in the vibrational state corresponding to wavelength range 2.5 to 1000 μm (FIR).
3. Changes in the rotational state corresponding to wavelengths above 1000 μm (microwaves).

In general, the food substances absorb FIR energy most efficiently through the mechanism of changes in the molecular vibrational state, which can lead to radiative heating. Water and organic compounds such as proteins and starches, which are the main components of food, absorb FIR energy at wavelengths greater than 2.5 μm [2].

The infrared absorption bands for chemical groups and relevant food components are summarized in following table [4]:

Sl. No.	*Relevant Food Component*	*Chemical Group*	*Absorption Wavelength (μm)*
1	Water, sugars	Hydroxyl group (OH)	2.7 to 3.3
2	Lipids, sugars, proteins	Aliphatic carbon hydrogen bond	3.25 to 3.7
3	Lipids	Carbonyl group (C=O) (ester)	5.71 to 5.76
4	Proteins	Carbonyl group (C=O) (amide)	5.92
5	Proteins	Nitrogen–hydrogen group (NH)	2.83 to 3.33
6	Unsaturated lipids	Carbon–carbon double bond (C=C)	4.44 to 4.76

Sources

1. *Krishnamurthy, K., Khurana, H., Sujin, J. and Irudayaraj, J., Infrared Heating in Food Processing: An Overview, Comprehensive Reviews in Food Science and Food Safety, 2008, 7(1), 2-13.*
2. *Sakai, N and Hanzawa, T., 1994, Applications and advances in far-infrared heating in Japan, Trends Food Sci Technol, 5: 357-62.*
3. *Decareau RV.1985. Microwaves in the food processing industry. Orlando, Fla.: Academic Press.*
4. *Rosenthal, I., 1992, Electro-magnetic radiations in food science, Berlin, Heidelberg: Springer-Verlag.*

IR Heat Transfer Modeling [1]

The modeling of IR heat transfer inside food materials has been a research intensive area because of the followings:

1. Complexity of optical characteristics,
2. Radiative energy extinction,
3. Combined conductive and/or convective heat transfer phenomena.

Some models for IR heat transfer are discussed below:

1. It was [2] assumed that most FIR radiation energy would be absorbed at the surface of a food system due to the predominant energy absorption of water. Energy would thereafter be transported by heat conduction in the food. Based on this assumption, a governing equation and boundary conditions to explain heat transfer derived from energy balance in a food system were solved using Galerkin's finite element method.
2. Researchers investigated following four mathematical drying models:
   i. An exponential model,
   ii. A Page model,
   iii. A diffusion model based on spherical grain shape,
   iv. An approximation of the diffusion model to address the thin layer infrared drying characteristics of rough rice. They found the Page model as most satisfactory for describing thin layer infrared radiation drying of rough rice
3. In general, numerical methods applied to solve the set of equations are finite elements, finite difference, and finite volume or the control volume method.
4. Scientist [3] investigated infrared drying of apple to create new suitable models, including combined effects of drying time and temperature. To explain the drying behavior of apple, 10 different drying models (Newton, Page, modified Page, Wang and Singh, Henderson and Pabis, logarithmic, diffusion approach, simplified Ficks diffusion [SFFD] equation, modified Page equation II, and Midilli equation) were developed and validated.
5. **Thermal death kinetics model**-Some examples are given below:
   i. A simple integrated model to predict the survivors of *E. coli* under predicted temperature distribution during FIR pasteurization, was developed [4].
   ii. An integrated model that combined the thermal death kinetics with the IR heat transfer model and could predict the survivors of fungal spores based on temperature prediction, was developed [5].

Sources

1. *Krishnamurthy, K., Khurana, H., Sujin, J. and Irudayaraj, J., Infrared Heating in Food Processing: An Overview, Comprehensive Reviews in Food Science and Food Safety, 2008, 7(1), 2-13.*

2. *Sakai, N and Hanzawa, T., 1994, Applications and advances in far-infrared heating in Japan, Trends Food Sci Technol, 5: 357-62.*

3. *Togrul H., 2005, Simple modeling of infrared drying of fresh apple slices. J Food Eng 71: 311-23.*

4. *Hashimoto A, Sawai J, Igarashi H, Shimizu M. 1992b. Farinfrared irradiation effect on pasteurization of bacteria on or within wet solid medium. J Chem Eng Japan 25(6): 666-71.*

5. *Jun S. 2002. Selective far infrared heating of food systems [PhD dissertation]. Pa.: The Pennsylvania State Univ.*

Food Processing Applications of Infrared [1]

Several novel IR based technologies have been developed recently by taking advantage of high heat transfer rate of IR radiation for blanching, dehydration and peeling of fruits and vegetables. It was found that IR heating achieved simultaneous blanching and dehydration eliminated the need for water or steam for blanching and reduced processing time and energy use [1a]. Infrared heating has been applied in drying, baking, roasting, blanching, pasteurization, and sterilization of food products. The details related to the applications of IR in Food processing are given below:

A. Drying and Dehydration [1c]

Application of FIR drying in the food industry is expected to represent a new process for the production of high-quality dried foods at low cost [1b]. The use of IR radiation technology for dehydrating foods has numerous advantages. FIR drying operations have been successfully applied in recent years for drying of fruit and vegetable products such as potatoes, sweetpotatoes, onions, kiwifruit, and apples. Drying of seaweed, vegetables, fish flakes, and pasta is also done in tunnel infrared dryers. Development of a continuous drying apparatus equipped with FIR heaters, NIR heaters, and hot air blast can reduce the economic costs, drying time, and operating temperature. However, vegetable size should be restricted to no more than 5 mm in thickness to improve drying efficiency [1b]. A continuous combined infrared and convective dryer was developed for vegetables. The synergistic effect of infrared and hot air led to rapid heating of the materials, resulting in a higher rate of mass transfer. The evaporation of water took 48 per cent less time and 63 per cent less energy consumption in combined mode drying as compared to convective drying [1c].

1. **Production of healthy French fries:** When it is applied to French fry production, the oil content in finished product and frying time were significantly reduced [1a].
2. **Roasting and pasteurization of almonds:** The Sequential IR and Hot Air (SIRHA) method produced high quality dry-roasted almonds while ensuring pasteurization and significantly reducing roasting time compared to the current hot air roasting method. When IR is used for pasteurizing raw almonds, the treatment retains the characteristics of the raw commodity [1a].
3. The Sequential IR and Freeze-Drying (SIRFD) method produced crispy strawberry pieces with a 40 percent reduction of freeze drying time [1a]. A combination of IR heating with freeze drying in sweetpotatoes could reduce the processing time by less than half [1c].

4. Short-(1µm) and intermediate- (5µm) wave IR heating modes have gained wide acceptance and have been applied for the rapid baking, drying, and cooking of foods of even geometry and modest thickness [1, 2].
5. **Bakery Products-**Short-wave IR has a penetration depth of several millimeters in many foods and has been successfully applied for the surface pasteurization of bakery products [1, 3].
6. Long-wave IR (30 µm) has been used for industrial cooking and drying applications, achieving shorter processing times (up to 70 per cent reduction) when compared to convective heating [1].

B. Enzyme Inactivation [1c]

IR heating also achieved simultaneous drying, disinfestation and partial enzyme inactivation of rough rice [1a]. Several studies have shown certain enzyme reactions involving the action of lipases, lipoxygenases, and α-amylases were affected by IR radiation [1c]. Lipooxygenase, an enzyme responsible for deterioration in soybeans, was inactivated 95.5 per cent within 60 s of IR treatment [1d]. FIR has been successfully used to inactivate enzymes responsible for the development of off-flavors in peas prior to the freezing process [1c], as well as other enzymes and bacteria in solution

C. Others

1. As a sustainable peeling method without using chemicals and steam, IR dry-peeling produced high quality peeled fruits and vegetables with reduced peeling loss [1a].
2. The technology has also been used for thawing, surface pasteurization of bread, and decontamination of packaging materials [1, 4].

Sources

1. *Hudaa Neetoo (Faculty of Agriculture, University of Mauritius, Réduit, Mauritius) and Haiqiang Chen (Department of Animal and Food Sciences, University of Delaware, Newark, Delaware, US), Alternative Food Processing Technologies.*

1a. *Zhongli Pan, Innovative applications of infrared heating for food processing, Journal of Food Processing and Technology, DOI: 10.4172/2157-7110. S1.023.*

1b. *Sakai, N and Hanzawa, T., 1994, Applications and advances in far- infrared heating in Japan, Trends Food Sci Technol, 5: 357-62.*

1c. *Krishnamurthy, K., Khurana, H., Sujin, J. and Irudayaraj, J., Infrared Heating in Food Processing: An Overview, Comprehensive Reviews in Food Science and Food Safety, 2008, 7(1), 2-13.*

1d. *Kouzeh KM, Van Zuilichem DJ, Roozen JP, Pilnik W. 1982. A modified procedure for low-temperature infrared radiation of soybeans. II. Inactivation of lipoxygenase and keeping quality of full fat flour. Lebensm Wiss Technol 15(3): 139-42.*

2. *Vicente A, Castro IA, 2007, Novel thermal processing technologies. In: Tewari G, Juneja VK (eds) Advances in Thermal and Non-thermal Food Preservation. Ames, IA: Blackwell, pp. 99-131.*

3. *Tewari G, 2007, Microwave and radiofrequency heating. In: Tewari G, Juneja V K (eds) Advances in Thermal and Non-thermal Food Preservation. Ames, IA: Blackwell Publishing, pp. 91-98.*

4. *Skjöldebrand C, 2001, Infrared heating. In: Richardson P (ed)Thermal Technologies in Food Processing. New York: CRC Press, pp. 208-227.*

IR Heating and Food Safety (Pathogen inactivation) [1]

IR heating can be effectively used for pathogen inactivation. It can be used for surface pasteurization applications. Its application in improving food safety is gaining popularity due to its higher thermal efficiency and faster heating rates and response time in comparison to conventional heating [1]. IR heating can be used to inactivate bacteria, spores, yeasts, and molds in both liquid and solid foods [1]. Inactivation of micro-organisms by IR heating may include inactivation mechanism similar to that of ultraviolet light (DNA damage) and microwave heating (induction heating) in addition to thermal effect, as infrared is located between ultraviolet and microwave in the electromagnetic spectrum.

Parameters Affecting Pathogen Inactivation

Efficacy of microbial inactivation by infrared heating depends on the following parameters [1]:

a. *Effect of Power [1]*

Increase in the power of infrared heating source produces more energy and thus total energy absorbed by micro-organisms increases, leading to microbial inactivation.

b. *Temperature of Food Sample [1]*

E. coli population was reduced by 0.76,0.90, and 0.98 $\log_{10}$ after 2 min exposure to IR radiation when the temperature of the bacterial suspension was maintained at 56, 58, and 61 °C, correspondingly [1a].

c. *Effect of Peak Wavelength and Bandwidth [1]*

The selected wavelength denatures the protein in micro-organisms, leading to a 40 per cent increase in inactivation of A. niger and F. proliferatum compared to normal IR heating. For instance, a 5-min treatment with nonselective and selective heating resulted in approximately 1.8 and 2.3 $\log_{10}$ CFU/g reduction of A. niger. Similarly, reductions of 1.4 and 1.95 $\log_{10}$ CFU/g of F. proliferatum were obtained with 5 min of nonselective and selective heating, respectively. Total energy decreases as the peak wavelength increases. Therefore, NIR radiation with short wavelength has a relatively higher energy level than FIR radiation with longer wavelength [1].

d. *Effect of Sample Depth [1]*

The penetration depth of IR radiation is very low. An increase in the sample depth slows down the bulk temperature increase of the food sample [1b]. A 90 per cent reduction in IR power was observed within a thin layer of 40 μm in bacterial suspension. Therefore, the effect of IR radiation on the microbial inactivation diminishes as the sample thickness increases. Decreasing the sample depth also accelerates the inactivation of spores [1] and *E. coli* and *S. aureus* [1, 4].

The following table shows the penetration depth of NIR energy into food products [1d]:

Product	*Special Peak (μm)*	*Depth of Penetration (mm)*
Dough wheat	1.0	4-6
Bread wheat	1.0	11-12
Bread Biscuit dried	1.0	4
Carrot	1.0	1.5
Raw potatoes	1.0	6
Tomato paste 70-80 per cent water	1.0	1

e. *Types of Micro-organisms [1]*

Resistance of different micro-organisms (bacteria, yeasts, and molds) to infrared heating may be different due to their structural and compositional differences. Spores are more resistant than vegetative cells. Upon infrared heat treatment, vegetative cells were inactivated followed by activation of spores. Then vegetative cells formed from spores will be activated and thus spores will be inactivated. Cereal surface is often contaminated with spore formers like Bacillus, Aspergillus, and Penicillium. Wheat was treated with infrared heating at 2.0 kW for 30 s, followed by cooling for 4 h, and again treated for 30 s with infrared heating to obtain a 1.56 $\log_{10}$ CFU/g reduction. Naturally occurring yeasts in honey were completely inactivated with an 8-min infrared heat treatment [1c]. The inactivation of pathogenic micro-organism by IR radiation is summerised below:

Inactivation of Pathogenic Micro-organisms by Infrared Heating

Sl. No.	*Pathogen*	*Food Materials*	*Temperature*	*Time*	*Log Reduction*
1	Monilia fructigena [1e]	Strawberry	Approx. 50 °C	10 s	2.5 to 5.2 log (estimated)
2	Natural microflora [1f]	Wheat	2.0 kW	63 s	Approx.2.0 log
3	Total bacterial count [1g]	Wheat or soybean surface	1.5 kW	10 s	Approx. 3.0 log
4	Total aerobic plate count [1h]	Onion	80 °C (average 2226 W/m^2)	Approx. 24 min	1.72 ± 0.45 $\log_{10}$ CFU/10g
5	Coliform counts [1h]	Onion			4.04 ± 0.47 $\log_{10}$ CFU/10g
6	Yeast and mold [1h]	Onion			1.26 ± 0.14 $\log_{10}$ CFU/10g
7	Yeast [1c]	Honey	0.2 W/cm^2	8 min	Approx 3.85 $\log_{10}$ CFU/mL

f. ***Effect of Moisture Content [1]***

Water molecules inside micro-organism readily absorb infrared radiation. These water molecules are attached to polar groups such as $-NH_2$, -COOH, and -COO within the cell. State and amount of water inside spores, bonding conditions of water molecules, and location of water molecule within M/Os affect their responses to infrared heating.

Mechanisms of Microbial Inactivation [1, 2]

IR heating is a thermal technology and therefore it is presumed to inactivate micro-organisms by heat, and the effect of heat on the internal contents of micro-organisms(thermal denaturation of proteins and nucleic acids) [2a]. However, as the IR spectrum lies between that of microwaves and UV light, an overlapping effect involving induction heating and damage to DNA has also been reported to be responsible for microbial inactivation [3].

- ☆ It was verified and demonstrated [3] the effect of IR heating on ***Staphylococcus aureus*** cells in milk using transmission electron microscopic and infrared spectroscopic observations. It was reported that IR heating resulted in both external and internal damage to the cells with the images revealing cell wall damage, shrinkage of cytoplasmic membranes, disintegration of mesosomes, and leakage of internal cellular contents. The rise in temperature effected by IR heating is expected to have effects similar to those of conventional heating methods, *i.e.* damage to cell walls and leakage of intracellular components.
- ☆ A more pronounced effect of IR radiation was observed [4] in experiments with ***Escherichia coli*** and ***S. aureus*** in a phosphate buffer medium. FIR radiation was found to perform better than both conductive water-bath heating and convective hot-air drying. This may be due to the increased energy levels of FIR concentrated on the surface of the phosphate buffer and agar, resulting in more inactivation compared with the distributed energy in other modes of heating.

Inactivation of *Staphylococcus aureus* (Pathogen) in Milk [1]

The efficacy of infrared heating for inactivation of *Staphylococcus aureus* in milk was studied to investigate the potential of this technology for milk pasteurization. S. aureus population was reduced from 0.10 to 8.41 log 10 CFU/ml, depending upon the treatment conditions. The effects of infrared lamp temperature (536, 619°C), volume of the treated milk sample (3, 5, and 7 ml), and treatment time (1, 2, and 4 min) were found to be statistically significant ($p<0.05$). Complete inactivation of S. aureus was obtained in two cases within 4 min at a 619°C lamp temperature, resulting in 8.41 log 10 CFU/ml reduction. Enrichment resulted in growth as some of the injured cells were able to repair. Further investigation of infrared heat treatment for longer treatment times (> 4 min) indicated that there was no growth observed following enrichment in most cases for treatment at a 619°C lamp temperature.

Sources

1. *Krishnamurthy, K., Jun, S., Irudayaraj, K.and Demirci, A., Infrared Heat Treatment for Inactivation of Staphylococcus aureus, Journal of Food Process Engineering, July 2008, 31(6): 798-816.*

1a. *Sawai J, Sagara K, Hashimoto A, Igarashi H, Shimizu M. 2003, Inactivation characteristics shown by enzymes and bacteria treated with far infrared radiative heating. Int J Food Sci Technol 38: 661-7.*

1b. *Sawai J, Sagara K, Igarashi H, Hashimoto A, Kokugan T, Shimizu M., 1995, Injury of Escherichia coli in physiological phosphate buffered saline induced by far infrared irradiation. J Chem Eng Jap 28(3): 294-9.*

1c. *Hebbar HU, Nandini KE, Lakshmi MC, Subramanian R., 2003, Microwave and infrared heat processing of honey and its quality. Food Sci Technol Res 9: 49-53.*

1d. *Ginzburg AS.1969, Application of infrared radiation in food processing, chemical and process engineering series. London, U. K.: Leonard Hil.*

1e. *Tanaka F, Verboven P, Scheer-linck N, Morita K, Iwasaki K, Nicolaý B., 2007, Investigation of far infrared radiation heating as an alternative technique for surface decontamination of strawberry, J Food Eng 79: 445-52.*

1f. *Uchino T, Hamanaka D, Hu W., 2000, Inactivation of micro-organisms on wheat grain by using infrared irradiation. Proceedings of Intl. Workshop Agricultural Engineering and AgroProducts Processing Toward Mechanization and Modernization in Agriculture and Rural areas.*

1g. *Daisuke H, Toshitaka U, Wenzhong H, Yaunaga E., 2001, The short time infrared ray sterilization of the cereal surface. Proceedings of IFAC control applications in postharvest and processing technology, Tokyo, Japan. p 195-201. 1h-Gabel MM, Pan Z, Amaratunga KSP, Harris LJ, Thompson JF, 2006, Catalytic infrared dehydration of onions. J Food Sci, 71(9): E351-7.*

2. *Ramaswamy, R. and Jun, S., in Microbial Decontamination in the Food Industry, 2012.*

2a. *Sakai, N.and Hanzawa, T., Application and advances in far-infrared heating in Japan, Trends in Food Science and Technology, Nov.1994, 5(11): 357-362.*

3. *Krishnamurthy, K., Microbial decontamination of milk and water by Piulswed UV and IRB heating, Ph D thresis, Pennslyvania satate University, Pennslyvania.*

4. *Hashimoto, A and Shimuzu, M., Effect of Far IR radiation of pasteurization of bacteria suspended in phosphate buffered saline, Kagaku Kogaluc Ronbunshu, 17, 627-633.*

Effect of Infrared on the Sensory and Nutritional Quality of Foods [1]

Several researchers have studied the quality and sensory changes of food materials during IR heating. The Effect of infrared on the sensory and nutritional quality of foods are discussed on next page.

1. The drying characteristics of carrots and garlic dried with hot air, microwaves or IR were compared [2]. Infrared-dried samples displayed the highest rehydration capacity (8.95 g H_2O/g) followed by microwave-dried (8.38 g H_2O)/g) and hot-air dried samples (7.96 gH_2O)/g).
2. The color parameters L, a, and b were determined, where L spans 0-100 (L = 0 represents total darkness (black) and L = 100 represents total lightness (white)), a runs from –a (green) to + a (red), and b runs from –b (blue) to + b (yellow). L and a values of dehydrated carrot samples were not significantly different. However, the b value of IR-dried carrots was significantly different from the fresh and air-dried products, indicating higher color protection by air drying.
3. It was concluded that the choice of processing methods would ultimately depend on the desired characteristics of the product.
4. The feasibility of IR for drying of shrimps was investigated [3]. It was demonstrated that the product quality index measured in terms of its thiobarbituric acid (TBA) value – a direct indicator of lipid oxidation – was highest at plate temperature of 357°C, air temperature of 43°C, and plate distance of 12.5 cm.
5. IR applied to barley resulted in an improvement in quality parameters such as germination and bulk density [4].
6. It was demonstrated [5] that intermittent IR treatment of cereals, *i.e.* IR heating at 2.0 kW for 30 sec, followed by cooling for 4 h, and treating again for 30 sec with infrared heating, resulted in minimal color changes; continuous treatment longer than 50 sec resulted in the discoloration of the wheat surface [5]. Application of infrared radiation in a stepwise manner by slowly increasing the power, with short cooling between power levels, resulted in less color degradation than with intermittent infrared heating [5a, 10]. Reductions in overall color change of 37.6 and 18.1 per cent were obtained for potato and carrot, respectively.
7. The irradiation of oysters by IR prior to freezing was found to result in higher moisture retention. Cooking of in-shell eggs using IR not only displayed a more attractive and brighter color, but also reduced the risk of cracking because the eggs were not in contact with each other [6].
8. Finally, the use of IR for bread baking was shown to be highly promising, as IR conditions controlled the crumb thickness, texture, and color of the finished product [7].
9. Longer infrared heat treatments may darken the color of onion due to browning [8, 10]. Chlorophyll content of dehydrated onions treated by infrared increased with an increase in irradiation power [8a, 10].
10. The evaluation of fullfat flour made from IRheat treated soybeans maintained freshness similar to fresh flour for 1 y. However, untreated samples resulted in rancidity development [9, 10].

11. The effect of IR treatment on nutritional quality of various food products, is summarised in following table. The available literature reveal that IR heating does not change the quality attributes of foods significantly, such as vitamins, protein, and antioxidant activities [10].

Effect of Infrared Treatment on Nutritional Quality of Food Products [10]

Food Products	*Parameters Effecting Nutritional Quality*	*Effect of Treatment*
Barley [4]	Germination rate at 55 °C	25 per cent increase by combination of IR heating and convectional heating
Full fat soybeans [10a]	Protein solubility	Infrared heating: 84 per cent Spouted bed drying: 82 per cent Extrusion: 73 per cent
Wheat [11]	Germination rates (heat treatment for 63s each)	Convectional heating: 90 per cent to 97 per cent Intermittent IR heating: 80 per cent to 86 per cent Continuous IR heating: 78 per cent to 85 per cent

Sources

1. *Hudaa Neetoo (Faculty of Agriculture, University of Mauritius, Réduit, Mauritius) and Haiqiang Chen (Department of Animal and Food Sciences, University of Delaware, Newark, Delaware, US), Alternative Food Processing Technologies.*

2. *Baysal T, 2003, Effects of microwave and infrared drying on the quality of carrot and garlic. European Food Research and Technology 218: 68-73.*

3. *Fu WR, Lien WR, 1998, Optimization of far infrared heat dehydration of shrimp using RSM. Journal of Food Science 63: 80–83.*

4. *Afzal TM, Abe T, Hikida Y, 1999, Energy and quality aspects during combined FIR–convection drying of barley. Journal of Food Engineering 42: 177-182.*

5. *Hamanaka D, Dokan S, Yasunga E, Kuroki S, Uchino T, Akimoto K, 2000, The Sterilization Effect of Infrared Ray on the Agricultural Products Spoilage Micro-organisms. ASABE Paper No. 006090. St Joseph, MI: American Society of Agricultural and Biological Engineering.*

5a. *Chua KJ, Chou SK. 2005. A comparative study between intermittent microwave and infrared drying of bioproducts, Int J Food Sci Technol, 40, 23-39.*

6. *Sakai N, Hanzawa T, 1994, Application and advances in far-infrared heating in Japan. Trends in Food Science and Technology 5: 357-362.*

7. *Regier M, Rother M, Schuchmann HP, 2010, Alternative heating technologies. In: Ortega-Rivas E (ed) Processing Effects on Safety and Quality of Foods. Boca Raton, FL: CRC Press, pp. 188-245.*

8. *Gabel MM, Pan Z, Amaratunga KSP, Harris LJ, Thompson JF. 2006. Catalytic infrared dehydration of onions, J Food Sci, 71(9): E351-7.*

8a. *Mongpreneet S, Abe T, Tsurusaki T. 2002. Accelerated drying of welsh onion by far infrared radiation under vacuum conditions. J Food Eng 55: 147-56.*

9. *Kouzeh KM, Van Zuilichem DJ, Roozen JP, Pilnik W. 1982. A modified procedure for low temperature infrared radiation of soybeans. II. Inactivation of lipoxygenase and keeping quality of full fat flour, Lebensm Wiss Technol, 15(3): 139-42.*

10. *Kathiravan Krishnamurthy, Harpreet Kaur Khurana, Jun Soojin, Joseph Irudayaraj, Ali Demirci, Infrared Heating in Food Processing: An Overview, Comprehensive Reviewsin food safety, Volume7, Issue1, 2008, 2-13.*

10a. *Wiriyaumpaiwong S, Soponronnarit S, Prachayawarakorn S. 2004. Comparative study of heating processes for fullfat soybeans, J Food Eng, 65, 371-82.*

11. *Uchino T, Hamanaka D, Hu W. 2000. Inactivation of micro-organisms on wheat grain by using infrared irradiation. Proceedings of Intl. Workshop Agricultural Engineering and AgroProducts Processing Toward Mechanization and Modernization in Agriculture and Rural areas.*

Chapter 15

Ohmic Heating Technology

Ohmic heating **[1]** is sometimes called as joule heating, electrical resistance heating, direct electrical resistance heating, electroheating or electroconductive heating. It is the process of passing electric currents through foods or other materials to heat them. Ohmic heating is often placed among the **so-called 'novel food processing technologies** and this method is one of the several electromagnetic based methods such as capacitive dielectric, radiative dielectric, inductive and radiative magnetic heating. Ohmic heating is somewhat similar to microwave heating but with very different frequencies [1a, 1b]. Ohmic heating (direct resistance heating) is a process in which food liquids and solids are heated simultaneously by passing an electric current through them [1b]. In this technique, heat is internally generated in a sample due to electrical resistance when electric current is passed through it. It is a novel technique which provides rapid and uniform heating, resulting in less thermal damage to the food product and this heating(OH) has immense potential for achieving rapid and uniform heating in foods, providing microbiologically safe and high quality foods [1b].

The defining characteristics of ohmic heating compared to other electrical methods (MW and RF heating) lie in the frequency and waveforms of the electric field, and the presence of electrodes that contact the material [1]. Ohmic heating apparently offers an attractive alternative because it heats materials through internal heat generation [4].

Ohmic heating is an alternative heating process in which the food product is heated internally due to inherent resistance [3]. Using electric current, food can be pasteurized, fermented, or sterilized in a manner that is equally comparable, if not better, than the current methods of processing. Ohmic heating for the food industry consists of using electrical energy to heat foods as a method of preservation, which can in turn be used for microbial inactivation or several other processes **such as pasteurization, extraction, dehydration, blanching or thawing**.

High Temperature Short Time (HTST) processes rely on rapid convection heat transfer and are thus well suited to liquid foods. HTST processes are, however, limited in application to particulates since, for particles more than a couple of millimeters thick, the processing time is insufficient for heat to transfer to the center to give sterility. **In ohmic heating processes**, foods are made part of an electric circuit through which alternating current flows, causing heat to be generated within the foods due to the electrical resistance of the foods. Therefore, in a liquid-particulate food mixture, if the electrical conductivity of the two phases are comparable, heat could be generated at the same or comparable rate in both phases in ohmic heating [4].

The concept of ohmic heating of foods is not new. In the nineteenth century, several processes were patented that used electrical current for heating flowable materials [4]. Ohmic heating is one of the newest alternative processing techniques to emerge in the last 15 years.

1. In the early twentieth century, 'electric' pasteurization of milk was achieved by passing [5] milk between parallel plates with a voltage difference between them [4, 5]. Six states in the United States had commercial electrical pasteurizers in operation [6].
2. The food industry has shown a renewed interest in ohmic technology in recent years, with new systems being designed since the early 1990's [2].
3. Schade described a blanching method of preventing the enzymatic discoloration of potato using ohmic heating [5]. It was thought at that time that lethal effects could be attributed to electricity. The technology virtually disappeared in succeeding years apparently due to the lack of suitable inert electrode materials and controls.
4. Within the past two decades, new and improved materials and designs for ohmic heating have become available. The Electricity Council of Great Britain has patented a continuous-flow ohmic heater and licensed the technology to APV Baker [7].

Sources

1. *Hudaa Neetoo (Faculty of Agriculture, University of Mauritius, Réduit, Mauritius) and Haiqiang Chen (Department of Animal and Food Sciences, University of Delaware, Newark, Delaware, US), Alternative Food Processing Technologies.*

1a. *De alwis AAP, Fryer PJ. The use of direct resistance heating in the food industry. J Food Eng. 1990; 11: 3-27.*

1b. *K. Shiby Varghese, M. C. Pandey, K. Radhakrishna, and A. S. Bawa, Technology, applications and modelling of ohmic heating: a review, J Food Sci Technol, 2014 Oct; 51(10): 2304-2317.*

2. *Destinee R. Anderson, A report on Ohmic heating as an alternative food processing technology, Food Science Institute, College of Agriculture, Kansas State University, Manhattan, Kansas.*

3. *Ranvir Kaur, Khalid Gul and A. K. Singh, Nutritional impact of ohmic heating on fruits and vegetables-A review, Cogent Food and Agriculture, Volume 2, 2016, Issue 1.*

4. *R. Ruan, X. Ye, P. Chen and C. J. Doona, University of Minnesota and I. Taub, Ohmic heating US Army Natick Soldier Center.*

5. *Schade, A. L., Prevention of enzymatic discoloration of potatoes, 1951, US patent 2, 569,075.*

6. *Sastry S. K., Palaniappan, S., Ohmic heating of liquid-particle mixtures, Fd Technol, 1992, 46, 64-7.*

7. *Skudder, P. J., Ohmic heating: new alternative for aseptic processing of viscous foods, Food Engineering, 1988, 60, 99-101.*

Principle and Working of Ohmic Heating [1, 1a]

The principles of ohmic heating are very simple. Ohmic heating is an advanced thermal process where food acts as an electrical resistor and it is based on the passage of alternating electrical current (AC) through a body such as a liquid-particulate food system which serves as an electrical resistance in which heat is generated [1a]. AC voltage is applied to the electrodes at both ends of the product body. The rate of heating is directly proportional to the square of the electric field strength, E, and the electrical conductivity [1a].

The experimental design usually consists of electrodes that contact the food, whereby electricity is passed through the substance using a variety of voltage and current combinations. The substance is heated by the dissipation of electrical energy. Electrical resistance heating allows particles and liquids to heat at the same rate and permits the rapid heating of mixtures with high solids fractions [2]. When **compared to conventional heating, where heat is conducted from the outside in using a hot surface, ohmic heating conducts heat throughout the entire mass of the food uniformly.** The success of ohmic heating depends on the rate of heat generation in the system, the electrical conductivity of the food, residence time and the method by which the food flows through the system [2a].

Ohmic heating has internal energy generation within its system. As per theory, there is no upper limit to the temperature that can be produced. The following other factors will influence the temperature achieved by the system [2a]:

1. The electrical conductivities of the food substance-The most important factor is the electrical conductivity of the product and its temperature dependence. The amount of energy generated as heat per second (the heating rate) depends on the voltage gradient and on the electrical conductivity of the product (σ) [4]. The most critical property affecting ohmic heating rate is σ.
2. The design of the system.
3. The time that the substance is subjected to the heating.

4. The thermophysical properties of the food- If the product has more than one phase such as in the case of a mixture of liquid and particulates, the electrical conductivity of all the phases has to be considered
5. The electric field strength and constancy.
6. The temperature dependence of electrical conductivities-The electrical conductivity increases with rising temperature, suggesting that ohmic heating becomes more effective as temperature increases.
7. The extent of the interstitial motion in the food [3].

Sources

1. *Destinee R. Anderson, A report on Ohmic heating as an alternative food processing technology, Food Science Institute, College of Agriculture, Kansas State University, Manhattan, Kansas.*

1a. *R. Ruan, X. Ye, P. Chen and C. J. Doona, University of Minnesota and I. Taub, Ohmic heating US Army Natick Soldier Center.*

2. *K. Shiby Varghese, M. C. Pandey, K. Radhakrishna, and A. S. Bawa, Technology, applications and modelling of ohmic heating: a review, J Food Sci Technol, 2014 Oct; 51(10): 2304-2317.*

2a. *Leizerson, S. and Shimoni, E. 2005a. Stability and Sensory Shelf Life of Orange Juice Pasteurized by Continuous Ohmic Heating. J. Agric. Food Chem. 53: 4012-4018.*

3. *Food and Drug Administration-Center for Food Safety and Applied Nutrition (FDA CFSAN), 2000, Kinetics of Microbial Inactivation for Alternative Food Processing Technologies-Ohmic and Inductive Heating. http: //www.cfsan.fda.gov/~comm/ift-ohm.html.*

4. *Halden, K., de Alwis, A. and Fryer, P., 1990, Changes in the electrical conductivity of foods during ohmic heating. Int. J. Food Sci. Technol., 25, 9-25.*

Ohmic Heating Equipment [1, 5a]

There are several possibilities for design of an ohmic heating system, but there are several key elements that will be present in each one. **A power supply** (generator) is needed to produce the electricity. **Electrodes** connected to the power supply must be in physical contact with the substance in order to pass the electric current through. **The electrode gap** (distance between the electrodes in the system) can fluctuate depending on the size of the system, but by changing this distance, the electric field strength, expressed in volts per centimeter [V/cm], can be varied. Ohmic heating equipment typically consists of the followings [1]:

1. Electrodes [5a]

A range of electrode materials from graphite to aluminum or stainless steel were used. In the early designs the electrolytic effect that causes the dissolution of the metallic electrodes was completely neglected. When foods (electrolytes) are subjected to a direct current, electrochemical reactions of reduction and oxidation will occur at the cathode and anode respectively. Under alternating current conditions, cathodes and anodes interchange places according to the frequency. The redox reactions therefore occur alternately at the same electrode site. In order to avoid such electrolytic effects and possible dissolution of the electrode material into the food, electrodes should be coated. This may be achieved by using food-compatible materials such as metal-coated titanium electrodes [3]. Therefore, in food processing, high standards of hygienic design are required. To overcome this problem, following steps are taken in recent technologies:

a. In APV ohmic process, the use of a food-compatible electrode electrical material and the correct current density has eliminated contamination problems. Specially-developed electrode material is used to eliminate polarization and contamination.

b. Utilisation of high power frequency, since at alternating frequencies above 100 kHz, there is no apparent metal dissolution.

The importance of controlling the ohmic heating process was emphasized [8]. In continuous processing, problems can result if a single electrode pair is used to heat food material through a large change in temperature; substantial changes in liquid conductivity and thus in heating rate, may result along the length of the electrode. Multiple sets of electrodes are easier to control [5a]. The use of multiple electrodes gives a much greater degree of control than is possible in other techniques, together with a uniform electric field in the pipe sections.

Arrangement of Electrodes in Flow Path [1a]

Currently, commercially available designs include electrodes that are located at various positions along the length of the product flow path **(in-line field),** or those located perpendicular to the flow **(cross-field),** differing principally in distribution of electric field strength [5a].

The details related to the aforesaid arrangements of electrodes are given below:

i In-line field system: Electrodes are placed at various positions along the length of the product flow path. The material upstream experiences higher field strength than the material downstream due to the drop in voltage throughout the system.

ii. Cross-field system*:* Electrodes are placed perpendicular to the flow path. The electric field strength is constant throughout.

2. Source of Power/Heaters [1, 5a]

A power supply of 60 Hz and an alternator was used for the experiment. Voltage and current transducers were used to change from Volts to Amperes and vice-versa [1a]. Ohmic heaters may run on either a static (batch) or continuous mode.

a. A typical batch ohmic heater: It consists of a horizontal cylinder with one electrode placed at each extremity.

b. Continuous ohmic systems: These systems call for more flexibility and important considerations in the design of ohmic heaters include electrode configuration the distance between electrodes, electrolysis, heater geometry, frequency of AC current, power requirements, current density, applied voltage, product velocity, and velocity profile [3]. Currently ohmic heaters are specifically engineered to interface with continuous process lines. The following two possible geometries can be considered [3a]:

i. ***The Transverse field mode:*** With the transverse mode the applied electric field and current flux are at right angles to the mass flow. The product flows parallel to the electrodes and perpendicularly to the electric field. In this case the electrodes are generally plane or coaxial. The localised overheating, boiling and electrode erosion have been observed. Development of the transverse mode heater has therefore been restricted to fluids containing no particles.

ii. ***Collinear field mode:*** With the collinear mode they are parallel. The product flows from one electrode to the other one and the flow is parallel to the electric field. A collinear heater is assembled from two basic units–the electrode housing and the spacer tube.

3. Temperature Measurement Devices

Such as an electrically isolated thermocouple *etc.* that do not interfere with the electric field produced by the unit should also be used [1]. A Teflon coated k-type thermocouple was used to monitor the geometric center temperature of the sample [1a].

4. A Data Logger

A data logger connected to the system would record essential information such as temperature, voltage, current and time [1]. The voltage, current and temperature were continuously measured using a CR 10× data logger which was connected to a computer [1a].

5. Additional Equipment

It may include a pH probe, a spectrophotometer for measuring optical density, a refractometer for measuring °Brix (soluble solids), and sampling ports (in a closed system) [5].

6. Auxiliary Components [1]

In addition to the heating device and aforesaid additional equipment, the unit may also consist of various auxiliary components, including feeding, holding, and cooling equipment, in addition to a tank for storage or direct aseptic filling [4].

Commercial Ohmic Heating System [5, 5a]

Ohmic heating systems made for commercial applications can vary greatly. The details related to an example of this system is given below:

a. This system (Figure 2) has 7 electrode columns, each comprised of polytetra-fluoro ethylene (PTFE) coverings and a single stainless steel cantilever electrode. These columns are structured vertically to allow for an upward flow of product, and are connected with insulated stainless

Electrode
Food Product
Voltage
Applied

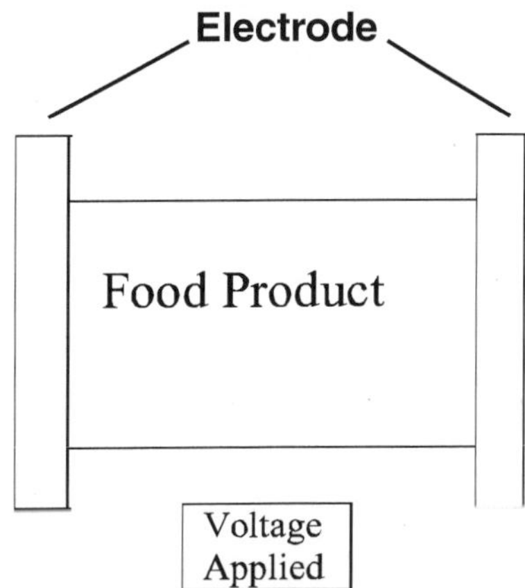

Figure 1. Schematic of Ohmic Heating Unit [2].

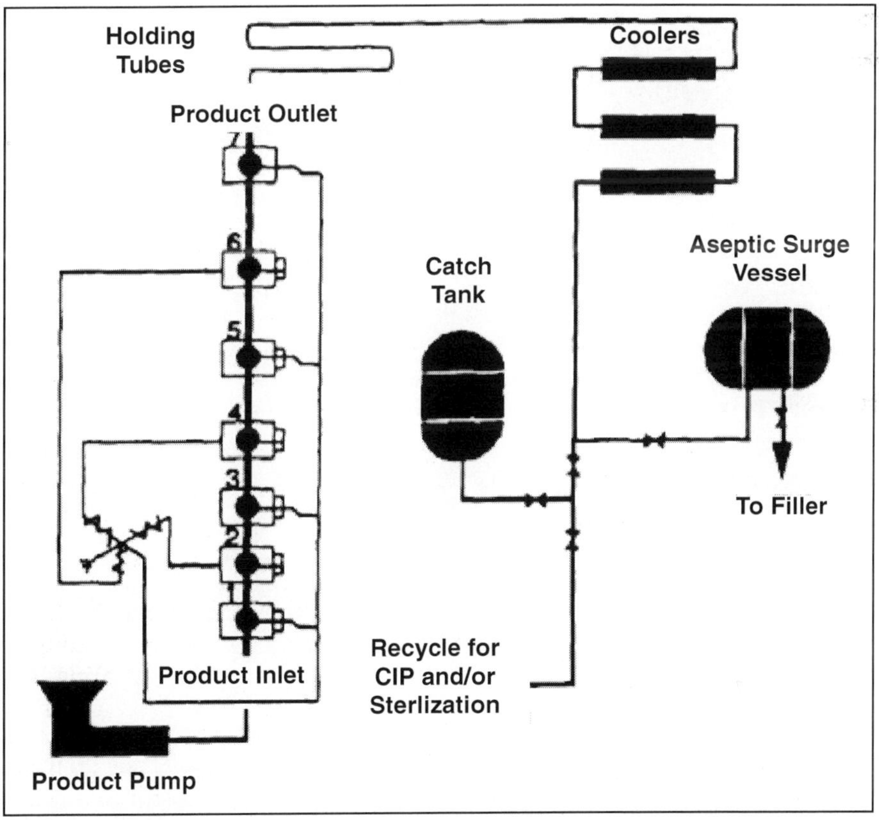

Holding
Tubes
Coolers
Product Outlet
7
6
5
4
3
2
1
Catch
Tank
Aseptic Surge
Vessel
To Filler
Product Inlet
Recycle for
CIP and/or
Sterlization
Product Pump

Figure 2. Schematic of Commercial Ohmic Design [6].

steel tubes. Due to the increased electrical conductivity of products as heat increases, the connecting tubes increase in length throughout the system to maintain the same electrical impedance. A temperature control system

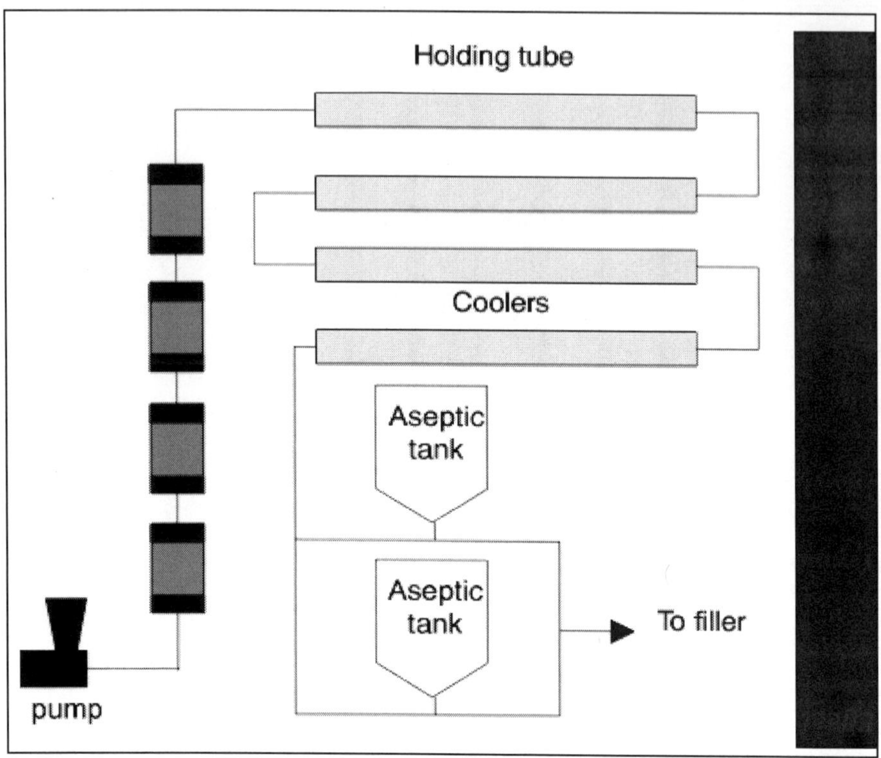

Figure 3. Schematic of a Continuous-Flowohmic Heating Process [5a].

constantly monitors the temperature, flow rate, heat capacity and specific heat of a product to calculate the electrical power needed for the system. The processing capacities of such systems can range from 3 to 6 tons an hour.

b. Most ohmic heating system configurations consist of three modules [5a]:

1. Heater assembly
2. Power supply
3. Control panel.

In continuous flow ohmic heating process (Figure 3), a viscous food product containing particulates enters the continuous-flow ohmic heating system via a feed pump hopper, the product then flows past a series of electrodes in the ohmic column, where it is heated to process temperature and then the product enters the holding tubes for a fixed time to achieve commercial sterility. Next, the product flows through tubular coolers and into storage tanks, where it is stored until filling and packaging [5a]. Equipment design is a critical factor that should be considered.

The most successful of ohmic heating systems have been for the processing of whole strawberries and other fruits for yogurt in Japan, and low acid ready-to-eat meals in the United States [5a, 5b]. Currently there are two commercial manufacturers of ohmic heating equipment [5a]. The names are given on next page.

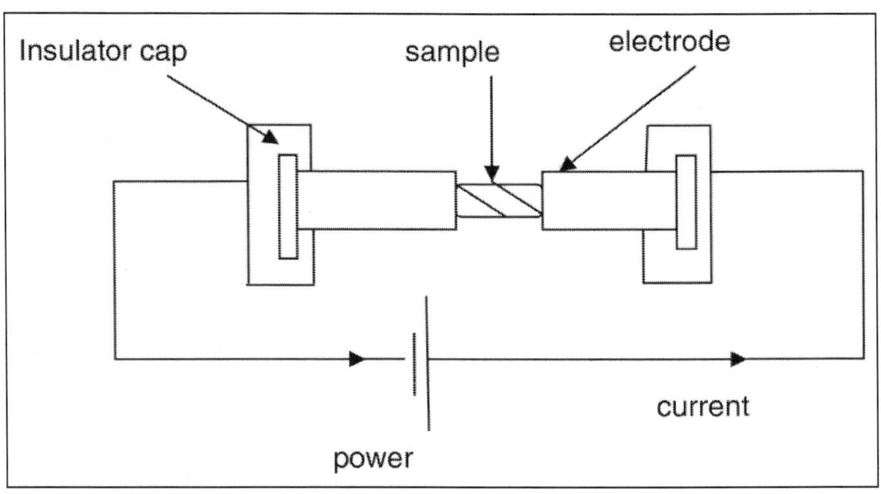

Figure 4. Schematic Diagram of Ohmic Heating Process [1a].

1. APV Baker, Ltd., Crawley, UK
2. Raztek Corp, Sunnyvale, CA, US.

APV Baker 'Ohmic Heating' Process

In this process, food is pumped through a vertical pipe containing a series of cylindrical electrodes connected to a 50Hz three-phase supply. Electric current thus flows through the food in the pipes connecting the electrodes. The food material is rapidly heated to sterilization temperature, then passes to a holding section and finally to an aseptic packaging plant. Specially-developed electrode material is used to eliminate polarization and contamination. The process allows food products containing particulates up to 25mm to be heated to sterilization temperatures up to 140°C in less than 90 seconds, after which they are cooled back down to ambient temperatures within 15 minutes. These processing times are significantly shorter than the typical two-hour process cycles for in-can sterilization techniques. The whole process is automatically controlled using a programmable logic controller [5a].

The informations on different commercial Ohmic heating systems are tabulated below [5]:

Sl. No.	*Countries*	*Company/ Scientist Name*	*Ohmic Heating System*
1	**United Kingdom**	Thomas R. Parker	An ohmic heating system to produce Japanese style breadcrumbs (Panko). Parker's device was designed to place the bread dough inside a container between two electrodes. The ohmic heating system bakes the dough without a crust, and the chamber constrains the dough, which forms the desired grain structure. The product is then cooled, ground and dried.

Sl. No.	*Countries*	*Company/ Scientist Name*	*Ohmic Heating System*
2	**Japan**	Yanagiya Machinery	An ohmic heating system for the production of tofu in 1995 [7]. The "Big-J", named for joule heating, can process up to 4, 800 pieces of tofu an hour. Yanagiya developed in 1999, the "Mini-J", a tabletop version of their "Big J". The "Mini J" has titanium electrodes, with two models ranging from 100-200 V.
		Wildfruit Products, a division of Nissei Co. Ltd. Of Japan	A 75-kilowatt system to process whole Fruits.
3	**England**	Sous Chef Ltd	A 75-kilowatt unit processes low acid meats and vegetables in bags.
4	**Italy**	Emmepiemme SRL, Piacenza	Uses ohmic heating to process many varieties of foods including- baby food, artichokes, carrots, mushrooms, ketchup, fruit nectars, fruit juice, peppers, cauliflower, tomato paste, sausages, pate, and fruit purée.
5	**United States**	A consortium {including food Industries (processors, equipment manufacturers, and ingredient suppliers), academia and government}	Evaluated a A 5kW pilot-scale continuous-flow ohmic system manufactured by APV Baker, Ltd., Crawley, UK. A wide variety of shelf-stable low and high-acid products, as well as refrigerated extended-shelf-life products were developed. The Consortium found to have texture, color, flavor, and nutrient retention that matched or exceeded those of traditional processing methods such as freezing, retorting, and aseptic processing and concluded that the technology was viable [5a].

Sources

1. *Hudaa Neetoo (Faculty of Agriculture, University of Mauritius, Réduit, Mauritius) and Haiqiang Chen (Department of Animal and Food Sciences, University of Delaware, Newark, Delaware, US), Alternative Food Processing Technologies.*

1a. *K. Shiby Varghese, M. C. Pandey, K. Radhakrishna, and A. S. Bawa, Technology, applications and modelling of ohmic heating: a review, J Food Sci Technol, 2014 Oct; 51(10): 2304-2317.*

2. *Vicente A, Castro IA, 2007, Novel thermal processing technologies. In: Tewari G, Juneja VK (eds) Advances in Thermal and Non-thermal Food Preservation. Ames, IA: Blackwell, pp. 99-131.*

3. *Bengtsson R, Birdsall E, Feilden S, Bhattiprolu S, Bhale S, Lima M, 2010, Ohmic and inductive heating. In: Hui YH (ed) Handbook of Food Science, Technology and Engineering, volume 3. Boca Raton, FL: CRC Press, pp. 121-127.*

3a. *Stirling R. Ohmic heating-a new process for the food industry. J Power Eng. 1987, 6, 365-371.*

4. *Regier M, Rother M, Schuchmann HP, 2010, Alternative heating technologies. In: Ortega-Rivas E (ed) Processing Effects on Safety and Quality of Foods. Boca Raton, FL: CRC Press, pp. 188-245.*

5. *Destinee R. Anderson, A report on Ohmic heating as an alternative food processing technology, Food Science Institute, College of Agriculture, Kansas State University, Manhattan, Kansas.*

5a. *R. Ruan, X. Ye, P. Chen and C. J. Doona, University of Minnesota and I. Taub, Ohmic heating US Army Natick Soldier Center.*

5b. *Anon, Ohmic heating garners 1996 industrial achievement award, Fd Technol., 1996, 20, 114-15.*

6. *Parrott, David L.1992, Use of ohmic heating for aseptic processing of food particulates, Food Technology. 68-72.*

7. *Noguchi, A., 2004, Potential of ohmic heating and high pressure cooking for practical use in soy protein processing. Japan International Research Center for Agricultural Sciences.*

8. *Getchell, B. E., Electric pasteurization of milk, Agric. Engng., 1935, 16, 408.*

Advantages of Ohmic Heating [1, 1a]

Ohmic heating is specially advantageous in processing semi solid, particulate foods. It has **proven advantages over conventional thermal processing and novel thermal alternative technologies like microwave heating, radio-frequency heating and induction heating**. Some key advantages are as follows:

1. Better product quality,
2. Less cooking time,
3. Lower capital cost,
4. Better energy efficiency,
5. An environment friendly process
6. It requires lower capital cost compared to other electro heating methods [1b].

Ohmic heating could be an effective future processing technique employed by food manufacturers. The particular interest in this technology stems from ongoing food industry interest in aseptic processing of liquid-particulate foods [1a]. The additional control over the system, the rapid and uniform heating, higher quality product, and ability to regulate the electrical conductivity of a food is a more cost effective and less damaging way to kill microbes in food [1]. Unlike most previous techniques, the advantage of ohmic heating in giving rapid sterility is not lost due to the packaging process [1a].

Special advantage is that ohmic heating can improve food quality and saves cost and energy to processors. Ohmic heating can produce safe, high quality food and to validate any commercial process by experimentally demonstrating its application. The shelf life of ohmically processed foods is comparable to that of canned and sterile, aseptically processed products. The details related to the advantages of ohmic heating are given on next page.

1. Ohmic heating systems are advantageous due to an optimization of investment (increased efficiency), instant shutdown of the system(Ease of process control with instant switch-on and shut-down) and reduced maintenance costs because of the lack of moving parts. Product safety as a result of high solids loading capacity [1, 1a]. **High energy efficiency because 90 per cent** of the electrical energy is converted into heat [1a]. Typical efficiencies of ohmic heaters are above 95 per cent.
2. Ohmic heating processes the particles and surrounding liquid simultaneously, preventing overcooking [2]. It can heat particulate foods and Liquid–particle mixtures. Higher temperature in particulates than liquid can be achieved, which is impossible for conventional heating [1a, 7].
3. Ohmic heating is the rapid and uniform heating process [1]. Its ability to heat materials rapidly and uniformly is its principal advantage. Better controllability.
4. As there are no hot surfaces for heat transfer, there is low risk of product damage due to burning. Reducing risks of **fouling** on heat transfer surface and burning of the food product, resulting in minimal mechanical damage and better nutrients and vitamin retention [1a]. The skim milk fouling was investigated during ohmic heating [1b, 1c]. The heating of the milk solution resulted in deposit formation on both electrodes. These deposits provided additional electrical resistance and the current passing through the milk solution dropped by up to 45 per cent within 4 h of operation. The applied voltage increased slightly during this time period. The deposit formation was reasonably uniform at any cross-section. Lowering the milk temperature at the inlet of the ohmic heater enhanced the rate of fouling. In contrast, increasing the milk solution flow rate by a factor of two was found to have almost no effect on the fouling rate.
5. Ambient-temperature storage and distribution when combined with an aseptic filling system [1a, 7].
6. A quiet environmentally friendly system [1a].
7. Heating food materials by internal heat generation without the limitation of conventional heat transfer. Heating takes place volumetrically and the product does not experience a large temperature gradient within itself as it heats [1a, 7].
8. Less damages caused to shear-sensitive food components- once shear rates are comparable to those existing in conventional pipe work with little additional disturbance to the flow; this leads to low pressure drops in the system.
9. Possibility of handling slurries with high solid contents (products with up to 80 per cent particulates can be processed).
10. Maintaining the color and nutritional value of food, shorter processing times, and higher yields [2a].

Microbial Inactivation

Ohmic heating can be considered a High Temperature Short Time (HTST) aseptic process, therefore the microbiological safety of the products is not affected and may even be improved. Microbial inactivation in relation to ohmic heating is primarily thermal in nature, much like conventional heating. It is generally believed that destruction of micro-organisms by ohmic heating is due to the thermal effect and electricity plays no role on inactivation [3, 3a]. However, the thermal requirement for inactivation is reduced when sub-lethal electrical treatment is applied [3a] and recent results clearly show that there is a non-thermal (electrical) effect of the application of ohmic heating on microbial growth [3b, 5]. It was demonstrated [4] that electrical pretreatment by ohmic heating can reduce the intensity of additional thermal applications for subsequent microbial inactivation. Microbial inactivation curves of ohmic heating are similar to conventional heating curves except for a difference in the slope, which can most likely be explained by the presence of the electric field [4a]. Some examples of Microbial inactivation by ohmic heating are given below:

1. An experiment using *Bacillus subtilis* revealed that a two stage ohmic treatment (ohmic heating, holding period, ohmic heating) resulted in accelerated death rates [5].
2. Another experiment involving *Saccharomyces cerevisae* showed that ohmic heating enhanced the leakage of intracellular components as compared with the conventional method of boiling in water [6].

Electroporation

It also known as electropermeabilization, is a significant increase in the electrical conductivity and permeability of the cell plasma membrane caused by an externally applied electrical field.

All living cells (prokaryote or eukaryote) contain cell membranes. These membranes are comprised of lipids and proteins. Prokaryotes have an additional layer called **Cell wall,** outside the membrane. At low frequencies (50-60 Hz) and high field strengths (>100V/cm) most commonly associated with ohmic heating, the naturally porous cell walls can allow the cell membrane to build up charges, forming disruptive pores [3]. Electroporation occurs because the cell membrane has a specific dielectric strength, which can be exceeded by the electric field. The dielectric strength of a cell membrane is related to the amount of lipids present in the membrane itself. The pores formed can vary in size depending on the strength of the electric field, and can reseal after a short period of time. Excessive exposure causes cell death due to the leakage of intracellular components through the pores [6]. Therefore, electroporation is highly damaging to a cell and would enhance the lethal effects of thermal abuse already present from the ohmic heating.

Sources

1. *Destinee R. Anderson, A report on Ohmic heating as an alternative food processing technology, Food Science Institute, College of Agriculture, Kansas State University, Manhattan, Kansas.*

1a. R. Ruan, X. Ye, P. Chen and C. J. Doona, University of Minnesota and I. Taub, Ohmic heating US Army Natick Soldier Center.

1b. K. Shiby Varghese, M. C. Pandey, K. Radhakrishna, and A. S. Bawa, Technology, applications and modelling of ohmic heating: a review, J Food Sci Technol, 2014 Oct; 51(10): 2304-2317.

1c. Bansal B, Chen XD, Lin SXQ, 2005, Skim milk fouling during ohmic heating. Heatexchanger fouling and cleaning: challenges and opportunities. Engineering Conferences Intl.; Kloster Irsee, Germany; 2005 June 5-10.

2. Parrott, David L., 1992, Use of ohmic heating for aseptic processing of food particulates. Food Technology, 68-72.

2a. Castro, I., Teixeira, J. A., Salengke, S., Sastry, S. K., and Vicente, A. A., 2004a, Ohmic heating of strawberry products: Electrical conductivity measurements and ascorbic acid degradation kinetics, Innovative Food Science and Emerging Technologies, 5, 27-36.

3. Palaniappan, S., Sastry, S. K.and Richter, E. R., 1990, Effects of electricity on micro-organisms: a review, J. Food Proc. Pres., 14, 393-414.

3a. Palaniappan, S., Sastry, S. K. and Richter, E. R., 1992, Effects of electroconductive heat treatment and electrical pre-treatment on thermal death kinetics of selected micro-organisms. Biotechnol. Bioeng., 39, 225-232.

3b. Pereira, R., Martins, J., Mateus, C., Teixeira, J. A., Vicente, A. A., 2007, Death kinetics of Escherichia coli in goat milk and Bacillus licheniformis in cloudberry jam treated by ohmic heating. Chemical Papers 61(2), 121-126.

4. Cho, H. Y., Yousef, A. E., Sastry, S. K., 1996, Growth Kinetics of Lactobacillus acidophilus Under Ohmic Heating. Biotechnol Bioeng, 49: 334-340.

4a. Food and Drug Administration-Center for Food Safety and Applied Nutrition (FDACFSAN)-2000, Kinetics of Microbial Inactivation for Alternative Food Processing Technologies-Ohmic and Inductive Heating.

5. Cho, H. Y., Yousef, A. E., Sastry, S. K, 1999, Kinetics of inactivation of Bacillus subtilis spores by continous or intermittent ohmic and conventional heating. Biotechnol Bioeng, 62(3): 368-372.

6. Lee, C. H. and Yoon, S. W. 1999, Effect of ohmic heating on the structure and permeability of the cell membrane of Saccharomyces cerevisae, 1999, IFT Annual Meeting. Chicago. July 24-28 1999.

7. Kim HJ, Choi YM, Yang TCS, Taub IA, Tempest P, Skudder P J, tucker G, Parrott DL, Validation of ohmic heating for quality enhancement of food products, Fd Technol, 1996, 253-61.

Disadvantages to Ohmic Systems [1]

Despite aforesaid advantages, there are potential problems with ohmic heating. The followings are the disadvantages of Ohmic heating:

1. The costs of commercial ohmic heating systems, including installation is high which is a costly investment for a manufacturing facility.

2. The main disadvantage is that it is not working properly with products with low electrical conductivity values. In this case, ohmic heating might just not be applicable due to the high power input required to process such products. A food that has fat globules can be troublesome to effectively heat ohmically, as it is non-conductive due to lack of water and salt [2]. If these globules are present in a highly electrical conductive region where currents can bypass them, they may heat slower due to lack of electrical conductivity. Any pathogenic bacteria that may be present in these globules may receive less heat treatment than the rest of the substance [3].
3. Another slight disadvantage relates to the electrical conductivity of a substance [1]. The limitation imposed by the minimum and maximum values of the electrical conductivity allowed for a given throughput, which in turn will limit the type of products that can be processed by ohmic heating.
4. The heat generation rate may be easily affected by **the electrical heterogeneity** of the particle, heat channeling, complex coupling between temperature and electrical field distributions, and particle shape and orientation. All these make the process complex and contribute to non-uniformity in temperature [1a].

As the temperature of a system rises, the electrical conductivity also increases due to the faster movement of electrons. Thus, this creates the possibility of '**runaway**' heating [4]. An ohmic heating system that has not been cleaned thoroughly enough may result in electrical arcing due to protein deposits on the electrodes.

By utilizing the knowledge of the aforementioned issues in designing an ohmic heating system, these disadvantages may be more easily controlled.

Problem with Ohmic Heating [1b]

Fouling is a common problem during ohmic heating. The research of fouling during ohmic heating of food concentrates revealed that the deposit creation on electrodes is pertinent to the problem of the electrodes corrosion. Effect of the corrosion of electrodes can be eliminated using higher frequencies (about 10 kHz) from a pulsed power supply. The corrosion effects can be suppressed either by using noble material like gold or platinum for electrodes. The positive effect of higher frequencies was known for a long time [1b].

Fouling creation of milk was studied [5]. Although the mechanism and reactions between all milk components are not yet fully understood, a relationship has been described between the denaturation of native β-lactoglobulin and fouling of heat exchangers [6]. The rate of the denaturation and agglomeration of proteins depend upon temperature and composition of milk. Roughness of electrode surface–sharp edges form singularities of electric field, with the possible risk of local overheating. Thus a microlayer of deposits can be formed by thermal denaturation of proteins.

Electrochemical reactions in the system electrode and electrolyte (redox reactions and electrochemical corrosion processes) start immediately after the power supply is connected. A primary micro layer can be formed on electrodes owing to oxidation or reduction reactions at relatively high values of the current density. This layer forms a part of electric circuit as an additional electric resistance. Because the relative electric resistance value of the primary micro layer is quite high, more electric energy is dissipated into heat in this layer, thus promoting biochemical changes in whey proteins.

Sources

1. *Destinee R. Anderson, A report on Ohmic heating as an alternative food processing technology, Food Science Institute, College of Agriculture, Kansas State University, Manhattan, Kansas.*

1a. *R. Ruan, X. Ye, P. Chen and C. J. Doona, University of Minnesota and I. Taub, Ohmic heating US Army Natick Soldier Center.*

1b. *K.shiby Varghese, M. C. Pandey, K. Radhakrishna, and A. S. Bawa, Technology, applications and modelling of ohmic heating: a review, J Food Sci Technol, 2014 Oct; 51(10): 2304-2317.*

2. *Rahman, M. Shafiur, 1999, Handbook of Food Preservation. CRC Press.*

3. *Sastry, S. K., 1992, A model for heating of liquid-particle mixtures in continuous flow ohmic heater. J Food Process Eng.15: 263-278.*

4. *Food and Drug Administration-Center for Food Safety and Applied Nutrition (FDACFSAN), 2000, Kinetics of Microbial Inactivation for Alternative Food Processing Technologies-Ohmic and Inductive Heating.*

5. *Ayadi MA, Leuliet JC, Chopard F, Berthou M, Lebouche M. Continuous ohmic heating unit under whey protein fouling. Innovat Food Sci Emerg Tech. 2004; 5: 465-473.*

6. *De Jong P. Modeling and optimization of thermal processes in the dairy industry. Ede: NIZO Verslag; 1996. p. 341.*

Effects of Different Factors on Ohmic Heating [1]

The different factors *e.g.* Temperature, Applied frequency, Field strength, Nature of food and ionic content effect, Particle size, concentration, and location are affecting the ohmic heating. Ohmic heating rate is dependent on the temperature, applied voltage gradient, frequency, particle size, concentration, and concentration of electrolytes. The success of ohmic heating depends on the rate of heat generation in the system, the electrical conductivity of the food, electrical field strength, residence time and the method by which the food flows through the system.

The efficiency of ohmic heating is dependent on the conductive nature of the food to be processed. The electrical conductivity of various fruits and meats increased linearly with the temperature during ohmic heating at constant voltage gradient. Lower electrical conductivity may be observed for highly porous materials like apples. Ohmic heating is appropriate for processing of particulate and protein rich foods [1a]. The details related to the factorwise effect on ohmic heating, are given on next page.

i. Temperature and Electrical Conductivity

Ohmic heating rate linearly increases with increase in temperature. Electrical conductivities (EC) increases with temperature. The temperature dependency of the EC liquid products follows linear or quadratic relations, depending on the product type tested such as strawberry pulps, sour cherry juice namely apple, orange, and pineapple juices, pomegranate, orange juice, lemon juice and grape juice. There was a linear increase in the EC of fruit purees with temperature rise. The most critical property affecting ohmic heating rate is food electrical conductivity (EC). It is a function mainly of food chemistry and structure and temperature. The electrical conductivity is affected by parameters such as ionic strength, free water and material microstructure [1a, 1b]. The presence of ionic substances such as acids and salts increases conductivity, while the presence of non polar constituents like fats and lipids decreases it. The electrical conductivity depends on the temperature, ionic breakup and microstructure setup of the food material undergoing heating and field strength of the experimental setup. It was also discussed the effect of temperature and sugar content on the EC values of strawberry-based products [2]. It was also noted the increase in EC of beetroot with temperature was sharper when processed electrically than the smoother increase seen when processed conventionally [3]. The observed and published conductivity values for various food samples during ohmic heating are given in following table [1a]:

Sl. No.	*Products*	*Conductivity during Ohmic Heating (At 25° C, S.m^{-1})*
1	Full Cream Milk	0.52
2	Desserts (Vanilla, rice custard)	0.50-0.38
3	Margarine	0.027
4	Fruits	0.067-0.186
5	Fruit Pieces	0.15-0.05
6	Vegetable pieces	0.1-0.06
7	Various soups	1.4-1.8
8	Chicken	0.8-0.37

ii. Field Strength

Ohmic heating rate linearly increases with increase in voltage gradient. The increase in field strength results in increasing fluid motion through the capillaries, which is directly proportional to EC [3]. The effect of field strength on EC was studied during ohmic heating of strawberry products [4]. An increase in EC with field strength was observed for two strawberry pulps and strawberry filling but not for strawberry topping or strawberry apple sauce. It was also measured the EC of fresh strawberries at different field strengths from 25 to 70 V/cm, and found that the EC increased almost linearly with the field strength [2].

iii. Applied Frequency

With increase in frequency, ohmic heating rate gets decreased. Researchers conducted studies on viscosity and EC of fruit juices and reported that EC was more at 10 kHz than at 1 kHz in all the juices indicating that heating rate increases with increasing frequency [5]. The ohmic heating of turnip as a function of frequency (4, 10, 25, and 60 Hz) and wave shape of alternating current were investigated [6] and it was shown that heating rate increased with decreasing frequency. The lower the frequency, the faster the sample reached elevated temperatures.

iv. Nature of Food and Ionic Content Effect

Ohmic heating rate linearly increases with increase in concentration of ionic constituents. Since electrical conductivity is influenced by ionic content, it is possible to adjust the electrical conductivity of the product (both phases) with ion (*e.g.* salts) levels to achieve effective ohmic heating. Generally, solid vegetable particles have lower EC than liquids. Researchers observed the heating behavior when there was no liquid phase in the ohmic heating cell during ohmic heating behavior of two-phase system. The OH cell was filled with carrot puree (100 per cent carrot particles) and observed much longer time (twice as much) was required to raise the temperature from 20 to 80°C than when the solid concentration was less [7]. During study of ohmic heating behavior of two-phase food systems, it was found that as particle concentration increased, values of EC decreased uniformly. Increase in the EC during heating of biological tissue occurs due to increase in the ionic mobility because of structural changes in the tissue like cell wall protopectin breakdown, expulsion of non-conductive gas bubbles, softening, and lowering in aqueous-phase viscosity [8].

v. Particle Size, Concentration, and Location

With increase in particle size, ohmic heating rate gets decreased. Researchers observed that EC values decreased as the particle size increased in carrot–starch mixture during ohmic heating behavior and EC of two-phase systems. They also observed that with the increase in particle size, the heating time to achieve the same temperature rise also increased [7]. It was noted that particle concentration is a critical factor in determining the heating rate of the two phases [9]. Investigations on the role of particle concentration in orange and tomato juices revealed that overall EC decreased as the percentage of solid constituents dispersed in the liquid phase increased. Solids of low conductivity relative to the fluid will heat slower than the fluid if they are in low concentration [10]. During the study of ohmic heating behavior of two-phase food systems, it was found that as particle concentration increased, values of EC decreased uniformly [7].

vi. Viscosity

Foods having high densities and high specific heat values are conducive to slower heating [1a, 10]. Fluid viscosity also influences ohmic heating; higher viscosity fluids tend to result in faster ohmic heating than lower viscosity fluids [11].

Sources

1. *Ranvir Kaur, Khalid Gul and A. K. Singh, Nutritional impact of ohmic heating on fruits and vegetables-A review, Cogent Food and Agriculture, Volume 2, 2016, Issue 1.*

1a. *K.shiby Varghese, M. C. Pandey, K. Radhakrishna, and A. S. Bawa, Technology, applications and modelling of ohmic heating: a review, J Food Sci Technol, 2014 Oct; 51(10): 2304-2317.*

1b. *Lima M, Sastry SK. The effects of ohmic heating frequency on hot-air drying and juice yield. J Food Eng. 1999.*

2. *Castro, I., Teixeira, J. A., Salengke, S., Sastry, S. K., and Vicente, A. A., 2003, The influence of field strength, sugar and solid content on electrical conductivity of strawberry products, Journal of Food Process Engineering, 26, 17-29.*

3. *Halden, K., De Alwis, A. A., and Fryer, P. J., 1990, Changes in the electrical conductivity of foods during ohmic heating, International Journal of Food Science and Technology, 25, 9-35.*

4. *Castro, I., Macedo, B., Teixeira, J. A., and Vicente, A. A., 2004, The effect of electric field on important food-processing enzymes: Comparison of inactivation kinetics under conventional and ohmic heating, Journal Food Science, 69, 696-701.*

5. *Singh, H., Singh, S. P., and PS, T., 2008, Study on viscosity and electrical conductivity of fruit juices, Journal of Food Science and Technology, 45, 371-372.*

6. *Lima, M., Heskitt, B., and Sastry, S., 1999, The effect of frequency and wave form on the electrical conductivity-temperature profiles of turnip tissue, Journal of Food Process Engineering, 22, 41-54.*

7. *Zareifard, M. R., Ramaswamy, H. S., Trigui, M., and Marcotte, M., 2003, Ohmic heating behaviour and electrical conductivity of two-phase food systems, Innovative Food Science and Emerging Technologies, 4, 45-55.*

8. *Bean, E. C., Rasor, J. P., and Porter, G. C., 1960, Changes in electrical conductivities of avocados during ripening, Year Book of California Avocado Society, 44, 75-78.*

9. *Sastry, S. K., 1991, Ohmic sterilisation and related safety issues (Paper No. 916616), International winter meeting of ASAE, Chicago, IL.*

10. *Palaniappan, S., and Sastry, S. K., 1991a, Electrical conductivities of selected solid foods during ohmic heating, Journal of Food Process Engineering, 14, 221-236.*

11. *Marcotte M, Trigui M, Ramaswamy HS. Effect of salt and citric acid on electrical conductivities and ohmic heating of viscous liquids. J Food Process Preserve. 2000; 24: 389-406.*

Modeling and Monitoring of Ohmic Heating [1]

Modern process systems engineering offers a number of tools (namely computer-aided simulation, optimisation and control), which can be used to design new processes or to improve existing ones. These process engineering tools are based on mathematical models and offer a powerful, rational and systematic way of designing and operating food processes [9, 10].

Mathematical Modeling

The utilization of ohmic heating by the Food industry will depend on development of adequate safety and quality assurance protocols. One crucial component in understanding the process lies in the development of mathematical models, which can then be used to simulate various effects of critical factors.

Mathematical modeling is a useful tool for designing an ohimc heating unit for pasteurization of solid foods. Pasteurization of a solid food undergoing ohmic heating has been analyzed using a mathematical model based on a previously validated [1b] multi-physic model solved with COMSOL 3.3. Modelling ohmic pasteurization involved simultaneous solution of the followings [1a]:

1. Laplace's equation which describes the distribution of electrical potential within a food
2. Heat transfer equation using a source term involving the displacement of electrical potential
3. Kinetics of inactivation of micro-organisms likely to be contaminating the product.

In the model, thermo-physical and electrical properties as function of temperature are used. The model predicted that, when temperature gradients established in the proximity of the outer curved surface, more cold areas were present at junctions of electrodes with lateral sample surface. For these reasons, colder external shells were the critical areas to be monitored, instead of internal points (typically geometrical center) as in classical pure conductive heat transfer [1a].

The following two different modeling approaches currently have been used to undrstand the ohmic heating:

i. De Alwis-Fryer Model

The initial model uses the solution to Laplace's equation to calculate heat generation rate together with transient energy balance equations to model a single particulate immersed in a fluid medium without convection. This model has been extended [3] to include multiple spheres uniformly distributed on a lattice within a non-convective fluid. In a typical situation, wherein a cylindrical particle is stationed in the middle of a tube filled with stationary, non-convective fluid. An electric field is applied along the length of the tube.

The electric field or voltage distribution can be developed from Maxwell's equations, or by combining Ohm's law and the continuity equation for electrical current.

$$\blacktriangledown . (\sigma i \blacktriangledown V) = 0$$

where, V-Voltage, $\blacktriangledown$ - gradient, σi- electrical conductivity of phase I which can take on different values for the particles and liquid.

Ignoring convection effects, the heat transfer problem is one of pure conduction with internal energy generation:

$$\blacktriangledown . (k_i \blacktriangledown T) + \mathring{u}_i = piCpi \frac{\delta T}{\delta t}$$

where, ***i***- again represents the phase, ***k***- thermal conductivity, ***ů***-specific internal energy generation rate, ***p***- density, ***Cp***- specific heat capacity, ***T*** -temperature and ***t***-time.

The system of aforesaid equations can be solved by the Galerkin-Crank-Nicolson algorithm, a hybrid-spatially finite element, temporally finite difference scheme [1].

ii. Sastry-Palaniappan Model

They [4] used circuit analogy to approximate electrical conductivity and thus the heat generation for a static heater with a particle immersed in a well-mixed fluid. Sastry [5] extended this approach to a continuous flow ohmic heater for high solid concentration. The effective electrical resistance can be determined using the circuit analogy. The effective resistance of the cell is calculated by the following formula:

$$R = R_{fs1} + R_{fs2} + \frac{R_{fp}R_{sp}}{R_{fp} + R_{sp}}$$

where, the resistance *R*'s can be calculated from the electrical conductivity and the geometry. The average voltage gradients ($\blacktriangledown V$) in the liquid and particulate can simply be calculated by the following formula:

$$\blacktriangledown V = \frac{V}{L} \; or \; \blacktriangledown V = \frac{IR}{L} \text{ and } I = \frac{V}{R}$$

where, V - the applied voltage and L- the distance between the two electrodes.

This system of equations can be solved by forward differences for the liquid phase, and the Galerkin-Crank-Nicolson method for the solid phase [1].

For **any aseptic process, safety deserves the highest priority**. The limitations of these two models and the lack of understanding of the temperature distribution mandate that the conservative approaches be used. Any low temperature regions that result will extend the process time needed to ensure satisfactory commercial sterility and will make it more likely that other regions of the food will be overcooked. Ohmic heating thus will lose its most attractive advantage. The effects of convection were over simplified in the two models. In reality, the convection heat transfer rate is neither zero nor infinity. A more general model is thus required and convection effects must be included to model the temperature profile [1].

Monitoring (Experimental and Instrumental) [1]

Invasive and Non-invasive methods are being used for monitoring the ohmic heating. The details related to the aforesaid methods are given below:

i. Invasive

Locating the coldest spot within an ohmically heated food system is of significant concern for food engineers and processors. For a high temperature short time thermal process such as heating, the determination of the spatial and temporal distribution of temperature within the particulate is necessary and important. It will not only provide information for calculation of lethality and cook value, but will also provide a guide to mathematical modeling and process control.

Thermal couples were used for modeling verification and process monitoring. Temperature could only be obtained from selected points and the integrity of the process could be disturbed. Moreover, it is difficult to monitor the temperature of a flowing particle.

ii. Non-invasive

Magnetic resonance imaging (MRI) is applied under this method. MRI has contributed greatly to a diversity of scientific disciplines in recent years. Applications in the food area have increased significantly as researchers have discovered the power and flexibility of this technique [6]. Food-related MRI research has gone beyond static imaging experiments to experiments involving dynamic processes, diffusion, water mobility, flow, water and oil distribution, and temperature distribution [6]. The major advantages of MRI are given below:

1. Nondestructive and Non-invasive to the material being imaged and the complexity of foods can be probed without disturbing the sensitive balance.
2. Can provide high spatial resolution.
3. Can provide diverse information such as proton density (which is related to moisture or fat concentration), internal structure, chemical shift, diffusion, temperature and flow.

MRI temperature thermometry is emerging as an attractive and promising temperature mapping method [7, 8]. It was pointed out [9] that the true potential of MRI lies not in static structure determination but in non-invasive, real time and dynamic changes as foods are processed, stored, packaged and distributed. When using MRI to map temperature distribution in a dynamic process like ohmic heating, the data acquisition should be as fast as possible so that a real-time measurement can be accomplished. If this temporal resolution cannot be achieved, the measurement will be inaccurate or perhaps even meaningless.

MRI is an extension of nuclear magnetic resonance (NMR) spectroscopy. NMR spectroscopy is based on magnetic behavior of certain nuclei (such as protons in water and fat) in a sample placed in an external bulk magnetic field, and subjected to a radiofrequency (RF) pulse. The nucleus of an atom possesses a positive charge, and is considered as spinning about an axis. The spinning of the nucleus generates a magnetic field, which is similar to that generated by a simple bar magnet, and is termed the nuclear magnetic dipole. MRI is based on a function of spatial position of magnetic fields instead of using a uniform static field.

Examples-Ohmic Heating of Potato, Carrot, and Beef [1]

Potato, carrot, and beef were cut into about 2.5 cm^3 cubes. The carrier medium was made up of 50 g/kg starch and 1 g/kg NaCl. An experimental ohmic heating device was constructed of glass. It consisted of a 38 mm dia and 318mm long glass tube with a rubber stopper at each end. A 30 mm diameter stainless still electrode was fixed to each of the stoppers and connected to a transducer corresponding to a 50Hz AC power supply. A hole was made in one of the stoppers to allow release of pressure built up within the tube during heating. Two fiber optic temperature sensors could be inserted into the tube through this hole, one near the end of the tube and the other in the center. A constant 140 volts was applied.

The MRI images show that the temperature in the particulates and the corresponding rate of heating the particulates increased with heating time. The increased heating rate was consistent with the rise in electrical conductivity of the particulates with temperature. These images indicate differences in temperature among different particulates in the same carrier fluids. Beef appeared to heat faster than the others. The differences in heating rate among different materials could be due to differences in electrical conductivity and/or non-uniformity in the electric field. The MR images also show that the temperatures in the center regions of a particulate were higher than in the outer regions, **indicating that the particulate heated intensively and transferred heat to the colder carrier liquid**. Variation in temperature was observed within a particle, probably due to spatial variation in electrical conductivity.

Sources

1. *R. Ruan, X. Ye, P. Chen and C. J. Doona, University of Minnesota and I. Taub, Ohmic heating US Army Natick Soldier Center.*

1a. *K. shiby Varghese, M. C. Pandey, K. Radhakrishna, and A. S. Bawa, Technology, applications and modelling of ohmic heating: a review, J Food Sci Technol, 2014 Oct; 51(10): 2304-2317.*

1b. *Marra F, Zell M, Lyng JG, Morgan DJ, Cronin DA. Analysis of heat transfer during ohmic processing of a solid food. J Food Eng. 2009; 91: 56-63.*

2. *De alwis AAP, Fryer P J, A finite element analysis of heat generation and transfer during ohmic heating of foods, Chem. Eng. Sci., 1990, 45, 1547-59.*

3. *Zhang L, Fryer P J, Models for the electrical heating of solid-liquid food mixtures, Chem. Eng. Sci., 1993, 48, 633-43.*

4. *Sastry SK, Palaniappan S, Mathematical modeling and experimental studies on ohmic heating of liquid-particle mixtures in a static heater, Journal of Food Process Engineering, 1992, 15, 241-61.*

5. *Sastry SK, Application of ohmic heating to continuous sterilization of food, Indian Food Industry, 1992, 11, 28-30.*

6. *Ruan R R, Chen P L, Water in foods and biological materials: A nuclear magnetic resonance approach, Lanaster: Technomic Publishing Co., 1998.*

7. *Chang K, Ruan RR, Chen P L, Fulcher RG, Bastian E, Moisture, fat and temperature mapping of cheese block during cooling using MRI, Paper No. 956535, 1995, presented at ASAE Summer meeting.*

8. *Sun X, Schmidt S J, Litchfield JB, Temperature mapping in a potato using half Fourier transform MRI of diffusion, J. Fd Proc. Engr., 1994, 17, 423-37.*

9. *Hill, S B, Food processing: an MRI perspective, Trends in Food Science and Technology, 6, 111-17.*

9. *Singh, R. P., and Banga, J. R., 1994, Recent advances in food process optimization. Chem. Ind., 13, 511-514.*

10. *Fryer, P., 1994, Mathematical models in food processing. Chem. Ind., 13, 515-518.*

Applications of Ohmic Heating in Food Processing [1]

Ohmic heating is an excellent alternative food processing technique that shows much promise in the food manufacturing industry. Since the early 1990's, the technology involving ohmic heating has increased dramatically. As time goes on, this technology will become even more streamlined and efficient. In the future, it is possible that even **'green' energy, such as water, solar, or wind power could create the electricity needed for ohmic heating systems [1a]**. A large number of actual and potential applications exist for ohmic heating which include **blanching, evaporation, dehydration, fermentation, sterilization, pasteurization, and heating of foods** to serving temperature [2a]. Despite all the efforts made so far, the application of ohmic heating to food processing is still not fully characterised and not all of its potentialities have been exploited due to the complexity of both the food materials and the phenomena occurring during ohmic heating processing. The application of ohmic heating in the future may be defined as follows [1f]:

1. Aseptic Process using Ohmic heating
2. Inactivation of Food Enzymes
3. Pasteurization and Sterilization
4. Ohmic Thawing.

The major applications of this technology are for sterilization of high proteinaceous materials, seafood processing, pretreatments for water removal, thawing, detection of starch gelatinization and extraction enhancement [1f]. The some applications of ohmic heating in food processing are given below:

1. Fruit and Vegetable Processing

a. Ohmic heating can be applied to a wide variety of foods, **including liquids, semi-solid slurries or solid foods** accompanied by a suitable carrier liquid.

b. Ohmic heating has been used commercially in the US to pasteurize liquid egg products and in the UK and Japan for processing whole fruits [2].

c. Ohmic heating has been successfully applied to a wide range of commodities such as fruits and vegetables, juices, meats, seafood, soups, crèmes, and pasta dishes [3]. Ohmic heating is especially useful for very sticky (viscous) materials or fluids containing solid particles. It may also be used wherever non-uniform heating must be avoided or where mechanical agitation to improve heat transfer is not recommended. The problems occurred during traditional thermal processing of fruit purees and jams (essentially due to their rheological properties), the production process of canned soups for instance, which normally integrate food solid particles (*e.g.* peas, carrot cubes) and pet food production can be solved by ohmic heating. Ohmic heating has following plus points on fruit processing:

i. Effect of Ohmic Heating on Sensory Properties of Orange Juices [1a]

The results of the study [3a] suggest that continuous ohmic heating can be used to effectively pasteurize orange juice, and yet retain the flavor and sensorial appeal of freshly squeezed juice. The ohmic heating technique is just as efficient in the required microbial and enzymatic inactivation as conventional pasteurization, but retains the flavor compounds associated with fresh orange juice. The study indicated that a thermal treatment such as ohmic heating could be used to extend the sensorial shelf life of pasteurized fresh squeezed orange juice. Ohmic heating can be performed without overheating, due the heat transfer aspects of electrical current through the juice, which results in a high retention of the sensorial attributes of the fresh juice.

ii. Increased Juice Yield

The researchers studied **[1b]** the effects of ohmic heating on juice yield from potato and apple tissues in the study, "**Ohmically Heated, Enhanced Expression of Juice from Apple and Potato Tissues".** The study showed that treating of plant tissue with electric fields less than 100 V/cm provides a dramatic increase in membrane destruction and mechanical softening. Previous studies [1c, 1d] showed that thermal damage is not significant for treatment times of 20 s and temperatures less than 60°C, so the changes in the mechanical properties of the tissues can be explained by electroporation. In the range of 50-60 V/cm and treatment times of 20 s, the threshold of electroporation is exceeded, and the majority of the cells are destroyed, releasing the juice. The final temperature of the tissues can greatly affect the juice yield also. For this phase of the experiment, different end point temperatures and electric fields strengths were used in conjunction to determine the greatest yield possible.

The informations related to applications of Ohmic heating in fruit and vegetable processing are tabulated below [1e]:

Fruit and Vegetables	*Applications*
1. Fruits	
Apple	Ohmic heating was found to enhance the extraction of apple juice from red delicious apples [2b]. The effect of ohmic heating frequency on apple juice yield and hot-air drying rate was evaluated. The extraction rates of ohmically treated apple tissue with respect to non-treated sample and the effect of frequency on extraction yield was determined [2c].
Grape and guava	The effects of voltage gradient, temperature, and holding time on the polyphenoloxidase activity were investigated for grape juice ohmic heating [2d]. The effects of ohmic heating application on mechanical extraction and sensory characteristics of guava juice was studied [2e].

Fruit and Vegetables	*Applications*
Orange	The effect of ultrahigh temperature continuous ohmic heating treatment on fresh orange juice in comparison to the conventional pasteurization was evaluated [2f]. Viscosity and EC of orange juice was measured over temperature range (25–70°C) during ohmic heating. The kinetics of ascorbic acid degradation during ohmic heating of orange juice by applying electric field strength of 42 V/cm was studied [2g].
Peach and pear	The EC of peach puree during ohmic heating was evaluated. The ohmic heating rate of peaches was evaluated at fixed electric field strength of 60 V/cm, and frequencies varying within 50 Hz-1 MHz. EC of peach and pear was determined over 25-140°C temperature range [2h]. The effect of ohmic heating on the osmotic dehydration kinetics and microstructure of pears was investigated [2i].
Pineapple and pome-granate	EC of pineapple was determined over 25-140°C temperature range [2h]. Viscosity and EC of pineapple juice was measured over temperature range (25-70°C) during ohmic heating. The effect of ohmic heating technique on EC, heating rate, system performance, and pH of pomegranate juice was investigated [2j].
Strawberry	EC of strawberry was determined over 25-140°C temperature range [2h]. The influence of ohmic heating on osmotic dehydration kinetics and microstructure of strawberries was studied [2k].
Apricot	The effect of ohmic heating processing on the quality and shelf life of apricots in syrup using a continuous pilot scale ohmic unit was studied [2l]. The EC of apricot puree during ohmic heating was also evaluated. Both the studies have successfully reported enhanced shelf life and quality of apricots.
2. Vegetables	
Beetroot and broccoli	The effect of temperature on EC of beetroot when processed ohmically and conventionally was studied [3b]. The leaching of soluble solids during blanching of beetroot by ohmic heating was examined [3c]. Similarly, the effect of ohmic and conventional heating process on textural properties of cylindrical pieces of red beet was investigated [3d].
Carrot	The effects of insoluble solids and applied voltage on EC of pre-pasteurized carrot juices during ohmic heating was studied [3e]. The researchers carried out the inactivation kinetics of peroxidase during ohmic heating of carrot [3f]. Similarly, the effect of ohmic and conventional heating process on textural properties of cylindrical pieces of carrot and golden carrot were investigated [3g].
Cauliflower and cabbage	Processing and stabilization of cauliflower by ohmic heating technology was investigated [4a]. The influence of precooking by ohmic heating on the firmness of cauliflower was studied [4b].
Pea and potato	Peroxidase inactivation and color changes during ohmic blanching of pea puree were examined [4c]. The effect of ohmic heating on cell membranes of potato was investigated by measurement of dielectric spectra from 100 Hz to 20 kHz. The effect of ohmic heating on juice yield from potato tissues was determined.
Radish, tomato, and turnip	Viscosity and EC of tomato juice was measured over temperature range (25–70°C) during ohmic heating [4d]. The effect of temperature and EC of turnip tissue at 4 frequencies (4, 10, 25, and 60 Hz) under ohmic heating was studied [4e]. The effect of the electric field frequency on heating rate of Japanese white radish in the range of frequencies 50 Hz–10 kHz was observed [4f].

2. Sterilisation

Ohmic heating is very often used in pasteurization/sterilization of food products resulting in excellent quality. It can be used for ultra high temperature (UHT) sterilization of foods, and especially those that contain large particles (up to 2.5 cm) that are difficult to sterilize by other means.]1f]. Ohmic heating can be used for sterilization purposes to produce high-quality shelf-stable low-acid foods such as ready-prepared meals and high-acid foods such as tomato-based sauces. More extensive studies into pathogenic bacteria inactivation, such as ***E.coli* O157: H7** or **Salmonella,** are also recommended [1a].

3. Enhancement of Process Efficiency of different Processes

In 2001, the FDA reported that "A large number of potential future applications exist for ohmic heating, including its use in blanching, evaporation, dehydration, fermentation, and extraction" [2]. Ohmic heating can increase process efficiency in blanching [4] as well as enhancing the drying of vegetable tissue such as potato and yam by 16 per cent and 43 per cent, respectively [5].

4. Oil Industry

Ohmic treatment was also found to enhance the extraction of valuable components such as rice bran oil from rice bran by up to 74 per cent [6]. It was reported [1f] that ohmic heating was an alternative method for stabilisation of rice bran because it increased the oil yield of rice bran to a maximum of 92 per cent, whilst only 53 per cent of oil was extracted from the control samples.

5. Dairy Industry [1a]

In the early nineties electric pasteurization of milk was carried out by passing it between two plates with an electrical potential difference [1f]. Fermentation is the process of releasing energy from a carbohydrate without oxygen by producing alcohol or lactic acid. An ohmic heating system may be particularly useful in this important process of food manufacturing, particularly the dairy industry, where fermentation by ***Lactobacillus acidophilus*** is necessary for the production of **cheese and yogurt.** The dairy industry's most important organism, ***Lactobacillus acidophilus*** is being used for fermentation of dairy products. A study [7] showed that ohmic heating of a fermentation vessel containing ***L. acidophilus*** reduced the lag period of the bacteria in the early stages of growth and because of this, a dairy is utilizing the ohmic heating process in the early stages of fermentation for shorten the total processing time of a dairy product which save time and labor costs. The final product of the ohmically heated sample had a higher final pH and lower bacteriocin levels but had little effect on glucose consumption and lactic acid production by ***L. acidophilus.***

6. Wine Industry

Fermented products like beer or wine would benefit from ohmic heating by decreasing the processing time, using the findings [7] that ohmic heating may decrease the lag period of fermentative bacteria.

7. Bakery [1a]

Baked goods like meringues and cookie dough may use ohmic heating as a method of drying and processing the eggs present in both products without affecting the final taste or structure.

8. Meat Industry [1a]

Lean meat products like chicken or rabbit may perhaps benefit from ohmic heating, as there is very little fat associated with these products, lessening the possibility of under processed fat globules.

9. Seafood Processing

Surimi is stabilized fish myofibrillar protein and is a major ingredient for seafood products; such as, surimi crab, shrimp, lobster and scallop products. The microbiological quality of these seafood products is very high. Ohmic heating experiments were conducted for fish mince [1f, 1g]. Pacific whiting surimi paste and stabilized mince in the 20-70 °C range were tested at frequencies from 55 Hz to 200 kHz. Sample impedance decreased slightly with frequency. The DC electrical conductivity and absolute dielectric loss of Pacific whiting surimi paste increased with temperature and salt concentration and those of the stabilized mince increased with temperature.

10. Ohmic Thawing [1f]

Ohmic heating can be used to thaw frozen foods placed between two electrodes and applying an alternating current to it. The advantages of this process include, water and wastewater is not generated, thawing can be relatively uniform due to volume heating and the process can be easy to control. As the frozen product thaws current passes through the thawed portion of the block more rapidly since frozen shrimp has 100 times lower conductivity than thawed shrimp [8]. With the current flowing through the thawed portion of the block, shrimp in that portion may cook, while the rest of the block is still frozen. **This is called runaway heating, or formation of hot spots.**

Sources

1. *Hudaa Neetoo (Faculty of Agriculture, University of Mauritius, Réduit, Mauritius) and Haiqiang Chen (Department of Animal and Food Sciences, University of Delaware, Newark, Delaware, US), Alternative Food Processing Technologies.*

1a. *Destinee R. Anderson, A report on Ohmic heating as an alternative food processing technology, Food Science Institute, College of Agriculture, Kansas State University, Manhattan, Kansas.*

1b. *Praporsic, I., Lebovka, N. I., Ghnimi, S., Vorobiev, E. 2006, Ohmically Heated, Enhanced Expression of Juice From Apple and Potato Tissues. Biosystems Engineering 93 (2): 199-204.*

1c. *Lebovka, N. I., Praporscic, I., and Vorobiev, E. 2004b. Combined treatment of apples by pulsed electric fields and by heating at moderate temperature. J. Food Eng. 65, 211-217.*

1d. *Lebovka, N. I., Praporsic, I., Ghnimi, S., Vorobiev, E. 2005. Does Electroporation Occur During the Ohmic Heating of Food? Journal of Food Science Vol 70, No. 5.*

1e. *Ranvir Kaur, Khalid Gul and A. K. Singh, Nutritional impact of ohmic heating on fruits and vegetables-A review, Cogent Food and Agriculture, Volume 2, 2016, Issue 1.*

1f. *K.shiby Varghese, M. C. Pandey, K. Radhakrishna, and A. S. Bawa, Technology, applications and modelling of ohmic heating: a review, J Food Sci Technol, 2014 Oct; 51(10): 2304-2317.*

1g. *Wu H, Kolbe E, Flugstad B, Park JW, Yongsawatdigul J. Electrical properties of fish mince during multi-frequency ohmic heating. J Food Sci. 1998; 63(6): 1029-1032.*

2. *FDA-2001.*

2a. *USA FDA-2000.*

2b. *Wang, W. C., 1995, Ohmic heating of foods: Physical properties and applications. Columbus, OH: The Ohio State University.*

2c. *Lima, M., and Sastry, S. K., 1999, The effects of ohmic heating frequency on hot-air drying rate and juice yield, Journal of Food Engineering, 41, 115-119.*

2d. *Icier, F., Yildiz, H., and Baysal, T., 2008, Polyphenol- oxidase deactivation kinetics during ohmic heating of grape juice, Journal of Food Engineering, 85, 410-417.*

2e. *Srikalong, P., Makrudin, T., Sampa-vamontri, P., and Kovitthaya, E., 2011, Effect of ohmic heating on increasing guava juice yield, 2nd International Conference on Biotechnology and Food Science, Vol.7, Bali Island.*

2f. *Leizerson, S. and Shimoni, E., 2005, Effect of ultrahigh - temperature continuous ohmic heating treatment on fresh orange juice, Journal of Agricultural and Food Chemistry, 53, 3519-3524.*

2g. *Vikram, V. B., Ramesh, M. N., and Prapulla, S. G., 2005, Thermal degradation kinetics of nutrients in orange juice heated by electromagnetic and conventional methods, Journal of Food Engineering, 69, 31-40.*

2h. *Sarang, S., Sastry, S. K., and Knipe, L., 2008, Electrical conductivity of fruits and meats during ohmic heating, Journal of Food Engineering, 87, 351-356.*

2i. *Moreno, J., Simpson, R., Estrada, D., Lorenzen, S., Moraga, D., and Almon-acid, S., 2011, Effect of pulsed-vacuum and ohmic heating on the osmo-dehydration kinetics, physical properties and microstructure of apples, Innovative Food Science and Emerging Technologies, 12, 562-568.https: //www.tandfonline.com/doi/full/10.1080/23311932. 2016.1159000*

2j. *Darvishi, H., Khostaghaza, M. H., and Najafi, G., 2013, Ohmic heating of pomegranate juice: Electrical conductivity and pH change, Journal of the Saudi Society of Agricultural Sciences, 12, 101-108.*

2k. *Moreno, J., Simpson, R., Baeza, A., Morales, J., Muñoz, C., Sastry, S., and Almonacid, S., 2012, Effect of ohmic heating and vacuum impregnation on the osmo-dehydration kinetics and microstructure of strawberries, LWT-Food Science and Technology, 45, 148-154.*

2l. *Pataro, G., Donsì, G., and Ferrari, G., 2011, Aseptic processing of apricots in syrup by means of a continuous pilot scale ohmic unit. LWT-Food Science and Technology, 44, 1546-1554.*

3. *Bengtsson R, Birdsall E, Feilden S, Bhattiprolu S, Bhale S, Lima M, 2010, Ohmic and inductive heating. In: Hui YH (ed) Handbook of Food Science, Technology and Engineering, volume 3. Boca Raton, FL: CRC Press, pp. 121-127.*

3a. *Leizerson, S. and Shimoni, E. 2005a. Stability and Sensory Shelf Life of Orange Juice Pasteurized by Continuous Ohmic Heating. J. Agric. Food Chem. 53: 4012-4018.*

3b. *Halden, K., De Alwis, A. A., and Fryer, P. J., 1990, Changes in the electrical conductivity of foods during ohmic heating. International Journal of Food Science and Technology, 25, 9-35.*

3c. *Mizrahi, S., 1996, Leaching of soluble solids during blanching of vegetables by ohmic heating. Journal of Food Engineering, 29, 153-166.*

3d. *Farahnaky, A., Azizi, R. and Gavahian, M., 2012, Accelerated texture softening of some root vegetables by ohmic heating. Journal of Food Engineering, 113, 275-280.*

3e. *Palaniappan, S., and Sastry, S. K., 1991a, Electrical conductivities of selected solid foods during ohmic heating. Journal of Food Process Engineering, 14, 221-236.*

3f. *Jakob, A., Bryjak, J., Wojtowicz, H., Illeova, V., Annus, J., and Polakovic, M., 2010, Inactivation kinetics of food enzymes during ohmic heating. Food Chemistry, 123, 369-376.*

3g. *Farahnaky, A., Azizi, R., and Gavahian, M., 2012, Accelerated texture softening of some root vegetables by ohmic heating, Journal of Food Engineering, 113, 275-280.*

4. *Wigerstrom K., 1976, Passing an electric current of 50-60 Hz through potato pieces during blanching. US Patent No.3, 997, 678.*

4a. *Goullieux, A., Zuber, F. and Godereaux, S. C. E., 2001, Processing and stabilisation of cauliflower by ohmic heating technology. Innovative Food Science and Emerging Technologies, 2, 279-287.*

4b. *Eliot, S. C., Goullieux, A. and Pain, J. P., 1999a, Processing of cauliflower by ohmic heating: Influence of precooking on firmness. Journal of the Science of Food and Agriculture, 79, 1406-1412.*

4c. *Icier, F., Yildiz, H., and Baysal, T., 2006, Peroxidase inactivation and colour changes during ohmic blanching of pea puree. Journal of Food Engineering, 74, 424-429.*

4d. *Singh, H., Singh, S. P. and PS, T., 2008, Study on viscosity and electrical conductivity of fruit juices. Journal of Food Science and Technology, 45, 371-372.*

4e. *Lima, M., Heskitt, B., and Sastry, S., 1999, The effect of frequency and wave form on the electrical conductivity-temperature profiles of turnip tissue, Journal of Food Process Engineering, 22, 41-54.*

4f. *Imai, T., Uemura, K., Ishida, N., Yoshizaki, S., and Noguchi, A., 1995, Ohmic heating of Japanese white radish Rhaphanus sativus, International Journal of Food Science and Technology, 30, 461-472.*

5. *Wang W, Sastry, S., 2000, Effects of thermal and electro-thermal pretreatments of hot air drying rate of vegetable tissue. Journal of Food Process Engineering 23: 299-319.*

6. *Lakkakula N, Lima M, Walker T., 2004, Rice bran stabilization and rice bran oil extraction using ohmic heating. Bioresource Technology 92: 157-161.*

7. *Cho, H. Y., Yousef, A. E., Sastry, S. K. 1996. Growth Kinetics of Lactobacillus acidophilus Under Ohmic Heating. Biotechnol Bioeng 49: 334-340.*

8. *Luzuriaga DA, Roberts JS, Balaban MO. Electrical conductivity of frozen shrimp and flounder at different temperatures and voltage levels. J Agric Food Proc Technol. 1996; 3(5): 41-63.*

Effect of Ohmic Heating on the Sensory and Nutritional Quality of Foods [1, 5]

Several researchers have shown that food samples processed by ohmic heating exhibited better quality attributes than their traditionally processed counterparts, with respect to nutrient retention, texture, color, and flavor. The properties such as viscosity, pH, rheology, and color were also examined. With increase in temperature, viscosity decreases leading to increase in ohmic heating rate. pH decreases with increase in voltage gradient.

1. Effect on Flavour and Heat Sensitive Compounds

i. Effect on Flavour

It was observed [2] that higher retention of flavor compounds, limonene and myrcene, in ohmically treated orange juice than in conventionally heated juice. Moreover, sensory evaluation showed that ohmically treated and fresh orange juices were indistinguishable.

ii. Effect on Heat Sensitive Compounds [5]

The results of several studies, showed that degradation of heat sensitive compounds (anthocyanins, ascorbic acid, and vitamin C) increases with both increasing voltage and solid contents. Some examples related to the degradation of Anathocyanin, Vitamin C *etc.* are given :

Anthocyanin

The available literature indicates for anthocyanin degradation in fruit products vary considerably. Anthocyanain stability is influenced by the intrinsic properties of the product and the process characteristics causing these differences to occur. Anthocyanin degradation in blueberry pulp was evaluated after thermal treatment using ohmic and conventional heating [2a]. Degradation increased with both increasing voltage and solids content. The comparison between ohmic and conventional heating showed that when lower voltage levels were used, the percentage of degradation was lower to those obtained during conventional heating. The pulp processed during ohmic heating exhibited higher anthocyanin degradation with high electric fields.

Vitamin C [5]

A higher level of degradation of vitamin C during ohmic heating using high voltages relative to conventional heating was studied [2b]. It was concluded that during ohmic heating, in addition to the degradation caused by heat, there is also electrochemical degradation due to number of reactions, including electrode reactions and electrolysis of the solution; in addition, reactions between the electrode materials and the electrolysis products may influence the degradation reaction mechanisms and the kinetic parameters. The effects of voltage and solid contents on vitamin C and ascorbic acid degradation in acerola pulp [2c], the degradation of vitamin C in strawberry products [2d] and temperature impact on ascorbic acid degradation in ground cashew apples [2e] were studied. Studies on the vitamin retention of ohmic and conventionally heated strawberry pulps and orange juices have presented similar ascorbic acid degradation kinetics [4].

2. Moisture Retention and Mechanical Properties [1]

Meat products cooked by ohmic heat were also indistinguishable from traditionally cooked samples with respect to moisture retention and mechanical properties.

3. Textural Strength and Colour [5]

i. However, it was observed [3] a decrease in textural strength of ohmic-treated sausages and suggested the use of binders to circumvent this problem. Ohmic blanching also appears to be a very promising application, causing minimal deleterious effects on the color of various vegetables and vegetable products [1].

ii. The color changes in acerola pulp during ohmic heating was investigated [3a]. Acerola has a good amount of anthocyanins and carotenoid pigments, their gradual degradation results in color changes in the final product. The studies reported that color degradation kinetics follow a first order reaction kinetics.

4. Inactivation of Enzymes [5]

There has been limited research on the effect of ohmic heating on enzymes. Electrical fields applied during ohmic heating leads to faster deactivation of enzymes and micro-organisms. The deactivation rate of different enzyme samples heated with ohmic or conventional heating was compared and postulated that the electrical field applied during ohmic heating caused the faster deactivation than the conventional heating [5a]. Some examples related to the inactivation of enzymes are given below:

i. PME

It is an enzyme that has been found in essentially every plant tissue, several fungi, and bacteria. PME has no prosthetic group and catalyzes de-esterification of galactosyluronate methylesters of pectins, releasing protons and methanol into the media. The study conducted on orange juice showed that ohmic heating reduced PME activity by 98 per cent [5b].

ii. Peroxidases

These are known to be the most heat stable enzymes in vegetables, and their inactivation is usually used to indicate the adequacy of blanching [6]. Ohmic blanching of pea puree at four different voltage gradients (20-50 V/cm) was performed by researchers [6a] and they observed that the inactivation time of peroxidase enzyme during ohmic blanching was less than water blanching by using 30 V/cm and above voltage gradient. As the voltage gradient increases the critical inactivation time decreases during ohmic blanching.

iii. Lipoxygenase and Polyphenoloxidase

These enzymes inactivated at a faster rate due to electric fields, applied during ohmic heating than conventional heating [5a]. Similarly, the effects of voltage gradient, temperature, and holding time on the polyphenoloxidase activity were investigated for grape juice ohmic heating [7]. There was an increase in ohmic heating rate with increase in voltage gradient. The critical deactivation temperature at higher voltage gradient was lower than at low voltage values because of the faster increase in EC at higher voltage gradients causing higher deactivation in polyphenol oxidase.

iv. Pepsin

The deactivation of pepsin activity was reported for grape juice as a function of electric field strength, EC and pH [8]. The researchers predicted that higher EC increased the deactivation of pepsin by pulsed electric field at the same temperature. Similar results for the deactivation of trypsin for fruit juice at temperatures ranging from 55 to 70°C was reported [8a].

5. Rheological Characteristics [5]

The rheological characteristics of quince nectar during ohmic heating was studied [9]. Ohmic heating was applied to quince nectar by changing the voltage gradient (10-40 V/cm) at 50 Hz. The rheological behavior of quince nectars was examined by application of different mathematical models. **Herschel-Bulkley model** fitted the experimental data best, for all temperatures. Non-Newtonian shear thinning behavior was obtained for quince nectar at the pasteurization conditions (temperature range of 65-70°C and holding time of 0-30 min. It was also reported that ohmic heating did not cause different effect on the rheological properties of quince nectar compared to conventional heating. Researchers also performed similar study comparing the available data in literature of fruit pulps (guava, raspberry, pineapple, apricot, apple, mango, tamarind, black currant) and found that these fruit pulps showed shear thinning behavior [10]. Researchers also studied the rheological behavior of ternary mixtures of mango pulp, and juices of carrot and orange, where the ternary mixtures presented the non-Newtonian fluid behavior [11].

Sources

1. *Hudaa Neetoo (Faculty of Agriculture, University of Mauritius, Réduit, Mauritius) and Haiqiang Chen (Department of Animal and Food Sciences, University of Delaware, Newark, Delaware, US), Alternative Food Processing Technologies.*

2. Leizerson S, Shimoni E, 2005, *Effect of ultrahigh-temperature continuous ohmic heating treatment on fresh orange juice. Journal of Agricultural and Food Chemistry 53: 3519-3524.*

2a. *Sarkis, J. R., Mercali, G. D., Tessaro, I. C., Marczak, L. D. F., 2013, Evaluation of key parameters during construction and operation of an ohmic heating apparatus, Innovative Food Science and Emerging Technologies, 18, 145-154.*

2b. *Assiry, A., Sastry, S. K., and Samaranayake, C., 2003, Degradation kinetics of ascorbic acid during ohmic heating with stainless steel electrodes, Journal of Applied Electrochemistry, 33, 187-196.*

2c. *Mercali, G. D., Jaeschke, D. P., Tessaro, I. C., and Marczak, L. D. F., 2012, Study of vitamin C degradation in acerola pulp during ohmic and conventional heat treatment. LWT-Food Science and Technology, 47, 91-95.*

2d. *Castro, I., Macedo, B., Teixeira, J. A. and Vicente, A. A., 2004, The effect of electric field on important food-processing enzymes: Comparison of inactivation kinetics under conventional and ohmic heating, Journal Food Science, 69, 696-701.*

2e. *Lima, J. R., Elizondo, N. J., and Bohuon, P., 2010, Kinetics of ascorbic acid degradation and colour change in ground cashew apples treated at high temperatures (100-180°C), International Journal of Food Science and Technology, 45, 1724-1731.*

3. *Piette L, 2004, Ohmic cooking of processed meats and its effects on product quality. Journal of Food Science 69: 71-78.*

3a. *Mercali, G. D., Schwartz, S., Marczak, L. D. F., Tessaro, I. C., and Sastry, S., 2014, Ascorbic acid degradation and color changes in acerola pulp during ohmic heating: Effect of electric field frequency, Journal of Food Engineering, 123, 1-7.*

4. *Vikram VB, Prapulla SG, Ramesh MN, 2005, Thermal degradation kinetics of nutrients in orange juice heated by electromagnetic and conventional methods, Journal of Food Engineering 69: 31-40.*

5. *Ranvir Kaur, Khalid Gul and A. K. Singh, Nutritional impact of ohmic heating on fruits and vegetables-A review, Cogent Food and Agriculture, Volume 2, 2016, Issue 1.*

5a. *Castro, I., Macedo, B., Teixeira, J. A. and Vicente, A. A., 2004, The effect of electric field on important food-processing enzymes: Comparison of inactivation kinetics under conventional and ohmic heating, Journal Food Science, 69, 696-701.*

5b. *Leizerson, S. and Shimoni, E., 2005, Effect of ultrahigh-temperature continuous ohmic heating treatment on fresh orange juice, Journal of Agricultural and Food Chemistry, 53, 3519-3524.*

6. *Akyol, C., Bayindirli, A. and Alpas, H., 2004, July 12-16, Effect of combined treatment of high hydrostatic pressure and mild heat on peroxidise inactivation in green beans, peas and carrots. Presented at IFT Annual Meeting, Las Vegas, NV.*

6a. *Icier, F., Yildiz, H., and Baysal, T., 2006, Peroxidase inactivation and colour changes during ohmic blanching of pea puree, Journal of Food Engineering, 74, 424-429.*

7. *Icier, F., Yildiz, H. and Baysal, T., 2008, Polyphenol - oxidase deactivation kinetics during ohmic heating of grape juice, Journal of Food Engineering, 85, 410-417.*

8. *Yang, R., Li, S. Q., and Zhang, Q. H., 2004, Effects of pulsed electric fields on the activity and structure of pepsin, Journal of Agricultural and Food Chemistry, 52, 7400-7406.*

8a. *Margot, A., Flaschel, E., and Renken, A., 1997, Empirical kinetic models for tryptic whey-protein hydrolysis. Process Biochemistry, 32, 217-223.*

9. *Bozkurt, H., and Icier, F., 2009, Rheological characteristics of quince nectar during ohmic heating, International Journal of Food Properties, 12, 844-859.*

10. *Krokida, M. K., Maroulis, Z. B., and Saravacos, G. D., 2001, Rheological properties of fluid fruit and vegetable puree products: Compilation of literature data, International Journal of Food Properties, 4, 179-200.*

11. *Branco, I. G., Gasparetto, C. A., 2003, Response surface methodology applied to the study of temperature effect on the rheological behaviour of ternary mixtures with mango pulp and orange and carrot juices, Ciencia e Tecnologia de Alimentos, 23, 166-171.*

Chapter 16

Sous-vide Processing Technology

Sous-vide **[1]** is a French term literally meaning "under vacuum." This technology allows food to be thermally processed using vacuum packaging in heat-stable, high barrier or air-impermeable multilaminate plastics. This form of processing is especially useful for food consisting of partially cooked ingredients alone or combined with raw foods that require low-temperature storage until the food is thoroughly heated immediately prior to serving [2]. In short, **sous-vide is an "assemble package- pasteurize-cool-store" process [3].** Sous-vide implies that food should be sealed in a plastic bag with all the air removed. An alternative method is to place the food in an open-sided plastic bag and partially submerge the bag into the water, forcing out the air. This method involves clipping the open side of the bag to the side of the pot to keep water from leaking into the opening [4].

Sous-vide is a method of cooking in vacuumized plastic pouches at precisely controlled temperatures. Precise temperature control gives more choice over doneness and texture than traditional cooking methods. Cooking in heat-stable, vacuumized pouches improves shelf-life and can enhance taste and nutrition [4a].

Sous-vide cooking differs from traditional cooking methods in following two fundamental ways:

1. Raw food is vacuum-sealed in heat-stable, food-grade plastic pouches.
2. The food is cooked using precisely controlled heating.

Vacuum-sealing has following several benefits:

1. It allows heat to be efficiently transferred from the water (or steam) to the food,
2. It increases the food's shelf-life by eliminating the risk of recontamination during storage,

3. It inhibits off-flavors from oxidation and prevents evaporative losses of flavor volatiles and moisture during cooking [4a, 4b]
4. Reduces aerobic bacterial growth. This results in especially flavorful and nutritious food [4c].

Developmental Steps (History) [4]

Sous-vide cooking/processing has following three characteristics, which are developed separately:

1. Low-temperature cooking,
2. Containerized cooking that separates the food from its heating environment.
3. Pressurized enclosure using full or partial vacuum.

The history related to the development of aforesaid characteristics are tabulated below:

Low-temperature Cooking	*Containerized Cooking and Pressurized Enclosure*
i. First described by Benjamin Thompson and Count Rumford in 1799.	i. Preparation of food under pressure, with or without heat, was developed by American and French engineers in the mid-1960s.
ii. Used air as the heat transfer medium while attempting to see if this could roast meat in a machine which had created to dry potatoes [5, 6].	ii. As this method was pioneered, applying pressure to food through vacuum sealing was sometimes called **Cryovacking**.
—	iii. The pressure notably concentrated the flavors of fruits, even without cooking [7].

- ☆ The sous-vide method of preserving and cooking in laminated plastic film sachets is generally credited to Chef Georges Pralus, a French chef. He adopted this method in 1974 for the Restaurant Troisgros in Roanne, France. He discovered that when foie gras was cooked in this manner (packaging the foie gras in a vacuum sealed plastic pouch and submersing it in a heated water bath with precisely controlled temperatures), it kept its original appearance (shrinkage reduces from 40 per cent to 5 per cent), did not lose excess amounts of fat, and had better texture [7].
- ☆ Bruno Goussault, chief scientist of Sterling, Virginia-based food manufacturer Cuisine Solutions, researched the effects of temperature on various foods and became well known for training top chefs in the method and developed the parameters of cooking times and temperatures for various foods [7].
- ☆ Goussault and Pralus independently worked on development of sous-vide in the 1970s, and eventually became collaborators. Goussault pioneered the marriage of vacuum sealing with low-temperature cooking. Pralus, considered the father of modern sous-vide, cooked at higher temperatures [7].

Sources

1. *Hudaa Neetoo (Faculty of Agriculture, University of Mauritius, Réduit, Mauritius) and Haiqiang Chen (Department of Animal and Food Sciences, University of Delaware, Newark, Delaware, US, Alternative Food Processing Technologies.*
2. *Ghazala S, 2004, Development in cook-chill and sous-vide processing. In: Richardson P (ed) Improving the Thermal Processing of Foods. Cambridge: Woodhead Publishing, pp. 152-173.*
3. *Juneja VK, 2003, Sous-vide processed foods: safety hazards and control of microbial risks. In: Novak JS, Sapers GM, Juneja VK (eds) Microbial Safety of Minimally Processed Foods. Boca Raton, FL: CRC Press, pp. 97-124.*
4. *Wikipedia.*

4a. *Baldin, D. E., Sous-vide cooking review, International journal of Gastronomy and Food sciences, Volume 1, Issue 1, January 2012, Pages 15-30.*

4b. *Church, I. J. and Parsons, A. L., The sensory quality of chicken and potato products prepared using cook-chill and sous-vide methods International Journal of Food Science and Technology, 35, 2000, pp. 155-162.*

4c. *Ghazala, S., Aucoin, J. and Alkanani, T. Pasteurization effect on fatty acid stability in a sous-vide product containing seal meat (Phoca groenlacdica), Journal of Food Science, 61 (3) (1996), pp. 520-523.*

5. *Sous-vide Historical Note: Count Rumford, Medellitin. Retrieved 14 October 2012.*
6. *Benjamin Count of Rumford, Essay X, On the construction of kitchen fire-places and kitchen utensils together with remarks and observations relating to the various processes of cookery; and proposals for improving that most useful art, Essays, Political, Economical, and Philosophical, vol. 3 (London, England: T. Cadell jun. and W. Davies, 1802), pp. 18-20.*
7. *Hesser, Amanda (2005-08-14), Under Pressure, The New York Times.*

Basic Techniques and Pathogens of Interest [1]

Sous-vide cooking typically takes following two forms:

i. Cook-Hold or Cook-Serve [1]

a. Technique

Cook-hold or cook-serve sous-vide cooking consists of preparing for packaging, vacuum packaging, heating or pasteurizing, finishing, and serving. It is a simplest and safest method of sous-vide cooking is cook-hold or cook-serve. The raw (or partially cooked) ingredients are vacuum sealed, pasteurized, and then held at 54.4°C or above until served.

b. Pathogens of Interest

- ☆ For cook-hold sous-vide, the main pathogens of interest are the Salmonella species and the pathogenic strains of *E. coli*.

- ☆ There are, of course, many other food pathogens but these two species are relatively heat resistant and require very few vegetative bacteria per gram to make an immuno-compromised person sick.
- ☆ The time and temperature combinations in FDA (2009) correspond to less than a 3-log10 reduction in the Salmonella species (sufficient for healthy person).
- ☆ For immuno-compromised people, most experts recommend a 6.5-7-log10 reduction of the *Salmonella* species and a 5-log10 reduction of pathogenic strains of *E. coli.*

ii. Cook Chill or Cook-Freeze [1]

a. Technique

Cook-chill or cook-freeze sous-vide cooking consists of preparing for packaging, vacuum packaging, pasteurizing, rapid chilling, refrigerating or freezing, reheating or rethermalizing, finishing, and serving. The most popular methods of sous-vide cooking are cook-chill and cook-freeze. The raw (or partially cooked) ingredients are vacuum sealed, pasteurized, rapidly chilled and either refrigerated or frozen until reheating for service.

b. Pathogens of Interest

- ☆ For cook-chill sous-vide, *L. monocytogenes* and the spore forming pathogenic bacteria are the pathogens of interest. That is because Listeria is the most heat resistant non-spore forming pathogen and can grow at refrigerator temperatures.
- ☆ While keeping the food sealed in its plastic pouches prevents recontamination after cooking, spores of *C. botulinum, C. perfringens,* and *B. cereus* can all survive the mild heat treatment of pasteurization. Therefore, after rapid chilling, the food must either be frozen.

Sous-vide cooking can be divided into following three broad categories:

1. Raw or unpasteurized,
2. Pasteurized
3. Sterilized

The following flow diagram of the main types of sous-vide cooking is helpful in understanding the Sous-vide cooking:

1. The branches in red and green (the rightmost three) are common in both restaurant and home kitchens.
2. While industrial food processors only use the branches in blue and green (the leftmost three).
3. The branches in red (the rightmost two) should only be served to healthy, immuno-competent people.
4. In the rightmost branch, they should understand and accept the risks.

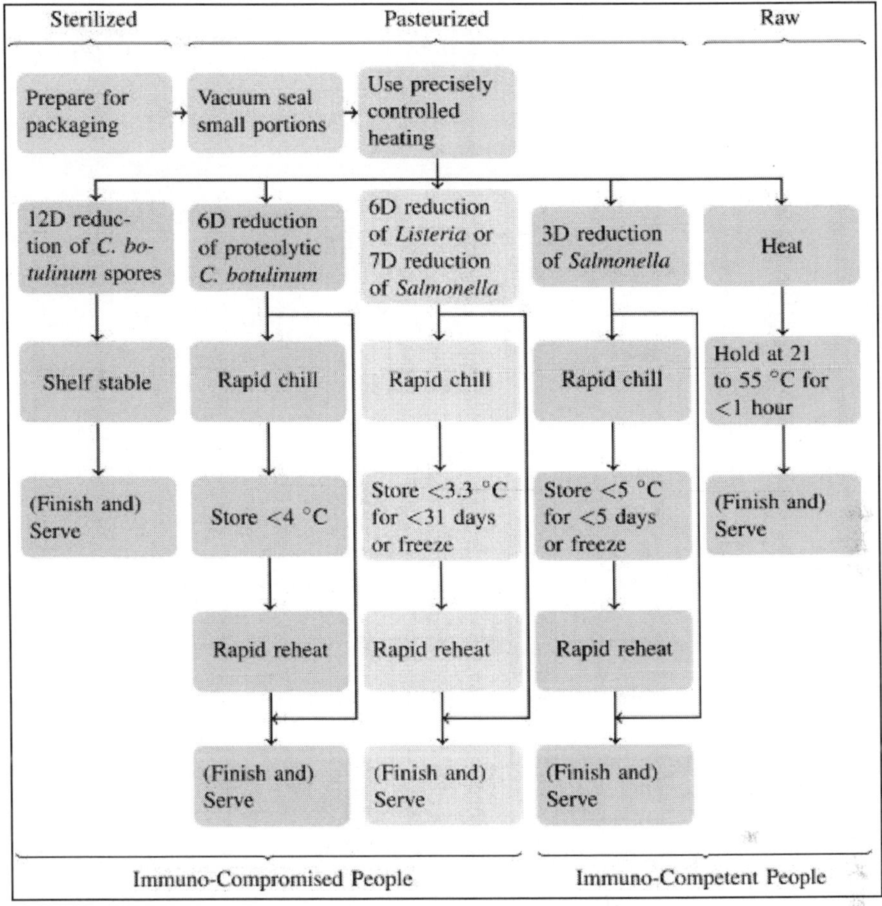

A Flow Diagram of Sous-vide Cooking.

Source

1. *D. E. Baldin, Sous-vide cooking review, International journal of Gastronomy and Food sciences, Volume 1, Issue 1, January 2012, Pages 15-30.*

Sous-vide Processing and Equipment [1]

Equipment [1]

Filling Pump, sous-vide pouches containers, thermoformer, conveyor-fed machine, water tank or "bain marie, steam combination ovens, low-pressure steam injection, iced water bath system, air-blast cabinets and cold rooms or chill cabinets.

Below are a few types of equipment for ultimate sous-vide setup [2b]:

i. Sous-vide Immersion Circulator

The Anova Sous-vide Precision Cooker is a stand alone immersion circulator that heats water and circulates it around the pot to maintain precise temperatures evenly. Stand-alone sous-vide devices don't require additional equipment to get started because they clamp on and adjust to any pot. Other examples include ChefSteps Joule, Nomiku, and Sansaire.

ii. Sous-vide Water Oven

Water ovens are often referred to as countertop water baths. These ovens are fully-contained, sous-vide devices that are about the size of a microwave. Sous-vide water ovens heat water, but unlike immersion circulators, these ovens do not circulate the water. Examples of water ovens include SousVide Supreme, AquaChef, and Gourmia.

iii. DIY Sous-vide Hacks

Cooler, rice cooker, and slow cooker hacks are great options for exploring sous-vide cooking.

Processing

Sous-vide is a method of cooking in which food is placed in a plastic pouch or a glass jar and then placed in a water bath or steam environment for longer than normal cooking times (usually 1 to 7 hours, up to 48 or more in some cases) at an accurately regulated temperature. The temperature is much lower than normally used for cooking, typically around 55 to 60 °C for meat, higher for vegetables. The intent is to cook the item evenly, ensuring that the inside is properly cooked without overcooking the outside, and to retain moisture [1a].

a. Processing Temperature and Time [1a]

The degree of accuracy and constancy of cooking temperature required varies with the food cooked [1a]. Cooking is the application of heat to change food for eating: some of these changes happen quickly and others happen slowly. Most traditional cooking is only concerned with fast changes because it is hard to hold food at a temperature (below boiling) with traditional heat sources for long enough that these slow changes become important. The precise temperature control in sous-vide cooking controls both the fast and the slow changes [1c]. Some examples of effect of Temperature and Time on sous-vide cooking are given below:

- ☆ A 15-millimeter-thick (0.6 in) piece of fish will cook in 17 to 18 minutes at any temperature from 44 °C to 61 °C [1b]. Such food can be cooked in a switched-off slow cooker filled with hot water and a thermometer.
- ☆ For an egg, which has proteins that denature at different temperatures, maintaining precise, constant temperature is more critical. Confit egg yolks are usually cooked at 63 degrees, this is hot enough to cook the white without setting the yolk. Typically the temperature of the bath is set to 64 degrees, meaning that the actual cooking temperature is 63 degrees.

  The fast changes happen quickly when the temperature of the food exceeds a certain threshold. For example, if a shelled chicken egg is heated until the temperature equalizes (say, for 30-60 min) then at:

i. 61.5 °C- the protein conalbumin denatures and causes the egg white to form a loose gel,

ii. 64.5 °C- the protein livetin denatures and causes the egg yolk to form a tender gel,

iii. 70 °C-the protein ovomucoid denatures and causes the egg white to form a firm gel (the egg yolk also coagulates around this temperature)

iv. 84.5 °C- the protein ovalbumin denatures and causes the egg white to become rubbery.

- Non-sous-vide cooking times are determined by when the center of the cooked item reaches a few degrees below the targeted temperature. Then heating should be stopped immediately. If the heating continues, the food will be overcooked. Sous-vide cooking continues until the center of the food has reached its target temperature. If it continues after this, the food will not be overcooked, and it will not cook more after it stops being heated.
- The time taken for the center of food to reach the target temperature depends on the initial temperature, the thickness and shape of the food and the temperature of the bath [1b].

The Mathematics of Sous-vide Cooking [1c]

The modeling for how long it takes the food to come up to temperature and how long it takes to pasteurize the food, is of great interest. Many simplifications and assumptions are necessary.

The transfer of heat (by conduction) is described by the partial differential equation:

$$T_t = \nabla\cdot(\alpha'' T)$$

where, αa" k/(ρCp) is thermal diffusivity (m²/s), *k* is thermal conductivity (W/m-K), ρ is density (kg/m³), and C_p is specific heat (kJ/kg-K).

Since the temperature at the slowest heating point of the food (typically the geometric center of the food), is in interest hence we can approximate the three dimensional heat equation by the one dimensional heat equation:

$$\rho Cp(T)T_t = k(T)\{T_{rr}+\beta T_r/r\},$$

$$T(r,0)=T0,\ T_r(0, t)=0,$$

$$k(T)T_r(R, t) = h\{T_{Water}-T(R, t),$$

where, $-0 \le r \le R,\ t \ge 0,\ 0 \le \beta \le 2$ is a geometric factor, T_0 is the initial temperature of the food, T_{Water} is the temperature of the fluid (air, water, steam) that the food is placed in, and *h* is the surface heat transfer coefficient (W/m²-K).

The geometric factor allows us to approximate any shape from a large slab (β=0) to a long cylinder (β=1) to a sphere (β=2). Indeed, a cube is well approximated by taking β=1.25, a square cylinder by β=0.70, and a 2: 3: 5 brick by β=0.28.

For Thawed Foods

k, ρ and C_p are essentially constant. It was reported [1e] that beef with above average fatness had followings:

1. Thermal conductivities- 0.48 W/m-K at 32 °F (0°C) and 0.49 W/m-K at 86 °F (30 °C);
2. Specific heat -3.81 kJ/kg-K at both 32 °F (0 °C) and 86 °F (30 °C)
3. A density - 1077 kg/m^3 at 41 °F (5 °C) and 1067 kg/m^3 at 86 °F (30 °C).

This is much less than the difference between beef sirloin (α=1.24×10^{-7}m^2/s) and beef round (α=1.11×10^{-7}m^2/s) [1e]. Therefore, we can model the temperature of thawed foods by

$$T_t=\alpha\{T_{rr}+\beta T_r/r\},$$

$$T(r,0)=T0,\ T_r(0,\ t)=0,$$

$$T_r(R,\ t)=(h/k)\{T_{Water}-T(R,\ t)\}$$

for $0 \le r \le R$ and $t \ge 0$. Since h is large even large deviations in h/k cause only minor deviations in the core temperature of the food [1f].

Most foods have a thermal diffusivity between 1.2 and 1.6×10^{-7} m^2/s [1g]. Thermal diffusivity depends on many things, including meat species, muscle type, temperature, and water content [1g]. The details related to the thermal diffusivity (at 0 °C to 65 °C) of various types of food reported in the literature, are tabulated below [1c]:

Sl. No.	Food	Thermal diffusivity (10^{-7} m^2/s)
1	Beef	1.11-1.30 [1e], 1.18-1.33 [1h].
2	Pork	1.17-1.25 [1e], 1.18-1.38 [1h].
3	Chicken	1.08-1.39 [1e], 1.36-1.42 (White) and 1.28-1.33 (Dark) [1i].
4	Fish	1.09-1.60 [1e], 1.22-1.47 [1h].
5	Fruits	1.12-1.40 (Apple), 1.42 (Banana), 1.07 (Lemon), 1.39 (Peach), 1.27 (Strawberry) [1h].
6	Vegetables	1.68 (Beans), 1.82 (Peas), 1.23-1.70 (Potato), 1.71 (Squash), 1.06-1.91 (Sweet Potato), 1.48 (Tomato) [1h].

Using the above models for the temperature at the slowest heating point of the meat, the classical model for the log reduction in pathogens is:

$$LR=1/D_{Ref}+\int^{0}_{t} 10(T(t2)-T_{Ref})/z\ dt2$$

where, D_{Ref} is the time required for an one decimal reduction in the pathogen at the reference temperature T_{Ref} and the z-value is the temperature increment needed for a 10-fold decrease in D.

Processing Steps

The sous-vide processing of food involves the following steps [1, 1a]:

i. Preparing for Packaging

Before packaging/filling, foods need to be treated. Tougher cuts of meat are frequently marinated, tenderized, or brined before vacuum packaging. Mechanical tenderizing has become quite common and is accomplished by inserting hundreds or thousands of thin blades into the meat to cut some of the internal fibers. Brining and curing has become increasingly popular in modern cooking, especially when cooking pork and poultry [1c]. Filling of the product is first achieved by dispensing it via a pump from the container into sous-vide pouches or containers on a thermoformer or a conveyor-fed machine [2].

ii. Vacuum Packaging and Packaging Materials

Vacuum-sealing's main benefit is that it allows heat to be efficiently transferred from the water bath or steam oven to the meat. For cook-chill or cook-freeze sous-vide cooking, vacuum packaging eliminates the risk of recontamination during storage and inhibits off-flavors from oxidation [1c].

a. Vacuum Packaging

The vacuum is carefully controlled to prevent damage to the contents. It has been generally recommended that the strongest vacuum possible (typically 10-15 mbar for firm food and 100-120 mbar for liquids) should be used to reduce the ballooning of the pouches during cooking and to reduce aerobic bacterial growth but it was found that pulling a 10-15 mbar vacuum significantly degraded the taste and texture of fish and poultry [2a]. Usually, sous-vide cooking in food service operations uses a water tank or "bain marie, " heated to the required temperature with a paddle to agitate the pouches [2].

b. Packagaing Materials [2b]

Some of the best types of sous-vide packaging are given below:

Sl. No.	*Packaging Materials*	*Details*
1	Reusable Silicone Bags	These bags made from silicone, like these Stasher Reusable Silicone Sous-vide Bag bags, make it easy to enjoy the same quality results night after night.
2	Resealable Bags or Jars	These bags are very versatile, and can be used with the water immersion method to remove air from the bag. Heavy-duty, BPA-free bags, like Ziplock's freezer bags, are recommended.
3	Vacuum Sealing Bags	With these bags, a vacuum sealer is not needed and these bags work well for batch cooking. Foodsaver and Oliso are great options, and both are pretty affordable.
4	Canning Jars	Several different types of foods can also be cooked in glass canning jars. Beans and grains both work well in jars, as do desserts such as cakes and custards.

iii. Cooking

In almost all cases, the cooking medium is either a water bath or a convection steam oven. Convection steam ovens allow large quantities of food to be prepared,

Different Types of Packaging Materials for Sous-vide [2b].

but do not heat uniformly enough to use. Circulating water baths heat very uniformly and typically have temperature swings of less than 0.1°C [1c]. A slightly different method uses steam combination ovens, which use either convected hot air (dry heat) or low-pressure steam injection with convection heating (wet heat), to achieve temperatures below 100°C [1a]. To prevent under-cooking, it is very important that the pouches are completely submerged and are not tightly arranged or overlapping [1c, 1d].

Industrial heating equipment for SVP can take several forms including air/vapor, water immersion or steaming water [4].

iv. Chilling

For cook-chill and cook-freeze sous-vide cooking, the food is rapidly chilled in its vacuum sealed pouch and refrigerated or frozen after pasteurizing. Before finishing for service, the food is then reheated in a water bath at or below the temperature it was cooked in [1c]. Chilling can be done by the following ways:

- ☆ Following cooking, products are then chilled in an iced water bath system, using paddles to accelerate the cooling effect by ensuring the rapid flow of water around the bags in conjunction with a heat exchanger to maintain a chilled temperature [1].
- ☆ In addition, products may be cooled in air-blast cabinets, in which the products are placed in trays mounted on trolleys.

v. Storage

The products are then stored chilled in standard cold rooms or chill cabinets [3].

Sous-vide Cooking [1a].

Sous-vide Cooking Process [1a].

Compact Vacuum Packing/ Sealing Device [1a].

Lower-cost Sous-vide WiFi Stick [1a].

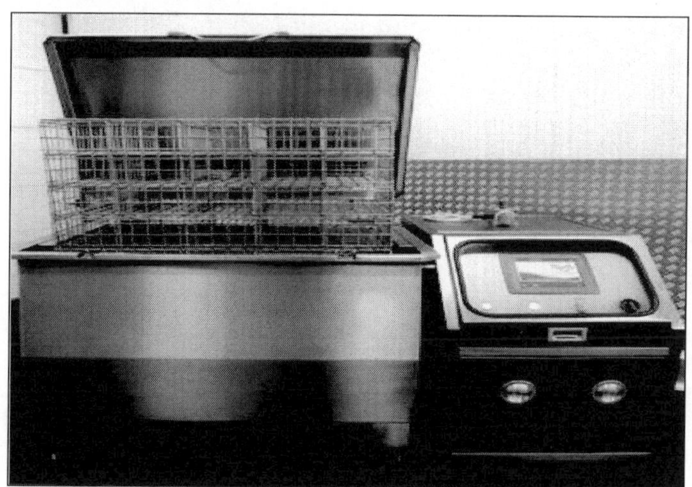

Sous-vide Processing [5].

The figure provides a simplified flow diagram that outlines the basic steps in sous-vide processing (SVP) [1]:

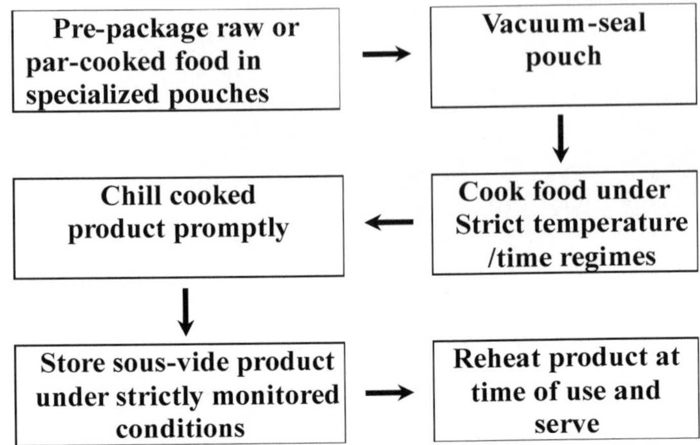

Generic Flow Diagram for a Sous-vide Processing Line.

Industrial sous-vide is a cooking technique that involves cooking food under vacuum in airtight plastic pouches which are submerged in a Cook Tank/Water Bath. Products are slow cooked in the cook tank which is held at a precisely controlled low temperature ensuring foods are cooked evenly and to perfection every time. The sous-vide method of cooking eliminates guesswork and allows commercial producers to cook foods with incomparable taste and texture; perfectly cooked steak, tender chicken breasts and ribs with the meat falling off the bone. With industrial sous-vide, products can also be cooked in a sauce or marinade giving food manufacturers additional recipe choices [5].

Sources

1. *Hudaa Neetoo (Faculty of Agriculture, University of Mauritius, Réduit, Mauritius) and Haiqiang Chen (Department of Animal and Food Sciences, University of Delaware, Newark, Delaware, US), Alternative Food Processing Technologies.*

1a. *Wikipedia.*

1b. *Baldwin: A Practical Guide to Sous-vide Cooking.*

1c. *D. E. Baldin, Sous-vide cooking review, International journal of Gastronomy and Food sciences, Volume 1, Issue 1, January 2012, Pages 15-30.*

1d. *Rybka-Rodgers, S., Developing a HACCP plan for extended shelf-life cook-chill ready-to-eat meals Food Australia, 51, 1999, pp. 430-433.*

1e. *Sanz, P. D., Alonso, M. D. and Mascheroni, R. H., Thermophysical properties of meat products: general bibliography and experimental valuesTransactions of the ASAE, 30, 1987, pp. 283-289.*

1f. *Nicolai, B. M. and Baerdemaeker, J. D., Sensitivity analysis with respect to the surface heat transfer coefficient as applied to thermal process calculations. Journal of Food Engineering, 28, 1996, pp.21-33.*

1g. *Baerdemaeker, J. D.and Nicolai, B. M., Equipment considerations for sous-vide cooking Food Control, 6 (4), 1995, pp. 229-236.*

1h. *Singh, R. P., Thermal diffusivity in food processing Food Technology, 36 (2), 1982, pp.134-137.*

1i. *Siripon, K., Tansakul, A. and Mittal, G. S., Heat transfer modeling of chicken cooking in hot water Food Research International, 40, 2007, pp. 923-930.*

2. *Cole JC, 1993, The User's Guide to Cuisine Sous-Vide. Buckingha-mshire: JC Group Associates, pp.63.*

2a. *Norén, N., Arnold, D., 2009, Cooking Issues-The French Culinary. Institute's Tech'n Stuff Blog. URL0.*

2b. *What is sous-vide-Every thing you need to know- Anova culinary.*

3. *Tansey FS, Gormley TR, 2005, Sous-vide/freezing technology for ready meals. In: Barbosa-Canovas GV, Tapia MS, Cano MP(eds) Novel Food Processing Technologies. Boca Raton, FL: CRC Press, pp.477-491.*

4. *Schellekens W, Martens T, 1992, "Sous-Vide"Cooking Part I: Scientific Literature Review. Brussels: Commission of theEuropean Committees Directorate General XII, Research and Development.*

5. *Sous-vide, D C Norris, America, https: //www. dcnorrisna. com/products/sous-vide.*

Production of Sous-vide Foods [1]

For the sous-vide method, manufacturers of specialised equipment for vacuum packing, cooking, cooling, chilled storage, reheating and packaging have co-operated in putting complete systems together in order to encourage their adoption. Applications for sous-vide techniques are split into following two general categories:

1. In-house, craft-based producers for small- to medium-scale production supplying, for example, hotels and restaurants, generally producing meal components for uninterrupted in-house usage.
2. Food manufacturing for large-scale industrial production of products to be sold to the food service industry or to the consumer through retail sale. Centralized manufacturing plants which fabricate hundreds of thousands of high quality frozen meals under factory conditions intended for later reheating

i. Sous-vide In-house, Small-scale Food Production [1]

The in-house producer (using a restaurant or hotel kitchen) will already have much of the necessary equipment such as:

a. A combination steam and forced convection cooker for heat treatment and reheating.

b. Bains-marie for reheating.

c. Chilled storage rooms.

The in-house producer will need to invest in vacuum-packing equipment, air blast chillers or iced water immersion chillers for rapid cooling of sous-vide packs and temperature monitoring and recording equipment.

Specialised equipment for reheating is becoming available. These are based on using pouches surrounded by hot water at a controlled temperature into which chilled sous-vide packs can be placed and removed when the built-in timers provide a signal [2, 3].

ii. Sous-vide Large-scale Food Production [1]

Industrial scale machinery is currently available for the preparation of raw materials by dicing, slicing and grating, and for pre-preparation steps such as blanching or for browning meats before vacuum packaging.

- ✰ For vacuum packing, the move from manual single or multi-station machines using pouches for small- to medium-scale producers requires investment in highly automated continuous packing systems where the base of the pack is formed on-line followed by filling and vacuum sealing under a top layer of packaging such as the Darfresh system [4].
- ✰ The heat treatment step for pasteurising sous-vide products on a large scale can be accomplished by a variety of equipment which often incorporate the rapid-chilling process. Several methods are used-most making use of computers for precise temperature control and controlling the sequence and timing of operations. Some exampes are given below:
- ✰ The Armor Inox Thermix system uses immersion in hot water for pasteurisation which is replaced by iced water for rapid chilling. The circulation system aims to minimise energy consumption.
- ✰ The Barriquand Steriflow system uses cascading water to heat and then chill sous-vide packs in an enclosed vessel. The water circulates at a high rate through a closed loop to minimise the risk of contamination and uses a heat exchanger for efficient heat transfer.
- ✰ The Lagarde system pasteurises packs stacked on trolleys using a mixture of water vapour and compressed air followed by cooling with streams of water in the same vessel.

Sources

1. *Ghazala S. (ed.), 1998, Sous-vide and Cook Chill Processing for the Food Industry. Gaithersburg, MD., USA: Aspen Publishers Inc., ISBN 0-7514-0433-0/978-07514-0433-3, pp. 25-56.*

2. *Baird, B., 1990, Sous-vide: What's all the excitement about? Food Technology, 44 (11), 92, 94, 96.*

3. *Lehr, H., 1990, Vie de France: creating a new market for sous-vide. Food Engineering International, 15 (5), 45-48.*

4. *Pre, G., 1992, Trends in processing and packaging technologies, in Proceedings of 'Packaging of Food', London, March, Euro Food Pack Group, PIRA (Packaging Industry Research Association), Leatherhead, Surrey, UK, 12 pp.*

Advantages and Disadvantages of Sous-vide Foods [1, 2]

Sous-vide foods and Sous-vide processing has several advantages and disadvantages. Sous-vide food preparation is used worldwide by professional chefs and is gaining popularity within the food processing industry as its advantages are being recognized. The details related to the advantages and disadvantages of Sous-vide/Cook-Chill processing are tabulated below [1]:

Parameters	*Advantages*	*Disadvantages*
Cost [1, 2]	i. Adding value to basic raw materials such as meat and fish by producing chilled prepared meals, ii. Cutting food costs by reducing materials for enhancing flavours, iii. Tighter portion control, iv. Lower weight loss through retention of moisture in packaging, v. Lower wastage due to quicker response to consumer demand, vi. Extension of shelf life [1, 2] leading to lower wastage during storage, vii. Economies of scale for variable food costs using a centralised production unit to supply several outlets, viii. Better Inventory Control [2] and ix. Eliminates Over-Production [2].	**Extra costs [1]** i. Cost associated with equipment and systems required for more stringent quality management systems to ensure food safety through monitoring raw materials, heat treatment storage, distribution, *etc.* ii. Higher capital costs of equipment for preparation, vacuum packing, pasteurising, chilling and storage, iii. Higher costs for packaging and iv. Higher costs for using devices for detecting temperature abuse.
Labour cost [1]	i. Using less-skilled or fewer skilled staff, ii. Easier staff recruitment, iii. More rapid service of food to the consumer iv. Higher productivity in terms of food produced per man-hour and v. Easy incorporation into a meal assembly system.	i. Costs of training for staff as part of the more stringent quality management systems, ii. Overcoming staff resistance relating to deskilling and iii. Costs of staff to liaise with suppliers and customers as part of the QMS.
Quality of Food and Service [1, 2]	i. Reducing problems due to oxygen by vacuum packing and Preserves Food Integrity [1, 2], ii. Reducing risk of contamination through use of vacuum packing and retards bacteria growth [1, 2], iii. Claimed superior sensory qualities, iv. Enabling production of pre-cooked foods otherwise needing long processing times (*e.g.*dried vegetables). v. Provide consistent good quality, tastier and Healthier food and vi. Complete Control Over Product-No over cooked and undercooked foods.	**Quality costs [1]** i. Increased risk of food poisoning if sous-vide packs are subjected to temperature abuse.

Parameters	Advantages	Disadvantages
Marketing cost [1]	i. Flexibility offered by a wide menu range, ii. Innovative concept for the use of technology and iii. Wide range of sizes from individual to multi portion for food service.	i. Ensuring reliable delivery of chilled foods over a wide geographical area, i. Overcoming fears of the consumer on food safety through education and iii. Overcoming confusion between pasteurised sous-vide products and sterilised 'boil-in-the-bag' products.

- ✰ Higher quality of service due to less time on repetitive food preparation work.
- ✰ Higher perceived quality of food due to more time for presentation of food.
- ✰ Better career development for chefs.

Sources

1. *Ghazala S. (ed.), 1998, Sous-vide and Cook Chill Processing for the Food Industry. Gaithersburg, MD., USA: Aspen Publishers Inc., ISBN 0-7514-0433-0/978-07514-0433-3, pp.25-56.*
2. *Sous-vide, D C Norris, America, https: //www.dcnorrisna.com/products/sous-vide.*

Reasons for Success and Failure [1]

There were several reasons which are contributing to success and failure of Sous-vide processing/foods.

Reasons Contributing to Success [1, 2]

A. Reasons Related to Quality [1, 2]

1. Supplying foods meeting customer requirements whether sous-vide processed or not [1]
2. **Tenderness**: It is one of the most noticeable benefits of sous-vide cooking over traditional methods is the tenderness that it brings to food. Because the food is cooked at a low temperature, the cell walls in the food don't burst, and the tough collagen which is found in the connective tissue of meat can be broken down without the protein in the meat being heated up so much that it toughens and loses its moisture. Sous-vide cooking allows vegetables to be cooked at below boiling temperature, ensuring that vegetables are properly cooked, without losing their firmness and crispness [2].
3. **Lack of Shrinkage**: One of the main benefits that sous-vide holds over traditional methods of cooking is that it yields far less shrinkage when cooking meat [2].

3. Waste reduction-Traditionally prepared food dries out and results in waste. For example, on average, traditionally cooked steak loses up to 40 per cent of its volume due to drying out. Steak cooked via precision cooking, loses none of its volume [3].
5. Moisture retention [2]
6. **Consistency [3].**

B. Reasons Related to other Aspects [1, 2]

1. A realistic attitude towards likely sales volume.
2. Emphasising advantages of portion control, reduction of labour costs and increased convenience as opposed to the 'haute cuisine' image.
3. Simplicity-Simply seal ingredients in a vacuum pouch and place it in water bath [2]
4. Selling the product at a realistic price [1].
5. **Flexibility**: Traditional cooking can require your constant attention. Precision cooking brings food to an exact temperature and holds it. There is no worry about overcooking [3].

Reasons Contributing to Failure [1]

1 Low market acceptance due to reservations of users and retailers on food safety.
2. An emphasis on marketing the technology and not its benefits.
3. Incorrect price/quality ratio giving poor profit margins.
4. Lack of profitability due to high start-up investment costs linked to expected short pay-back periods
5. Professional resistance to pre-cooked foods.
6. Costs and reliability of distribution over a wide area.
7. Consumer concerns on food safety and lack of product knowledge.

Sources

1. *Ghazala S. (ed.), 1998, Sous-vide and Cook Chill Processing for the Food Industry. Gaithersburg, MD., USA: Aspen Publishers Inc., ISBN 0-7514-0433-0/978-07514-0433-3, pp. 25-56.*
2. *Sousvidetools.com.*
3. *https: //anovaculinary. com/anova-precision-cooker/.*

Food Safety and Sous-vide Processing [1]

Food safety is a function of both time and temperature. A temperature usually considered insufficient to render food safe may be perfectly safe if maintained for long enough. People with compromised immunity should never eat food that has not been properly pasteurized. Some examples related to the food safety concerns with Sous-vide processing/cooking are given on next page.

1. Women eating food cooked **sous-vide** while pregnant expose themselves and/or their unborn children to risk and thus may choose to avoid unpasteurized recipes [2].
2. Sous-vide cooking must be performed under carefully controlled conditions to avoid botulism poisoning [3] as *Clostridium botulinum* bacteria can grow in food in the absence of oxygen and produce the deadly botulinum toxin.
3. The botulism bacteria are killed by pasteurization, but the possibility of hardy botulism spores surviving and reactivating once cool remains a concern. For that reason, Baldwin's treatise specifies precise chilling requirements for "cook-chill", so that the botulism spores do not have the opportunity to grow. Pasteurized food can then be stored for up to two weeks at around 3°C sealed within the vacuum pack [2].
4. Some plastic bags may have plasticizers with endocrine disruptor properties that are released when heated and posed danger for health. The plasticizers are now less common than in earlier uses because of improving standards for food-grade plastics [4]. The plastic pouches made for sous-vide cooking must be food-grade to limit the migration of plastic constituents into the food [5].

Steps to be Taken to Ensure Safety of Sous-vide Foods [6]

By identifying following critical control points and establishing hurdles to microbial growth, all of the safety concerns related to vacuum packaging and sous-vide cooking may be virtually eliminated:

1. Only the freshest, highest-quality ingredients must be used when preparing sous-vide packages. This can significantly lower initial microbial levels, extending shelf life and product freshness.
2. It is also critical to calibrate equipment on a daily basis and quality-check all seals and packages for leaks.
3. Raw packets must not be stored for more than two days before pasteurization above 60°C, and must be cooled below 3°C within two hours
4. Pasteurized inventory should be stored below 3°C and consumed or frozen within a specified time period.
5. The date and time of packaging, pasteurization, and expiration must be documented and labeled on the package.
6. The entire production process must be governed by a Hazards Analysis Critical Control Point (HACCP) system to ensure compliance and corrective actions [7].

Sources

1. *Wikipedia.*
2. *D. E., Baldwin, A Practical Guide to Sous-vide Cooking.*

3. *Hyytia-Trees E, Skytta E, Mokkila M, et al., 2000, "Safety evaluation of sous-vide-processed products with respect to nonproteolytic Clostridium botulinum by use of challenge studies and predictive microbiological models". Appl. Environ. Microbiol, 66 (1), 223-9.*

4. *Sous-Vide Cooking in Plastic- Is It Safe? Chow.com.*

5. *Baldwin, Douglas, 2010, Sous-vide for the home cook. Incline Village, NV: Paradox Press. ISBN 978-0-9844936-0-9.*

6. *Jamie Tiampo-Seal Appeal: The Nutrition, Food Safety, and Operational Benefits of Sous-vide.*

7. *Schellekens, W. and T. Martens, 1993, "Sous-vide Cooking, "Food-Linked Agro-Industrial Research (FLAIR).*

Effect of SVP on the Sensory and Nutritional Quality of Foods [1]

Sous-vide is an extremely delicate and healthy (healthful) method of preparing food. Most of the sensory and nutritional benefits are directly related to the fact that the food is placed into an evacuated package (reduced oxygen environment), sealed and cooked with mild heat [2]. Apart from ensuring uniform cooking, sous-vide cooking facilitates development of desired organoleptic flavors and limits off flavors due to oxidation [1a].

1. The overall effect is to achieve a tight control over heat, oxygen content, and presence of moisture–three primary factors that contribute to a decreased nutritional content of conventionally prepared foods.
2. The presence of the plastic barrier also significantly reduces the extent and rate of oxidation, there by preserving the qualities of essential polyunsaturated fatty acids [3, 4]. The plastic films also lock in moisture. Sealing the food in sturdy plastic bags retains juices and aroma that, otherwise, would be lost in the process [1a].
3. The plastic films also lock desirable flavors, consequently, sensory quality is retained(flavors are amplified), requiring less seasoning and spices [3] *i.e.*fewer spices and less salt is required, lowering the overall sodium content of sous-vide foods. The pouch also eliminates mineral and water loss, unlike other traditional cooking methods [5].
4. In addition, vitamins are also protected from destabilization or loss of activity during sous-vide cooking. Research has shown that vitamin C retention was higher after SVP than following pasteurization. Between different cooking methods, sous-vide provides the highest retention of vitamins versus steaming and boiling [5b].
5. The desirable organoleptic attributes of sous-vide cooked products characterized by their fresh-like textures and vivid flavors, together with their wholesomeness, make them increasingly more appealing to consumers [5]. During cooking of vegetables, extreme tenderness or softness is seen as undesirably overcooked where as in sous-vide procesing, vegetables are cooked at a temperature below the boiling point

of water allows vegetables to be thoroughly cooked while maintaining a firm or somewhat crisp texture. While the cell walls will generally not burst, the de-polymerization of the pectic polysaccharides that connect the vegetable cells together and/or the gelatinisation of starch in the vegetable can be achieved without overcooking [5a].

6. In the case of meat cooking, tough collagen in connective tissue can be hydrolysed into gelatin, without heating the meat's proteins high enough that they denature to a degree that the texture toughens and moisture is wrung out of the meat [1a].
7. Examples of successful sous-vide products include beefsteaks, meat loaves, fish fillets, and fruits and vegetables.

Critical Factors that Limit the Scope of SVP [1]

Unfortunately, there are a following few caveats to recognize in SVP:

a.

Only top-quality ingredients should be used and preparation should be accomplished in a clean environment to minimize initial contamination.

b. Microbiological Threats

The time-temperature regime of the cooking step as well as the storage conditions need to be strictly monitored as they would affect the type of micro-organisms that may survive and grow in the final product.

- ☆ The anaerobic conditions, together with the relatively low temperature heat treatment of sous-vide cooking creates an atmosphere in which the most deadly food-borne pathogen, *Clostridium botulinum,* thrives. The ubiquitous nature of spores of *C. botulinum, C. perfringens,* and B.cereus in the environment, possible or even probable contamination of ingredients and raw materials and difficulty in maintaining the cold chain at the retail display and in the domestic refrigerator can all limit the prospects of SVP technology.
- ☆ If improperly handled, the following conditions [4a] can exacerbate normal bacterial growth and spoilage:
    1. Sous-vide foods lack chemical preservatives, which normally arrest biological activity in processed foods.
    2. Gentle heat treatment only kills vegetative forms of pathogens if the core temperature reaches 60°C. It is impossible to monitor core temperature without an invasive probe.
    3. The reduced oxygen packaging extends the shelf life by retarding aerobic micro-organisms, but can facilitate the growth of anaerobic pathogens such as *Clostridium botulinum, Clostridium perfringens, Bacillus cereus,* and *Listeria monocytogenes.*
    4. Sufficient cooling must be rigorously applied to prevent bacterial reproduction at all stages of preparation, storage, and regeneration.

c. *Limitation in Browning of Foods [1a, 1b]*

Another limitation of sous-vide cooking is that during this process, browning of food could not be done. Browning **(Maillard reactions)** of food happens at temperatures above the boiling point of water (100°C). The Maillard or browning reaction is a very complex reaction between amino acids and reducing sugars. After the initial reaction, an unstable intermediate structure is formed that undergoes further changes and produces hundreds of reaction by-products. The Maillard reaction can be increased by adding a reducing sugar (glucose, fructose or lactose), increasing the pH (*e.g.*, adding a pinch of baking soda), or increasing the temperature [1c]. The flavors and 'crust' texture developed by browning are generally seen as very desirable in the cooking of certain types of meat, such as a **steak**. The flavors and texture produced by browning cannot be obtained with only the sous-vide technique. In many cases, meats and other foods will be browned before placing these foods in the water bath or after the sous-vide cooking, using techniques such as grilling or searing.

Sources

1. *Hudaa Neetoo (Faculty of Agriculture, University of Mauritius, Réduit, Mauritius) and Haiqiang Chen (Department of Animal and Food Sciences, University of Delaware, Newark, Delaware, US), Alternative Food Processing Technologies.*

1a. *Wikipedia.*

1b. *Sousvidetools.com.*

1c. *D. E. Baldin, Sous-vide cooking review, International journal of Gastronomy and Food sciences, Volume 1, Issue 1, January 2012, Pages 15-30.*

2. *Ghazala S, 2004, Development in cook-chill and sous-vide processing. In: Richardson P (ed) Improving the Thermal Processing of Foods. Cambridge: Woodhead Publishing, pp., 152-173.*

3. *Creed PG, 1998, Sensory and nutritional aspects of sous-vide processed foods. In: Ghazala S (ed) Sous-Vide and Cook-Chill Processing for the Food Industry. Gaithersburg, MA: Aspen Publishers, pp.57-88.*

4. *Ghazala S, Aucoin EJ, (1996), Optimization of pasteurization processes for a sous-vide product in rectangular thin profile forms. Journal of Food Quality 19: 203-215.*

4a. *Ghazala, S. and R. Trenholm, 1998, Concepts in sous-vide and cook-chill products. Sous-vide and cook chill processing for the Food Industry. S. Ghazala. Gaithersburg, MD, Aspen Publishers: 295-309.*

5. *Tansey FS, Gormley TR, 2005, Sous-vide/freezing technology for ready meals. In: Barbosa-Canovas GV, Tapia MS, Cano MP (eds) Novel Food Processing Technologies. Boca Raton, FL: CRC Press, pp. 477-491.*

5a. *Baldwin, A Practical Guide to Sous-vide Cooking.*

5b. *Petersen, M. A., 1993, "Influence of sous-vide processing, steaming and boiling on vitamin retention and sensory quality in broccoli florets." Zeitschrift fur Lebensmittel Untersuchung und Forschung 197(4): 375-380.*

Food Processing Applications of Sous-vide [1]

Sous-vide has been used widely in the processing of various types of raw or par-cooked meat, poultry, fish, and even vegetable-based products to enhance sensory characteristics, improve microbiological safety and extend shelf-life.

A. General Applications

Some examples of general applications are given below:

i. In Europe, recommendations, guidelines, and codes of practice have been developed to ensure the safe production of sous-vide foods [2].

ii. The following table shows the proportion of published studies on the application of SVP for various food commodities:

Sl. No.	*Product*	*Use of Sous-vide Cooking*
1	Pork products	13 per cent
2	Poultry products	10 per cent
3	Fish products	12 per cent
4	Seafood products	1 per cent
5	Staple food	5 per cent
6	Fruit products	3 per cent
7	Vegetable products	15 per cent
8	Gravies Sauce/Soups	8 per cent
9	Beef products	33 per cent

iii. It was demonstrated [3] that the capacity of sous-vide cooking to reduce the counts of *Staphylococcus aureus, Bacillus cereus, Clostridium perfringens,* and *Listeria monocytogenes* on rainbow trout and salmon to below detectable levels and extend the shelf-life to >45 days during storage at 2 °C.

iv. Similarly, study showed [4] an improvement in the microbiological quality of sous-vide fish cakes during chilled storage (3°C) with an eight-fold increase in its shelf-life compared to their conventionally cooked counterparts.

v. Current research has also focused on spore forming micro-organisms that constitute a safety risk in sous-vide products, such as *Clostridium botulinum, C. perfringens,* and *Bacillus* spp.

vi. The study observed [5] the growth potential of *C. perfringens* in sous-vide cooked turkey products formulated with 0-3 per cent salt and stored at temperatures of 4-28°C. Overall, storage at 4°C and a salt level of 3 per cent proved to be most effective in controlling spore outgrowth.

vii. Similarly, it was demonstrated [6] that addition of calcium lactate (1.5 per cent) and sodium lactate (3 per cent) completely inhibited *B. cereus* outgrowth in beef goulash samples.

viii. The sous-vide method is used in many high-end gourmet restaurants. Sous-vide has become common on cooking TV shows. It has also been used to quickly produce significant quantities of meals [1a].

Sources

1. *Hudaa Neetoo (Faculty of Agriculture, University of Mauritius, Réduit, Mauritius) and Haiqiang Chen (Department of Animal and Food Sciences, University of Delaware, Newark, Delaware, US), Alternative Food Processing Technologies.*

1a. *Wikipedia.*

2. *Juneja VK, Snyder OP, 2007, Sous-vide and cook-chill processing of foods: concept development and microbiological safety. In: Tewari G, Juneja VK (eds) Advances in Thermal and Non-Thermal Food Preservation. Ames, IA: Blackwell Publishing, pp.145-166.*

3. *Gonzalez-Fandos E, Garcia-Linares MC, Villarino-Rodrýguez A, Garcia-Arias MT, Garcia-Fernandez MC, 2004, Evaluation of the microbiological safety and sensory quality of rainbow trout (Oncorhynchus mykiss) processed by the sous-vide method. Food Microbiology 21: 193-201.*

4. *Shakila JR, Jeyasekaran G, Vijayakumar A, Sukumar D, 2009, Microbiological quality of sous-vide cook-chill fish cakes during chilled storage (3°C). International Journal of Food Science and Technology 44: 2120-2126.*

5. *Juneja VK, Marmer BS, 1996, Growth of Clostridium perfringens from spore inocula in sous-vide turkey products. International Journal of Food Microbiology 32: 115-123.*

6. *Aran N, 2001, The effect of calcium and sodium lactates on growth from spores of Bacillus cereus and Clostridium perfringens in a sous-vide beef goulash under temperature abuse. International Journal of Food Microbiology 63: 117-123.*

B. Industrial Applications

The details related to the industry wise applications are given below:

i. Meat Processing [1]

Vacuum packing has long been a useful packaging method for large joints of fresh and cured meats. It has also been used with low-temperature water baths to produce cooked joints for use in food service or for retail sale of sliced cooked meats [2]. The economics of the continuous production of sausages pasteurised in vacuum packs was described [3]. The use of vacuum-packing as part of the sous-vide system has been suggested for eliminating problems of oxidation found with the conventional preparation of many meat-based prepared meals [4] and adding value to meat by selling it as prepared meals of higher quality at a higher profit [5, 6].

ii. Fruit and Vegetable Processing [1]

The preparation of Fruits and vegetables by using the sous-vide method, is considered to be an improvement over frozen or sterilised versions as long as processing temperatures are not too low [7].

Vegetables [12]

Vegetables, which are a rich source of vitamins and minerals, when boiled or steamed lose nutrients to cooking water [12a] whereas, Sous-vide cooked vegetables, in comparison, retain nearly all their nutritive value [12b]. The retention of nutrients also intensifies the flavor inherent in the vegetable. Sous-vide can cause some vegetables(turnips and rutabaga), to have a flavor that is too pronounced for some people [12c]. Vegetables that are boiled, steamed, or microwaved lose their nutrients because the cell walls are damaged by heat and allow the water and nutrients in the cells to leach out [12a]. Sous-vide vegetables leave the cell walls mostly intact and make the vegetables tender by dissolving some of the cementing material that holds the cells together [12d].

Fruits [12]

Fruits are often eaten raw, sometimes apples and pears are cooked until they are tender. Tart (high acid) apples soften faster than sweet (low acid) apples, because the acid lowers the temperature at which the cementing material dissolves [12a].

iii. Seafood Processing [1]

- ✰ The use of Barriqu and Steriflow equipment was noted [8] for producing consumer packs of sous-vide processed salmon with pasta for retail sale as part of a low-calorie range of products in France. Vacuum packing salmon slices so that the end user can then cook them 'in the bag' is one idea suggested by packaging manufacturers [9].
- ✰ It was suggested that the sous-vide can extend the shelf life of crab meat [10].
- ✰ **Fish and shellfish [11, 12]**-It was considered [11] that the sous-vide method offered the fish industry many opportunities for providing chilled convenience products, providing that stringent working practices and pasteurisation levels are in place. Fish is cooked to change its texture, develop flavor, and destroy food pathogens. Traditionally, fish is considered to be cooked when it flakes. Fish flakes when the collagen separating the flakes is converted into gelatin at around 46-49 °C [12, 14]. This temperature is too low, however, to destroy any food pathogens. FDA (2011) generally recommends pasteurizing fish which will reduce all non-spore forming pathogens and parasites to a safe level, it will not reduce the risk of hepatitis A virus (HAV) or norovirus infection from shellfish. Since the spores of non-proteolytic *Clostridium* botulinum are not inactivated by pasteurization, the fish should be stored at below 3.3 °C for less than 4 weeks [12]. It was found that fish cooked sous-vide **retains more healthful omega-3 fatty acids** and nutrients than traditionally cooked fish-11].

iv. Milk Pasteurisation [12]

Sous-vide processing can be used for milk pasteurisation. It is very easy to pasteurise milk. There are many types of pasteurisation. Of all the different methods, people would choose the following, because it is the best suited for the SmartVide system and low temperature:

First, place the milk into a vacuum bag or other container that allows a vacuum to be made. Next, heat to 65° for 30 minutes. It is very important that the internal temperature reaches 65 degrees and the temperature is constant during the 30 minutes. Once the time is up, cool it as quickly as possible to 4°C. The milk can now be used to make yogurt using a traditional artisan method.

v. Poultry [12]

Cooking poultry is very similar to cooking meat. Like lean meat, lean poultry is only plump and juicy if it does not exceed 60-65 °C. Tougher cuts of poultry can be cooked to higher temperatures and remain juicy, because the melted fat lubricates the lean meat/poultry [13].

Traditional cooking methods make poultry safe by cooking the coldest part to 74 °C or above. Poultry can also be made safe at lower temperatures, it just takes longer. Indeed, cooking chicken and turkey breasts at 60 °C for the prescribed times is just as safe as cooking them to 74 °C.

For example, chicken and turkey breasts are moist, plump, and juicy when pasteurized between 58 °C and 63 °C for the prescribed times and duck breasts are usually pasteurized at 57 °C/135 °F for the prescribed times. Dark poultry meat, such as legs and thighs, is usually cooked well done at 70-80 °C until it is fall-apart tender, about 4-6 h at 80 °C or 8-12 h at 70 °C.

Sources

1. *Ghazala S. (ed.), 1998, Sous-vide and Cook Chill, Processing for the Food Industry. Gaithersburg, MD., USA: Aspen Publishers Inc., ISBN 0-7514- 0433-0/978-07514-0433-3, pp. 25-56.*
2. *Vann der Leest, A. J., 1985, Pasteurised meat products in co-extruded bags, in Proceedings of International Symposium - Trends in Modern Meat Technology, 30 October-1 November 1984, Wageningen, Netherlands, Pudoc, Wageningen, pp.93-95.*
3. *Wiesel, V., 1987, Einsatz kontinuierlich arbeitender Koch-, Kuhl- und Pasteurisieranlagen für Wurst- Kaliberware und Wurstchen-Packungen. Fleischwirtschaft, 67 (4), 386, 388-390, 452.*
4. *Anon., 1987, Sealed with a hiss. Meat Processing, July, 66.*
5. *Beauchemin, M., 1990, Sous-vide technology. Q. C. R. and D. Research Bulletin, 17 (1), 1-11.*
6. *Proller, T., 1990, Frisch-Gar-System fur die Fleischerei: Sous-vide - Frische, die uberzeugt. Fleischerei, 41(5), 346-348.*
7. *Varoquaux, P. J. A. and Nguyen-The, C., 1994, Vacuum processing: a new concept for precooked fruit and vegetables. Food Science and Technology Today, 8 (1), 42-49.*
8. *Urch, M., 1991, Processing developments on show in France. Seafood Processing and Packaging International, 5 (3), 26-27.*

9. *Varney-Burch, A., 1991, Macropak reflects 'green' concerns. Seafood Processing and Packaging International, 5 (3), 18-19.*

10. *Hackney, C. R., Rippen, T. E. and Ward, D. R., 1991, Principles of pasteurisation and minimally processed seafoods, in Microbiology of Marine Food Products (eds. D. R. Ward and C. R. Hackney), Van Nostrand Reinhold, New York, pp. 355-371.*

11. *Ghazala, S., 1994, New packaging technology for seafood preservation: shelf life extension and pathogen control, in Fisheries Processing: Biotechnological Applications (ed. A. M. Martin), Chapman and Hall, London, pp. 82-110.*

12. *Baldin, D. E., Sous-vide cooking review, International journal of Gastronomy and Food sciences, Volume 1, Issue 1, January 2012, Pages 15-30.*

12a. *Charley, H., Weaver, C., Foods: A Scientific Approach (third ed.), Prentice-Hall, Inc, 1998.*

12b. *Creed, P. G., The sensory and nutritional quality of 'sous-vide' foods Food Control, 6 (1), 1995, pp. 45-52.*

12c. *Baldwin, D. E., Sous-vide for the Home Cook Paradox Press, 2010.*

12d. *Georget, D. M. R., Smith, A. Ng, A. C.and Waldron, K. W., The thermomechanical properties of carrot cell wall material, Journal of the Science of Food and Agriculture, 78, 1998, pp. 73-80.*

13. *McGee, H., On Food and Cooking: The Science and Lore of The Kitchen Scribner, New York, 2004.*

14. *Belitz, H., D., W. Grosch, P. Schieberle Food Chemistry (third ed.), Springer, 2004.*

Section-C

Other Latest Processing Technologies

Chapter 17

Ozone Technology in the Food Industry

Interest in application ozone has enhanced in recent years in response to consumer demands for 'greener' food additives, regulatory approval and the increasing acceptance that ozone is an environmentally friendly technology [2]. Ozone is a safe, powerful disinfectant and hence it can be used to control biological growth of unwanted organisms in products and equipment used in the food processing industries. Some important aspects of Ozone are given below:

- ☆ Ozone is particularly suited to the food industry because of its ability to disinfect micro-organisms without adding chemical by-products to the food being treated, or to the food processing water or atmosphere in which food are stored [1].
- ☆ Ozone, a powerful oxidant, is effective against various kinds of micro-organisms on fruits, vegetables, meat grains and their products. The multi-functionality of ozone makes it a promising food processing agent [2]. Ozone as an oxidant is used in water treatment, sanitising, washing and disinfection of equipment, odour removal, and fruit, vegetable, meat and seafood processing [2].
- ☆ In aqueous solutions, ozone can be used to disinfect equipment, process water, and some foodstuff. Ozone is an effective sanitizer with great potential applications in the food industry. It decomposes into simple oxygen with no safety concerns about consumption of residual by-product.
- ☆ Due to its high oxidation capacity and microbial inactivation potential, ozone has prevented various kinds of microbial spoilages usually encountered in fruits and vegetables [2].

☆ Ozone can act as a preservative for certain food products and can also sanitize food packaging materials. Some products currently being preserved with ozone include eggs during cold storage, fresh fruits and vegetables, and fresh fish [1].

Excess ozone auto decomposes rapidly to produce oxygen and thus leaves no residues in foods from its decomposition. The only result of ozone, when it decomposes, is oxygen; so, food products treated with ozone are free of disinfectant residue. Spontaneous decomposition without forming hazardous residues in the treatment medium makes ozone safe in food applications. If improperly used, ozone can cause some deleterious effects on products, such as losses in sensory quality [2]. Ozone treatment assures the retention of sensory, nutritional and physico-chemical characteristics of food. Treatment conditions should be specifically determined for all kinds of products for effective and safe use of ozone [2].

Year-wise History

Brief history of application of ozone for water and food products is tabulated below [5, 6]:

Year	*Work*
1906	Ozone used to provide safe drinking water in Nice,France.
1910	First use of ozone in a German meat packing plant.
1918	Ozone used to sanitize swimming pools in the United States.
1936	Ozone used to treat shellfish in France.
1942	Ozone used in egg storage rooms and in cheese storage facilities in the United States.
1972	Ozone used to purify process water in Germany.
1977	Ozone used to reduce Salmonella in shell eggs in Russia.
1982	Ozone declared GRAS (Generally Recognized as Safe) for bottled water in the United States - Reaffirmed Gras in 1995.
1997	Expert Panel convened by EPRI declared ozone GRAS in food processing in the United States.
2000	Food Additive Petition filed with the FDA, August 15, 2000.
2001	FDA recognizes ozone as a secondary direct food additive to kill foodborne pathogens. This approval opened the floodgates for food processors to begin utilizing ozone in their plants. FDA officially approved media containing ozone for use in the food industry, also for direct contact with food products, including fish, meat and poultry (Federal Register, Vol. 66, no. 123, Tuesday, June 26, 2000. Rules and Regulations); 2001: Food Safety and Inspection Service (FSIS) declares acceptable in poultry and meat products.

Today, the use of ozone is steadily replacing conventional sanitation techniques such as chlorine, steam or hot water. It's gaining momentum in the food processing industry as the safest, most cost-effective and chemical-free way of dealing with food safety management [7].

Ozone-Properties, Characteristics and its Bactericidal Effect [2]

Ozone (O_3) is the second strongest common oxidizing agent after fluorine [8] and because of this, it has a strong capacity of disinfection and sterilization. It was discovered by Scheonbein in 1839 and was first used commercially to treat drinking water in France more than a century ago [9]. Disinfecting agents have widespread applications to assure safety and quality in the food industry. Some of these agents (chlorine *etc.*), are inefficient against some organisms and chlorine can react to form trihalomethanes, which are of concern for both human dietary safety and as environmental pollutants. Therefore, the food industry is in search of applications that are:

- Effective in inactivation of common and emerging pathogens, and removing toxic contaminants.
- Leading to less loss in product quality and ensure 'freshness'.
- Environmental friendly.

1. Properties and Characteristics

Ozone (O_3) is tri-atomic oxygen formed by addition of a free radical of oxygen to molecular oxygen. The three atoms of oxygen in the ozone molecule are arranged at an obtuse angle, whereby a central oxygen atom is attached to two equidistant oxygen atoms; the included angle is approximately116° 492 and the bond length is 1.278 Å [2]. Ozone, a bluish gas with pungent smell, is an extremely reactive and unstable allotrope of oxygen (O_2) possessing a high oxidation potential of **-2.07 V** that conveys broad-spectrum antimicrobial properties [9].

- Ozone exists in the gaseous state at room and refrigeration temperature and it is partially soluble in water.
- In normal conditions of pressure and temperature, ozone is unstable. This instability increases with the temperature and the humidity, arriving to be total over 200°C. On the contrary, its greater degree of stability reaches to – 50°C and 38 mm Hg of pressure, in other words, a twentieth part of the atmospheric pressure. At room temperature, ozone is an unstable gas. At room temperature, ozone attacks saturated organic compounds slowly. This offensive capability increases at temperatures of and even inferior 78°C [8]. Ozone readily degrades but has a longer half-life in the gaseous state than in aqueous solution [4].
- Ozone is relatively stable in air but highly unstable in water, decomposes in a very short time. It cannot be stored and must be generated continuously.
- Purity of water usually affects ozone stability. Although ozone in pure water degrades rather quickly to oxygen, it degrades even more rapidly in impure solutions.
- Ozone solubility in water is 13 times that of oxygen at 0-30°C and it is progressively more soluble in colder water [4].
- ozone has capacity of absorption of flavors and strange smells in the water. This must simply to the fast destruction of organic compounds responsible

for the smell. In the same way, ozone has a deodorization role of the air [4a]. In waters also, it is useful for the elimination from heavy metals like iron and manganese that precipitate quickly in oxide form [4b].

- ✰ Ozone is a toxic gas, toxicity is dependent on concentration and length of exposure. Ozone at high concentration is corrosive and toxic even at lower ppm.
- ✰ Ozone causes the formation of free radicals with pH>8. At lowest pH, the mechanism of ionic reaction predominates (ozonolysis) and generates the peroxide production [8].
- ✰ Some other characteristics of Ozone are tabulated below [2, 4, 8, 9]:

Sl. No.	*Properties/Characteristics*	*Values*
1	Boiling point	-111.9 ± 0.3 °C
2	Melting point	-192.5 ± 0.4 °C
3	Critical temperature	-12.1°C
4	Critical pressure	54.6 atm
5	Readily detectable level	0.01–0.05 ppm
9	Molecular weight	48 g.mol^{-1}
10	Density (kg/m^3)	2.14
11	Colour	Light Blue

2. Bactericidal Effects

Ozone is a strong oxidant and potent disinfecting agent. The bactericidal effects of ozone have been documented on a wide variety of organisms *e.g.* Gram positive and Gram negative bacteria, spores and vegetative cells. Ozone has many application area in the industry such as food surface hygiene, sanitation of food plant equipment, reuse of wastewater, treatment and lowering biological oxygen demand (BOD) and chemical oxygen demand (COD) of food plant waste.

Health and Safety Aspects of Ozone Application [9]

The efficacious systems for the detection and catalytic or thermal destruction of ozone are reasonably required for the safety of personnel in food processing plants [10]. As exposure to ozone at low concentrations of around 0.1 mg/L only causes irritation to the eyes, throat and nose and ozone levels as high as 95 mg/L can even have irreversible lethal effects on humans [11]. This is especially important if ozone is used in the gaseous form. In such cases, a continuous ozone analyser must be installed which triggers a general alarm (*i.e.* both acoustic and visual warning signals) as soon as the concentration of ozone exceeds 0.1 ppm (equalling 0.2 mg/m^3) in the atmosphere of the ozonation room [11, 12].

Ozone Advantages [1]

The followings are the advantages of application of ozone in food processing:

1. Most powerful oxidizer available,

2. Instantly destroys microbes,
3. Eliminates chemical storage,
4. Environmentally friendly,
5. Stops mold spores,
6. Does not affect product taste,
7. Can be used in air and water
8. No harmful by-products.

Ozone is easy to use and install. Dissolved ozone can be added to your existing washing, or spraying application. Ozone can also be used as a gas for extending shelf life and eliminating cross contamination [3].

Sources

1. *Ranjan Saha, New and Emerging Non-Thermal Technologies in Food Preservation, April 2015.*

2. *Vithu Prabha, Rahul Deb Barma, Ranjit Singh, Aditya Madan, Ozone Technology in Food Processing: A Review, Trends in Biosciences 8(16), Print: ISSN 0974-8, 4031-4047, 2015.*

3. *www.ozonesolutions.com.*

4. *Rice, R. G., 1986. Application of ozone in water and wastewater treatment. In: Rice, R. G., Browning, M. J. (Eds.), Analytical Aspects of Ozone Treatment of Water and WasteWater. Syracuse, The institute, New York, pp.726.*

4a. *Rice, R. G.; Farquhar, J. W. and Bollyky, L. J, 1982, Review of the applications of ozone for increasing storage times of perishable foods. Ozone: Science and Engineering, 4, 147-163.*

4b. *Tech Brief, 1999, A National Drinking Water Clearing house Fact Sheet, Tech Brief Twelve, December, 4 p.*

5. *Tiwari B K and Rice R G, 2012, Regulatory and legislative issues. In Ozone in Food Processing, pp 7-17. O'Donnell C, Tiwari B K, Cullen P J and Rice R G, eds. Oxford: Wiley-Blackwell.*

6. *Alex Augusto Gonçalves, Ozone- an emerging technology for the seafood industry, Braz. arch. biol. technol. vol.52 no.6 2009.*

7. *Vaz-Velho, M. et al., 2006, Inactivation by ozone of Listeria innocua on salmon-trout during cold-smoke processing, Food Control,* **17**, *609-616.*

8. *Guzel-Seydim Z B, Greene A K and Seydim A C, 2004, Use of ozone in the food industry, LWT-Food Science and Technology 37, 453-460.*

9. *Laszlo Varga and JenoSzigeti, Use of ozone in the dairy industry: A review, International journal of Dairy Technology, 2016, 69, 2, Pages 157-168.*

10. *Kim J G, Yousef A E and Dave S, 1999, Application of ozone for enhancing the microbiological safety and quality of foods: a review. Journal of Food Protection 62, 1071-1087.*

11. *Khadre M A, Yousef A E and Kim J G, 2001, Microbiological aspects of ozone applications in food: a review. Journal of Food Science 66, 1242-1252.*

12. *Damez F, Langlais B, Rakness K L and Robson C M, 1991, Operating an ozonation facility. In Ozone in Water Treatment: Application and Engineering, pp 469-490. Langlais B, Reckhow D A and Brink D R, eds. Chelsea: Lewis Publishers.*

Regulatory Approval and Regulations for Ozonation

A. Regulations for Ozonation

Ozonation applications in food processing have been legally approved, although to varying degrees, in North America, Australia, New Zealand, Japan and several European countries [1]. The specific rules and regulations issued by the US FDA with respect to ozonation of food products, including bottled water, have triggered increased global interest in the use of ozone for food treatment and processing purposes (United States Department of Health and Human Services, Food and Drug Administration 2001). In many parts of the world, ozonation is becoming more and more widely accepted in the food industry as an ecofriendly green technology [2, 3].

B. Regulatory Approval, Legislation and Reference Exposure level of Ozone

a. Regulatory Approval of Ozone Technology [4]

A Brief History of Regulatory Approval of Ozone Technology is given below:

- ☆ **Prior to mid-1997**: There were few or no commercial applications of ozone in food processing or treatment in the U. S. The reason was entirely regulatory in nature.
- ☆ **In 1980**: The FDA approved IBWA's petition for ozone in bottled water, and in 1982, published in the ***Code of Federal Regulations (CFR)*** a formal FDA regulation affirming GRAS Status (*maximum* dosage of ozone of 0.4 mg/L over four minutes contact time, and that the water to be treated must already meet the potable water requirements of the U. S. Environmental Protection Agency -EPA) for use of ozone [5].
- ☆ **In 1982**: The GRAS approval for ozone disinfection of bottled water in 1982 contained the additional statement: "All other food additive applications for ozone must be the subject of appropriate Food Additive Petitions (FAP)" [21 *CFR* 184.1(b) (2)]. This statement effectively mandates the filing of FAPs in order to gain FDA approval for other uses of ozone that come into direct contact with foods.
- ☆ **In June 1997**: An expert panel of food scientists convened by the Palo Alto, CA-based Electric Power Research Institute (EPRI) concluded the following: "The available information supports the safety of ozone when used as a food disinfectant or sanitizer, and further that the available information supports a GRAS classification of ozone as a a disinfectant or

sanitizer for foods **when used at levels and by methods of application consistent with good manufacturing practices"** [6].

- ☆ **In mid-1999**: EPRI's GRAS affirmation gave a clear green light to food processors to test and use ozone for a variety of food processing applications. In mid-1999, the FDA suggested to the EPRI that a single FAP that would provide FDA with specific data showing the antimicrobial properties of ozone in a number of food processing applications could be reviewed quickly, EPPJ developed such an FAP and formally filed it with the FDA in August 2000 [7]. FDA approval of this FAP was published June 26, 2001 in the **Federal Register**.
- ☆ Later in the year, USDA/FSIS approved ozone for use on meat and poultry products, including treatment of ready-to-eat meat and poultry products just prior to packaging, with no labeling issues in regard to treated product.
- ☆ Formal regulatory approval by the FDA (and by the USDA/FSIS) for the use of ozone as an antimicrobial agent in direct contact with foods clears away the regulatory hurdle that has impeded application of ozone to foods in the U. S., and will reassure food processing firms wishing to improve the qualities of their products by approaches involving ozone.

b. Legislation in different Countries [8]

The Legislations being used in different countries for regulation on the application of Ozone in food processing, are given below:

i. **In the United States**: **a.** The FDA has approved the use of ozone like antimicrobial agent for the direct contact with all foods and food products (June 26th, 2001) and **b.** The Ministry of Agriculture of the United States approved it like antimicrobial agent for the direct contact with meats, poultry, fish, mollusks and crustaceans (December, 2001). Ozone was given Gras approval for direct contact on food in 2001 by the FDA Final Rule 21 CFR Part 173.336.

ii. **In Japan**: The government of Japan in 1996 admitted its use for the direct contact with all types of food. Ozone appears in a similar list of FDA as "Generally Recognized as Safe" (GRAS). The Japanese boats of fishing use ozonized water routinely to wash the fresh fish, to make the ice with ozonized water and to pack it on board.

iii. **In Canada**: The Canadian Food Inspection Agency (CFIA) has approved the use of ozone for the cleaning of the contact surfaces of food.

iv. **In Australia**: The government of Australia (also in 1996) authorized the use of ozone for the contact with all foods - similar to the Japanese approval.

v. **In Germany**: The fishing boats German also use ozone in the water and ice.

vi. **In Norway**: The farms of aquaculture and the plants of processing use ozone for the preservation of the fish.

c. Reference Exposure Level [5]

The ozone itself is not toxic in low concentrations. Ozone is an irritating gas with poisonous effects for human. Ozone is toxic by inhalatory via (Following table) but never noticed an undesired effects if provided correctly by other routes (secondary effects due to errors in the technique of supplying in most of the cases) and in the suitable doses (1-40 mg/ml. O_2).

Sl. No.	*Institution*	*Maximum Concentration Permitted (ppm) in Air*	*Exposure Time for Human in Ozonated Air*
1	Food and Drug Administration (FDA)	0.05	8 h
2	Environmental Protection Agency (EPA)	0.08	8 h
3	Ministry of Labour and Employment (Brazil) – Portaria 321478	0.08	48 h/week
4	Occupational Safety and Health Administration (OSHA)	0.10	8 h
5	National Institute of Occupational Safety and Health (NIOSH)	0.10	Permanent

Sources

1. *Tiwari B K and Rice R G, 2012, Regulatory and legislative issues. In Ozone in Food Processing, pp 7-17. O'Donnell C, Tiwari B K, Cullen P J and Rice R G, eds. Oxford: Wiley-Blackwell.*
2. *Pascual A, Llorca I and Canut A, 2007, Use of ozone in food industries for reducing the environmental impact of cleaning and disinfection activities. Trends in Food Science and Technology 18 S29-S35.*
3. *O'Donnell C, Tiwari B K, Cullen P J and Rice R G, 2012a, Status and trends of ozone in food processing. In Ozone in Food Processing, pp 1-6, O'Donnell C, Tiwari B K, Cullen P J and Rice R G, eds. Oxford: Wiley-Blackwell, 8-16.*
4. *Rip G. Rice, Ph. D., Dee M. Graham And Matt T. Lowe, Recent Ozone Applications in Food Processing and Sanitation, Food Safety Magazine, October/November 2002.*
5. *USFDA. GRAS status of ozone. Federal Register, 47, 215, 50209-50210, 1982.*
6. *EPRI. Technical Report-i 08026, Vols.1, 2, 3: Evaluation of the history and safety of ozone in processing food for human consumption. EPRI. Palo Alto, CA., 1997, www.epri.com.*
7. *EPRI. "Ozone as an Antimicrobial Agent for the Treatment, Storage and Processing of Foods in Gas and Aqueous Phases." Food Additive Petition filed with FDA. June 2000.*
8. *Alex Augusto Goncalves, Ozone - an emerging technology for the seafood industry, Braz. arch. biol. technol. vol. 52 no.6 2009.*

Ozone System/Apparatus for Ozone Treatment [1]

This system has to perform the activities given on next page.

1. Takes up air or feed gas from the atmosphere.
2. Concentration of oxygen for ozone generation.
3. Analysis of concentration.
4. Passing of ozone to the treatment chamber for microbial inactivation.
5. Release of excess ozone to prevent chance of health hazards.

Hence a complete ozonation system (gaseous ozone) for use in food processing plants consists of the following subunits:

1. Oxygen Concentrator [1]

Oxygen-enriched air (>90 per cent O_2) is provided simply and conveniently by means of oxygen concentrators for food processing industries. These oxygen concentrators operate on the principle of pressure swing adsorption (PSA) drying and these devices take in ambient air, automatically filter (remove dust particles), then separate and remove nitrogen, which is also dried to below the desired maximum dew point ("4°C), all at the same time in one small device.

- ☆ An air compressor pressurises the airflow and sends it through a molecular sieve bed that adsorbs or traps nitrogen and moisture, while providing oxygen-enriched air to the supply output of the concentrator.
- ☆ The molecular sieve bed loaded with nitrogen and moisture, desorb to waste in vapour form to the environment, recovering the adsorption capacity of the sieve bed.
- ☆ Most ozone equipment manufacturers have optimised their equipment for food processing plants to operate on oxygen feed gas.

2. Ozone Generator [1]

Ozone is formed by a high energy input that splits the oxygen (O_2) molecule in the air into free radical oxygen. Single oxygen (O) molecules rapidly combine with available O_2 to form ozone. The ozone production takes place generally by the ventilation of electrical discharges of high voltage in the air or pure oxygen [2, 3]. This radiation affects a common oxygen molecule that is found in atmosphere which produces the split of the molecule and separation of free oxygen atom. These atoms collide with other oxygen molecules, forming therefore ozone molecules [4]. The step wise details of above reaction are given below [2]:

i. The energy absorbed by an oxygen molecule break it in two oxygen atoms.

$$O_2 + hv \rightarrow O + O$$

ii Each one of these atoms is joined to an oxygen molecule to give another one of ozone.

$$O + O_2 \rightarrow O_3$$

iii. Finally, the ozone molecule is destroyed again absorbing more ultraviolet radiation.

$$O_3 + hv \rightarrow O + O_2$$

iv. Ultraviolet energy is absorbed in a closed cycle of formation and destruction of the ozone.

In order to generate ozone, a diatomic oxygen molecule must first be split. The resulting free radical oxygen is thereby free to react with other diatomic oxygen to form the triatomic ozone molecule. However, in order to break the O–O bond, a great deal of energy is required [3]. Ultraviolet radiation (188nm wave length) and corona discharge methods can be used to initiate free radical oxygen formation and, thereby generate ozone. In order to generate commercial levels of ozone, the corona discharge method is usually used [2, 3].

Each ozone generator is designed to operate at an optimal flow rate depending upon size of ozone generator and such information should be stated in the supplier's equipment manual. With moisture present in the feed gas, the very corrosive nitric acid (HNO_3) is readily formed. Ozone generators are of the following types:

a. Corona Discharge (CD) Generators

CD generators use **Corona discharge or plasma technique** and require an air preparation prior to ozone generation because the quality of gas fed to an ozone generator is a critical factor. The gas feeding an ozone generator must be very dry (maximum "54 °C dew point), because the presence of moisture also affects ozone production (output slightly increased by increasing the oxygen content or drying the air), as well as leading to the formation of nitric acid. This very corrosive acid can destroy the internal parts of a CD ozone generator, which can cause premature system failure and will increase the frequency of required maintenance. Advantages of Corona Discharge are as follows:

i. High ozone concentrations.
ii. Best for water applications.
iii. Fast organic (odor) removal.
iv. Equipment can last for years without maintenance.

The two common types of gas preparation to feed CD ozone generators are oxygen and dry air. If an air dryer is selected to feed a CD ozone system, make sure that the air preparation equipment is matched and sized to the ozone.

The most commercially significant technique is by corona discharge (CD) or plasma technique (the so-called "silent electrical discharge" procedure). In a CD ozone generator, the feed gas (dried air, oxygen, or mixtures thereof), passes between two closely spaced electrodes (one of which is coated with a dielectric material) under a nominal applied potential of ~10 kV. A silent or barrier discharge occurs when the gas becomes partially ionized, resulting in a characteristic violet glow when air is the feed gas (with high purity oxygen the violet coloration is seldom observed). Efficiency of ozone production by corona discharge depends on the strength of micro-discharges which are influenced by the following factors:

1. The gap width.
2. Gas pressure.

3. Properties of the dielectric and metal electrodes.
4. Power supply.
5. The presence of moisture.

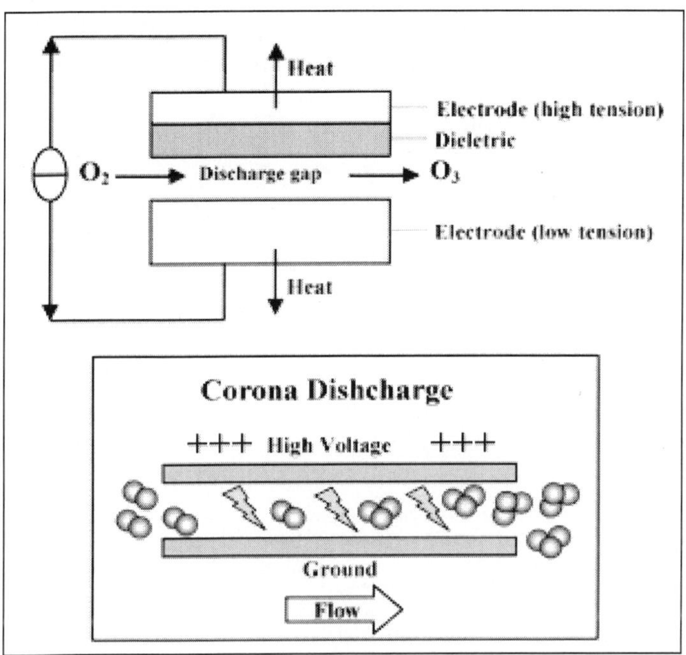

Scheme of Corona Discharge Method: Oxygen is forced between high voltage plates to simulate corona discharge. The oxygen is broken apart and recombines into ozone (Adopted from Ozone Solutions, 2007) [2].

The process of electrically rupturing oxygen atoms to produce oxygen ions or atoms that combine with more oxygen to produce ozone liberates considerable heat. This heat generated must be removed from the generator to avoid the reverse reaction (ozone reverting to oxygen). Hence cooling of the corona discharge ozone generator becomes a critical component in generator design. Normally cooling is provided by water.

There are air cooled CD- ozone generators commercially available, especially on smaller scale, appropriate for many food applications.

b. UV Generators

These generators use the ultraviolet radiation for ozone generation. UV generators of ozone do not require any special air preparation. oxygen atoms formed by the photo-dissociation of oxygen by short wavelength.

UV radiation react with oxygen molecules to form ozone. Although the theoretical quantum yield of ozone by this technique is 2 per cent, in practice the actual yield is more on the order of 0.5 per cent, because the low pressure mercury lamps produce not only the 185 nm radiation responsible for the production of ozone, but also the 254 nm radiations that destroy ozone.

- ☆ An advantage of generating ozone by UV radiation is that ambient air can be used efficiently as the feed gas.
- ☆ The low concentrations of ozone available from UV generators limit their applicability for water treatment to special applications.

c. Other Ozone Generators

These ozone generators use electrolysis technique. High current density electrolysis of aqueous phosphate solutions at room temperature produces ozone and oxygen in the anodic gas. Electrolysis of 68 wt- per cent sulfuric acid can produce 18-25 wt- per cent ozone in oxygen when a well-cooled cell is used.

- ☆ Although electrolysis of water can produce high concentrations of ozone, the output is low, and the cost is several times more than that of the corona discharge process.
- ☆ However, small electrolytic units are being used commercially for treatment of ultra-high purity waters in pharmaceutical and electronic industries.

Some Special Features of Ozone generators [1]

When using oxygen as the feed gas, the concentration of ozone produced is increased two to three times over ozone concentrations in dried air for the same energy expenditure. For example:

- ☆ Dry air-fed CD generators (of the size used in food processing plants) produce 1-2 wt per cent ozone, but when fed high-purity oxygen, ozone concentrations of 3-6 wt per cent are produced for the same energy expenditure. Most ozone equipment manufacturers have optimised their equipment for oxygen feed gas.
- ☆ In some instances, ozone generators can produce as high as 20 wt per cent ozone (from oxygen), but the energy and other requirements are higher to produce ozone at this concentration.
- ☆ Properly designed and operated CD-ozone generators commercially available today are capable of producing kg/h quantities of ozone in gas phase concentrations of 1-5 per cent by weight in air and up to 14 per cent by weight in high purity oxygen. The advantage of a higher weight per cent ozone product is economics.

3. Flow Meters [1]

For accurate measurement of total oxygen gas flow from the oxygen concentrator to ozone generator, the Mass flow meter is used. Ozone flow meter use small gas stream (less than 2 LPM) to measure ozone concentration and for regulation of gas flow through ozone analyser.

4. Treatment Chamber [1]

The treatment chamber is air tight and avoids leakages. It is the collector where ozone enters from the top and returns though the bottom outlet after completing the exposure time. The outlet is connected to ozone analyser and then to ozone destructor. The dimension and type will depend on the followings:

i. Mode of application of ozone.
2. Type of product being processed.

5. Ozone Analyser [1]

This device will be used for measuring the ozone concentration in per cent by weight, or g/m that enters and leaves the treatment chamber.

6. Ozone Destructor [1]

Destructor is an important part that assures non-hazardous by-product or waste in ozone treatment. In this device, the excess ozone is allowed to release into the air such that the contact of both is controlled. Since ozone is unstable in air, decomposition of ozone takes place and oxygen is given out through the outlet. At the outlet of the destructor, excess ozone is destroyed by broking down into oxygen and send to the atmosphere in order to prevent any harm for the worker.

7. Ozone Analyser for Data and Analysis [1]

The data such as exposure time, ozone concentration can be acquired using ozone analyser and used for analysis of results.

Sources

1. *Vithu Prabha, Rahul Deb Barma, Ranjit Singh, Aditya Madan, Ozone Technology in Food Processing: A Review, Trends in Biosciences 8(16), Print - ISSN 0974-8, 4031-4047, 2015.*

2. *Alex Augusto Gonçalves, Ozone - an emerging technology for the seafood industry, Braz. arch. biol. technol. vol.52 no.6 2009.*

3. *Guzel-Seydim, Z. B.et al., 2004, Use of ozone in the food industry. Lebensm. Wiss. U. Technol. 37, 453-460.*

4. *Kogelschatz, U., Eliasson, B. and Hirth, M., 1988, Ozone generation from oxygen and air: Discharge physics and reaction mechanisms. Ozone Science and Engineering, 10, 367-378.*

Advantages and Limitations [1]

The advantages and limitations of ozone disinfection methods proposed for fresh-cut organic vegetables, fruits and meat products applications [2] are tabulated on next page.

Sl. No.	Advantages	Limitations
1	High antimicrobial activity compared to non-oxidative biocins (chlorine) in terms of concentration and time.	Ozone is toxic; when inhaled it cause throat and nasal problems, even lead to asthma.
2	Short contact time for disinfection compared to other disinfection methods.	Ozone is highly unstable gas so controlled release on requirement is to be established.
3	No residue problem as it is completely utilized and get reduced.	Recontamination problems in clean in process pipes as ozone decomposes completely within a short duration.
4	Non-hazardous at low ppm (lower than 4 ppm) and effective in bactericidal uses.	Corrosive at high ppm (higher than 4 ppm),care should be taken in using ozone and releasing to treatment chamber.
5	No need to store hazardous substance compared to other sanitation methods.	It requires regular monitoring in indoor applications for any leakages.
6	Lower running costs, cost matters only to filling of oxygen cylinders and power supply.	Higher initial investment for the generation Equipment.
7	No heat requirement and no heat generation in treatment (applicable to heat sensitive foods) and thus saves need of input energy.	Onsite generation is required as it is unstable and not suitable for storage.
8	Saves transport of disinfectant chemicals and storing of gas cost.	Storage of ozone is not possible as it decomposes quickly.
9	Eco-friendly and economically feasible technology.	It can be mostly surface treatment as ozone decomposes in short time and it is liable to oxidation with organic matter.

Sources

1. *Vithu Prabha, Rahul Deb Barma, Ranjit Singh, Aditya Madan, Ozone Technology in Food Processing: A Review, Trends in Biosciences 8(16), Print: ISSN 0974-8, 4031-4047, 2015.*

2. *H. Olmez and U. Kretzschmar, LWT- Food Science and Technology, 42(3), 686-693, 2009.*

Modes of Application of Ozone [1]

The mode of application may be selected based on the type of food and the need of ozone treatment. There are many modes of application of ozone to the food product. The mode of application of ozone [2] is explained as follows:

1. Application in the Gas Phase

Reaction rates of contaminants in air are much lower (slower) in the gas phase than in aqueous phase.

a. Pumping ozone gas from the generator into an air space to be treated is the simplest approach, and is most effective when the air contaminants to be treated are rapidly reactive with ozone. For example- Odoriferous compounds such as hydrogen sulfide, molds, spores, and some airborne micro-organisms.

b. For those air contaminants that are only slowly affected by ozone, the accepted procedure is to draw contaminated air into an enclosed structure in which ozone is mixed with the contaminated air for such period of time as is necessary to destroy (or inactivate) the contaminants.

ii. Application in Aqueous Phase

Ozone forms true solution with water. Half-life of ozone in aqueous solution is less than that of gaseous ozone; so decomposes rapidly. The decomposition of ozone is very rapid in the presence of impure water. Ozone is only partially soluble in water; efficient transfer of ozone into solution requires the dispersion of gaseous ozone into small bubbles. This is accomplished in various types of positive pressure ozone contractors such as bubble diffuser/bubble columns, mechanically agitated vessels, turbine mixers, tubular reactors, in-line static mixers as well as negative pressure reactors (venturi) and injectors.

- ☆ In positive pressure devices, ozone gas under pressure exiting the ozone generator is forced through small apertures supported under the water.
- ☆ With negative pressure devices, ozone gas is drawn into the flowing water stream to be treated. Under these mixing conditions, bubbles are sheared and mixed thoroughly with the aqueous fluid, decreasing the liquid film thickness but increasing both interfacial area and contact time.

iii. Application of Adjuncts with Ozone [1]

The antimicrobial efficacy can be enhanced considerably when ozonation is combined with other chemical (*e.g.*, hydrogen peroxide) or physical (*e.g.*, ultraviolet radiation) treatments. The coupling ozone with either hydrogen peroxide or with ultraviolet radiation are reported to be used for effective food processing. Some recent reports indicate what appears to be a synergy - or an increased amount of microbial inactivation by applying these combinations to certain foodstuffs over what is obtained by applying ozone or UV radiation alone. The synergistic sporicidal activities of gaseous ozone and UV irradiation was investigated [1a]. The details related to the combined effects of Ozone+Hydrogen peroxide and Ozone + U. V. radiation are given below:

- ☆ **Ozone+Hydrogen Peroxide:** With this combination of oxidants in aqueous solution, both agents destroy each other, and both agents give rise, eventually, to hydroxyl free radicals. Ozone reacts immediately with hydrogen peroxide in solution, and if the amount of ozone dosed in the contactor is always greater than the amount of peroxide initially added to solution, at the outlet of the contact chamber it will not be possible to measure a level of residual ozone in solution.

  It is advisable when evaluating the use of this advanced oxidation process to first determine which polluting constituents of the water or wastewater requiring treatment are present, and then to determine experimentally the optimum range of peroxide to ozone weight ratio required for their destruction. Scientists [1b] have shown that if the weight ratio of peroxide to ozone rises above 1: 1, the rates of oxidation of organics in water actually slowly down. This means that if excess hydrogen peroxide is present over the amount of ozone added, at least some of the advantages of advanced oxidation are lessened. It also means that there will be no molecular ozone present at any time during ozone contacting for microbial disinfection.

- **Ozone + UV Radiation:** With this combination of agents, it is customary to place a UV bulb (or multiple bulbs) in the ozone contacting chamber. As water flows through the chamber, first the UV bulb(s) is (are) turned on and ozone is added. As long as the amount of UV radiation dosed is in excess of the amount of ozone present, all ozone will be converted instantaneously to decomposition products, ending rapidly as hydroxyl free radicals.
- Several reports have been made of an apparent increase in antimicrobial activity in some food applications when ozone is combined in water with either peroxide or UV radiation.

Sources

1. *Vithu Prabha, Rahul Deb Barma, Ranjit Singh, Aditya Madan, Ozone Technology in Food Processing: A Review, Trends in Biosciences 8(16), Print: ISSN 0974-8, 4031-4047, 2015.*

1a. *Naitoh, S. 1992. Synergistic sporicidal effect of gaseous ozone and UV on Bacillus and Clostridium spores. Bokin Bobai, 20: 293-300.*

1b. *Glaze, W. H.; Kwang, J. W. and Chapin, D. H., 1987, Chemistry of water treatment processes involving ozone, hydrogen peroxide and ultraviolet radiation. Ozone Science and Engineering 9, 335-352.*

2. *Graham, D. M. 1997, Use of ozone for food processing, Food Technology, 51: 72-5.*

Factors Affecting Efficacy of Ozone Processing [1]

There are following different extrinsic and intrinsic parameters that affect the disinfection ability of ozone in liquid processing treatment:

a. Extrinsic Parameters [1]

These include flow rate, concentration of ozone, temperature.

i. Flow Rate

Depending on the gas flow rate applied for ozone production, different bubble sizes are produced. Bubble size has been shown to have an effect on ozone's solubilisation rate and disinfection efficacy. It was reported [1a] that ozone mass transfer and disinfection efficacy increased as bubble size decreased. The higher interfacial area available for mass transfer at the smaller bubble size may have been responsible for this effect. Free suspended bacteria migrate toward the ozone bubbles due to their surface active properties and are preferentially inactivated by comparatively high ozone concentrations at the gas liquid interface of the bubble.

ii. Concentration

Ozone concentration present or available in the medium is determining ozone efficacy. Increased ozone concentration causes saturation and thus makes addition of further ozone to the reactor ineffective, resulting in longer times to achieve the same log-reduction values.

iii. Temperature

As temperature increases, ozone becomes less stable and less soluble but more reactive. Therefore, temperature should be carefully controlled during ozonation in order to maintain a balance in solubility, stability and sanitising efficacy [6, 7, 9]. Ozone solubility in water is 13 times that of oxygen at 0-30° C and it is progressively more soluble in colder water [2]. The solubility ratio for ozone increases as the temperature of water decreases. As temperature increases, ozone becomes less soluble and less stable, with an increase in the decomposition rate [3].

iv. Humidity

Microbial destruction rates are greater when ozone is used in an atmosphere having an increased relative humidity [7, 9].

b. Intrinsic Parameters [1]

These include pH, and presence of solid contents (organic matter).

i. pH

It is widely accepted that ozone is less stable at high than at low pH values. However, when ozone auto decomposes at high pH values, the radicals formed evidently contribute to its efficacy [6, 9]. The effect of pH on ozone inactivation is mainly attributed to the fact that ozone decomposition rate changes substantially with changes in pH. It was observed [4] that ozone inactivation of *E. coli* was much faster at the lower pH. The ozone treatment duration required for achieving a 5-log reduction was 4 min at the lowest pH and 18 min at the highest pH studied.

ii. Presence of Solid Contents (Organic Matter)

Dissolved organic matter reduces the disinfection activity by consuming ozone to produce compounds with little or no microbiocidal activity, thereby reducing the concentration of active species available to react with micro-organisms [1]. Hence the absence of organic compounds in ozonated water to be used in food processing is highly desirable [6, 8]. Researchers [5] studied the inactivation of *E. coli* in orange juice, and found that the efficacy of ozonation was reduced in the presence of ascorbic acid and organic matter.

Sources

1. *Vithu Prabha, Rahul Deb Barma, Ranjit Singh, Aditya Madan, Ozone Technology in Food Processing: A Review, Trends in Biosciences 8(16), Print: ISSN 0974-8, 4031-4047, 2015.*

1a. *M. Ahmad; S. Farooq, "Influence of Bubble Sizes on Ozone Solubility, Utilization, and Disinfection, " Water Sci. Technol. 17: 1081-1090, 1983.*

2. *Rice, R. G., 1986. Application of ozone in water and wastewater treatment. In: Rice, R. G., Browning, M. J. (Eds.), Analytical Aspects of Ozone Treatment of Water and WasteWater. Syracuse, The institute, New York, pp. 726.*

3. *Rice, R. G., Robson, C. M., Miller, G. W., Hill, A. G., 1981. Uses of ozone in drinking water treatment. J. Am. Water Works Ass. 73 (1), 4457.*

4. *Patil, S., Valdramidis, V. P., Cullen, P. J., Frias, J., Bourke, P., 2010a. Inactivation of Escherichia coli by ozone treatment of apple juice at different pH levels. Food Microbiol. 27 (6), 835840.*

5. *Williams, R. C., S. S. Sumner, and D. A. Golden, 2004."Survival of Escherichia coli O157: H7 and Salmonella in Apple Cider and Orange Juice as Affected by Ozone and Treatment Temperature", J. Food Prot., 67(11): 2381-2386.*

6. *Khadre M A, Yousef A E and Kim J G., 2001, Microbiological aspects of ozone applications in food: a review. Journal of Food Science 66 1242-1252.*

7. *Kim J G, Yousef A E and Dave S., 1999, Application of ozone for enhancing the microbiological safety and quality of foods: a review. Journal of Food Protection 62 1071-1087.*

8. *Cullen P J and Norton T, 2012, Ozone sanitisation in the food industry. In Ozone in Food Processing, pp 163-176. O'Donnell C, Tiwari B K, Cullen P J and Rice R G, eds. Oxford: Wiley-Blackwell.*

9. *Laszlo Varga and JenoSzigeti, Use of ozone in the dairy industry: A review, International journal of Dairy Technology, 2016, 69, 2, Pages 157-168.*

Microbial Inactivation by Ozone [1]

Microbial inactivation by ozonation is a complex process. Ozone destroys micro-organisms by the progressive oxidation of vital cellular components. The antimicrobial activity of ozone is based essentially on its powerful oxidizing effect, which causes irreversible damage to the fatty acids in the cell membrane and to cellular macromolecules, such as proteins, and DNA. This action is particularly effective in air at high relative humidity, the bacteria being killed by ozone more readily in the swollen state than when dry.

Micro-organisms are inactivated by disruption of the cell envelope or disintegration leading to cell lysis. Two following major mechanisms of ozone destruction of the target organisms were identified:

1. Ozone oxidises sulfhydryl groups and amino acids of enzymes, peptides and proteins to smaller peptides.
2. Ozone oxidises polyunsaturated fatty acids to acid peroxides.

Some authors concluded that molecular ozone is the main inactivator of micro-organisms (direct oxidation), while others emphasize the antimicrobial activity of the reactive by-products of ozone decomposition (indirect oxidation) such as OH, O_2^-, and HO_3.

It has not been well established whether molecular ozone or the radical species produced by ozone breakdown, are responsible for inactivation of micro-organisms.

Thus there are two main reactions happening on incidence of ozone on the microbial surfaces ***viz.* direct and indirect reactions**. Ozone takes any of the pathway or both for the oxidation of sulfhydryl group of enzymes, amino acids of peptides, proteins and enzymes and polyunsaturated fatty acids.

i. Direct Reaction

Direct reaction with molecular ozone, is the predominant mechanism for inactivation of micro-organisms. It is the direct oxidation of target groups in microbial cell by ozone. Because of the molecular structure of ozone, it can act as an electrophilic or nucleophilic agent during reactions, with these types of reaction occurring in solutions containing organic pollutants (with microbes).

- ☆ Generally, electrophilic reactions will occur with organic water contaminants with a high electron density and will act faster in solutions consisting of high levels of aromatic compounds.
- ☆ Nucleophilic reactions take place mainly when there is a shortage of electrons and particularly at carbon compounds that contain electron-withdrawing groups such as –COOH and $-NO_2$.

Overall, the direct oxidation of organic matter by ozone involves a quite selective reaction mechanism. Moreover, it is important to note how the pH value of the water system can influence ozone decomposition; with pH > 7 causing an increase in the rate of ozone decomposition. Also, in strongly acidic solutions (pH < 3) the OH radicals do not influence the decomposition of ozone.

ii. Indirect Reaction

Indirect reactions with radicals are responsible for inactivation for some groups of microbes. Ozone decomposition has been explained as occurring in following three stages:

- ☆ **Initiation:** During the initiation step, free radicals are generated, such as superoxide radical ions and hydroperoxide radicals, which lead to formation of the highly reactive hydroxyl radical. These hydroxyl radicals are one of the factors contributing to ozone decomposition.,
- ☆ **Promotion:** The promotion step involves regeneration of the hydroperoxide and superoxide radicals through reactions involving participation of promotors such as formic acid, glyoxylic acid, primary alcohols and aryl groups.
- ☆ **Inhibition:** In contrast, in the inhibition step the consumption of hydroxyl radicals occurs via ions like bicarbonate, carbonate, tertiary alcohols and alkyl groups, without regeneration of the superoxide radical [2]. Bicarbonate ions are generally present in microbial cells, which could act as scavengers of radicals otherwise responsible for inactivation of micro-organisms.

Additionally, factors promoting ozone decomposition in the system can lead to faster dissipation of ozone, resulting in a requirement for increased ozone concentration in order to achieve the desired inactivation level. The resultant disruption or lysis of cell walls (probably by oxidative destruction) associated with ozone is a faster inactivation mechanism than that of other disinfectants, which require the disinfecting agent to permeate through the cell membrane in order to be effective. Generally, with regard to the spectrum of microbial action, each micro-organism has an inherent sensitivity to ozone. **Bacteria are more sensitive than yeasts and fungi. Gram-positive bacteria are more sensitive to ozone than Gram-negative organisms, and spores are more resistant than vegetative cells.** Moulds are more resistant than yeasts, and yeasts are more resistant than bacteria, with gram-negative bacteria being even more sensitive than gram positives. Ozone is less effective against both fungal and bacterial spores than vegetative cells [2, 3, 4]. Due to the mechanism of ozone action, which destroys the micro-organism through cell lysis, the development of resistance to ozone disinfection is not found.

Sources

1. *Vithu Prabha, Rahul Deb Barma, Ranjit Singh, Aditya Madan, Ozone Technology in Food Processing: A Review, Trends in Biosciences 8(16), Print: ISSN 0974-8, 4031-4047, 2015.*
2. *Khadre, M. A., Yousef, A. E.and Kim, J.2001. Microbiological aspects of ozone applications in food: a review, Journal of Food Science, 6: 1242-52.*
3. *Moore G, Griffith C and Peters A, 2000, Bactericidal properties of ozone and its potential application as a terminal disinfectant. Journal of Food Protection 63 1100-1106.*
4. *Laszlo Varga and JenoSzigeti, Use of ozone in the dairy industry: A review, International journal of Dairy Technology, 2016, 69, 2, Pages 157-168.*

Different Target Sites of Ozone Activity [1]

Inactivation of bacteria by ozone is a complex process because ozone attacks numerous cellular constituents Ozone is capable of attacking various constituents in cell membranes, cell walls, the cytoplasm, endospore coats, virus capsids and viral envelopes [2, 3]. Details are given below:

a. Peptidoglycans in Cell Envelopes

Ozone may oxidize various components of cell envelope including polyunsaturated fatty acids, membrane- bound enzymes, glycoproteins and glycolipids leading to leakage of cell contents and eventually causing lysis. When the double bonds of unsaturated lipids and the sulfhydryl groups of enzymes are oxidized by ozone, disruption of normal cellular activity including cell permeability and rapid death ensues.

b. Proteins, Unsaturated Lipids and Respiratory Enzymes in Cell Membranes

Several authors referred to enzyme inactivation as an important mechanism by which ozone kills cells.ozone acted as a general protoplasmic oxidant. Inactivation of enzymes by ozone is probably due to oxidation of sulfhydryl groups in Cysteine residues. It is noteworthy that the double bonds of unsaturated fatty acids are especially vulnerable to ozone attack [3, 4].

c. Enzymes and Nucleic Acids in the Cytoplasm

Reaction of aqueous ozone with nucleic acids in vitro found that it may damage nucleic material inside the cell. Ozone modified nucleic acids in vitro, with thymine being more sensitive than cytosine and uracil. In another study, ozone opened the circular plasmid DNA and reduced its transforming ability, produced single double-strand breaks in plasmid DNA and decreased transcription activity. It was [5] found that ozone may induce mutations. Compared to other known mutagens, ozone was found to be a weak mutagen on *Saccharomyces cerevisiae*.

d. Proteins and Peptidoglycan in Spore Coats and Virus Capsids

Bacterial Spore Coats

It was [6] found that bacterial spores (*Bacillus cereus*) with coat proteins removed were rapidly inactivated by ozone, compared to intact spores. The researcher concluded that the spore coat is a primary protective barrier against ozone.

Recently, it was found that spores of Bacillus subtilis treated with aqueous ozone showed heavily disrupted outer spore coats. Transmission Bacillus spores treated with ozone suggest that ozone inactivates spores by degrading the outer spore component, thus exposing the cortex and core to the action of ozone [2].

Capsids

It was [7] found that aqueous ozone inactivated bacteriophages (f2 and T4) by attacking capsid protein, with liberation and inactivation of the nucleic acid. The RNA from bacteriophage (f2) was partially inactivated prior to release from the capsid. It was suggested that ozone breaks the protein capsid into subunits liberating RNA and disrupting virus adsorption to the host pili, and that the RNA may be secondarily inactivated. The DNA released from bacteriophage (T4) was rapidly inactivated by ozone at about the same rate as that in the intact phage. The damage to viral envelopes is the main cause of inactivation of viruses by ozone. Viruses are similar to bacteria in sensitivity to ozone with bacteriophages showing the least resistance [2, 3].

Sources

1. *Vithu Prabha, Rahul Deb Barma, Ranjit Singh, Aditya Madan, Ozone Technology in Food Processing: A Review, Trends in Biosciences 8(16), Print: ISSN 0974-8, 4031-4047, 2015.*
2. *Khadre, M. A., Yousef, A. E. and Kim, J.2001. Microbiological aspects of ozone applications in food: a review, Journal of Food Science, 6: 1242-52.*
3. *Laszlo Varga and JenoSzigeti, Use of ozone in the dairy industry-A review, International journal of Dairy Technology, 2016, 69, 2, Pages 157-168.*
4. *Guzel-Seydim Z B, Greene A K and Seydim A C, 2004, Use of ozone in the food industry. LWT - Food Science and Technology 37, 453-460.*
5. *L'Hérault P, Chung YS, Mutagenicity of ozone in different repair-deficient strains of Escherichia coli, Mol Gen Genet. 1984; 197(3): 472-7.*

6. *P. M. Foegeding. Ozone inactivation of Bacillus and Clostridium spores and the importance of the spore coat to resistance, Food Microbiology, 1985, 2(2), 123-134.*

7. *Kim CK, Gentile DM, Sproul OJ, Mechanism of Ozone Inactivation of Bacteriophage f2, Appl Environ Microbiol. 1980, 39(1): 210-8.*

Effect of Ozone on Quality and Nutrition of Food Products [1]

The effects of ozone treatment on quality and physiology of various foods are reported. Applying ozone at doses that are large enough for effective decontamination may change the sensory qualities of food. Ozone is not universally beneficial and in some cases may promote oxidative spoilage in foods. Surface oxidation, discolouration or development of undesirable odours may occur in substrates from excessive use of ozone [2]. It was reported [2a] that there was no detrimental change in the quality attributes of apple cider when it was treated with ozone. Several studies were conducted to observe the effect of ozone treatment on the Chemical attributes, Texture, Visual quality and sensory quality of food products.

a. Chemical Attributes

i. No change in onion chemical composition and sensory quality was reported [2b].

ii. Ozone-containing water treatment resulted in no significant difference in total sugar content of celery and strawberries [3] during storage.

iii. Ozone washing treatment was reported to have no effect on the final phenolic content of fresh-cut iceberg lettuce [4].

b. Texture (Rheological Property) of Products

Fruits and vegetables with a firm, crunchy texture are highly desirable because consumers associate these textural attributes with freshness and wholesomeness. Textural changes in fruits and vegetables could be due to various enzymatic and non-enzymatic processes. Ozone treatment of fresh fruit and vegetables either by washing or in storage consisting of ozone gas is reported to have significant effects on texture. Change in texture during ozonation and subsequent storage may possibly be due to postharvest changes in cellulose and hemicellulose contents due to ozone application during MAP. This could be due to polymerisation and epimerisation of cellulose and hemicelluloses contents of cell walls inducing thickening of the cell walls, causing textural changes in fresh-cut green asparagus during storage after ozone treatment [5].

- ✰ Firmness of fresh coriander leaves was reported to decrease through washing with ozone-containing water compared to control.
- ✰ Another study [6] reported no significant changes in firmness of fresh-cut cantaloupe irrespective of gaseous ozone concentration (5000 or 2000 ppm) for 30 minutes during storage compared to control.
- ✰ Ozonation of fruits has been reported to enhance firmness of citrus fruits and cucumbers compared to controls [7].

- ✰ Ozone is a strong oxidizing agent which can cause oxidation of feruloylated cross linkages or phenolic cross linkages among cell-wall pectin, structural proteins or other polymers, and thereby change the firmness of the product [8].

c. Visual Quality

i. It was observed [9] that ozone fumigation with up to 10 000 ìl/L for up to 2 hours helps to control postharvest grey mould of table grapes caused by *Botrytis cinerea*. However, grapes stored in ozone-rich atmospheres may develop thin longitudinal darkened lesions. This injury is reported to be irregular and was not always associated with an ozone dose or cultivar.

ii. It was investigated [10] that the effect of several sanitisers on the visual quality and colour of rocket leaves during storage in air and low O_2 (1-3 kPa) + high CO_2 (11-13 kPa) for 15 days at 4 °C. They observed that ozone effects were comparable with other sanitisers, except for lactic acid treated samples.

d. Sensory Quality of Product

The most notable effect of ozone on the sensory quality of fruits reported is the loss of aroma.

- ✰ Ozone-enriched cold storage of strawberries resulted in reversible losses of fruit aroma [11]. This behaviour is probably due to the oxidation of volatile compounds.
- ✰ Scientists [12] did not observe any significant changes in tomato fruit weight, antioxidant status, CO_2/H_2O exchange, ethylene production or in organic acid, vitamin C (pulp and seed) or total phenolic content when exposed to ozone concentrations ranging between 0.005 and 1.0 ìmol/mol at 13 °C and 95 per cent RH.

Sources

1. *Vithu Prabha, Rahul Deb Barma, Ranjit Singh, Aditya Madan, Ozone Technology in Food Processing: A Review, Trends in Biosciences 8(16), Print: ISSN 0974-8, 4031-4047, 2015.*

2. *Khadre, M. A., Yousef, A. E.and Kim, J. 2001. Microbiological aspects of ozone applications in food: a review, Journal of Food Science, 6: 1242-52.*

2a. *Dock, L. L., Development of thermal and non-thermal preservation methods for producing microbially safe apple cider. Thesis, Purdue University, West Lafayette, 1999.*

3. *Zhang, L., Lu, Z., Yu, Z. and Gao, X. 2005 Preservation Fresh-cut Celery by Treatment of Ozonated Water. Food Control, 16, 279-283.*

4. *Beltrán, D., Selma, M. V., Marin, A. and Gil, M. I. 2005a, Ozonated water extends the shelf life of fresh-cut lettuce, Journal of Agricultural and Food Chemistry, 53: 5654-63.*

5. *An, J., Zhang, M., Lu, Q., 2007, Changes in some quality indexes in fresh-cut green asparagus pretreated with aqueous ozone and subsequent modified atmosphere packaging. J. Food Eng. 78, 340-344.*

6. *Selma MV, Ibanez AM, Cantwell M, Suslow T. 2008. Reduction by gaseous ozone of Salmonella and microbial ora associated with fresh-cut cantaloupe. Food Microbiol, 25, 558-65.*

7. *Skog, L. J.and Chu C. L., Effect of ozone on qualities of fruits and vegetables in cold storage, Canadian Journal of Plant Science, 2001, 81(4), 773-778.*

8. *Heun Hong, J. and Gross, K. C., 1998, Surface sterilization of whole tomato fruit with sodium hypochlorite influences subsequent postharvest behavior of fresh-cut slices, Postharvest Biology and Technology, 13(1), 51-8.*

9. *Gabler, F. M., Mnasour, M. F. and Karaka, H, Influence of fumigation with concentration of ozone gas on postharvest gray mold and fungicide residues on table grapes, 2010, Postharvest Biology and Technology 55(2).*

10. *Martínez-Sánchez, A., Allende, A., Bennett, R. N., Ferreres, F. and Gil, M. I., 2006, Microbial, nutritional and sensory quality of rocket leaves as affected by different sanitizers, Postharvest Biology and Technology, 42(1): 86-97.*

11. *Nadas, A., Olmo, M., and Garcia, J. M., 2003, Growth of Botrytis cinerea and strawberry quality in ozone-enriched atmospheres. J. Food Sci. 68: 1798-1802.*

12. *Tzortzakis, N. G., Borland, A., Singleton, I., Barnes, J. D., 2007b. Impact of atmospheric ozone-enrichment on quality-related attributes of tomato fruit. Postharvest Biol. Technol. 45, 317-325.*

Applications in Food Processing

Ozone is an accepted commercial technology in many aspects of the foods/ Food processing industry such as water and air treatment, food decontamination and disinfection, safe packaging and storage *etc.* In general, ozone finds wide application in the food industry, including surface decontamination of meat, fruits and vegetables, drinking water disinfection, pesticide removal, safe storage and wastewater treatment [1]. Ozone can reduce contamination on food and increase storage life. The age-enhancing hormone ethylene is instantly destroyed. This reduces spoiling and keeps the food looking fresh [1].

The use of ozone in the processing of foods has recently come to the forefront as a result of stories of food contamination reported in the media. These serve to highlight the need for the food industry to seek better, more effective methods of ensuring food safety [2].

- ☆ The use of ozone in food processing applications has gained popularity in the past few years due to the approval by the U. S. Food and Drug Administration approving the use of ozone as an anti-microbial agent for food treatment, storage, and processing.
- ☆ It is worthy of noting, the use of ozone in food processing has been allowed and accepted in Japan, Australia, France and other countries for some time.

- ☆ Samples treated with 2.3 PPM Dissolved ozone had an average reduction of aerobic bacteria plate counts of 97.82 per cent.
- ☆ *E. coli* was reduced by 98.22 per cent on samples Treated with 2.3 PPM Dissolved ozone.

Flume using Ozonated Water.

Potatoes Sprayed with Dissolved Ozone.

Waterzone Ozone Injection System [2].

Sources

1. *Vithu Prabha, Rahul Deb Barma, Ranjit Singh, Aditya Madan, Ozone Technology in Food Processing: A Review, Trends in Biosciences 8(16), Print: ISSN 0974-8, 4031-4047, 2015.*
2. *www.ozonesolutions.com.*

A. Application of Ozone in Food Industries [1]

Some of the ozone uses (given by Bharathraj -2000) in food industries are discussed below:

i. Removal of Contaminants [1]

The surface of food can be contaminated by fungicides, pesticides and other chemicals being used during farming practises. This can be potentially dangerous to the human health. Ozone can be used to oxide the chemicals and remove the contaminants and make it safe for sale or for further processing.

ii. Cleaning in Process (CIP) [1]

Ozone charged water is used for CIP for cleaning pipes, tanks floors, surface equipment *etc.* Use of ozone system to food processing is that it provides the opportunity to reuse the water that could bring about a lot of savings in terms of availability and water cost. CIP washing of plant processing equipment and drains with ozone-containing water is now common practice.

iii. Sanitation [1]

Food for consumption must be free from pathogenic micro-organisms. Microbial load of raw material, improper handling and storage, use of contaminated wash water, processing equipment, and transportation facilities, as well as cross-contamination from other products contribute to the microbial hazards associated with foods *e.g.* meat, fruits and vegetables.

- ✰ Usually **non- oxidative biocides such as chlorine** were widely used for 2-log unit reduction of micro-organisms. Chlorine, the most common used disinfecting agent, selectively destroys certain intracellular enzyme systems, whereas ozone causes widespread oxidation of internal cellular proteins causing rapid cell death.
- ✰ **Ozone is an oxidative biocide** and is better than non- oxidative to avoid undesirable taste or carcinogenic effects. Ozone is 3000 times powerful than chlorine and regarded as "Generally Recognised as Safe (GRAS)". Commercial processing applications of ozone expanded in processing water treatments especially in fish, beverage-producing plants, and wineries.

iv. Stop Ripening and Spoilage

Ethylene formation on the food surface is responsible for the food ripening and spoilage. By virtue of its chemical properties, ozone prevents ethylene formation and thereby retards ripening and spoilage by micro-organisms. This property extends the shelf life of food.

Ozone oxidises ethylene completely and leaves carbon dioxide, water and oxygen. by following the below mentioned reaction:

$$C_2H_4 + 6O_3 \rightarrow 2CO_2 + 2H_2O + 6O_2$$

B. Application in Seafood and Shell Fish Processing [1]

Few examples are discussed below:

- ☆ Researchers [1a, 1b] developed the use of ozone for shell fish depuration from laboratory stage to full commercial installations in Southern Europe. Depuration is a process whereby shellfish, freshly harvested from their natural environments, are placed for several days in storage chambers, through which clean, pathogen-free water is passed. Ozonated water was used for depuration.
- ☆ In the fresh fish and bivalve mollusk, ozone application suppresses the smell characteristic which sometimes can be disagreeable, giving a healthful aspect to these seafood. It is advisable to consider that ozone, in this case, does not have to be used to mask the low quality avoiding, thus, the economic fraud [6].
- ☆ Chilled tilapias were stored at zero and 5°C after short ozone (6 ppm) pretreatment of live fish [6].
- ☆ The ozonized water for dipping and washing fish or fish fillets showed an effective reduction of microbiological flora and simultaneously had no effect on the product [7, 8].

c. Application in Meat Processing [1]

The use of ozone (in gas and aqueous phases) for the preparation of complete meals in a central food processing plant was demonstrated [1c]. The meals were then packaged and sealed in a sterilised manner.

Ozone is reported to be effective against contamination of natural casings, which are generally contaminated with several micro-organisms.

- ☆ It was suggested [1d] that a combination of treatments, such as washing with ozonated water to whiten casings with another treatment, such as irradiation, might prove effective for enhancing the casing value while ensuring the destruction of potential food borne pathogens.
- ☆ Scientists [1e] evaluated trimming techniques and beef carcass washing techniques to improve the microbial quality of meat.
- ☆ Intervention treatments included knife trimming, washing with water and rinsing with ozone (0.3-2.3ppm) or hydrogen peroxide (5 per cent). Ozone treatment reduced carcass surface contamination by 1.30 CFU/cm^2, whereas hydrogen peroxide reduced aerobic plate counts by 1.14 CFU/cm^2.
- ☆ Ozone has also been used as a pre-treatment before cooking to determine any synergistic activity on reducing micro-organisms.

D. Ozone in Grain Processing [1]

Storage grains are very susceptible to a number of insects, such as Tribolium, Sithophilus and moths, which cause considerable damage and which could potentially develop resistance to the currently used insecticides. Ozone that can be used in fumigation is an interesting alternative to applied chemicals for the control of insect development.

- The efficiency of ozone fumigation in a corn grain mass against adult insects was evaluated [2], Results demonstrated a significant insect mortality increase (92–100 per cent compared to 3-10 per cent in the control) when the insect species in grain samples were treated with 50 ppm for 3 days.
- Due to its inactivating action on fungi, ozone can also be considered as helping to reduce **mycotoxin accumulation** during grain storage. Furthermore, its oxidant properties could also be used for **mycotoxin degradation and detoxification.** It should also be noted that ozone was successfully used to detoxify commonly occurring mycotoxins by either completely degrading them or causing chemical modifications, thus considerably reducing their bioactivity [4, 5].

E. Application in Microbial Reduction and Sanitation [1]

Ozone destroys micro-organisms by the progressive oxidation of vital cellular ozonation. Numerous studies were conducted to investigate the effectiveness of ozone against various kinds of micro-organisms. Studies on microbial inactivation by ozone in fruits and vegetables are summarized in following Table [1]:

Microbial Inactivation by Ozone

Sl. No.	*Type of Product*		*Treatment*	*Findings*
1	Fruits	Orange and Lemon	Gas Zone (0.3ppm) for 4 weeks.	Spore was reduced of *P. Italicum.*
		Apple [3]	Ozone bubbling and dipping in pre-ozonated water, 3min	Decrease in counts of *E. coli* (< 1 log CFU).
		Strawberry	Storage for 3 days at 2°C with 1.5 ppm ozone.	Mycelial growth developed more slowly.
		Date fruit	0.002,0.006, and 0.01 g/m^3 gaseous ozone for 1 hour.	Reduction of fungal count (total mycoflora).
2	Grains	Barley	Gaseous ozone	Reduction mycelia grow and spores germinates.
		Wheat	1.4 g/m^3 gaseous ozone	Reduction of fungal counts (Micromycetes).
		Maize [2]	Gaseous ozone	Reduction of 63 per cent fungal growth counts after3 days.
3	Vegetables	Onion	Gaseous ozone	Reduction on spore germination, change of colony colour (Aspergillus).
		Tomato	Washing with ozonated water (3.8ppm) for 10 min.	Spores of *B. cinereaon* surface were inactivated.
4	Milk Products	Cheese	Gaseous ozone at 3 different levels	Reduction of fungal counts on ripening room and on cheese surface.

Sources

1. *Vithu Prabha, Rahul Deb Barma, Ranjit Singh, Aditya Madan, Ozone Technology in Food Processing: A Review, Trends in Biosciences 8(16), Print: ISSN 0974-8, 4031-4047, 2015.*

1a. *Voille, H., On the sterilization of seawater by ozone. Application of this method to purification of contaminated shellfish. Rev. d'Hyg. Prevent. Med. prevent. [Jan.], 1929.*

1b. *Salmon, J.and J. Le'Gall, Application of ozone to maintain the freshness and to prolong the durability of conservation of fish. Rev. Gen. duFroid. Nov. 1936, p. 317.*

1c. *C. Steffen and R. G. Rice, Ozone - Science and Engineering, 2010; 32(2): 144-149).*

1d. *H. Benli, B. S. Hafly, J. Keeton and L. M. Lucia, Biomechanical and microbiological changes in natural hog casings treated with ozone, Meat Science 79(1), 155-62, 2008.*

1e. *J. O. Reagan, G. R Acuff, D. R. Buege and M. J. Buyck, Trimming and Washing of Beef Carcasses as a Method of Improving the Microbiological Quality of Meat, Journal of food protection, 59(7), 51-756, 1996.*

2. *Kells, S. A., Mason, L. J., Maier, D. E. and Woloshuk, C. P.2001. Efficacy and fumigation characteristics of ozone in stored maize, J of Stored Products Research, 37: 371-82.*

2a. *V. Bharath Raj, Ozone Technologies and Systems (India) Pvt. Ltd. Chennai.*

3. *Achen, M. and Yousef, A. E. 2001. Efficacy of ozone against Escherichia coli O157: H7 on apples, J Food Sci, 66(9): 1380-4.*

4. *LaszloVarga and JenoSzigeti, Use of ozone in the dairy industry: A review, International journal of Dairy Technology, 2016, 69, 2, Pages 157-168.*

5. *McKenzie K S, Sarr A B, Mayura K et al., 1997, Oxidative degradation and detoxification of mycotoxins using a novel source of ozone. Food and Chemical Toxicology 35 807-820.*

6. *Alex Augusto Gonçalves, Ozone - an emerging technology for the seafood industry, Braz. arch. biol. technol., 52 no.6 2009.*

7. *Ravesi, E. M.; Licciardello, J. J. and Racicot, L. D, 1988, Ozone Treatments of Fresh Atlantic Cod, Gadus morhua. Marine Fisheries Review, 49 (4), 37-42.*

8. *Tapp, C.and Sopher, C. D., 2002, Ozone Applications in Fish and Seafood Processing-Equipment Suppliers Perspective-Summary Paper. In: Ozone Applications in Fish Farming, EPRI, Palo Alto, CA: 2002, 4 pages.*

F. Application of Ozone in Dairy Industry [1]

Implementation of good hygienic practices on dairy farms is a prerequisite for the production of high-quality and microbiologically safe raw milk and ozone helps in doing so. Ozone treatment has also been found to be a promising method for reducing the concentrations of pollutants in dairy wastewaters [2].

a. Dairy Farm [2]

The use of ozone can considerably lower chemical costs and completely eliminate hot water costs on dairy farms. Ozone treatment can be applied on dairy farms for a variety of reasons and purposes.

- If the hind legs, udder and teats of cows are thoroughly washed off with ozonated water prior to milking, many hygienic problems may easily be prevented (Ozone Systems 2014).
- Ozone has even been successfully used to treat bovine mastitis [3]. When 6-30 mg of ozone was infused into each inflamed udder quarter of cows with acute clinical mastitis, 60 per cent of sick animals completely recovered without administration of antibiotics. Ozone therapy is a safe, effective and inexpensive method of curing mastitis without leaving antibiotic residues in raw milk [3].

b. Dairy Plants [1]

Ozone has successfully been used for the removal of milk residues and biofilm forming bacteria from stainless steel surfaces and in milk processing, including fluid milk, powdered milk products and cheese [1, 2]. Ozone can be used in dairies as a gas which sterilises the air, or in ozonized water as a more effective and healthier alternative to chlorinated water. Details are given below [1]:

i. Air Disinfection [1]

Ozone has been compared with chemical fogging and UV exposure as different alternatives to air disinfection. Ozone was shown to be both effective and reproducible in its effects. UV was effective, but shadowing posed a problem in a factory environment. Fogging was shown to be the least effective method of air disinfection.

ii. Cleaning and Sanitation [1]

The effectiveness of warm water (40°C) and ozonated cold water (10°C) as a prerinse for removing dairy soil from stainless steel plates was quantified and visualized [4]. Ozonated water removed 84 per cent of milk residues from plates, whereas the nonozonated warm water treatment removed only 51 per cent of dairy soil materials, but the two values did not differ significantly ($P > 0.05$) [2]. There are many different lab reports, which show the effectiveness of ozone over routine cleaning methods. Swabs taken from food processing tables, warehouses or area after routine cleaning with a chemical cleaning solution, compared with swabs from ozonated area. These tests show ozone has a kill rate of 99 per cent. Bacterial counts after routine cleaning methods showed their ineffectiveness in the prevention of cross- contamination and spoilage. Ozonated water has been shown as an effective alternative to chlorinated water in the disinfection of dairy equipment. Not only does it destroy surface-attached bacteria, but requires no heat, and therefore uses less energy than sanitising systems which use steam or hot water [1]. The use of ozonated water can only be recommended to replace warm water and chlorine for cleaning and sanitising purposes, respectively, if the surfaces of dairy processing equipment are not adversely affected [2].

iii. Sanitisation of Processing Hall [1]

- ☆ Using ozone in the production area to control the growth of yeast and mould.
- ☆ The ozone generator runs for 3 hours, flooding the area overnight with the help of 2 fans.
- ☆ The ozone is allowed to degrade for 2-3 hours before the staff come in for the morning shift.
- ☆ Have had great success in keeping down the levels of mould and bacteria.

iv. Quality of Milk and Milk Products [1, 2]:

a. Milk

Bacteria levels on the upper surface of milk are reduced significantly [1].

- ☆ A method was patented [5] for the mild ozone treatment of liquids, including milk and fluid dairy foods, thereby minimizing their possible quality deterioration.
- ☆ The pressurised ozone (5-35 mg/L for 5-25 min) was used [6] to preserve skim milk by decreasing its microbial populations. The treatment was shown to reduce the number of psychrotrophs by more than 99 per cent.

b. Proprietary Powdered Milk

Lab test reports revealed that the use of ozone on proprietary powdered milk showed the control sample bacteria levels 1000 times higher than on the sample treated with ozone [1].

Milk Powders [2]

Cronobacter spp., previously known as Enterobacter sakazakii are the organisms, frequently isolated from milk powder and the environments of dried milk production facilities [2, 7, 8].

1. Ozonation was shown [2, 8] to be an effective method of destroying Cronobacter sakazakii ATCC 51329 cells in milk powders, especially in dried skim milk. Whole and skim milk powder samples were exposed to gaseous ozone at concentrations of 2.8 mg/L or 5.3 mg/L for 0.5-2 h. Both ozone levels reduced Cronobacter counts in skim milk powder by approximately 3 $\log_{10}$ orders following 120 min of exposure.

   As fat effect the effectiveness of ozone and therefore a reduction of only <2 $\log_{10}$ units was observed in the viability of C. sakazakii in whole milk powder after 2 h of gaseous ozonation.
2. Ozone treatments may also influence the chemical, physical, functional and organoleptic properties of dried milk products:
   - ☆ Spray-dried skim milk powders manufactured under a background ozone level of 32 ppb received significantly lower sensory scores from

a trained taste panel than those produced in air containing 2 ppb of ozone [2, 9].

- ✰ Whole milk powders were shown to suffer more ozone damage than skim milk powders, suggesting that reactions between milk fat and ozone were responsible for the off-flavour produced [9].

Whey Protein Isolates [2]

Whey protein isolates were treated with gaseous and aqueous ozone [10] and found that ozonation substantially enhanced the foaming capacity and foam stability of proteins. Gaseous ozone treatments decreased the solubility of protein samples to a greater degree than did aqueous ozonation.

c. *Cheese*

Ozonation has been shown to prevent mould growth on cheese and inactivate airborne moulds in cheese ripening and storage facilities. Exposure to low levels of ozone prevents the development of new mould growths on the sides and ends of unwaxed cheeses. Free mould spores significantly reduced in rooms treated with ozone. The use of ozone does not affect the flavour of any of the cheeses. 0.1mg/ M3 of ozone in a cheese ripening room inactivates 80 per cent of the mould spores without affecting the organoleptic qualities of the cheese. Using ozone to flood the production area over night in order to kill *staphylococcus aureus* [1].

- ✰ The high-ozone and low-ozone treatments also reduced the mean mould spore counts in the curing rooms by 94 per cent and 88 per cent, respectively [2].
- ✰ Two ozone concentrations were employed to combat a well-established mould growth and inhibit mould development on Cheddar cheese [2, 11].
- ✰ The usability of various plastic films for cheese packaging purposes was tested [12]. It was found that thick films (37-480 lm) were not permeable to ozone, whereas thin films (12 lm), because of their ozone permeability, were suitable for ozone-based surface sterilisation of packaged cheese.
- ✰ The study on the effect of ozone ttreatment on different types of cheeses *e.g.* Russian- and Swiss type cheeses, Italian cheese (*i.e.* Ricotta Salata di Pecora, Taleggio PDO and Gorgonzola PDO), Brazilian fresh cheese-Minas Frescal, mozzarella cheese, were also made.

d. *Yoghurt*

Use an ozone generator to control bacteria spore levels on equipment in the processing area overnight. Significant reduction in spore counts. Flooding the premises with ozone helps in reduction of micro-organism.

e. *Storage*

Ozone treatment helps in storage of commodities such as cheese, bacon, butter and vegetables for long periods in the same transport compartment or area without any cross- contamination.

v. ***Wastewater Treatment in Dairy Processing [2]***

Dairy processing plants use huge amounts of water and, as a consequence, generate large quantities of wastewater, the latter being characterized by high organic contents. Dairy wastewaters are conventionally purified with physico-chemical and biological methods but now ozonation was employed.

- ☆ Ozonation was shown to decrease the initial fat content of 80-230 mg/L by 96-98 per cent, thereby enabling the lightly polluted effluent to be discharged into natural water courses [13].
- ☆ Researchers [14] studied the possibility of using ozone for oxidising biologically pretreated cheese whey wastewaters to obtain effluents dischargeable to natural aquatic systems.
- ☆ It was also [2] found that ozone treatment to be a very promising method for reducing the concentrations of organic pollutants in dairy wastewaters. Owing to its microflocculation effect, ozonation increased the removal of COD from dairy wastewater samples during subsequent nanofiltration.

Sources

1. *Ozone Use in the Dairy, Eco-Logical Technology, Ozone Systems-Domestic and Industrial, Durban, South Africa.*
2. *Laszlo Varga and JenoSzigeti, Use of ozone in the dairy industry: A review, International journal of Dairy Technology, 2016, 69, 2, Pages 157-168.*
3. *Ogata A and Nagahata H, 2000, Intra-mammary application of ozone therapy to acute clinical mastitis in dairy cows. Journal of Veterinary Medical Science 62 681-686.*
4. *Guzel-Seydim Z B, Wyffels J T, Greene A K and Bodine A B, 2000, Removal of dairy soil from heated stainless steel surfaces: use of ozonated water as a prerinse. Journal of Dairy Science 83 1887-1891.*
5. *Sander M, 1985, Process for the gentle ozone treatment of liquids, such as fruit juices, milk, liquid milk products, wine, oils, liquid medicaments, blood and/or similar products (In German). Patent No. DE 3325 568A1.*
6. *Rojek U, Hill A R and Griffiths M, 1995, Preservation of milk by hyperbaric ozone processing. Journal of Dairy Science 78(suppl.1), 125.*
7. *Kandhai M C, Reij M W, Gorris L G M, Guillaume-Gentil O and van Schothorst M, 2004, Occurrence of Enterobacter sakazakii in food production environments and households. Lancet 363 39-40.*
8. *Torlak E and Sert D, 2013, Inactivation of Cronobacter by gaseous ozone in milk powders with different fat contents, International Dairy Journal 32 121-125.*
9. *Kurtz F E, Tamsma A, Selman R L and Pallansch M J, 1969, Effect of pollution of air with ozone on flavor of spray-dried milks. Journal of Dairy Science 52 158-161.*
10. *Uzun H, Ibanoglu E, Catal H and Ibanoglu. S, 2012, Effects of ozone on functional properties of proteins. Food Chemistry 134 647-654.*

11. *Gibson C A, Elliott J A and Beckett D C, 1960, Ozone for controlling mold on Cheddar cheese. Canadian Dairy and Ice Cream Journal 39 24-28.*

12. *Volodin V I and Shiler G G, 1978, Method for determination of ozone permeability coefficient of polymer films (In Russian). Trudy, Vsesoyuznyi Nauchno-issledovatel'skii Institut Maslodel'noi i Syrodel'noi Promyshlennosti 24 100-103, 110.*

13. *Loorits K A, Munter R R, Siirde E K and Lisenkova L L, 1975, Use of ozone for oxidation of major milk components in effluent (In Russian). Molochnaya Promyshlennost' 4 27-30.*

14. *Martins R C and Quinta-Ferreira R M, 2010, Final remediation of postbiological treated milk whey wastewater by ozone. International Journal of Chemical Reactor Engineering 8 A142.*

G. Application of Ozone in other Industries [1]

The other applications are given below:

- ☆ **Air Treatment:** To purify the atmospheres contaminated with volatile organic compounds and micro-organisms.
- ☆ **Water Treatment:** To reduce in great amount the chlorine use, without form chlorinated organ compound [2].
- ☆ **Medicine/dentistry:** Like active medicine and dental surgery, presenting and displaying viral inactivation, bactericidal and fungicide effect [3, 4].
- ☆ **Industrial Processes:** To reduce the use of chemical agents, as chlorine; to reduce the residual pesticides on food [5].

Sources

1. *Alex Augusto Gonçalves, Ozone - an emerging technology for the seafood industry, Braz. arch. biol. technol. vol.52 no.6 2009.*

2. *Rice, R. G., 1997, Applications of ozone for industrial wastewater treatment. Ozone: Science and Engineering, 18, 477-515.*

3. *Azarpazhooh, A. and Limeback, H. 2008, The application of ozone in dentistry: A systematic review of literature. Journal of Dentistry, 36, 104-116.*

4. *Bocci, V., 2005, Is it true that ozone is always toxico, The end of a dogma, Toxicol. Appl. Pharmacol., 216 (3), 493-504.*

5. *Brooks, G. M. and Pierce, S. W., 1990, Ozone applications for commercial catfish processing. Paper presented at 15th Annual Tropical and Subtropical Fisheries Technological Conference of the Americas, December 2-5, Orlando, Florida.*

Chapter 18

Nanotechnology in Food and Food Processing

Nanotechnology [1] is based on the prefix "nano", a Greek word meaning "dwarf". It is the understanding and control of matter at dimensions of roughly 1 to 100 nanometers [2]. To be more specific, nanotechnology is defined as the design, production and application of structures, devices, and systems through control of the size and shape of the material at the nanometer (10^{-9} of a meter) scale where unique phenomenon enable novel application [3].

- ☆ The term "Nanotechnology" was first used in 1974 by the late Norio Taniguchi and concepts were given by Richard Feynman in 1959 [2a].
- ☆ In 2006 Food and Drug Administration (FDA) defined nanomaterials as "Particles with dimensions less than 1 micrometre scale that exhibit unique properties" [2b]. As the size of the materials transit from micro-scale to nano-scale, there are dramatic increase in the surface area and number of energically unstable molecules which have significant relationship with functionality.
- ☆ United states Department of Agriculture (USDA) was first to address the usefulness of Nanotechnology in agricultural and food industries in its road map published in September 2003 [2c].
- ☆ Over the past few decades, nanotechnology has increasingly been considered as to be attractive technology that has revolutionized the food sector. It is a technology on the nanometer scale and deals with the atoms, molecules, or the macro-molecules with the size of approximately 1-100 nm to create and use materials that have novel properties. [3a].

- ✰ This technology has already revolutionized the health care, textile, information technology, and energy sectors *etc.* and has been well publicized [4].
- ✰ Several products enabled by nanotechnology are already in the market, such as antibacterial dressings, transparent sunscreen lotions, light-diffracting cosmetics, penetration enhanced moisturizers, stain and odour repellent fabrics, scratch free paints for cars, and self cleaning windows, dirt repellent coatings, long lasting paints and furniture varnishes, and even some food products [5].
- ✰ Tomorrow food will be designed by shaping molecules and atoms. Nanoscale biotech and nano-bio-info will have big impacts on the food and food-processing industries. The future belongs to new products, new processes with the goal to customize and personalize the products. Improving the safety and quality of food will be the first step [6].

Nanotechnology has been described as the new industrial revolution and both developed and developing countries are investing more in this technology. The government of India established the **Nanoscience and Technology Initiative** in the later part of the 2001 through Department of Science and Technology (DST), New Delhi and invested about Rs. 350 Crores (2002-06) and granted approval for the Nanomission worth Rs. 1000 crores for next five years [7].

According to a definition in a recent report, food is nanofood when nanaoparticles, nanotechnology techniques or tools are used during cultivation, production, processing, or packaging of the food. It does not mean atomically modified food or food produced by nanomachines. Although there are ambitious thoughts of creating molecular food using nanomachines, this is unrealistic in the foreseeable future [6]. Most of the food giants including Nestle, Kraft, Heinz, and Unilever support specific research programmes to capture a share of the nanofood market in the next decade [6].

The importance of nanotechnology in food processing can be evaluated by considering its role in the improvement of food products in terms of:

1. Food texture.
2. Food appearance.
3. Food taste.
4. Nutritional value of the food.
5. Food shelf-life.

It is a matter of fact that surprisingly nanotechnology not only touches all the above-mentioned aspects but has also brought about significant alterations in food products providing them novel qualities [3a].

The benefits of nanotechnology for the food industry are many and are expected to grow with time. Some of the important aspects of nanotechnology in food sector are given on next page [8]:

- This new, rapidly developing technology impacts every aspect of the food system **from cultivation to food production to processing, packaging, transportation, shelf life and bioavailability of nutrients**.
- Commercial applications of nanomaterials will continue to impact the food industry because of their unique and novel properties.
- Human exposure to nanomaterials is increasing and will continue to increase with time. Therefore, the health impact of nanomaterials in food is of public interest and concern.
- Public acceptance of food and food-related products containing nanomaterials will depend on their safety. Consequently, a uniform international regulatory framework for nanotechnology in food is necessary.

Sources

1. *Qureshi, Mehar Afroz, Karthikeyan, Swaminathan, Punita Karthikeyan, Khan, Pervez Ahmed, Uprit, Sudhir, and Mishra, Umesh Kumar, Application of nanotechnology in food and dairy processing: An overview, PAK. J. Food Sci., 22 (1), 2012: 23-31, ISSN: 2226-5899.*

2. *Pehanich, M. 2006, Small gains in processing, packaging. Food Processing, 46-48.*

2a. *Warad H. C. and Dutta J., 2005, Nano-technology for agriculture and food systems a view. Available: http: //www. nano. ait.ac. th/Download/AIT per cent 20 Papers/2005/ Nanotechnology/View.pdf.*

2b. *Miller G. and Senjen R. 2008, Out of the laboratory and on to our plates - Nano-technology in food and agriculture. Available: http: //www.foeeurope. org/activities/ nanotechnology/Documents/Nano_food_report. Pdf.*

2c. *Joseph T. and Morrison M., 2006, Nanotechnology in Agriculture Food: A Nano forum report published in 2006. Available: www. nanoforum.org.*

3. *Ravichandran, R. and Kala, P. Sasi. 2006, Nanoscience and nanotechnology: perspectives and overview. School Sci., 43-49.*

3a. *Trepti Singh, Shruti Shukla, Pradeep Kumar, Verinder Wahla, Vivek K. Bajpai, and Irfan A. Rather, Application of Nanotechnology in Food Science: Perception and Overview, Front Microbiol. 2017; 8: 1501.*

4. *Kumar, O. and T. Rai. 2009. Nanotechnology in Milk and its Products. Indian Dairyman, 37-40.*

5. *Miller, G. 2008. Nanotechnology- the new threat to food. Global Research cited from http: //www.globalresearch.ca/index.php?context=va and aid=10755.*

6. *Ranjan Saha, New and Emerging Non-Thermal Technologies in Food Preservation, April 2015.*

7. *Patra, Debasmita, Haribabu, Ejnavarzala and Prajit K. Basu.2009. Nano-science and Nanotechnology: ethical, legal, social and environmental issues. Current Science, 96(5): 651-657.*

8. *A. Wallace Hayes and Saura C. Sahu, Nano-technology in the Food Industry: A Short Review, NANOTECHNOLOGY, February/March 2017.*

Natural Nano Food [1]

Basic component of food (Protein, Fat, carbohydrates) are very useful in the engineering of nano foods through various approach [2]. Newly developed nano foods (Through nanotechnology) exhibit various nutritional, functional and sensory functions differ from previous one. The knowledge related to naturally occurring nanostructures in the food is much important to improve the functional behaviour of the food [3]. Some examples of Natural Nano foods are given below:

1. The crystalline structures in starch have influence in the gelatinization and nutritional quantity of processed starch-based foods which can be manipulated through Nano technology [4].
2. Denatured components of milk protein (β-lactoglobulin) obtained by various denaturing factors like pressure, heat, drastic change pH; are compiled and framed even larger gel networks and the best β-lactoglobulin.
3. α-lactalbumin is also used to develop nanotubes through hydrolyzation by bottomup approach [6].
4. Homogenization and fine grating processes create favourable environment for the synthesis of nano engineered products.
5. Decreased size of fat globule is very useful in synthesis of liposomes.
6. In nano technological storm, casein micelles, fat globules, whey proteins are very useful in dairy industry to formulate micro sized and nanosized structures to achieve desirable dairy products [5].
7. Milk is an example of an emulsion of nanoscale particles, and the industry uses nano-sized structures to create emulsions (butter), foams (ice cream), solids (cheese) and gel networks (yoghurt) [7].

Sources

1. *Chavada P. J, Chudasama M. M, Bariya A. R., Gadhvi Y. H., Deokar S. S.and Sadhu D. B, Novel Application of Nanotechnology in Dairy and Food industry- Nano Inside.*
2. *Morris V. J. and R. Parker R. 2008, Natural and designed self- assembled nanostructures in foods. The World of Food Science: Food Nanotechnology. Available: http: //www. worldfoodscience. org/cms.*
3. *Shekhon B. S. 2010, Nanotech. Sci. Appl., 3, 1.*
4. *Rudolp M. J. 2004, Food Technol, 58, 32-34.*
5. *Momin J. K., Chitra J. and Prajapati J. B. 2012, Emir. J. Food Agric., 25(1), 10-19.*
6. *Bugusu, Mejia B. C., Magnuson B. and Tafazoli S. 2009, Food Technology (Chicago), 63(5), 24-28.*
7. *FSANZ 2015, Nanotechnology and Food.*

Approaches for Application [1]

Nanotechnology can be applied by two different approaches either "bottom up" or "top down." in food and dairy processing [2].

1. **The topdown approach:** The "topdown" approach involves converting large size materials into nanometre size range by employing various mechanical processes such as grinding, milling, etching and lithography [1a, 3]. Some examples are given below:
   i. It involves a physical processing of the food materials such as dry-milling of wheat into fine flour that has a high water-binding capacity.
   ii. The antioxidant activity in green tea powder is improved by when the size of the powder is reduced to 1000 nm, the digestion and absorption resulted in an increase in the activity of an oxygen-eliminating enzyme [3].
2. **The Bottom Up approaches:** Bottom-up approach of nanotechnology was derived from the basic concept of biology like self-assembly and self-organization. The self-assembly and self-organization are concepts derived from biology that have inspired a bottom-up food nanotechnology. For example:
   i. Self-assembly structures through organization of casein micelles or starch.
   ii. The folding of globular proteins and protein aggregates which create stable entities to form nanometer scale via self organization [4].

Sources

1. *Qureshi, Mehar Afroz, Karthikeyan, Swaminathan, Punita Karthikeyan, Khan, Pervez Ahmed, Uprit, Sudhir, and Mishra, Umesh Kumar, Application of nanotechnology in food and dairy processing: An overview, PAK. J. Food Sci., 22 (1), 2012: 23-31, ISSN: 2226-5899.*

1a. *Chavada P. J, Chudasama M. M, Bariya A. R., Gadhvi Y. H., Deokar S. S.and Sadhu D. B, Novel Application of Nanotechnology in Dairy and Food industry- Nano Inside.*

2. *Ravichandran, R. 2010. Nanoparticles in drug delivery: potential green nanobiomedicine applications. Int. J. Nanotechnol. Biomed. 1: 108-130.*

2a. *Shibata T., 2002, Method for producing green tea in microfine powder." United States Patent US, 6416803B1.*

4. *Dickinson, E.and Van Vliet, T.2003. Food Colloids Biopolymers and Materials Royal Society of Chemistry, London.*

Food Safety and Regulation of Nanotechnologies [1]

There are few issues, particularly regarding the accidental or deliberate use of nanoparticles in food, or food-contact materials, that consumers are concerned about the potential negative effects of nanotechnology-based delivery systems on human health and also regulatory stands. There is an urgent need for regulation of nanomaterials before their incorporation into food and dairy processing including packaging. Nano-materials must not cause any health risks for consumers or to the environment [1].

a. Food Safety [1]

Consumers are thrilled at the exciting food and dairy products emerging through the application of nanotechnology. Consumers are exposed to nanomaterials by consumption of food and beverages containing these extremely small particles of large reactive surface area of unknown safety. Once absorbed in the gastrointestinal system, they may bioaccumulate in various organs of the body, leading to potentially adverse effects. Thus, application of nanotechnology by the food industry is of public concern. Public acceptance of food and food products containing nanomaterials depends on their perceived safety. There is a serious question about safety that will require attention of the policy makers as well as the food industry.

The nanomaterials (increased contact surface area), might have toxic effects in the body that are not apparent in the bulk materials [1a]. A wide variety of food and nutrition products containing nanoscale additives are already in the market (*e.g.* iron in nutritional drink mixes, micelles that carry vitamins, minerals, and phytochemicals in oil, and zinc oxide in breakfast cereals *etc.*) because of the lack of regulations and risk knowledge.

The nanoparticles are more reactive, more mobile, and likely to be more toxic. This toxicity is one of the important issues must be addressed. There is strong possibility that nanoparticles in the body can result in increased oxidative stress that, in turn, can generate free radicals, leading to DNA mutation, cancer, and possible fatality. The ingredients in these nanoparticles must undergo a full safety assessment by the relevant scientific advisory association before these are permitted to be used in the dairy and food products including packaging (U. K. RS/RAE, 2004).

b. Regulations [1]

The effect of food processing techniques that produce nanoparticles and nanoscale emulsions on human health, warrant the attention of food regulations. The potential for such foods to pose new health risks must be investigated in order to determine whether or not related new food safety standards are required [2]. The details related to the regulations being recommended/imposed in different countries are mentioned below:

- ✰ The European Union regulations for food and food packaging have recommended that for the introduction of new nanotechnology, specific safety standards and testing procedures are required [3].
- ✰ In March 2009, the scientific committee of the European Food Safety Agency published an opinion on nanoscience and nanotechnology regarding food and animal feed safety [3a, 3b].

- ☆ A guidance document on how to assess potential risks associated with certain food-related uses of nanotechnology followed in May 2011, providing practical recommendations to regulators on how to assess applications from industry to use engineered nanomaterials in food additives, enzymes, flavorings, food contact materials, novel foods, food supplements, feed additives and pesticides [3a].
- ☆ In the United States, nanofoods and most of the food packaging are regulated by the United States Food and Drug Administration (US FDA) [4].
- ☆ In Australia, nanofood additives and ingredients are regulated by Food Standards Australia and New Zealand (FSANZ), under the Food Standards Code [5].
- ☆ However, no internationally accepted standard protocols for toxicity testing of nanomaterials in food or feed are currently available. Such protocols are in the development stage by organizations such as the International Alliance for Nano Environment, Human Health and Safety Harmonization and the U. S. National Research Council [3a]. There is an urgent need for a common and uniform international regulatory system capable of managing any risks associated with nanofoods and the use of nanotechnologies in dairy and food industry. Governments must also respond to nanotechnology's broader social, economic, civil liberties and ethical challenges.
- ☆ To ensure democratic control of these new technologies in the important area of food and dairy, public involvement in nanotechnology decision making is essential (U. K. RS/RAE, 2004).

Sources

1. *Qureshi, Mehar Afroz, Karthikeyan, Swaminathan, Punita Karthikeyan, Khan, Pervez Ahmed, Uprit, Sudhir, and Mishra, Umesh Kumar, Application of nanotechnology in food and dairy processing: An overview, PAK. J. Food Sci., 22 (1), 2012: 23-31, ISSN: 2226-5899.*

1a. *Dowling, A. P.2004. Development of nano-technologies. Mater. Today. 7: 30-35.*

2. *Bowman, D. and G. Hodge. 2007. A Small Matter of Regulation: An International Review of Nanotechnology Regulation. Columbia Sci Technol Law Rev. 8: 1-32.*

3. *Halliday, J. 2007. EU Parliament votes for tougher additives regulation. FoodNavigator. com 12.07.2007.*

3a. *A. Wallace Hayes and Saura C. Sahu, Nanotechnology in the Food Industry: A Short Review, NANOTECHNOLOGY, February/March 2017.*

3b. *www.efsa.europa.eu/en/topics/topic/nanotechnology.*

4. *Badgley, C., J. Moghtader, E. Quintero, E. Zakem, L. M. Chappel, K. Aviles-Vazquez, A. Salon and I. Perfecto. 2007. Organic agriculture and the global food supply. Renew Ag. Food Sys. 22 (2): 86-108.*

5. *Bowman, D. and G. Hodge. 2006. Nanotechnology: Mapping the wild regulatory frontier. Futures. 38: 1060-1073.*

Nanotechnology in the Food Industry [1]

The use of nanotechnology provides many opportunities for innovation in the food sector, with potential applications in primary production, food manufacturing and preservation, and packaging. For the food sector, these include [1c]:

- ✰ Nanostructured food products (including emulsions).
- ✰ Nanoencapsulation of food additives (colours, preservatives, flavourings and supplements)
- ✰ Enhanced food packaging (improved mechanical, barrier and antimicrobial properties)
- ✰ Nanocoatings on food equipment (non-stick and easier to clean).

For example, silver has a long history of use as an antimicrobial substance. So it is not surprising that silver nanoparticles have been found to be potent agents against numerous species of pathogenic bacteria. Hence its potential for incorporation as a nanoparticle into polymers to form functional antimicrobial coatings and packaging.

The applications of nanotechnology in food sector can be summarized in following two main groups :

i. Food Nanostructured Ingredients

Food nanostructured ingredients encompass a wide area from food processing to food packaging. In food processing, theses nanostructures can be used as food additives, carriers for smart delivery of nutrients, anti-caking agents, antimicrobial agents, fillers for improving mechanical strength and durability of the packaging material, *etc.* The nanostructured food ingredients are being developed with the claims that they offer improved taste, texture, and consistency.

ii. Food Nanosensing

Food nanosensing can be applied to achieve better food quality and safety evaluation [1a].

Nanotechnology increasing the shelf-life of different kinds of food materials and also help brought down the extent of wastage of food due to microbial infestation [1b].

In food and dairy industries, the applications of nanotechnology include the followings:

1. Nanoparticulate Delivery Systems (Nanodispersions and Nanocapsules).
2. Packaging (Nanolaminates, Nanocomposites bottles, bins with silver nano-particles).
3. Food Safety and Biosecurity (Nanosensors) *etc.* [2].

The nanotechnology will play an vital role in the food and dairy processing in near future and would involve two forms of nano food applications *viz.*:

1. **Food additives (nano inside)**: The nanoscale food additives may be used to influence texture, flavour, nutritious improvement, provide functionally and even detect pathogens.
2. **Food packaging (nano outside):** Food packaging involves extend food shelf life, edible, nano wrapper which will envelope foods, preventing gas and moisture exchange, 'smart' packaging (containing nano-sensors and anti-microbial activators) for detecting food spoilage and releasing nano-anti-microbes to extend food shelf life [3].

A-The details related to the Nano inside and Nano outside application are tabulated below [1]:

Sl. No.	*Forms and Sub Form of Application*	*Details*
1	**Nano inside** **a. Nanodispersions and Nanocapsules**	Nowadays nanocarriers are being utilized as delivery systems to carry food additives in food products without disturbing their basic morphology. Particle size may directly affect the delivery of any bioactive compound to various sites within the body as it was noticed that in some cell lines, only submicron nanoparticles can be absorbed efficiently but not the larger size micro-particle[1a]. Functional ingredients (drugs, vitamins,antimicrobials, antioxidants, flavorings, colorants and preservatives *etc.*) come in different molecular and physical forms. These ingredients are rarely utilized directly in their pure form; instead, they are often incorporated into some form of delivery system. A delivery system must perform following different roles [4]- i. It serves as a vehicle for carrying the functional ingredient to the desired site of action. Ensure availability at a target time and specific rate. ii. It may have to protect the functional ingredient from chemical or biological degradation during processing, storage, and utilization(the functional ingredient in its active state). iii. It may have to be capable of controlling the release of the functional ingredient(the release rate or the specific environmental conditions-pH, ionic strength, or temperature). iv. The delivery system has to be compatible with the other components in the system, as well as being compatible with the physico-chemical and qualitative attributes of the final product. In order to achieve above said objectives, a number of potential delivery systems based on nanotechnology could be used as under: **a. Association colloids:** Surfactant micelles,vesicles, bilayers, reverse micelles, and liquid crystals are some examples of association colloids which have been used to encapsulate and deliver polar, nonpolar, and amphiphilic functional ingredients. **b. Biopolymeric nanoparticles:** Polylactic acid (PLA) a common biodegradable nanoparticle is often used to encapsulate and deliver drugs and micronutrients like iron, vitamin, protein *etc.* It has shown that the PLA need an associative compound such as polyethylene glycol for successful results and the functional ingredients can be encapsulated in nanoparticles and released in response to specific environmental triggers [5].

Sl. No.	Forms and Sub Form of Application	Details
		c. **Nanoemulsion:** Emulsions are often referred to as "nanoemulsions", when the use of high-pressure valve homogenizers or microfluidizers often causes emulsions with droplet diameters of less than 100 to 500 nm and functional food components can be incorporated within the droplets, the interfacial region, or the continuous phase. It is possible to develop smart delivery systems by engineering the properties of the nanostructured shell around the droplets. This interfacial engineering technology would utilize food-grade ingredients and processing operations that are already widely used in the manufacture of food emulsions [4].
	b. Nanofibers	Since nanofiberes are usually not composed of foodgrade substances, they have only a few potential applications in the food industry [4]. If in future,it will be possible to manufacture Nano fibres from food grade biopolymer, then its application may increase. Electrospinning is a manufacturing technology capable of producing thin, solid polymer strands (nanofiberes) from solution by applying a strong electric field to a spinneret with a small capillary orifice. The food industry can use electrospun microfibers in following ways: i. As a building/reinforcement element of composite green (that is, environmentally friendly) food packaging material, ii. As building elements of the food matrix for imitation/artificial foods. iii. As nanostructured and microstructured scaffolding for bacterial cultures.
	c Nanotubes	Carbon nanotubes have been used nonfood application. The structures have been used as low resistance conductors or catalytic reaction vessels among other uses. It has have reported [6] that certain globular proteins from milk (such as hydrolyzed α-lactalbumin) can be made to self assemble to form nanotubes under appropriate conditions. In general protein hydrolysis increases the digestibility of protein. Furthermore α-lactalbumin has important nutritional value. A nanotube made by food/dairy proteins or their derivatives have so far only been reported for α-lactalbumin.
	d. Nanocapsules	Because of the considerable demand,scientists were interested in exploring the potential of nanotechnology in encapsulation and delivery of biologically active substances into targeted tissues, enhance the flavour and other sensory characteristics of food and dairy products [1]. Nanoparticles have better properties for encapsulation and release efficiency than traditional encapsulation systems. Nanoencapsulations mask odors or tastes, control interactions of active ingredients with the food matrix, control the release of the active agents, ensure availability at a target time and specific rate, and protect them from moisture, heat, chemical, or biological degradation during processing, storage, and utilization, and also exhibit compatibility with other compounds in the system [4]. Casein micelle (CM) plays a role as natural nano-capsular vehicle for nutraceuticals. The CM is important due to their biological activity, good digestibility. The general approach is to develop nanosized carriers or nanosized materials, in order to improve the absorption and, hence, potentially the bioavailability of added materials such as vitamins, phytochemicals, nutrients, or minerals. The materials can be incorporated into solid foods, delivered as liquids in drinks, or even sprayed directly on to mucosal surfaces [1].

<table>
<tr><th>Sl. No.</th><th>Forms and Sub Form of Application</th><th>Details</th></tr>
<tr><td></td><td></td><td>Various synthetic and natural polymer-based encapsulating delivery systems have been elaborated for the improved bioavailability and preservation of the active food components. The details are given below [1d]:
<table>
<tr><th>Nano-technique</th><th>Characteristic Feature</th><th>Examples</th></tr>
<tr><td>Nano-emulsions</td><td>i. Greater stability to droplet aggregation and gravitational separation [14a],
ii. Higher optical clarity
iii. Inceased oral bioavailability [14b].</td><td>β-Carotene-based nano-emulsion [14a,14b]</td></tr>
<tr><td>Polymeric micelles</td><td>Solubilize water-insoluble compounds in the hydrophobic interior, high solubility, low toxicity.</td><td>PEO-b-PCL [poly (ethylene glycol)block-poly (caprolactone)] polymeric micelles[14c].</td></tr>
<tr><td>Edible coatings</td><td>To preserve the quality of fresh foods during extended storage.</td><td>Gelatin-based edible coatings containing cellulose nanocrystal [15], Chitosan/nano-silica coatings [15a], Chitosan film with nano-SiO_2 [15b].</td></tr>
<tr><td>Hydrogels</td><td>Can be easily placed into capsules, protects drugs from extreme environments and to deliver them in response to environmental stimuli such as pH and temp.</td><td>Protein hydrogels[16].</td></tr>
</table></td></tr>
<tr><td>2</td><td>Nano outside</td><td>Food Packaging (Nano Outside)-Food packaging is considered to be one of the earliest commercial applications of nanotechnology in the food sector. The significant purpose of nano-packaging is to set longer shelf life by improving the barrier properties of food pack-aging to reduce gas and moisture exchange and UV light exposure [7]. Du Pont has announced the release of a nano-titanium dioxide plastic additive (" light stabilizer 210"), which could reduce UV damage of foods in transparent packaging [8].</td></tr>
<tr><td></td><td>Nano-Coatings</td><td>Nanotechnology has enabled the development of nanoscale edible coatings as thin as 5 nm wide, which are invisible to the human eye. Edible coatings and films are currently used on a wide variety of foods, including fruits, vegetables, meats, chocolate, cheese, candies, bakery products, and French fries [9,10]. These coatings or films could serve as moisture, lipid, and gas barriers. Alternatively, they could improve the textural properties of foods or serve as carriers of functional agents such as colors, flavors, antioxidants, nutrients, and antimicrobials and could also increase the shelf life of manufactured foods.</td></tr>
<tr><td></td><td>Nanosensors</td><td>Packaging equipped with nano-sensors is also designed to track either the internal or external conditions of food products, pellets and containers, throughout the supply chain. United State used modified carbon nanotube as biosensor to detect micro-organisms, toxic substances and spoilage of foods or beverages [11].</td></tr>
</table>

Sl. No.	Forms and Sub Form of Application	Details
		Nanosensors in plastic packaging can detect gases given off by food when it spoils and the packaging itself changes color to alert you. These films are packed with "silicate nanoparticles" to reduce the flow of oxygen into the package and the leaking of moisture out of the package to stay food fresh. Nanosensors are being developed that can detect bacteria and other contaminates such as salmonella on the surface of food at a packaging plant. Nanotechnology is also enabling sensor packaging to incorporate cheap Radiofrequency Identification (RFID) tags. The nano-enabled RFID tags are much smaller, flexible and can be printed on thin labels. This increases the tags versatility and thus enables much cheaper production. United States company Oxonica Inc, has developed [12] nano-barcodes to be used for individual items or pellets, which must be read with a modified microscope for anti- counterfeiting purposes. Another trend in the application of nanopackaging is the nano-biodegradable packaging.
	Nanolaminates	Nanotechnology provides food scientists with a number of ways to create novel laminate films suitable for use in the food and dairy industry. A nanolaminate consists of 2 or more layers of materials with nanometer dimensions that are physically or chemically bonded to each other. It is reported that nanolaminates offer some advantages for the preparation of edible coatings and films over conventional technologies and may thus have a number of important applications within the food and dairy industry. A variety of different adsorbing substances could be used to create the different layers, including natural polyelectrolytes (proteins, polysaccharides), charged lipids (phospholipids, surfactants) and colloidal particles (micelles, vesicles, droplets).
	Clay nano particles and nano crystals	Nanoclays and nanocrystals help in improving the barrier properties of dairy and food packaging materials by incorporating as well as embedding. The plastic films and bottles containing these nanoparticles are able to block oxygen, carbon dioxide and moisture from reaching food products. The advantage of clay nanocomposite in the packaging material offers improved shelf life, shutter proof, light in weight and heat resistant [13].

The types of application include: smart packaging, on demand preservatives, and interactive foods. Building on the concept of "on-demand" food, the idea of interactive food is to allow consumers to modify food depending on their own nutritional needs or tastes. The concept is that thousands of nanocapsules containing flavour or colour enhancers, or added nutritional elements (such as vitamins), would remain dormant in the food and only be released when triggered by the consumer [14].

Sources

1. *Qureshi, Mehar Afroz, Karthikeyan, Swaminathan, Punita Karthikeyan, Khan, Pervez Ahmed, Uprit, Sudhir, and Mishra, Umesh Kumar, Application of nanotechnology in food and dairy processing: An overview, PAK. J. Food Sci., 22 (1), 2012: 23-31, ISSN: 2226-5899.*

1a. *Ezhilarasi P. N., Karthik P., Chhanwal N., Anandharama krishnan C., 2013, Nanoencapsulation techniques for food bioactive components: a review. Food Bioprocess Technol. 6, 628-647.*

1b. *Pradhan N., Singh S., Ojha N., Srivastava A., Barla A., Rai V., et al. 2015, Facets of nano-technology as seen in food processing, packaging, and preservation industry. BioMed Res. Int.2015: 365672 10.1155/2015/365672.*

1c. *FSANZ, 2015, Nano-technology and Food.*

1d. *Trepti Singh, Shruti Shukla, Pradeep Kumar, Verinder Wahla, Vivek K. Bajpai, and Irfan A. Rather, Application of Nanotechnology in Food Science: Perception and Overview, Front Microbiol. 2017, 8, 1501.*

2. *Chen, H., J. Weiss and F. Shahidi. 2006. Nano-technology in nutraceuticals and functional foods. Food Tech. 60(3), 30-6.*

3. *Richardson, N. S. M. and K. E. Piehowski. 2008, Nanotechnology in nutritional sciences. Minerva Biotechnol. 20: 17-126.*

4. *Weiss, J., Takhistov Paul, and D. Julian. Mc'clements. 2006. Functional Materials in Food Nanotechnology, Journal of food sci. 71(9): 107-116.*

5. *Riley, T., T. Govender, S. Stolnik, C. D. Xiong, M. C. Garnett, L. Illum and S. S. Davis. 1999. Colloidal stability and drug incorporation aspects of micellar like PLA-PEG nanoparticles. Colloids Surf. B.16: 47-159.*

6. *Graveland-Bikker, J. F., I. A. T. Schaap, C. F. Schmidt and C. G. de Kruif. 2006. Structural and mechanical study of a self assembling protein nanotube. Nano Lett. 6: 616-621.*

7. *Sorrentino, A., G. Gorrasi and V. Vittoria, 2007. Potential perspectives of bio-nano composites for food packaging applications. Trends Food Sci Technol, 2007, 18: 84-95.*

8. *El Amin, A, 2007, Nanoscale particles designed to block UV light. Food Production Daily.com Europe. 18 October.*

9. *Rhim, J. W. 2004. Increase in water vapor barrier property of biopolymer-based edible films and coatings by compositing with lipid materials. J Food Sci Biotechnol 13: 528-35.*

10. *Morillon, V., F. Debeaufort, G. Blond, M. Capelle and A. Voilley. 2002. Factors affecting the moisture permeability of lipid-based edible films: a review. Crit Rev Food Sci Nutr.42: 67-89.*

11. *Nachay, K.2007, Analyzing Nanotechnology. Food Technol. 1: 34-36.*

12. *Roberts, R.2007, The role of nanotechnology in brand protection, Packaging Digest January 2007.*

13. *Ravichandran, R.2010. Nano-particles in drug delivery: potential green nano biomedicine applications. Int. J. Nano-technol. Biomed. 1: 108-130.*

14. *Ranjan Saha, New and Emerging Non-Thermal Technologies in Food Preservation, April 2015.*

14a. *Kong M., Chen X. G., Kweon D. K., Park H. J., 2011, Investigations on skin permeation of hyaluronic acid based nanoemulsion as transdermal carrier. Carbohydr. Polym 86, 837-843.*

14b. Yuan Y., Gao Y., Zhao J., Mao L., 2008, *Characterization and stability evaluation of β-carotene nano-emulsions prepared by high pressure homogenization under various emulsifying conditions. Food Res. Int. 41, 61-68.*

14c. Ma Z. S., Haddadi A., Molavi O., Lavasanifar A., Lai R., Samuel J., 2008, *Micelles of poly(ethylene oxide)-b-poly (epsilon-caprolactone) as vehicles for the solubilization, stabilization, and controlled delivery of curcumin. J. Biomed. Mater. Res. A. 86 300-310.*

15. Fakhouri F. M., Casari A. C. A., Mariano M., Yamashita F., Mei L. I., Soldi V., et al., 2014, *"Effect of a gelatin-based edible coating containing cellulose nanocrystals (CNC) on the quality and nutrient retention of fresh strawberries during storage," in Proceedings of the IOP Conference Series: Materials Science and Engineering, Conference 1 2nd International Conference on Structural Nano Composites, NANOSTRUC, 2014, Vol. 64 Madrid.*

15a. Shi S., Wang W., Liu L., Wu S., Wei Y., Li W., 2013, *Effect of chitosan/nano-silica coating on the physico-chemical characteristics of longan fruit under ambient temperature. J. Food Eng.118, 125-131.*

15b. Yu Y., Zhang S., Ren Y., Li H., Zhang X., Di J., 2012, *Jujube preservation using chitosan film with nano-silicon dioxide. J. Food Eng., 113, 408-414.*

16. Qui Y., Park K., 2001, *Environment sensitive-hydrogels for drug delivery. Adv. Drug Devl. Rev., 53, 321-339.*

B. Other Important Applications

The applications of Nanotechnology can be easily understood by the following figure:

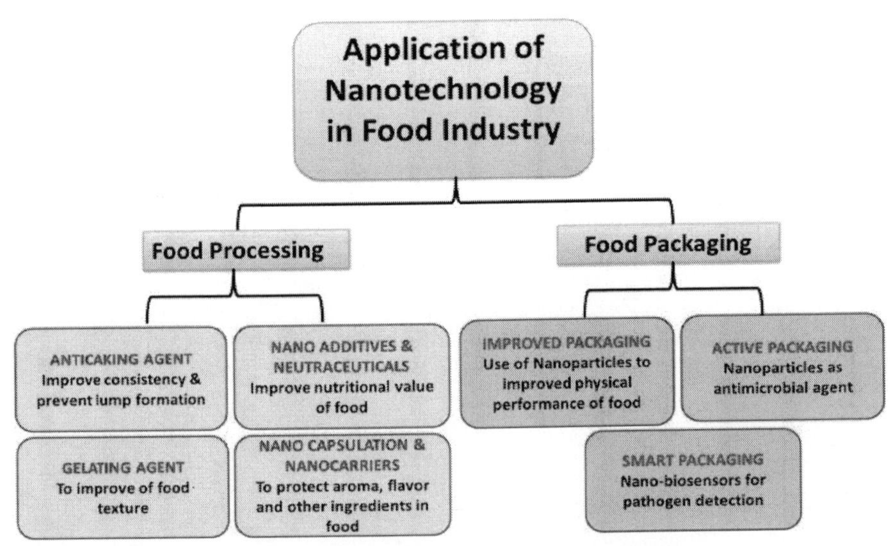

Diagram Showing Role of Nanotechnology in different Aspects of Food Sectors [5].

1. Nanoceuticals [1]

It is nano version of neutraceuticals. Neutraceuticals are framed by two words- Nutrition and Pharmaceutical. These are the substances which apart from nutritional quality, exhibit health promoting benefits. Nanoceuticals is gaining popularity and commercial dairy/food and food supplements containing nanoparticles are available [2, 3]. Some examples of food-related nanoproducts are given below:

i. Carotenoids nanoparticles can be dispersed in water, and can be added to fruit drinks for improved bioavailability.

ii. Canola oil based nanosized micellar system is claimed to provide delivery of materials such as vitamins, minerals, or phytochemicals,

iii. Patented "nanodrop" delivery systems, in the form of encapsulated materials, such as vitamins, transmucosally, rather than through conventional delivery systems such as pills, liquids, or capsules

iv. Chinese nanotea (nano-based mineral supplements) claimed to improve selenium uptake.

- ☆ A wide range of nanoceutical products containing nanocages or nanoclusters that act as delivery vehicles, *e.g.*, a chocolate drink claimed to be sufficiently sweet without added sugar or sweeteners.
- ☆ Nanosilver or nanogold are available as mineral supplements.
- ☆ To prevent the accumulation of cholesterol some of the nutraceuticals incorporated in the carriers include lycopene, beta-carotenes and phytosterols.
- ☆ A synthetic lycopene has been affirmed GRAS ("generally recognized as safe") under US FDA procedures.

2. Nanotechnology in Food Fortification [1]

The fortification of processed dairy and food products with nanoencapsulated nutrients, is being tried by Nanotech companies. The appearance and taste are boosted by nano-developed colours, their fat and sugar content removed or disabled by nano-modification, and 'mouth feel' improved. Food 'fortification' will be used to increase the nutritional claims. Some examples are given below:

- ☆ Nanotechnology will also enable junk foods like ice cream and chocolate to be modified to reduce the amount of fats and sugars that the body can absorb, by using nanoparticles to prevent the body from digesting or absorbing these components (fats and sugars) of the food.
- ☆ The nano industry could market vitamin and fibre-fortified, fat and sugar-blocked junk food as health promoting and weight reducing [4].
- ☆ The inclusion of 'medically beneficial' nano-capsules will soon enable chocolate chip cookies or hot chips to be marketed as health promoting or artery cleansing.

3. Application in Improvement of Texture, Taste, and Appearance of Food [5]

Nanotechnology gives several options to improve the food quality and to enhance food taste. Nanoencapsulation techniques have been used broadly to improve the flavor release and retention. It is also used to deliver culinary balance [6, 7] used the nanoencapsulation for the highly reactive and unstable plant pigment anthocyanins which have various biological activities.

- Rutin is a common dietary flavonoid with great important pharmacological activities. Due to poor solubility, its application in the food industry is limited.
- The ferritin nanocages encapsulation enhanced the solubility, thermal and UV radiation stability of ferritin trapped rutin as compared to free rutin [8].
- The use of nanoemulsions to deliver lipid-soluble bioactive compounds is much popular since they can be produced using natural food ingredients using easy production methods, and may be designed to enhance water-dispersion and bioavailability.
- The food industry is beginning to use nanotechnology to develop nanoscale ingredients to improve color, texture and flavor of food. Many metallic oxides such as titanium dioxide and silicon dioxide (SiO_2) have conventionally been used as color or flow agents in food items [9]. SiO_2 nanomaterials are also one of the most used food nanomaterials as carriers of fragrances or flavors in food products [10]. The nanoparticle amorphous silica [10a] is also used as food additives. TiO_2 is used as a coloring in the powdered sugar coating on doughnuts.

4. Nutritional Value [5]

The bioactive compounds such as lipids, proteins, carbohydrates, and vitamins are sensitive to high acidic environment and enzyme activity of the stomach and duodenum. Encapsulation of these bioactive compounds enables them to resist such adverse conditions and also allows them to assimilate readily in food products). The nanocomposite, nano-emulsification, and nanostructuration are the different techniques which have been applied to encapsulate the substances in miniature forms to more effectively deliver nutrients like protein and antioxidants for precisely targeted nutritional and health benefits. Polymeric nanoparticles are found to be suitable for the encapsulation of bioactive compounds (*e.g.*, flavonoids and vitamins) to protect and transport bioactive compounds to target functions [5].

5. Preservation or Shelf-Life [5]

If the bioactive components in functional foods gets degraded and eventually led to inactivation due to the hostile environment then helps and nanoencapsulation of these bioactive components extends the shelf-life of food products by slowing down the degradation processes or prevents degradation until the product is delivered at the target site.

☆ The edible nano-coatings on various food materials could provide a barrier to moisture and gas exchange and deliver colors, flavors, antioxidants, enzymes, and anti-browning agents and could also increase the shelf-life of manufactured foods, even after the packaging is opened [11]. Encapsulating functional components within the droplets often enables a slowdown of chemical degradation processes by engineering the properties of the interfacial layer surrounding them. For example, curcumin the most active and least stable bioactive component of turmeric (*Curcuma longa*) showed reduced antioxidant activity and found to be stable to pasteurization and at different ionic strength upon encapsulation [12].

6. Food Production and Packaging [13]

Nano-based "smart" and "active" food packagings confer several advantages over conventional packaging methods from providing better packaging material with improved mechanical strength, barrier properties, antimicrobial films to nanosensing for pathogen detection and alerting consumers to the safety status of food [5]. Nanomaterials used for food packaging provide following benefits [13]:

i. Improved mechanical barriers,

ii. Detection of microbial contamination and

iii. Potentially enhanced bioavailability of nutrients.

A number of nanocomposites(as an active material), polymers containing nano-particles, are used by the food industry for food packaging and food contact materials [13a]. The use of ZnO and MgO nanoparticles for food packaging has been reported [13b]. Many nanoparticles such as silver, copper, chitosan, and metal oxide nanoparticles like titanium oxide or zinc oxide have been reported to have antibacterial property [5]. Engineered water nanostructures generated as aerosols are very effective at killing foodborne pathogens such as ***Escherichia coli, Listeria* and *Salmonella*** on steel food production surfaces [13c]. As such food contact substances containing nano-materials have the potential of migrating from food packaging into food, so this technology must have regulatory compliance before it gains wide-spread acceptance in the industry.

☆ The application of nanoparticles is not limited to antimicrobial food packaging but nanocomposite and nanolaminates have been actively used in food packaging to provide a barrier from extreme thermal and mechanical shock extending food shelf-life [5].

☆ Antimicrobial nanocomposite films which are prepared by impregnating the fillers (having at least one dimension in the nanometric range or nanoparticles) into the polymers offer two-way benefit because of their structural integrity and barrier properties [13d].

7. Food Storage [13]

The antimicrobial properties of nanomaterials enable them to preserve food during storage and transport [14, 15]. Nanosensors can be used for a variety of applications. Commercial use of nanosensors has been reported to check storage conditions [14] and during food transport in refrigerated trucks for temperature control [15].

8. Nanosensors for Pathogen Detection [5]

Nanomaterials for use in the construction of biosensors offers the high level of sensitivity and other novel attributes. Nanosensors or Nanobiosensors are used for the following purposes:

i. Detection of pathogens in processing plants or in food material,

ii. Quantification of available food constituents,

iii. Alerting consumers and distributors on the safety status of food [5a]. The nanosensor works as an indicator that responds to changes in environmental conditions such as humidity or temperature in storage rooms, microbial contamination, or products degradation [5b]. Some examples are given below:

- ☆ A dimethylsiloxane microfluidic immunosensor integrated with specific antibody immobilized on an alumina nanoporous membrane was developed for rapid detection of foodborne pathogens *Escherichia coli* O157: H7 and *Staphylococcus aureus* with electrochemical impedance spectrum [5c].
- ☆ Biosensors based on carbon nanotubes also gained much attention due to their rapid detection, simplicity and cost effectiveness and have also been successfully applied for the detection of micro-organisms [5].

A summary of the kind of applications, industry and universities are working on nanotechnology, is tabulated below [16]:

Sl. No.	*Application*	*Status of work*
1	**Food** i. **Processed nanostructured or -textured food** (*e.g.* less use of fat and emulsifiers, better taste.	Nanostructured food ingredients and additives are understood to be in the R and D pipeline, *e.g.* Mayonnaise.
	ii. **Organic nanosized additives** for food, supplements and animal feed.	Materials range from colors, preservatives, flavorings to supplements and antimicrobials.
	iii. **Inorganic nanosized additives** for food, health food, and animal feed.	Inorganic additives (silver, iron, silica, titanium dioxide, selenium, platinum, calcium, magnesium) are available for supplements, nutraceuticals, and food and feed applications.
	iv. **Nanocarrier systems for delivery of nutrients and supplements** in the form of liposomes or biopolymer-based nanoencapsulated substances.	A number are commercially available in some countries.

Sl. No.	*Application*	*Status of work*
	V. Food packaging applications *e.g.* plastic polymers containing or coated with nanomaterials for improved mechanical or functional properties.	This area makes up the largest share of the current/short-term market for nanotech applications in the food sector (*e.g.* plastic polymers with nanoclay as gas barrier; nanosilver and nanozinc oxide for anti-microbial action, nanotitanium nitride for strength).
	vi. **Nanocoatings on food contact surfaces** for barrier or antimicrobial properties.	Nanomaterial-based coatings are available for food preparation surfaces and for coating food preparation machinery.
	vii. Surface-functionalized nanomaterials.	Main uses are currently in food packaging; possible uses emerging in animal feed.
	viii. Nanosensors for food labelling.	R and D stage
2	**Water decontamination**	Nano iron is already available in industrial-scale quantities. Companies thought to be using the technology in developing countries.
3	**Animal feed applications**	Nanosized additives specifically developed or are under development for feed include nanomaterials that can bind and remove toxins or pathogens

Sources

1. *Qureshi, Mehar Afroz, Karthikeyan, Swaminathan, Punita Karthikeyan, Khan, Pervez Ahmed, Uprit, Sudhir, and Mishra, Umesh Kumar, Application of nanotechnology in food and dairy processing: An overview, PAK. J. Food Sci., 22 (1), 2012: 23-31, ISSN: 2226-5899.*

2. *Chen, H., J. Weiss and F. Shahidi. 2006. Nano-technology in nutraceuticals and functional foods. Food Tech. 60(3): 30-6.*

3. *Mozafari, M., J. Flanagan, L. Matia, M. Merino, A. Awati, A. Omri, Z. Suntres and H. Singh. 2006. Recent trends in the lipid-based nanoencapsulation of antioxidants and their role in foods. J Sci. Food Ag. 86, 2038-2045.*

4. *Miller, G. 2008. Nanotechnology- the new threat to food. Global Research cited from http: //www. globalresearch.ca/index.*

5. *Trepti Singh, Shruti Shukla, Pradeep Kumar, Verinder Wahla, Vivek K. Bajpai, and Irfan A. Rather, Application of Nanotechnology in Food Science: Perception and Overview, Front Microbiol. 2017, 8, 1501.*

5a. *Cheng Q., Li C., Pavlinek V., Saha P., Wang H., 2006, Surface-modifi ed antibacterial TiO_2/Ag+nanoparticles: preparation and properties. Appl. Surface Sci. 252 4154-4160.*

5b. *Bouwmeester H., Dekkers S., Noordam M. Y., Hagens W. I., Bulder A. S., Heer C., et al., 2009, . Review of health safety aspects of nanotechnologies in food production. Reg. Toxicol. Pharmacol. 53 52-62.*

5c. Tan F., Leung P. H. M., Liud Z., Zhang Y., Xiao L., Ye W., et al., 2011, microfluidic impedance immunosensor for E. coli O157: H7 and Staphylococcus aureus detection via antibody-immobilized nanoporous membrane. Sensor. Actuat. B. Chem. 159 328-335.

6. Nakagawa K., 2014, "Nano- and micro-encapsulation of flavor in food systems, " in Nano- and Microencapsulation for Foods, Chap. 10 ed. Kwak H.-S., editor. (Oxford: John Wiley and Sons), 249-272.

7. Zhang T., Lv C., Chen L., Bai G., Zhao G., Xu C., 2014, Encapsulation of anthocyanin molecules within a ferritin nanocage increases their stability and cell uptake efficiency. Food Res. Int. 62 183-192.

8. Yang R., Zhou Z., Sun G., Gao Y., Xu J., Strappe P., et al., 2015, . Synthesis of homogeneous protein-stabilized rutin nanodispersions by reversible assembly of soybean (Glycine max) seed ferritin. RSC Adv. 5 31533-31540.10

9. Ottaway P. B. 2010, Nanotechnology in Supplements and Foods-EU Concerns. Available at: , February 26, 2010.

10. Dekkers S., Krystek P., Peters R. J., Lankveld D. X., Bokkers B. G., van Hoeven-Arentzen P. H., et al., 2011, Presence and risks of nanosilica in food products. Nanotoxicology 5 393-405.

10a. Uboldi, C et al., 2012. "Amorphous Silica Nanoparticles Do Not Induce Cytotoxicity, Cell Transformation or Genotoxicity in Balb/3T3 Mouse Fibroblasts."Mutat Res 745(1-2), 11-20.

11. Weiss J., Takhistov P., McClements J., 2006, Functional materials in food nanotechnology. J. Food Sci. 71 R107-R116.

12. Sari P., Mann B., Kumar R., Singh R. R. B., Sharma R., Bhardwaj M., et al, 2015, Preparation and characterization of nanoemulsion encapsulating curcumin. Food Hydrocol, 43 540-546.

13. A. Wallace Hayes and Saura C. Sahu, Nanotechnology in the Food Industry, A Short Review, NANOTECHNOLOGY, February/March 2017.

13a. Gerloff, K et al., 2009. "Cytotoxicity and Oxidative DNA Damage by Nano-particles in Human Intestinal Caco-2 Cells." Nanotoxicol 3(4): 355-364.

13b. Llorens, A et al., 2012. "Metallic-Based Micro- and Nanocomposites in Food Contact Materials and Active Food Packaging." Trends Food Sci Technol 24: 19-20.

13c. Pyrgiotakis, G et al., 2015, "Inactivation of Foodborne Micro-organisms Using Engineered Water Nanostructures (EWNS)."Environ Sci Technol 49(6): 3737-3745.

13d. Rhim J. W., Ng P. K.2007, Natural biopolymer-based nanocomposite films for packaging applications. Crit. Rev. Food Sci. Nutri. 47 411-433.

14. Bouwmeester, H et al., 2009, "Review of Health Safety Aspects of Nano-technologies in Food Production."Regul Toxicol Pharmacol, 53, 52-62.

15. Buzby, JC.2010, "Nano-technology for Food Applications: More Questions than Answers." J Consumer Affairs 44(3): 528-545.

16. Nanotechnology in Food, Nanowerk newspaper email digest.

Chapter 19

Processing Technologies for Prolong Shelf-Life Foods

Prolong Shelf-Life [1]

Shelf-life is the time it takes a food product to deteriorate to an unacceptable degree under specific storage, processing, and packaging conditions [2]. The longer food shelf life, the easier the shipping, storing and selling. Processing technologies, packaging technologies and proper storage of food products are significant factors for extended shelf-life of food products. The shelf life of a food depends on the three how's: how it is packaged, how it is processed, and how it is stored. The following methods are used alone or in combination to extend the normal biological life of foodstuffs and to enhance marketing:

1. Processing- Freezing, Heating, Drying *etc.* (Food-treatment methods).
2. Use of Chemicals (Antimicrobial compounds *etc.*, Next-generation additives).
3. Modifying product atmosphere (New packaging, Novel packaging solutions).

In addition to the processing technologies mentioned in earlier chapters, some other processing technologies like Freezing *etc.* play important role in extending the shelf-life of food products. The application of antimicrobial compounds and advanced packaging are also proved as significant tools for prolong shelf-life. The reduction in the rate of quality loss and increasing the shelf life of the product, are two important goals of packaging. Different packaging methods generally achieve an extended shelf life for food by removing or reducing oxygen from around the surface of food. Lowering the amount of oxygen slows down and inhibits the growth of any bacteria on food. There are many different types of packaging methods to improve the shelf life of products.

If food is stored in a warm environment, it will only last half the time it would last in a cool place. Good shelf life is achieved by storing products at 60°F. Optimum storage temperature for shelf life is 40°F or less. It is important to find a place where the temperature remains constant. Frequent temperature changes and excess moisture can shorten storage life.

There are many benefits of extending shelf life. Highly seasonal products can be stock piled. Product returns from retailers are reduced, and more extensive product distribution is possible.

Sources

1. *Brianna Jordan, Christian Brothers University, Extended shelf-life of products by advanced pacakaging, International Journal of Advanced Packaging Technology, Vol 8, No 1, 2020.*
2. *dictionary.org.*

1. Freezing

The Food is being preserved by Freezing from the time it is prepared to the time it is eaten [1]. Freezing is a quick and convenient way to preserve fruits and vegetables at home [1a]. Freezing food slows down decomposition by turning residual moisture into ice, inhibiting the growth of most bacterial species. Natural food freezing (using winter frosts) had been in use by populations in cold climates for centuries [1]. Freezing provides a significant extended shelf life and has been successfully employed for long-term preservation of many foods. Freezing is still one of the most widely used methods of food preservation even though several new technologies, such as high pressure, infrared irradiation, pulsed electric field, and ultrasound, are gaining importance [1b]. There are many feezing processes which are used in the food commodity industry. Mechanical [1] and Cryogenic (or flash freezing) [2] are some of the examples of these methods.

The freezing kinetics is important to preserve the food quality and texture. Preserving food in domestic kitchens during modern times is achieved using household freezers. It is to freeze food on the day of purchase. The Food Standards agency was reported as supporting the change, provided the food had been stored correctly up to that time [3]

The micro-organisms do not grow when the temperature of the food is below -9.5°C (15 °F) hence frozen products do not require any added preservatives. Freezing is sufficient on its own in preventing food spoilage. Long-term preservation of food may call for food storage at even lower temperatures. Carboxymethylcellulose (CMC), a tasteless and odorless stabilizer, is typically added to frozen food because it does not adulterate the quality of the products [4].

Freezing is an effective form of food preservation because the pathogens that cause food spoilage are killed or do not grow very rapidly at reduced temperatures.

- ☆ The process is less effective in food preservation than are thermal techniques, such as boiling, because pathogens are more likely to be able to survive cold temperatures rather than hot temperatures [5].
- ☆ Foods may be preserved for several months by freezing. Long-term frozen storage requires a constant temperature of -18 °C (0 °F) or less [6].
- ☆ One of the problems surrounding the use of freezing as a method of food preservation is the danger that pathogens deactivated (but not killed) by the process will once again become active when the frozen food thaws.

Sources

1. *Wikipedia.*

1a. *William Schafer, emeritus extension specialist and Suzane Driessen, Extension educator, The science of freezing foods, University of Minnesota Extension.*

1b. *Mohammad Shafiur Rahman and Jorge F. Velez-Ruiz, Food Preservation by Freezing, Handbook of Food Preservation, Second Edition page-635-665.*

2. *Sun, Da-Wen Advances in food refrigeration. Leatherhead Food Research Association Publishing. p.318. (Cryogenic refrigeration), 2001.*

3. *Smithers, Rebecca, "Sainsbury's changes food freezing advice in bid to cut food waste", The Guardian. Retrieved February 10, 2012.*

4. *Arsdel, Michael, Robert. Quality and Stability of Frozen Foods, TIme-Temperature Tolerance and its Significance. pp. 67-69.*

5. *Mathlouthi, M. Food Packaging and Preservation. pp. 112-115.*

6. *Tressler, Evers, Evers. Into the Freezer-and Out. pp. 56-82.*

History of Freezing

The history related to the development/application of freezing technique is tabulated below [1]:

Year	*Particular*
1861	The first freezing works in the world was established by Thomas Sutcliffe Mort at Darling Harbour in Sydney, Australia.
1868	The first trial shipment of frozen meat to London was in 1868.
1885	A small number of chickens and geese were being shipped from Russia to London in insulated cases using this technique.
1899	The "British Refrigeration and Allied Interests" reported that a food importing business, "Baerselman Bros", was shipping some 200,000 frozen geese and chickens per week from three Russian depots to New Star Wharf, Lower Shadwell, London over three or four winter months.
1929	Clarence Birdseye introduced "flash freezing" to the American public.
1934	The Icelandic Fisheries Commission was created to initiate innovation in the industry.
-	Other experiments involving orange juice, ice cream and vegetables, were conducted by the military near the end of World War II.

Source

1. *Wikipedia.*

Freezing Methods/Technologies [1, 1a]

The freezing technology itself, is developing to become faster, more efficient and more cost-effective [1]. There are different freezing methods. No single freezing system can satisfy all freezing needs because of a wide variety of food products and process characteristics [1a]. The selection criteria of a freezing method are the type of product, reliable and economic operation, easy cleaning ability, hygienic design, and desired product quality. In the food industry, plate contact, immersion, air blast, fluidized-bed, and cryogenic freezing are common methods [1b].

i. Mechanical Freezing

Mechanical freezers were the first to be used in the food industry and are used in the vast majority of freezing/refrigerating lines. They function by circulating a refrigerant, normally ammonia, around the system, which withdraws heat from the food product. This heat is then transferred to a condenser and dissipated into air or water. The refrigerant itself, now a high pressure, hot liquid, is directed into an evaporator. As it passes through an expansion valve, it is cooled and then vaporises into a gaseous state. Now a low pressure, low temperature gas again, it can be reintroduced into the system.

ii. Flash (Cryogenic) Freezing

Cryogenic or flash freezing of food is a more recent development, but is used by many leading food manufacturers all over the world. Cryogenic freezing is the quickest freezing technology available due to the ultra low liquid nitrogen temperature - 196 °C(-320 °F). Quicker freezing generates smaller ice crystals and maintains cellular structure [2]. Cryogenic equipment uses very low temperature gases, usually liquid nitrogen or solid carbon dioxide, which are applied directly to the food product [3]. In cryogenic freezing, liquefied gases are placed in direct contact with the foods. Food is exposed to an atmosphere below - 60°C through direct contact with liquid nitrogen or liquid carbon dioxide or their vapor [3a]. It also provides flexibility by being compatible with various types of food products and having low capital cost [3b]. The rapid formation of small ice crystals greatly reduces the damage caused by cell rupture, preserving color, texture, flavor, and nutritional value. The rapid freezing also reduces the evaporative weight loss from the products, provides high product throughput, and has low floor space requirements [3b].

Freezing methods are also categorized on the basis of contact with cooled solid, liqid and gas. The details are given below [1a]:

1. Freezing by Contact with a Cooled Solid (Plate Freezing) [1a]

In plate freezing, the product is sandwiched between metal plates and pressure is usually applied for good contact. Plate freezers are only suitable for regular shaped materials or blocks. When the product has been frozen, hot liquid is circulated to break ice seal and defrost.

2. Freezing by Contact with a Cooled Liquid (Immersion Freezing) [1a]

In Immersion Freezing, food is immersed in a low-temperature brine to achieve fast temperature reduction through direct heat exchange [3a]. The fluids usually used are salt solutions (sodium chloride), sugar solutions, glycol and glycerol solutions, and alcohol solutions. The solutes used must be safe to the product in terms of health, taste, color, and flavor, and the product must be denser than fluids. To ensure that the food does not come into contact with liquid refrigerants, flexible membranes can be used to enclose the food completely while allowing rapid heat transfer [3b].

- ☆ A mixture of glycerol and glycol is liquid–liquid medium, thus it can also be used since there is no eutectic point.
- ☆ Methanol or ethanol can also be used. Although the methanol will be removed during cooking, it is poisonous whereas ethanol is safe. Alcohols also pose a fire hazard in processing plants.

3. Freezing by Contact with a Cooled Gas [1a]

Cabinet Freezing and Blast Freezing are discussed below:

i. Cabinet Freezing

In this process, cold air is circulated in a cabinet where product is placed in a tray. The moisture pick-up from the product surface may with air velocity at least 5 m/s generates high heat-transfer rates [3a].

ii. Air-Blast Freezing

In this process, the temperature of food is reduced with cold air flowing at a relatively high speed. Air velocities between 2.5 and 5 m/s give the most economical freezing. This method can be further sub divided into following three types of freezing depending on how air interacts with the product [3a].

a. **Fluidized Bed Freezing:** A fluidized bed freezer consists of a bed with perforated bottom through which refrigerated air is blown vertically upward. The air velocity must be greater than the fluidization velocity. This freezing method is suitable for small particulate food bodies of fairly uniform size, *e.g.*, peas, diced carrots and potatoes, corns, and berry fruits.

b. **Belt Freezing:** The first mechanized air-blast freezers consisted of a wire mesh belt conveyor in a blast room for continuous product flow. Uniform product distribution over the entire belt is required to achieve uniform product contact and effective freezing. Temperatures used usually are -10°C to - 4°C in the precool section and -32°C to - 40°C in the freezing section [5].

c. **Spiral Freezing:** A spiral belt freezer consists of a long belt wrapped cylindrically in two tiers, thus requiring a minimal floor space. The spiral freezer uses a conveyor belt that can be bent laterally. It is suitable for products with a long freezing time (generally 10 min to 3 h), and for products that require gentle handling during freezing.

d. **Tunnel Freezing:** In this method, products are placed in trays or racks in a long tunnel and cool air is circulated over the product.

To maintain top quality, frozen fruits and vegetables should be stored at zero F or lower. Storing frozen foods at temperatures higher than zero F increases the rate at which deteriorative reactions can take place and can shorten the shelf life of frozen foods [4].

Emerging Freezing Techniques [1a]

The following emerging freezing techniques or combinations are being used for their potential benefits, technical and economical advantages, and quality enhancements:

a. Pressure Shift Freezing

Improved structures by pressure-shift freezing are reported for tofu (soybean curd) [5] and carrots [6]. In case of the tofu, the texture was almost same for pressure above 200 MPa as in the case of the untreated tofu [5]. Similarly, in case of carrots above pressure 100 MPa, the damage could be reduced significantly [6].

b. A Cryomechanical or Combined Freezing

A cryomechanical or combined freezing with cryogenic freezing followed by air-blast freezing was utilized to improve frozen food product quality. This is used for delicate products with poor mechanical resistance, such as shrimp, raspberries, strawberries, or chicken and mushrooms when significant changes occurred in the product [7]. In this combined process, a protective crust is formed through the immersion in the liquid N_2, which is characteristic of the products.

c. A Dehydrofreezing Method

It is a combined method of a controlled dehydration previously to the freezing stage, which is mainly promising for fruits and vegetables preservation due to the high moisture content. This technique reduces refrigeration loads and packaging, storage, and distribution costs; and provides a comparable quality [8].

Sources

1. *Wikipedia.*

1a. *Mohammad Shafiur Rahman and Jorge F. Velez-Ruiz, Food Preservation by Freezing, Handbook of Food Preservation, Second Edition page-635-665.*

1b. *ASHRAE Handbook, Refrigeration Systems and Applications, American Society of Heating, Refrigerating, and Air-conditioning Engineers, Atlanta, GA, 1994.*

2. *Sun, Da-Wen Advances in food refrigeration. Leatherhead Food Research Association Publishing. p.318. (Cryogenic refrigeration), 2001.*

3. *W. B. Bald, Food Freezing: Today and Tomorrow, J. P. Miller, The Use of Liquid Nitrogen in Food Freezing, p.157-170, Institute for Applied Biology, Springer-Verlag.*

3a. *Y. C. Hung and N. K. Kim, Fundamental aspects of freeze-cracking, Food Technol. 50: 59, 1996.*

3b. *R. M. George, Freezing processes used in the food industry, Trends Food Sci. Technol. 4: 134, 1993.*

4. *William Schafer, emeritus extension specialist and Suzane Driessen, Extension educator, The science of freezing foods, University of Minnesota Extension.*

5. *M. Fuchigami and A. Teramoto, Structural and textural changes in kinu-tofu due to high-pressure-freezing, J. Food Sci. 62: 828, 1997.*

6. *M. Fuchigami, K. Miyazaki, N. Kato, and A. Teramoto, Histological changes in high-pressure-frozen carrots, J. Food Sci. 62(4): 809-812, 1997.*

7. *M. E. Agnelli and R. H. Mascheroni, Quality evaluation of foodstuffs frozen in a cryomechanical freezer, J. Food Engng. 52: 257, 2002.*

8. *B. Li and D.-W. Sun, Novel methods for rapid freezing and thawing of foods: A review, J. Food Eng. 54: 175, 2002.*

New Freezing Technologies [1]

The following new freezing techniques or combinations are being developed and these have potential benefits, technical and economical advantages and quality enhancements:

- ✰ A cryomechanical or combined freezing with cryogenic freezing followed by air-blast freezing was utilized to improve frozen food product quality. It is recommended for delicate products with poor mechanical resistance, such as shrimp, raspberries, strawberries [2].
- ✰ Improved structures by pressure-shift freezing are reported for tofu (soybean curd) [3] and carrots [4]. In case of the tofu, the texture was almost same for pressure above 200 MPa as in the case of the untreated tofu [3]. Similarly, in case of carrots above pressure 100 MPa, the damage could be reduced significantly [4].
- ✰ A dehydrofreezing method is a combined method of a controlled dehydration previously to the freezing stage, which is mainly promising for fruits and vegetables preservation due to the high moisture content. This technique reduces refrigeration loads and packaging, storage, and distribution costs and provides a comparable quality [5].

Sources

1. *Mohammad Shafiur Rahman and Jorge F. Velez-Ruiz, Food Preservation by Freezing, Handbook of Food Preservation, Second Edition page-635-665.*

2. *M. E. Agnelli and R. H. Mascheroni, Quality evaluation of foodstuffs frozen in a cryomechanical freezer, J. Food Engng. 52: 257, 2002.*

3. *M. Fuchigami and A. Teramoto, Structural and textural changes in kinu-tofu due to high-pressure-freezing, J. Food Sci. 62: 828, 1997.*

4. *M. Fuchigami, K. Miyazaki, N. Kato, and A. Teramoto, Histological changes in high-pressure-frozen carrots, J. Food Sci. 62(4): 809-812, 1997, 67.*

5. *B. Li and D.-W. Sun, Novel methods for rapid freezing and thawing of foods: A review, J. Food Eng. 54: 175, 2002.*

Freezing as Hurdle Technology [1]

Freezing can be used as one of the hurdles in combined methods of food preservation. Some examples are given below:

- ✰ An osmotic dehydration treatment followed by freezing provided good quality and acceptance for apple cubes. The optimum conditions were found to be 55°Brix, 35°C for the solution, and 60 min of osmotic dehydration time with a fast freezing [2].
- ✰ The ultrasound with freezing was applied to potato tissues [3], which resulted in an improved freezing rate. Higher freezing rates produced better cellular structure, less intercellular void formation, and less cell disruption when studied by cryoscanning microscopy analysis. The application of power ultrasound was effective in improving the structure of frozen-then-thawed potatoes.
- ✰ Scientists [4]utilized a modified-atmosphere packaging, with different N_2/CO_2 ratios, combined with freeze chilling to extend the shelf life of raw whiting, mackerel, and salmon, finding that these combined technologies confer logistic benefits not only during frozen storage but also in product distribution and retailing.

Sources

1. *Mohammad Shafiur Rahman and Jorge F. Velez-Ruiz, Food Preservation by Freezing, Handbook of Food Preservation, Second Edition page-635-665.*
2. *A. Bunger, P. C. Moyano, R. E. Vega, P. Guerrero, and F. Osorio, Osmotic dehydration and freezing as combined processes on apple preservation, Food Sci. Tech. Int. 10: 163, 2004.*
3. *B. Li and D.-W. Sun, Effect of power ultrasound on freezing rate during immersion freezing of potato, J. Food Eng. 55: 277, 2002.*
4. *J. D. Fagan, T. R. Gormley, and M. M. Uí Mhuircheartaigh, Effect of modified atmosphere packaging with freeze-chilling on some quality parameters of raw whiting, mackerel and salmon portions, Innovative Food Sci. Emerg. Technol. 5: 205, 2004.*

Mode of Preserving Action [1]

The freezing reduces the random motion and rearrangement of molecules in the matrix [2]. Freezing involves the use of low temperatures, and reactions take place at slower rates as temperature is reduced. The presence of ice and an increase in solute concentration can have significant effects on the reactions and the state of the matrix [3]. The final influence of temperature on chemical reactions due to freezing could be grouped as follows:

1. Normal stability (a temperature decrease results in a slower reaction rate, thus better stability when foods are stored),
2. Neutral stability (the temperature has no influence on the reaction rate), or

3. Reversed stability (a temperature decrease results in an increased reaction rate) [4].

Regardless of the type of aqueous system, concentration during freezing causes the unfrozen portion to undergo marked changes in physico-chemical properties such as ionic strength, pH, viscosity, water activity, surface and interfacial tension, and oxidation-reduction potential. It is important to note that oxygen is almost totally expelled from ice crystals as they are formed [4].

The three types of cell damage due to freezing- **osmotic damage, solute-induced damage, and structural damage** were reviewed [5]. In slow cooling, ice forms slowly in the external cells. If there is sufficient time, water from the cells migrates out by osmotic pressure. This results in cell shrinkage and some membrane damage. This water does not return to the cells on thawing due to the damage to the cell wall, and the consequence is drip loss [5]. The concentration of the solute increases as freezing progresses. Thus, high solute concentrations of the unfrozen matrix, in particular high salt, can cause damage to many polymeric cell components and may kill the cell [6]. This concentration effect is present irrespective of whether freezing is fast or slow. Cryoprotectants, such as sugars, are usually added to the aqueous phase to reduce salt-induced damage [3].

In addition to the concentration effect, the formation of ice within the cell may cause damage to the delicate organelle and membrane structure of the cell. As a consequence, enzyme systems may be released, leading to a variety of effects, including off-flavor production. This can be prevented by blanching, a prefreezing heat treatment that denatures enzymes [3].

Sources

1. *Mohammad Shafiur Rahman and Jorge F. Velez-Ruiz, Food Preservation by Freezing, Handbook of Food Preservation, Second Edition page-635-665.*
2. *Y. C. Hung and N. K. Kim, Fundamental aspects of freeze-cracking, Food Technol. 50: 59, 1996.*
3. *D. S. Reid, Fruit freezing, Processing Fruits: Science and Technology. Vol.1. Biology, Principles and Applications (L. P. Somogyi, H. S. Ramaswamy and Y. H. Hui, eds.), Technomic Publishing, Lancaster, CA, 1996, p. 169.*
4. *K. Porsdal and F. Lindelov, Acceleration of chemical reactions due to freezing, Water Activity: Influences on Food Quality (L. B. Rockland and G. F. Stewart, eds.), Academic Press, New York, 1981.*
5. *D. S. Reid, Fruit freezing, Processing Fruits: Science and Technology. Vol.1. Biology, Principles and Applications (L. P. Somogyi, H. S. Ramaswamy and Y. H. Hui, eds.), technomic Publishing, Lancaster, CA, 1996, p. 169.*
6. *P. Mazur, The role of intracellular freezing in the death of cells cooled at superoptimal rates, Cryobiology 14: 251, 1977,*

Packaging [1]

Frozen food packaging must maintain its integrity throughout filling, sealing, freezing, storage, transportation, thawing, and often cooking [2]. As many frozen foods are cooked in a microwave oven, manufacturers have developed packaging that can go straight from freezer to the microwave. Some examples of the packaging materials for frozen foods are given below:

a. Differential Heating Container (DHC)

Initially the first DHC was sold in 1974. A DHC is a sleeve of metal designed to allow frozen foods to receive the correct amount of heat. Various sized apertures were positioned around the sleeve. The frozen dinner is put into the sleeve according to what needed the most heat. This ensured proper cooking [3].

b. Multiple Packaging Options

Now multiple options for packaging frozen foods *e.g.* Boxes, cartons, bags, pouches, Boil-in-Bags, lidded trays and pans, crystallized PET trays, and composite and plastic cans are being used [4].

c. Active Packaging

This new aspects of frozen food packaging offers a host of new technologies that can actively sense and then neutralize the presence of bacteria or other harmful species. Active packaging can extend shelf-life, maintain product safety, and help preserve the food over a longer period of time. The following functions of active packaging are being researched:

i. Oxygen scavengers,
ii. Time Temperature Indicators and digital temperature data loggers,
iii. Antimicrobials,
iv. Carbon Dioxide controllers,
v. Microwave susceptors,
vi. Moisture control: Water activity, Moisture vapor transmission rate, *etc.*,
vii. Flavor enhancers,
viii. Odor generators,
ix. Oxygen-permeable films
x. Oxygen generators [5].

Sources

1. *Wikipedia.*
2. *Decareau, Robert. Microwave Foods: New Product Development. pp. 45-48.*
3. *Whelan, Stare. Panic in the Pantry: Facts and Fallacies About the Food You Buy.*
4. *Russell, Gould. Food Preservatives. pp. 314.*
5. *Sun, Da-Wen. Handbook of Frozen Food Processing and Packaging. pp. 786-792.*

Microbial Growth in the Freezer [1, 2]

The detrimental effects of freezing on micro-organisms may be desirable or undesirable, depending on the types of food products.

Undesirable

In frozen foods without any added beneficial cultures, microbial growth or spoilage is not desirable,

Desirable

Whereas care must be taken to reduce the damage in cells during freezing of foods containing microbial cultures.

The maximum recommended storage temperature at which microbiological spoilage ceases is registered between -9°C and -12°C [1].

The micro-organisms which may be present on fruits and vegetables, are not actually destroyed by freezing process. While blanching destroys some micro-organisms and there is a gradual decline in the number of these micro-organisms during freezer storage, sufficient populations are still present to multiply and cause spoilage of the product when it thaws. For this reason it's necessary to carefully inspect any frozen products which have accidentally thawed by the freezer going off or the freezer door being left open [2].

The ability of bacteria to resist damage by freezing vary considerably [3]. Gram-negative bacteria are less resistant to freezing death than are Gram-positive bacteria. Nonsporulating rods and spherical bacteria are the most resistant, while bacterial spores, such as Clostridium and Bacillus, remain unaffected by freezing [1]. Bacteria in the stationary phase are more resistant than those in the log phase [3]. The principal site of damage in the bacterial cells during freezing has been shown to be the membrane.

- ☆ In fermented foods (yogurt) frozen storage should increase the viability of beneficial cultures incorporated for their potential health benefits and control of other spoiling micro-organisms. The potential beneficial roles of bifidobacteria in the human intestine reported include antagonistic effects on enteropathogenic bacteria, breakdown of carcinogenic N-nitrosamines, and suppression of liver tumorigenesis [4]. Thus, bifidobacteria are incorporated in dairy products.
- ☆ Yeast cells in bakery products do not withstand freezing well. This can be partially compensated for by increasing the amount of yeast used in the formulation or adding improved yeast strains having a better survival rate in freezing [5].

Sources

1. *Mohammad Shafiur Rahman and Jorge F. Velez-Ruiz, Food Preservation by Freezing, Handbook of Food Preservation, Second Edition page-635-665.*
2. *William Schafer, emeritus extension specialist and Suzane Driessen, Extension educator, The science of freezing foods, University of Minnesota Extension.*

3. G. P. Archer, C. J. Kennedy and A. J. Wilson, Position paper: Towards predictive microbiology in frozen food systems-A framework for understanding microbial population dynamics in frozen structures and in freeze-thaw cycles, Int. J. Food Sci. Technol. 30: 711, 1995.

4. K. M. K. Kebary, Viability of Bifido-bacterium bifidum and its effect on quality of frozen Zabady, Food Res. Int. 29: 431, 1996.

5. H. Symons, Frozen foods, Shelf Life Evaluation of Foods(C. M. D. Man and A. A. Jones, eds.), Blackie Academic and Professional, London, 1994, p. 298.

Effects of Freezing on Quality and Nutritional Values/Nutrients [1, 2a]

The freezing of foods slows down, but does not stop, the physico-chemical and biochemical reactions that govern the deterioration of foods [1a]. During storage, there is a slow progressive change in organoleptic quality, which does not become objectionable for some time [1b]. The loss of quality of frozen foods depends primarily on storage temperature, length of storage time, and thawing procedure. Microbial growth is completely stopped below –18°C, and both enzymatic and nonenzymatic changes continue at much slower rates during frozen storage [1c]. Freezing has following effects on quality and nutrients of foods:

A. Effect on Quality [1]

The speed of the freezing has a direct impact on the size and the number of ice crystals formed within a food product's cells and extracellular space. Fast freezing leads to smaller but more numerous ice crystals while slow freezing leads to fewer but larger ice crystals and these large ice crystals can puncture the walls of the cells of the food product which will cause a degradation of the texture of the product as well as the loss of its natural juices during thawing [2]. That is why there will be a qualitative difference observed between food products frozen by ventilated mechanical freezing, non-ventilated mechanical freezing or cryogenic freezing with liquid nitrogen. Over loading of freezer with unfrozen product will result in a long, slow freez and poor quality product [2a].

Fluctuating temperatures in the freezer can cause the migration of water vapor from the product to the surface of the container. This defect is sometimes found in commercially frozen foods which have been improperly handled [2a].

Recommended storage times ensure maximum quality. Food stored longer will be safe to eat but you may notice changes in flavor, color and texture. For best quality, use frozen fruits and vegetables within 8 to 12 months [2a].

a. Chemical Changes during Freezing [2a, 2b]

i. Colour Change

Fresh fruits and vegetables, when harvested, continue to undergo chemical changes which can cause spoilage and deterioration of the product. This is why these products should be frozen as soon after harvest as possible and at their peak degree of ripeness [2a].

The most important color changes in fruits and vegetables are related to following three biochemical or physico-chemical mechanisms [2b, 2e]:

i. Changes in the natural pigments of vegetable tissues (chlorophylls, anthocyanins, carotenoids),
ii. Development of enzymatic browning
iii. Breakdown of cellular chloroplasts and chromoplasts.

Color and flavor are important sensory attributes where as vitamin content is a functional attribute of frozen foods. The green color of vegetables is lost by chlorophyll degradation during freezing and frozen storage resulting from the conversion of chlorophyll to pheophytins or the destruction of both chlorophyll and pheophytins, giving a dull khaki color. Storage temperature and time, acidity and blanching time affect the loss of chlorophyll in frozen vegetable [2b]. The maximum stability of carotenes in frozen spinach was 2 years at -28.9°C, 1 year at -6.7°C and 7 days at 4.4°C [2f].

Enzymes cause loss of color, flavor changes and nutrient loss in frozen fruits and vegetables. These enzymes must be inactivated to prevent such reactions from taking place. The following processes are being used to inactivate enzyme present in fruits and vegetables:

Blanching of Vegetables

Blanching is used to inactivate enzyme in vegetable.

It is the exposure of the vegetables to boiling water or steam for a brief period of time. The vegetable must then be rapidly cooled in ice water to prevent it from cooking. Blanching also helps to destroy micro-organisms on the surface of the vegetable and to make some vegetables, such as broccoli and spinach, more compact.

Addition of Ascorbic Acid to Fruit to Control Enzymes

The major problem associated with enzymes in fruits is the development of brown colors and loss of vitamin C. Because fruits are usually served raw, they're not blanched like vegetables and hence enzymes in frozen fruit are controlled by using chemical compounds which interfere with deteriorative chemical reactions. The most common control chemical is ascorbic acid (vitamin C-Pure form or in commercial mixtures with sugars).

Less Effective Methods of Controlling Enzymes

Soaking the fruit in dilute vinegar solutions or coating the fruit with sugar and lemon juice are the temporary measures to control enzyme activated browning. These methods are not as effective as treatment with ascorbic acid.

ii. Change in Flavour/Development of Rancid Flavour [2a, 2b]

Change in Flavour [2b]

Freezing affects the flavor and aroma of frozen foods.

- For example, freezing of strawberries is usually associated with a reduction in aroma and the development of off-flavor. It was found [2g] that the development of off-flavor in frozen thawed strawberries was due to the chemical production of H_2S rather than enzymatic activity.
- In fish and seafood, FA is formed during cold storage by enzymatic decomposition of TMAO. It was mentioned [2h] that changes in the flavor of fish and seafood generally occur in following three distinct phases during frozen storage:
    i. Gradual loss of flavor due to the loss of or decrease in concentration of some flavour compounds.
    ii. Detection of a neutral, bland, or flat flavour.
    iii. Development of off-flavors due to the presence of compounds such as acids and carbonyl compounds that are products of lipid oxidation.

Development of Rancid Flavour

Another group of chemical changes that can take place in frozen products is the development of rancid oxidative flavors through contact of the frozen product with air. This problem can be controlled by using a wrapping material which does not permit air to pass into the product. Also, remove as much air as possible from the freezer bag or container to reduce the amount of air in contact with the product [2a]. Freezing results in a concentration of solutes, which catalyze the initiation of oxidative reactions and disrupts and dehydrate cell membranes, exposing membrane phospholipids to oxidation [2b].

- The degradation of lipids in frozen peas during storage at -18°C led to flavor damage due to the formation of hydroperoxides, thiobarbituric acid, and fatty acids, particularly in unblanched samples [2d].
- Oxygen availability and tissue composition play important roles in the acceleration of lipid oxidation frozen fish. Lipid hydrolysis occurs in fish during storage, which may affect lipid oxidation [2b].
- Poultry fat becomes rancid during very long storage periods or at extremely high storage temperatures. Rancidity in frozen whole poultry stored for 12 months is not a serious problem if the bird is packaged in essentially impermeable film and held at -18°C or below [2b].

B. Effect on Nutritional Values [1]

Freezing may potentially preserve the greatest quantity of nutrients. To maintain top nutritional quality in frozen fruits and vegetables, it's essential to follow directions prescribed for pretreatment of the vegetables, to store the frozen product at zero F and to use it within suggested storage times [2a]. Freezing has following effect on the vitamins of frozen foods:

Vitamin A (Carotene)

There is little loss of carotene during preparation for freezing and freezing of most vegetables. Much of the vitamin loss is incurred during the extended storage period [3].

Vitamin B [2a]

Vitamin B losses sometimes occur in frozen meat products. Vitamin B losses may be significant in frozen poultry products, but most losses are the result of the subsequent thawing and cooking treatments rather than of the freezing process [9].

Vitamin B1 (Thiamine)

A vitamin loss of 25 per cent is normal. Thiamine is easily soluble in water and is destroyed by heat [4].

Vitamin B_2 (Riboflavin)

Not much research has been done on it. One study found an 18 percent vitamin loss in green vegetables, while another determined a 4 percent loss [5]. It is commonly accepted that the loss of Riboflavin has to do with the preparation for freezing rather than the actual freezing process itself.

Vitamin C

The destruction of vitamin C (ascorbic acid) occurs during freezing and frozen storage. This loss is influenced by blanching conditions, types of freezing, package types, and time-temperature conditions [6b]. Usually lost in a higher concentration than any other vitamin [6]. There is a 10-fold increase in the rate of loss of ascorbic acid per 8.9°C rise in storage temperature of frozen vegetables [6a]. Generally, frozen vegetables stored at -24°C displayed better ascorbic acid retention than those at -12°C and -18°C, respectively [6b].

- ☆ A study was performed on peas to determine the cause of vitamin C loss. A vitamin loss of ten percent occurred during the blanching phase with the rest of the loss occurring during the cooling and washing stages [7]. The vitamin loss was not actually accredited to the freezing process.
- ☆ Another experiment was performed involving peas and lima beans (Frozen and canned vegetables). The frozen vegetables were stored at -23 °C (-10 °F) and the canned vegetables were stored at room temperature 24 °C (75 °F). After 0, 3, 6, and 12 months of storage, the vegetables were analyzed with and without cooking. The vitamin content of the two vegetables when they were ready for the plate of the consumer, there did not appear to be any marked advantages attributable to method of preservation, frozen storage, processed in a tin, or processed in glass [8].

C. Changes in Functional Properties during Freezing [2a, 2b]

Some Properties such as textural, mechanical, rheological (both flow and dynamic), consistency, appearance and sensory attributes and water-holding capacity have been correlated with freezing and thawing processes [2b].

Textural Changes

Texture is important in frozen vegetables. After freezing and thawing, firmness decreased, rupture strain increased, and consequently crispness decreased [2b]. The water (90 per cent of the weight) and other chemical substances give support structure and texture to the fruit or vegetable. Freezing fruits and vegetables actually consists of freezing the water contained in the plant cells and when the water freezes, it expands and the ice crystals cause the cell walls to rupture. Consequently, the texture of the produce, when thawed, will be much softer than it was when raw. This textural difference is especially noticeable in products which are usually consumed raw. For example, when a frozen tomato is thawed, it becomes mushy and watery. This explains why celery and lettuce are not usually frozen. In the partially thawed state, the effect of freezing on the fruit tissue is less noticeable. Textural changes due to freezing are not as apparent in products which are cooked before eating because cooking also softens cell walls. These changes are also less noticeable in high starch vegetables, such as peas, corn and lima beans. The extent of cell wall rupture (or Control texture change) can be controlled by freezing produce as quickly as possible. In rapid freezing, a large number of small ice crystals are formed. These small ice crystals produce less cell wall rupture than slow freezing which produces only a few large ice crystals [2a].

Rheological Properties [2b]

These properties of fresh and frozen thawed okra dispersions were significantly different when measured at 20°C– 80°C. The dispersions were pseudoplastic with both the consistency coefficient and the flow behavior index influenced by temperature. The consistency coefficient was higher for the unfrozen sample than the frozen thawed sample when measured at 20°C and 50°C, but the reverse was observed at 80°C. The flow behavior indices were not different at any temperature between 20°C and 80°C [2i].

Mechanical Property

Strawberries, apples, peaches, and citrus, contain thin-walled cells with a large proportion of intracellular water, which can freeze, resulting in extensive cell rupture and radical alteration of the mechanical properties of the material [2b, 2j].

D. Physical Changes and Quality

i. Moisture Loss/Defect in Appearance

Different types of water (free and unfreezable) are present in frozen foods. A major cause of product degradation is the amount of unfrozen water present in the frozen matrix. Unfrozen water is known to be reactive, particularly during frozen storage, rendering the product susceptible to deteriorative and enzymatic reactions and limiting its frozen shelf life [2b].

Dehydration or weight loss should be regarded as an important quality parameter for frozen unpacked foods, mainly in animal tissue. Foods lose moisture during the freezing process because their surface is exposed to heat and a moisture gradient exists within the environment [2c]. Ice recrystallization during frozen storage influences the product quality in different ways.

ii. Freezer Burn

Moisture loss, or ice crystals evaporating from the surface area of a product, produces **freezer burn** - a grainy, brownish spot where the tissues become dry and tough. This surface freeze-dried area is very likely to develop off flavors. Packaging designed specifically for freezing foods will prevent freezer burn. To prevent freezer burn, remove all air and seal airtight [2a].

iii. Glass Formation [2b]

The glass transition has a dramatic effect on frozen food quality. The product is most stable below Tg$, and moisture has little influence on Tg$. The presence of low molecular weight (LMW) solutes lowers the Tg$, and high molecular weight (HMW) exerts little effect. This means that with increasing maturity, many vegetables display a decrease in sugars and an increase in starch, thus increasing the Tg$ and the stability of frozen foods.

iv. Denaturation of Protein [2a]

Freezing causes protein denaturation and solubility changes but the practical significance of these changes is not clear [9]. During freezing, water molecules freeze out and migrate to form ice crystals, resulting in the disruption of the organized H-bonding system that stabilizes the protein structure, and the hydrophobic as well as the hydrophilic regions of protein molecules become exposed to a new environment. This may allow the formation of intermolecular cross-linkages either within a protein molecule or between two adjacent molecules [9a]. Some examples of protein denaturation during feezing are discussed below:

- ☆ Fish deterioration during frozen storage is associated with a decrease of protein solubility and extractability, diminishing the nutritional value [10].
- ☆ Quality loss during cold storage of meat is characterized by an increase in loss of waterholding capacity, a decrease in protein extractability, a decrease in sulfhydryl groups, and a slight loss in ATPase activity [9, 11].

Sources

1. *Wikipedia.*

1a. *R. M. George, Freezing processes used in the food industry, Trends Food Sci. Technol. 4: 134, 1993.*

1b. *B. Salvage, Frozen Food Rep. 3: 28-33, 1992.*

1c. *Y. G. Moharram and S. D. Rofael, Shelf life of frozen vegetables, Shelf Life Studies of Foods and Beverages (G. Charalambous, ed.), Elsevier Science Publishers B. V., Amsterdam, 1993.*

2. *W. F. Stoecker, Industrial Refrigeration Hand-book, 2000, Chapter 17 Refrigeration and freezing of foods, 17.10 The freezing process.*

2a. *William Schafer, emeritus extension specialist and Suzane Driessen, Extension educator, The science of freezing foods, University of Minnesota Extension.*

2b. *Mohammad Shafiur Rahman and Jorge F. Velez-Ruiz, Food Preservation by Freezing, Handbook of Food Preservation, Second Edition page- 635-665.*

2c. *L. A. Campañone, L. A. Roche, V. O. Salvadori and R. H. Mascheroni, Monitoring of weight losses in meat products during freezing and frozen storage, Food Sci. Tech. Int. 8: 229, 2002.*

2d. *H. M. Henderson, J. Kanhai, and N. A. M. Eskin, The enzymic release of fatty acids from phosphatidyl-choline in green peas (Pisum sativum), Food Chem. 13: 129, 1983.*

2e. *M. P. Cano, Vegetables, Freezing Effects on Food Quality (L. E. Jeremiah, ed.), Marcel Dekker, New York, 1996, p. 247.*

2f. *W. C. Dietrich, M. M. Boggs, M. D. Nutting and N. E. Weinstein, Time-tolerance of frozen foods. XVI. Quality retention of frozen foods. XIII. Quality changes in frozen spinach, Food Technol. 14: 155, 1960.*

2g. *H. Deng, Y. Ueda, K. Chachin, and H. Yamanaka, Off-flavor production in frozen strawberries, Postharv. Biol. Technol. 9: 31, 1996.*

2h. *E. E. M. Santos-Yap, Fish and seafood, Freezing Effects on Food Quality (L. E. Jeremiah, ed.), Marcel Dekker, New York, 1996, p.109.*

2i. *A. O. Olorunda and M. A. Tung, Rheology of fresh and frozen okra dispersions, J. Food Technol. 12: 593, 1977.*

2j. *J. M. DeMan, P. W. Voisey, V. F. Rasper and D. W. Stanley, Rheology and Texture in Food Quality, Van Nostrand Reinhold/Avi, New York, 1976.*

3. *Tressler, Evers. The Freezing Preservation of Foods. pp. 976-978.*

4. *Gould, Grahame. New Methods of Food Preservation. pp. 237-239.*

5. *Tressler, Evers. pp. 973-976.*

6. *Tressler, Evers. The Freezing Preservation of Foods. pp. 620-624.*

6a. *G. Bennett, J. F. Cone, M. L. Dodds, J. C. Garrey, N. N. Guerrant, J. G. Heck, J. F. Murphy, J. E. Nicholas, J. S. Perry, R. T. Piercf, and M. D. Show, Some Factors Affecting the Quality of Frozen Foods, Bull. of Penn. State Univ., State College, PA, No. 580, 1954.*

6b. *Y. G. Moharram and S. D. Rofael, Shelf life of frozen vegetables, Shelf Life Studies of Foods and Beverages (G. Charalambous, ed.), Elsevier Science Publishers B. V., Amsterdam, 1993. 7-Tressler, Evers. The Freezing Preservation of Foods. pp. 961-964 Tressler.*

8. *Tressler, Evers. The Freezing Preservation of Foods.p.627.*

9. *J. G. Sebranek, Poultry and Poultry Products, Freezing Effects on Food Quality (L. E. Jeremiah, ed.), Marcel Dekker, 1996, p.85.*

9a. *E. E. M. Santos-Yap, Fish and seafood, Freezing Effects on Food Quality (L. E. Jeremiah, ed.), Marcel Dekker, New York, 1996, p. 109.*

10. *C. Alvarez, A. Huidobro, M. Tejada, I. Vázquez, E. De Migel, and I. A. Gómez de Segura, Consequences of frozen storage for nutritional value of hake, Food Sci. Tech. Int. 6: 493, 1999.*

11. *D. S. Reid, Fruit freezing, Processing Fruits: Science and Technology. Vol. 1. Biology, Principles and Applications (L. P. Somogyi, H. S. Ramaswamy, and Y. H. Hui, eds.), Technomic Publishing, Lancaster, CA, 1996, p. 169.*

Processing and Packaging Factors [1]

The following factors are affecting the freezing as well as quality of frozen food:

1. Pretreatments for Freezing

Successful freezing will only retain the inherent quality present initially in a food item and will not improve quality characteristics, hence quality level prior to freezing is a major consideration. Microbiological quality prior to freezing remains a major determinant of post thaw quality. Although freezing can reduce some pathogens, there is also usually significant survival. The details related to the commonly used pretreatments are given below:

i. Blanching

Most vegetables and some fruits are blanched before freezing. Blanching destroys the permeability of cell membranes, destroys cell turgor, removes intercellular air, filling these spaces with water, and establishes a continuous liquid phase [1]. Properly blanched vegetables have a long shelf life at frozen food temperatures, enabling them to be exported all over the world and span the seasons [2]. Blanching has also other advantages, such as destruction of micro-organisms and wilting of leafy vegetables assisting packaging [3]. Blanching of fruits may be detrimental in many cases, resulting in a- rapid discoloration by enzymatic browning [2], b- loss of texture, c- formation of cooked taste, d- some loss of soluble solids, especially in water blanching and e- adverse environmental impact due to energy requirements and disposal of used water [3]. Blanching at 70°C-105°C is associated with the destruction of enzyme activity. Hot-water blanching is usually carried out between 75°C and 95°C for 1-10 min, depending on the size of the individual vegetable pieces. High-pressure steam blanching is more energy efficient than water blanching. It is important that cooling be carried out shortly after blanching, especially for products to be frozen [3].

ii. Heat Treatments

Low temperature and long-time pretreatment were useful in improving the texture of frozen vegetables by avoiding excessive softness. Carrots heated for 30min at 60°C and frozen above 5°C/min (optimum rate) should escape both cell damage and excessive softening [4]. It was found [5]that preheating followed by quick freezing was effective in improving excessive softness and cell damage. The optimum preheating occurred with 30 min at 60°C or 5 min at 70°C, and the optimum freezing was -5°C to -50°C. Preheated carrots retained a firmer texture than those blanched in boiling water; A greater variety of products cooked prior to freezing is becoming popular with consumers. These include breaded and fried portions, cured and smoked products, and items in marinades or broths [6].

iii. Dipping Pretreatments

Generally foods are dipped or soaked in different solutions before freezing and the type of solute used depends on the desired purpose. Apple slices are commonly treated by soaking it in 1 per cent salt solution to remove intercellular air. Fruits are also dipped in ascorbic acid and sugar solutions to minimize browning or blanched for a short time to inactivate enzymes [7]. The frozen product is protected by a suitable glazing compound. A glaze acts as a protective coating against the two main causes of deterioration during storage: dehydration and oxidation. It protects against dehydration by preventing moisture from leaving the product, and against oxidation by mechanically preventing air contact with the product. The different glazes available include inorganic salt solutions of sodium acid phosphate, sodium carbonate, and calcium lactate, alginate solution (also known as Protan glaze), antioxidants (such as ascorbic and citric acids, glutamic acid, and monosodium glutamate), and other edible coatings (such as corn syrup solids) [8].

iv. Bacterial Ice Nucleators or Antifreeze Proteins

The use of bacterial ice nucleators to the freezing of some model food systems and real elevates nucleation temperatures, reduces freezing times, and improves the quality specially flavor and texture. When bacterial cells were added to isotropic aqueous dispersions of hydrogels composed of proteins and polysaccharides, the bulk of the water was converted into directional ice crystals at subzero temperatures not lower than 5°C, and resulted in the formation of anisotropically textured products [9].

Antifreeze proteins, found in polar fish and cold tolerant insects and plants, can affect freezing in following several different ways:

1. By lowering the freezing temperature,
2. By retarding recrystallization on frozen storage
3. By promoting ice nucleation causing supercooled solutions to freeze more rapidly [10].

The antifreeze proteins depress the freezing point by attaching to ice crystals and interfering with water molecules joining the ice lattice. There are two groups of antifreeze proteins: antifreeze glycoproteins and antifreeze proteins.

v. Osmotic Concentration

Partial removal of water by osmotic treatment to freezing is recognized as convenient for reducing cellular damage of fruits and vegetables, which causes softness after thawing. Osmotic concentration of vegetables prior to freezing is a pretreatment that can improve final product quality [11]. The osmotic dehydration improves the product quality in terms of color, flavor, and texture. Partially concentrating fruit and vegetables prior to freezing saves packaging and distribution costs [12].

vi. Cryoprotection

Meat and fish muscle is susceptible to freeze denaturation, which decreases gel-forming potential, water holding capacity, and protein solubility. Cryoprotectants are generally added to protect fish myofibrillar proteins from freeze denaturation during frozen storage [13]. Freeze drying had a greater effect when no additive or NaCl was present. The blast-frozen and freeze-dried samples with no cryoprotectants had a very coarse structure with no obvious fine network system.

In the presence of sorbitol or sucrose, there was a finer meshwork for freeze-dried material, which was finer for frozen material. In the presence of sorbitol or sucrose with tripolyphosphate, the network was even finer but with less obvious spaces in the matrix for both freeze-dried and frozen material.

HMW polymer cryoprotectants have the following advantages over LMW cryoprotectants [14]:

- ☆ Do not generally penetrate the cell membrane and remain in the extracellular suspensions and/or outer surface of the cell.
- ☆ Do not produce a significant freezing point depression within the range of concentrations that can be applied in practice.
- ☆ There is no binary eutectic effect, *i.e.*, the hydrated polymer does not crystallize from aqueous solution as LMW agents do.
- ☆ Have the ability to keep a substantial portion of the solution from freezing.

Although the presence of HMW polymeric cryoprotectants is limited to the extra-cellular suspension medium, HMW additives affect the intracellular composition by the efflux of intracellular water due to chemical potential change across the membrane when extracellular ice is formed [14].

vii. Irradiation

High-dose irradiation can produce changes in the chemical composition and taste of fish and seafood. A combination of irradiation and freezing can be used in foods. A combination treatment involving freezing in conjunction with irradiation has recently been proposed as a means of retarding spoilage.

2. Thawing [1]

Thawing as a final and obligatory step of the freezing process is quite important. Thawing properly is essential to maximize quality retention and safety of frozen foods. Microbiologically safe thawing process includes the followings:

a. Inside a refrigerator at temperatures below 5°C,

b. Microwave oven, or

c. As part of the cooking treatment [15].

Scientists [15] found that sensory parameters, such as flavor and color, were more sensitive to thawing at refrigeration temperature (3°C-7°C) than nutritional properties, such as vitamin C and starch contents. High pressure, microwave, ohmic, and acoustic thawing are innovative applications and may require shorter thawing times, but some limitations have been found [16, 17]. Heterogeneous heating, controlled frequencies, and much more investigations need to be considered with these new thawing methods [17].

3. Storage and Display [1]

Packaging, storage, and display also affect the frozen food quality. Loss of quality in frozen foods is a gradual process; the changes being slow or very slow, cumulative, and irreversible [18]. Optimum quality requires care in every stage of processing, packaging, storage, and marketing sequence. Storage temperature is important for frozen food.

Packages

To provide the greatest protection, a package must be well evacuated of air (oxygen) using a vacuum or gas-flushed system and provide an adequate barrier to both oxygen and moisture [6]. The shelf life of frozen foods kept in open display cabinets at -15°C packed in 23 different types of plastic, cardboard, and laminate was studied. It was found that aluminium foil-laminated and metallized packages gave the best results. This is due to low levels of oxygen permeability, water vapor transmission, and light transparency, and less fluctuating temperatures [1]. Two terms used to describe the shelf life of frozen foods are **Practical storage life (PSL) and Just noticeable difference (JND).** PSL is the level of quality expected for the product by the ultimate consumer. JND is usually determined by a trained taste panel.

Storage Temperature

Quality losses of frozen food increase log linearly with the storage temperature when greater than -18°C [19]. The rate of quality loss increases about 2–2.5 times for every 5°Cincrease over -18°C [20, 21]. Some types of foods, such as fish, pork, animal organs, fried chicken, and spinach, can be maintained in a high-quality state for only 3-7 months at -20°C, whereas other foods, such as beef, sugared fruits, many bakery products, and many vegetables, can be maintained in a high-quality state for more than 12 months at -20°C [22].

Sources

1. *Mohammad Shafiur Rahman and Jorge F. Velez-Ruiz, Food Preservation by Freezing, Handbook of Food Preservation, Second Edition page- 635-665.*
2. *H. Symons, Frozen foods, Shelf Life Evaluation of Foods(C. M. D. Man and A. A. Jones, eds.), Blackie Academic and Professional, London, 1994, p. 298.*
3. *M. P. Cano, Vegetables, Freezing Effects on Food Quality (L. E. Jeremiah, ed.), Marcel Dekker, New Yor.k, 1996, p. 247.*
4. *M. Fuchigami, N. Hyakumoto, K. Miyazaki, T. Nomura, and J. Sasaki, Texture and histological structure of carrots frozen at a programmed rate and thawed in an electrostatic field, J. Food Sci. 59: 1162, 1994.*
5. *M. Fuchigami, K. Miyazaki, and N. Hyakumoto, Frozen carrots texture and pectic components as affected by low-temperature-blanching and quick freezing, J. Food Sci.60: 132, 1995.*
6. *J. G. Sebranek, Poultry and Poultry Products, Freezing Effects on Food Quality (L. E. Jeremiah, ed.), Marcel Dekker, 1996, p.85.*
7. *D. S. Reid, Fruit freezing, Processing Fruits: Science and Technology. Vol.1. Biology, Principles and Applications(L. P. Somogyi, H. S. Ramaswamy and Y. H. Hui, eds.), Technomic Publishing, Lancaster, CA, 1996, p.169.*

8. F. Wheaton and A. Lawson, Processing Aquatic Food Products, Wiley, New York, 1985.

9. J. Li and T. Lee, Bacterial ice nucleation and its potential application in the food industry, Trends Food Sci. Technol. 6: 259, 1995.

10. R. M. George, Freezing processes used in the food industry, Trends Food Sci. Technol, 4: 134, 1993.

11. A. Sormani, D. Maffi, G. Bertolo and D. Torreggiani, Texture and structural changes of dehydrofreeze thawed strawberry slices: Effects of different dehydration pretreatments, Food Sci. Tech. Int.5: 479, 1999.

12. R. N. Biswal, K. Bozorgmehr, F. D. Tompkins, and X. Liu, Osmotic concentration of green beans prior to freezing, J. Food Sci.56: 1008, 1991.

13. S. Park, M. S. Brewer, F. K. McKeith, P. J. Bechtel and J. Novakofski, Salt, cryo-protectants and preheating temperature effects on surimi-like material from beef or pork, J. Food Sci. 61: 790, 1996.

14. N. K. Kim, Mathematical modeling of cryogenic food freezing, Ph. D. Dissertation, Univ. of Georgia, Athens, 1993.

15. R. C. Martins and C. L. M. Silva, Frozen green beans (Phaseolus vulgaris, L.) quality profile evaluation during home storage, J. Food Eng. 64: 481, 2004.

16. B. Li and D.-W. Sun, Novel methods for rapid freezing and thawing of foods: A review, J. Food Eng. 54: 175, 2002.

17. B. Li and D.-W. Sun, Effect of power ultrasound on freezing rate during immersion freezing of potato, J. Food Eng. 55: 277, 2002.

18. H. Symons, Frozen foods, Shelf Life Evaluation of Foods(C. M. D. Man and A. A. Jones, eds.), Blackie Academic and Professional, London, 1994, p. 298.

19. T. P. Labuza, Drying food: technology improves on the sun, Food Technol. 30: 34, 1976.

20. A. Kramer, Effects of freezing and frozen storage on nutrient retention of fruits and vegetables, Food Technol. 33: 58, 1979,.

21. A. Kramer and J. W. Farquhr, Testing of time-temperature indicating and defrost devices, Food Technol. 30: 50, 1976.

22. O. Fennema, Frozen foods: Challenges for the future, Food Australia 45: 374, 1993.

Defrosting [1]

Many cooked foods that have been previously frozen require defrosting prior to consumption. Preferably, some frozen meats should be defrosted prior to cooking to achieve the best outcome: cooked through evenly and of good texture. Ideally, most frozen foods should be defrosted in a refrigerator to avoid significant growth of the pathogens. Food is often defrosted in one of following ways:

1. At room temperature: This is dangerous since the outside may be defrosted while the inside remains frozen [2].

2. In a refrigerator [2].
3. In a microwave oven [2].
4. Wrapped in plastic and placed in cold water [2] or undercold running water.

Frozen foods defrost at room temperature should be promptly consumed after cooking or discarded and never be refrozen or refrigerated since pathogens are not killed by freezing process.

Sources

1. *Wikipedia.*
2. *"Consumer Resources-NSF International". www. Nsf. org.*

2. Food Applications of Antimicrobial Compounds [1]

Many food products are perishable by nature and require protection from spoilage during their preparation, storage, and distribution to give them desired shelf life. The demand for minimally processed, easily prepared, and ready-to-eat fresh food products, globalization of food trade, and distribution from centralized processing pose major challenges for food safety and quality [1]. Traditional food preservation methods are less efficient in reducing the growth of food-borne pathogens in food products, and the ever-increasing demand for chemical-free food has paved the way for antimicrobials to be used in food industry [1a]. The use of antimicrobials is a new technology by the food industry to increase the shelf life of food and overcome the issues of food quality and safety [1a].

Many micro-organisms can cause undesirable reactions that deteriorate flavor, odor, color, sensory, and textural properties of foods. In packaged foods, growth and survival of common spoilage and pathogenic micro-organisms such as ***Listeria monocytogenes, Escherichia coli* O157, *Salmonella, Staphylococcus aureus, Bacillus cereus, Campylobacter, Clostridium perfringens, Aspergillus niger,*** **and** ***Saccharomyces cerevisiae*** are affected by a variety of following factors [1, 2, 3]:

i. Intrinsic factors- such as pH and presence of oxygen.
2. Or by extrinsic factors associated with storage conditions, including temperature, time, and relative humidity.

Numerous efforts are conducted to find natural alternatives to prevent bacterial and fungal growth in foods. Because of the great consumer awareness and concern regarding synthetic chemical additives, now foods preserved with natural additives have become very popular.

Natural antimicrobials from different sources are used to preserve food from spoilage and pathogenic micro-organisms. Plants (herbs and spices, fruits and vegetables, seeds and leaves) are the main source of antimicrobials and contain many essential oils (rosemary, sage, basil, oregano, thyme, cardamom, and clove) that have preservation effect against different micro-organisms. These essential oils are very effective against many pathogenic and spoilage micro-organisms like Salmonella, *Escherichia coli, Listeria monocytogenes, Campylobacter* spp., and *Staphylococcus aureus* and help to increase their quality and shelf stability. These antimicrobial compounds are also used in combination with edible food coatings and inhibit the ability of micro-organisms to grow on the surface of food and food products [1a].

Sources

1. *Annalisa Lucera, Cristina Costa, Amalia Conte and Matteo A. Del Nobile, Food Applications of Natural antimicrobial compounds, Front Microbiol. 2012; 3: 287.*

1a. *Muhammad Sajid Arshad and Syeda Ayesha Batool, Natural Antimicrobials, their sources and Food Safety, 2017, DOI; 10.5772/intechopen. 70197.*

2. *Singh, T. K., Drake, M. A.and Cadwallader, K. R., 2003, Flavour of Cheddar cheese: a chemical and sensory perspective. Comp. Rev. Food Sci. Food Saf., 2, 139-162.*

3. *Rydlo, T., Miltz, J., and Mor, A., 2006, Eukaryotic anti-microbial peptides: promises and premises in food safety, J. Food Sci., 71, 125-135.*

Natural and Synthetic Antimicrobial Compounds

These are used to preserve food products. The antimicrobials could be of natural or synthetic type, but natural antimicrobials are gaining much importance than synthetic ones. Even though synthetic preservatives are approved by government agencies for human use, many of these preservatives still threaten our health. Thus, researchers give more importance toward the potential of natural products for their antimicrobial activities [1a]. The details are given below:

A. Synthetic Antimicrobial Compounds

The sulphonamides were the first compounds found to suppress bacterial infections. Sulphanilamide seemed a likely metabolic breakdown product of Prontosil and was found to possess antibacterial properties. The more effective derivatives were searched.

- ✰ The first of these to come into general use was sulphapyridine (M and B 693), which in turn was replaced by compounds with less unpleasant side-effects.
- ✰ Several of these early sulphonamides are still in regular use, the best known being sulphadiazine and sulphadimidine.
- ✰ Sulphafurazole has quite similar biological properties, but has the advantage of higher solubility. Structures of these sulphonamides are shown.
- ✰ The sulphonamides act against a wide range of bacteria but their main success immediately following their discovery was in the treatment of streptococcal infections and pneumococcal pneumonia.
- ✰ More recently a combination of a sulphonamide with trimethoprim has opened a vast new area of use. The structural requirements for antibacterial activity in the sulphonamide series are relatively simple. Starting from sulphanilamide the modifications have been almost entirely variations in substitution on the nitrogen of the sulphonamide group. Substitution on the aromatic amino group causes loss of activity.

B. Natural Antimicrobial Compounds [1, 1a]

An antimicrobial is an agent that kill micro-organism or stops their growth. These agents have long been researched for their effectiveness to kill or to inhibit growth of micro-organism in and on food. This agent has been used in an effort to increase food safety for the consumer as well as to increase the shelf life of food products and it should be added to the food during processing. It is either removed, converted into normal food constituents or functional additives that leave insignificant nonfunctional residuals [2a]. The use of these agents depends on following factors [2a]:

1. Desired effect.
2. Legal limits of use.
3. Effect on the food.

The direct addition of natural compounds to food is the most common method of application, even if numerous efforts have been made to find alternative solutions to the aim of avoiding undesirable inactivation. Dipping, spraying, and coating treatment of food with active solutions are currently applied to product prior to packaging as valid options. To inhibit growth of undesirable micro-organisms in food, the anti-microbials can be directly added into the product formulation, coated on its surface or incorporated into the packaging material. Direct incorporation of active agents into food results in an immediate but short-term reduction of bacterial populations, while the antimicrobial films can maintain their activity for a long period of time [2].

Toxicologic Safety of Antimicrobial Compounds [2a]

1. It is essential that an additive for use as a food antimicrobial be safe for human consumption.
2. Antimicrobials occur in nature. It is often thought that naturally occurring antimicrobials are less toxic than synthetic compounds but it is not true. A naturally occurring antimicrobial must be shown to be nontoxic either by animal testing or by its continuous consumption by consumers as a food over a long period.
3. It is also important that the antimicrobial be metabolized and excreated by the body. The compound or its breakdown products should also not result in buildup of residues in body.

Types of Natural Antimicrobials

The following main natural compounds are used to preserve food products:

i. Essential Oils Derived from Plants (Basil, thyme, oregano, cinnamon, clove, and rosemary)

Plant essential oils (Generally Recognized as Safe-GRAS) are gaining a wide interest in food industry for their potential as decontaminating agents. The essential oil fractions have a wide spectrum of antimicrobial activity, against food-borne pathogens and spoilage bacteria [3]. The antimicrobial activity of plant essential oils is due to their chemical structure, in particular to the presence of hydrophilic functional groups, such as hydroxyl groups of phenolic components and/or lipophilicity of some essential oil components [4]. Allyl-isothiocyanate is the major antimicrobial component of mustard and horseradish oil. It has been found to be more effective against Gram-negative bacteria with less or no effect on lactic acid bacteria.

ii. Enzymes Obtained from Animal Sources (Lysozyme, Lactoferrin)

The enzymes, another group of natural compounds found application in food as valid preservatives. Lysozyme, a lytic enzyme found in foods (milk and eggs), can hydrolyze β-1, 4 linkages between N-acetylmuramic acid and N-acetylglucosamin [5]. It is well known that lysozyme is bactericidal against Gram-positive bacteria, whereas it is essentially ineffective against Gram-negative bacteria, owing to the presence of a lipopolysaccharide layer in the outer membrane.

iii. Bacteriocins from Microbial Sources (Nisin, Natamycin) [1]

The use of bacteriocin-producing lactic acid bacteria or their more or less purified bacteriocins has been also receiving increased interest. Bacteriocins are small bacterial peptides that show strong antimicrobial activity against closely related bacteria. Nisin is a polypeptide produced by *Lactococcus lactis* spp. It has been approved as a food additive with GRAS status. Use of nisin in conjunction with ethylenediamine tetra-acetic acid (EDTA) may increase the effectiveness. Moreover, the effect of nisin can be enhanced by using exposure to chelating agents, sub-lethal heat, osmotic shock and freezing, because these treatments make the cell wall of Gram-negative.

iv. Organic Acids (Sorbic, Propionic, Citric acid)

Organic acids and their salts are widely used as chemical antimicrobial agents. The most effective organic compounds are acetic, lactic, propionic, sorbic and benzoic acid. Their anti-microbial effect is based on the increase in proton concentration thereby lowering the external pH. Organic acids may affect the integrity of microbial cell membrane or cell macromolecules or interfere with nutrient transport and energy metabolism, causing bactericidal effect [6]. Mixtures of acids could exert a wider antimicrobial activity than a single organic acid [7].

v. Naturally Occurring Polymers (Chitosan)

Chitosan has been used in medical, food, agricultural, and chemical industry, mainly due to its high biodegradability and antimicrobial properties. The biological activity of chitosan depends on its molecular weight, degree of deacetylation and derivatisation, such as degree of substitution, length, and position of a substitute in glucosamine units of chitosan, pH of chitosan solution and the target organisms [8]. It is commercially produced from crab and shrimp shell wastes, with different deacetylation grades and molecular weights and, hence, it possess different functional properties, like emulsification ability, dye binding, and gelation.

Application of Antimicrobial Compounds [1]

The uses of natural compounds as antimicrobial compounds, to control microbiological and physico-chemical shelf life of following main food categories, are given below:

1. Minimally Processed Food and Vegetables Juices [1]

Minimally processed products *e.g.* fresh-cut fruit and vegetables are facing the difficulties in preserving their fresh-like quality during prolonged periods. Dipping, impregnation, coating, and spraying are the different ways of applications of active agents to fresh-cut fruit and vegetables but among them, the most recent results on the application of active compounds to ready-to-eat fruit and vegetables deal with coating systems. Some relevant examples are reported below:

i. Malic acid in combination with various stabilizing compounds was used for fresh-cut apples [9].

ii. The combined effect of chemical dip and/or edible coating and/or controlled atmosphere (CA) on quality of fresh-cut banana was investigated [10]. Banana slices were dipped into a solution containing 1 per cent (w/v) calcium chloride, 0.75 per cent (w/v) ascorbic acid, and 0.75 per cent (w/v) cysteine and/or combined with a carrageenan coating and/or combined with CA (3 per cent O_2 + 10 per cent CO_2). Dipping combined with CA treatment prevented product weight loss and increased polyphenol oxidase activity. The combined strategies prevented microbial growth after 5 days of storage at 5°C.

iii. The combined effects of malic acid and essential oils of cinnamon, palmarosa, and lemongrass (0.3 and 0.7 per cent) and their main active compounds (eugenol, geraniol, and citral, 0.5 per cent) on microbiological and physico-chemical shelf-life of fresh-cut "Piel de Sapo" melon (*Cucumis melo* L.) was investigated [11]. The incorporation of essential oils or their active compounds into the edible coating prolonged the microbiological shelf-life by more than 21 days.

iv. Natural volatile compounds such as methyl jasmonate, ethanol, tea tree oil, and garlic oil were applied on fresh-cut tomato stored at 5°C for 15 days. Ethanol combined with methyl jasmonate was more effective in suppressing microbial proliferation than each single compound. In addition, this combination preserved firmness and color better than the other antimicrobial preservatives [1].

v. The antimicrobial effect of essential oils against L. innocua inoculated into apple pieces before coating was also tested. A release system of antimicrobial volatiles was adopted by encapsulation of garlic oil in β-cyclodextrin and tested on microbial growth and sensory quality of fresh-cut tomato. Grape-fruit seed extract was used as antimicrobial compound into a coating of sodium alginate to prolong the shelf-life of minimally processed kiwifruits. The combination of an active compound to an alginate-based coating delayed microbial growth, whereas the sole dipping treatment was inefficient [1]. The combined use of modified atmosphere packaging (MAP) and coating treatments further prolonged the shelf-life up to 13 days [12].

vi. The effectiveness of chitosan incorporated in the edible methyl cellulose coating on the microbiological quality of fresh-cut cantaloupe was evaluated [13]. The coating application preserved the overall visual quality, the microbial proliferation and reduced surface whiteness during storage. While the content of total phenolics markedly increased in coated carrot sticks stored under moderate gas levels, it was controlled under low O_2 and high CO_2 levels. Film based on methyl cellulose incorporating chitosan and chitosan/methyl cellulose film incorporating vanillin were applied to fresh-cut cantaloupe and pineapple to control microbial quality during storage at 10°C [1]

vii. Juices are food very susceptible to yeasts attack. Clary, sage, juniper, lemon, and majoram essential oil were chosen to preserve apple juice, as being efficient in *in vitro* test and not containing phenolics but alcohol terpenoids, linalool (clary sage), and terpinen-4-ol (marjoram), the cyclic monoterpenes α-pinene (juniper) and limonene (lemon) [1].

2. Dairy Products [1]

Fresh dairy products are ready-to-eat foods easily contaminated by undesirable micro-organisms(spoilage micro-organisms -may produce unwanted visual appearance and diminish the commercial value of cheese, and pathogens - affect product safety). Fungal spoilage can also occur in dairy products. Recently, some studies have recorded the efficacy of natural compounds, alone or in combination with other preservation methods, when directly applied to milk [14] or to cheese by spraying, immersing, or dusting the products. Antimicrobials may also be spread onto the packaging materials that come in contact with the cheese or incorporated into the plastic films used for packaging [15].

i. Cheese

Some examples of applications of natural active agents to cheese are reported below:

☆ **Mozzarella cheese**: A study on the effectiveness of lysozyme and EDTA on microbiological shelf life of mozzarella cheese was made [16]. Mozzarella was packaged in a brine that contained lysozyme (0.25 mg mL^{-1}) and different amounts of EDTA (10, 20 and 50 mmol L^{-1}), and stored at 4 ± 1°C for 8 days. The packaging system significantly inhibited growth of coliforms and Pseudomonadaceae, without affecting the typical lactic acid bacteria.

The addition of different concentrations of lemon extract in the brine of mozzarella and in a gel solution made up of sodium alginate was evaluated [15]. Shelf-life tests were run at 15°C to simulate thermal abuse. An increase in shelf-life of all active packaged mozzarella cheese was observed

- ☆ **Ricotta cheese:** Edible coatings made of galactomannans incorporating nisin were tested against *L. monocytogenes* in Ricotta cheese. The system not only help in retarding the growth of *L. monocytogenes* but also help in the maintenance of water content, therefore reducing cheese weight loss [17].
- ☆ **Kashar cheese:** Natamycin was incorporated into wheat gluten and methyl cellulose biopolymers and tested against A. niger and P. roquefortii inoculated on surface of fresh kashar cheese [18].
- ☆ **Semi hard cheese [1]:** Natamycin was also used in combination with chitosan. In particular, the effects of the chitosan coating containing natamycin (0.50 mg/mL) on semi-hard cheese were assessed. The natamycin coated samples presented a decrease of moulds and yeasts compared to the control after 27 days of storage.

ii. Yoghurt

Three essential oils, thyme, marjoram and sage, were added to concentrated yoghurt (labneh) at concentrations of 0.2,0.5, and 1.0 ppm. The better concentration of each essential oil was 0.2 ppm that allowed obtaining a shelf-life up to 21 days [19].

The anise volatile oil and its oleoresin added to yogurt (prepared from buffalo's milk) (stored up to 20 days at $4 \pm 1°C$) at various concentrations (0.1–1.0 g L^{-1}) were effective in controlling spoilage micro-organisms [1].

iii. Cow Milk

Different concentrations of active substances were added during processing cow's milk. Lemon extract and chitosan showed a good compromise between the antimicrobial effectiveness and the sensory impact caused by their addition, thus promoting a satisfactory shelf-life increase (129 per cent) [1].

3. Meat-Based Products

Spoilage of meat products contributes to deterioration of texture and change in flavor and color. The use of natural antibacterial compounds, (s extracts of spices and herbs, essential oils, organic acids, salts, and bacteriocins) is reported to improve the shelf life of meat [20]. Some examples are given below:

Pork Sausages

The combined use of natural antimicrobials, such as lemon and thymol, and MAP was suggested [1] to improve the shelf life of reduced pork back-fat content sausages. In particular, the application of thymol and thymol-MAP limited the development of *Pseudomonas* spp., responsible for sausages' unacceptability.

Chitosan (0.5 per cent and 1 per cent) added individually or in combination with nitrites (150 ppm) as ingredients was tested to protect fresh pork sausages from microbial spoilage [1].

It was reported [21] that the addition of a mixture of organic acids (sodium lactate 90 per cent and sodium acetate 10 per cent) at different concentrations (from

0 to 20 g/Kg) on Marguez sausages, made with lamb and beef, significantly reduced microbial cell loads during storage at 8°C.

Pork Meat

Best shelf life values were obtained for pork meat added with garlic and marjoram oil [1].

Pork Chops

Effects of pork chops dipped in organic acids, such as ascorbic acid (500 ppm) and citric acid (250 ppm) individually or in combination, packaged under MAP and vacuum and stored at 1°C were studied [1]. Ascorbic acid dipping reduced psychrotrophic microbial count, while ascorbic and citric acids improved lipid stability. The obtained results were enhanced by packaging under MAP conditions.

Chicken Meat

The use of bay essential oil combined to MAP without oxygen (20 per cent CO_2-80 per cent N_2) was suggested to control *L. monocytogenes* and *E. coli* growth and also to extend the shelf life of naturally contaminated ground chicken meat [22].

A relevant preservation effect for fresh chicken breast meat, stored at 4°C, was obtained by dipping meat in oregano oil, prior to packaging under MAP [1].

Meat

Antioxidant and antibacterial effects of rosemary, orange, and lemon extracts was also investigated in cooked Swedish-style meat-balls. Results indicated that significant advantages were obtained using rosemary and citrus extracts in rancidity-susceptible meat products; however, only rosemary slightly reduced lactic acid bacteria [23].

Dipping of thyme and oregano oil in concentrations of 0.1 and 0.3 per cent were carried out to improve the shelf life of meat-based products [1]. The combination of dipping and MAP extended the shelf life to about 22 days, against 7 days of the control sample packaged in air.

4. Fish-Based Products

Because of the biological composition, fresh fish is a highly perishable product. The activity of spoilage seafood micro-organisms is the main cause of deterioration in quality of fish. These micro-organism provoke loss of essential fatty acids, fat-soluble vitamins and protein functionality, production of biogenic amines, and formation of off-odors [24]. The treatments with natural compounds are effective preservation methods for fish products. Some examples are provided below:

- ✰ The use of thyme (1 per cent) and laurel essential oil (1 per cent) is proposed to extend the shelf life of bluefish by about 3-4 days [25].
- ✰ It is demonstrated [1] that combination of MAP and thyme essential oil (0.1 per cent) resulted in a significant shelf life extension of fresh Mediterranean swordfish fillets. In particular, the addition of thyme essential oil extended product shelf life (13 days) if compared to the control (8 days), whereas

its combination to MAP conditions further prolonged product shelf life (about 20.5 days).

- ☆ Researchers [26] highlighted the possibility to extend the microbial acceptability limit of fresh fish burgers by using a mix of thymol, grapefruit seed extract, and lemon extract. The mixture of the three natural compounds prolonged the sensory quality without compromising the flavor of fish. In a subsequent work, thymol, grapefruit seed extract, and lemon extract were used in combination to MAP to demonstrate that MAP further enhanced the effects of the natural active compounds [1].
- ☆ It is reported [27] that the use of a coating with chitosan and cinnamon essential oil improved trout fillet shelf life (16 days vs. 10 days of the control) and in particular it enhanced texture, odor, and color.
- ☆ To control quality of northern snakehead fish fillets at refrigeration temperature cinnamon, coatings with nisin and EDTA, alone and in combination were used [1].
- ☆ Acetic acid, glucono-delta-lactone and chitosan were tested, individually and in combination, to inhibit microbial growth in surimi.

5. Cereal-Based Products

The shelf life of cereal based products specially bread is generally limited due to the staling phenomenon and fungi spoilage, in particular moulds. The treatments with natural compounds are effective preservation methods for enhancing the shelf life of cereal based products.

i. Bread

Some examples are given below:

- ☆ It was reported that chitosan coating improved bread quality by inhibiting microbial growth and retarding oxidation and staling. A reduced microbial proliferation was obtained for bread coated with chitosan during storage at room temperature [28].
- ☆ Different applications of citrus peel essential oils in bread, were reported [29]. Results demonstrated that the oils influenced sensory characteristics and delayed microbial growth. Maximum inhibitory effect against moulds and bacteria was achieved by spraying peel essential oil. The combination of MAP and mustard oil was proposed for wheat and rye bread artificially inoculated with moulds [30].
- ☆ The studies on colony count in bread at different storage time showed that treatment containing 0.32 per cent of suhanjna, 3 per cent of lecithin, and 0.1 per cent of ascorbic acid proved to be most effective. The different combination of the three selected natural additives improved bread shelf life. In particular, lower cell loads of yeasts and moulds were observed for bread with lecithin and ascorbic acid [31].

ii. Pasta

Natural active compounds were also applied to fresh pasta that is a product easily perishable for its high water content. Some examples are given below:

- ✰ Researchers [32] proposed the use of different natural antimicrobial compounds such as thymol, lemon extract, chitosan, and grapefruit seed extract at different concentrations (2000 mg/kg and 4000 mg/kg) to improve the microbiological stability of refrigerated amaranth-based fresh pasta. Results pointed out that chitosan were the most successful among the investigated compounds in slowing down the spoilage, whereas lemon extract was the less effective. In a subsequent work, the antimicrobial activity of chitosan in combination with different MAP was tested.

iii. Noodles

To improve the shelf life of yellow alkaline noodle, scientists [33] tested the antibacterial activity of ethanol and water extracts of six types of leaves against the principal spoilage microbial groups of this product. It was found that ethanol extracts of aromatic leaf, *Murraya koenigii* (curry leaves), added in yellow alkaline noodle contributed to improve shelf life better than the other extracts.

Sources

1. *Annalisa Lucera, Cristina Costa, Amalia Conte and Matteo A. Del Nobile, Food Applications of Natural antimicrobial compounds, Front Microbiol. 2012; 3: 287.*

1a. *Muhammad SajidArshad andSyeda Ayesha Batool, Natural Anti-microbials, their sources and Food Safety, 2017, DOI; 10.5772/intechopen. 70197.*

2. *Appendini, P., and Hotchkiss, J. H., 2002, Review of antimicrobial food packaging. Inno. Food Sci. Emerg. Technol, 3, 113-126.*

2a. *Antimicrobial agents used in foods, S. Sharma, D. Singh, R. Singh and A. Pathania, slideshare.*

3. *Gutierrez, J., Barry-Ryan, C., and Bourke, P., 2008, The antimicrobial efficacy of plant essential oil combinations and interactions with food ingredients. Int. J. Food Microbiol. 124, 91-97.*

4. *Dorman, H. J. D., and Deans, S. G., 2000, Antimicrobial agents from plants: antibacterial activity of plant volatile oils. J. Appl. Microbiol. 88, 308-316.*

5. *Cunningham, F. E., Proctor, V. A., and Goetsch, S. J., 1991, Egg white lysozyme as food preservative: an overview. Worlds Poult. Sci. J. 47, 141-163.*

6. *Ricke, S. C., 2003, Perspectives on the use of organic acid and short chain fatty acid as antimicrobials. Poult. Sci. 82, 632-639.*

7. *Theron, M. M., Lures, J., and Rykers, F., 2010, "Application of organic acid in food preservation" in Organic Acids and Food Preservation, eds Taylor and Francis Group (Boca Raton, FL: CRC Press), 51-95.*

8. *No, H. K., Meyers, S. P., Prinyawiwatkul, W., and Xu, Z., 2007, Application of chitosan for improvement of quality and shelf life of foods: a review. J. Food Sci. 72, 100-187.*

9. *Raybaudi-Massilia, R. M., Mosqueda-Melgar, J., Sobrino-Lòpez, A., Soliva-Fortuny, R., and Martín-Belloso, O., 2009, Use of malic acid and other quality stabilizing compounds to assure the safety of fresh-cut "Fuji" apples by inactivation of Listeria monocytogenes, Salmonella enteritidis and Escherichia coli O157, H7. J. Food Saf. 29, 236-252.*

10. *Bico, S. L. S., Raposo, M. F. J., Morais, R. M. S. C., and Morais, A. M. M. B., 2009, Combined effects of chemical dip and/or carrageenan coating and/or controlled atmosphere on quality of fresh-cut banana. Food Contr. 20, 508-514.*

11. *Raybaudi-Massilia, R. M., Mosqueda-Melgar, J., and Martín-Belloso, O., 2008., Edible alginate-based coating as carrier of antimicrobials to improve shelf-life and safety of fresh-cut melon. Int. J. Food Microbiol. 121, 313–327.*

12. *Mastromatteo, M., Mastromatteo, M., Conte, A., and Del Nobile, M. A., 2011a, Combined effect of active coating and MAP to prolong the shelf life of minimally processed kiwifruit (Actinidia deliciosa cv. Hayward). Food Res. Int. 44, 1224-1230.*

13. *Krasaekoopt, W., and Mabumrung, J., 2008, Microbiological Evaluation of Edible Coated Fresh-Cut Cantaloupe. Kasetsart J. (Nat. Sci.) 42, 552-557.*

14. *Cava, R., Nowak, E., Taboada, A., and Marin-Iniesta, F., 2007, Antimicrobial activity of clove and cinnamon essential oils against Listeria monocytogenes in pasteurized milk. J. Food Prot. 70, 2757-2763.*

15. *Conte, A., Scrocco, C., Sinigaglia, M. and Del Nobile, M. A., 2007, Innovative active packaging systems to prolong the shelf life of Mozzarella cheese. J. Dairy Sci. 90, 2126.*

16. *Sinigaglia, M., Bevilacqua, A., Corbo, M. R., Pati, S., and Del Nobile, M. A., 2008, Use of active compounds for prolonging the shelf life of mozzarella cheese. Int. Dairy J. 18, 624-630.*

17. *Martins, J. T., Cerqueira, M. A., Souza, B. W. S., Avides, M. D. C., and Vicente, A. A., 2010, Shelf life extension of ricotta cheese using coatings of galactomannans from nonconventional sources incorporating nisin against Listeria monocytogenes. Agric. Food Chem. 58, 1884-1891.*

18. *Ture, H., Eroglu, E., Ozen, B., and Soyer, F., 2011, Effect of biopolymers containing natamycin against Aspergillus niger and Penicillium roquefortii on fresh kashar cheese. Int. J. Food Sci. Technol. 46, 154-160.*

19. *Otaibi, M. Al., and Demerdash, H. El., 2008, Improvement of the quality and shelf concentrated yoghurt (labneh) by the addition essential oils. Afr. J. Microbiol. Res. 2, 156-161.*

20. *Jamilah, M. B., Abbas, K. A., and Rahman, R. A., 2008, A review on some organic acids additives as shelf life extenders of fresh beef cuts. Am. J. Agric. Biol. Sci.3, 566-574.*

21. *Ayachi, B. E., Daoudi, A., and Benkerroum, N., 2007, Effectiveness of commercial organic acids' mixture (Acetolac™) to extend the shelf life and enhance the microbiological quality of Merguez sausages. Am. J. Food Technol. 2, 190-195.*

22. *Irkin, R., and Esmer, O. K., 2010, Control of Listeria monocytogenes in ground chicken breast meat under aerobic, vacuum and modified atmosphere packaging conditions with or without the presence of bay essential oil at 4°C. Food Sci. Technol. Res. 16, 285-290.*

23. *Fernàndez-Lòpez, J., Zhi, N., Aleson-Carbonell, L., Perez-Alvarez, J. A., and Kuri, V., 2005, Antioxidant and antibacterial activities of natural extracts: application in beef meatballs. Meat Sci. 69, 371-380.*

24. *Gram, L., and Dalgaard, P., 2002, Fish spoilage bacteria-problems and solutions. Curr. Opin. Biotechnol. 13, 262–266.*

25. *Erkan, N., Tosun, S. Y., Ulusoy, S., and Uretener, G., 2011, The use of thyme and laurel essential oil treatments to extend the shelf life of bluefish (Pomatomus saltatrix) during storage in ice. J. Consum. Protec. Food Saf.6, 39-48.*

26. *Corbo, M. R., Speranza, B., Filippone, A., Granatiero, S., Conte, A., Sinigaglia, M., and Del Nobile, M. A., 2008, Study on the synergic effect of natural compounds on the microbial quality decay of packed fish hamburger. Int. J. Food Microbiol. 127, 261-267.*

27. *Ojagh, S. M., Rezaei, M., Razavi, S. H., and Hosseini, S. M. H., 2010, Effect of chitosan coatings enriched with cinnamon oil on the quality of refrigerated rainbow trout. Food Chem. 120, 193-198.*

28. *No, H. K., Meyers, S. P., Prinyawiwatkul, W., and Xu, Z., 2007, Application of chitosan for improvement of quality and shelf life of foods: a review. J. Food Sci. 72, 100-187.*

29. *Rehman, S., Hussain, S., and Nawaz, H., 2007, . Inhibitory effect of citrus peel essential oils on the microbial growth of bread. Pak. J. Nut. 6, 558-561.*

30. *Suhur, K. I., and Nielsen, P. V., 2005, Inhibition of fungal growth on wheat and rye bread by modified atmosphere packaging and active packaging using volatile mustard essential oil. Inst. Food Technol. 70, 37-44.*

31. *Latif, A., and Masud, T., 2006, Use of natural preservative in bread making. Am. J. Food Technol. 1, 34-42.*

32. *Del Nobile, M. A., Di Benedetto, N., Suriano, N., Conte, A., Lamacchia, C., Corbo, M. R., and Sinigaglia, M., 2009c, Use of natural compounds to improve the microbial stability of Amaranth-based homemade fresh pasta. Food Microbiol. 26, 151-156.*

33. *Rosyid, T. A., Karim, R., Adzahan, N. M. and Ghazali, F. M., 2011, Antibacterial activity of several Malaysian leaves extracts on the spoilage bacteria of yellow alkaline noodles. Afr. J. Microbiol. Res. 5, 898-904.*

3. Modern Packaging System [1, 2, 3]

Packaging is the science, art and technology of enclosing or protecting products for distribution, storage, sale, and use. Packaging also refers to the process of designing, evaluating, and producing packages. Packaging can be described as a coordinated system of preparing goods for transport, warehousing, logistics, sale, and end use. Packaging contains, protects, preserves, transports, informs, and sells [1a]. Packaging maintains the benefits of food processing after the process is complete, enabling foods to travel safely from point of origin to the point of consumption [1c].

Packaging plays a crucial role in reducing the rate of quality loss in foods. Vacuum packaging, modified atmosphere packaging, and post-pasteurization shrink bag packaging are just a few of the many methods used to package food products.

Food packages are required to have multi-functions in terms of chemical, physical, and biological alterations of the food to prolong its shelf life. Chemically, the packages should be able to control and/or prevent oxidation and some other chemical reactions(hydrolytic rancidity and Maillard reaction), in food products. The packages should maintain the moisture level in the products which is a critical parameter controlling bread staling and changes in texture and physical appearance. Most importantly, microbial growth in food during storage is a serious problem and a major cause to shorten food shelf life. Packages that can inhibit microbial proliferation are currently in demand [4, 5]. Nowadays, novel food packaging technologies, such as active packaging, aseptic packaging, intelligent packaging, nano-packaging, and bioactive packaging intentionally associated with the food products have proved to be best technological research areas [6b].

1. Aseptic Packaging [1b]

Aseptic packaging consists of filling and sealing a sterilized packaging material with a sterilized product. These packaging systems are filled with pre-sterilized packaging and then sealed under strictly sterile conditions. To select the appropriate thermal stabilization conditions, the pH value of the product is of fundamental importance. Useful properties of systems with aseptic packaging are given below:

i. A more short-term exposure to high temperature, allows to save nutritional advantages.

ii. Taste.

iii. Packaging in a sterile area using sterile air overpressure.

iv. Long shelf life of products without refrigerated storage.

v. Obtaining additional protection of products by introducing inert gas into the packaging.

vi. High level of disinfection packaging material or packaging.

The versatility of aseptic packaging systems for products determines their ability to transport and store for long periods without using reduced temperatures.

Packaging Materials [2a]

Aseptic packaging material not only has to assure sterile conditions within the package and protect the product from physical damage, but also maintain the quality of the product inside the packaging [2c]. To achieve this, a laminate material is formed from the following components [2b]:

i. Semi-rigid paper: Paper (70 per cent) provides the stiffness, strength, and the efficient brick shape to the package.

ii. Aluminum: Aluminum (6 per cent) is located on the inside of the aseptic package, forming a barrier against light and oxygen, thereby eliminating the need for refrigeration and preventing spoilage without using preservatives [2d].

iii. Plastic: Low-density polyethylene (24 per cent), the most common plastic used for aseptic packaging, located on the innermost layer forms the seals that make the package liquid-tight.

Most packaging material used in aseptic packaging is made from plastics instead of metal or glass containers due to the relatively low cost of producing plastic material when compared to metal and glass. Plastics are lighter than metal or glass making them cheaper and easier to transport. Plastics also required much less energy to produce than metal and glass [2b]. These factors have made plastic the packaging material of choice for use in aseptic processing.

Advantages and Disadvantages of Aseptic Packaging

The following are the advantages and disadvantages of the aseptic packaging:

Advantages

i. **Convenience:** it is portable and light.
ii. **Food safety:** the process ensures that the food is safe from harmful bacteria.
iii. **No refrigeration is required:** which saves energy in the transportation and storage of the food product.
iv. **Less energy is required:** less energy to heat and sterilise the product and the container.
v. **Protection of nutritional content:** the food product contains more nutrients due to less heat damage.
vi. **Low packaging to product ratio:** there is more than 90 per cent food product to 10 per cent package.
vii. **Storage efficiency:** brick shapes are easier to pack in bulk than cans or jars.

Disadvantages

i. Recycling in some countries (Australia) is difficult, though other countries reuse aseptic carton packaging for paper towels, tissues and writing paper
ii. **Production costs:** due to a more complicated system, production costs increase, resulting in increased costs for the consumer

2. Active Packaging (AP) [1b, 6]

This is another packaging technique. Active packaging is a booming area of food packaging process that has preservation benefits on many food products. Because of timely advances in packaging process, growth in biotechnology and new consumer demands, active packaging has developed a trust in the consumers. This technology extends shelf life; maintain sensory quality besides ensuring quality and safety of food products [6b].

In this the packages not only act as a protective barrier but interact with the product in order to protect food from adulteration by, for example, moisture, oxygen, ethylene, and micro-organisms [6]. Active packaging emerged with the objective of consumer satisfaction in regard to demand for natural, recyclable, and bio-degradable packaging materials [6a]. In this packaging system, intrinsic conditions of the package are changed due to interaction of packaging material, food product, and the environment [6b]. The system of "active packaging" has a positive effect on the product, ensuring its high quality and storage time, and actively protects the product. Active packages are packaged in a modified atmosphere (MAP) or a modified gaseous medium (MAP) [1b]. Active packaging is a packaging system that changes the condition of the packaging and maintains these conditions throughout the storage period to extend shelf-life or to improve safety or sensory properties while maintaining the quality of packaged food [7]. Active packaging may be defined as "packaging that interferes with the internal condition of a packaged food product and changes the condition to extend the shelf life or improve food safety or sensory properties in order to maintain the quality". Active packaging techniques for food preservation can be divided into following three categories-

i. **Scavengers (absorbers)-**Absorbing systems remove undesirable compounds such as oxygen, carbon dioxide, ethylene, excessive water, taints and other compounds.
ii. **Releasing agents** – Releasing systems actively add or emit compounds to the packaged food or into the headspace of the package such as carbon dioxide, antioxidants and preservatives.
iii. **Others-** Other systems include self-heating, self-cooling systems, *etc.*

Gas absorbents and emitters control the atmospheric balance inside packaging. The controlled supply of active substances into the food via packaging helps to extend shelf life, and avoids unnecessary flavor [6b]. Examples of the common active packaging applications that are used within the food industry are oxygen scavengers, carbon dioxide scavengers/emitters, ethylene scavengers, preservative releasers, ethanol emitters, moisture absorbers, flavor/odor absorbers, and temperature control packaging (ICAR Online e-courses). Nowadays, oxygen absorbers can be added as small separate sachets in the package to have the most commercial significance in active food packaging [6b].

There are following two types of AP, have been developed [6] :

i. Sachet-based Packing

Sachets, containing active ingredients such as oxygen absorber, ethylene scavenger, and moisture absorber, were developed in the late 1970s in Japan. Although this technique is considerably practical for packaging industries, some disadvantages are inevitable. They cannot be used in liquid products or in the tight fitting film as their functionalities would be stifled. Furthermore, risks of accidental ingestion of these sachets are a serious concern [6]. Sachets are not well accepted by consumers since there is a possibility of accidental consumption with package contents by children [6b].

ii. Plastic Film-based Packaging

The aforesaid drawbacks can be alleviated if those functionality components are incorporated to the plastic film itself [6]. In the coming decades, the use of active packaging systems in the form of thin films would be becoming universal [6b].

Active Package Compounds

Some examples of active package compounds are tabulated below:

Sl.No.	*Active Package Compounds*	*Examples*
1	Oxygen absorbing	Chemical system-Catechol, Ascorbic acid oxidation , Ferrous carbonate. Enzymatic system-Glucose oxidase- glucose ,alcohol oxidase ethanol vapour.
2	Moisture absorbing	Silica gel, propylene glycol.
3	CO_2 absorbing / emitting	Iron powder-calcium hydroxide.
4	Ethanol emitting	Encapsulated ethanol.

3. Modified Atmosphere Packaging (MAP) [1]

This is food packaging in which the composition of the air surrounding the food inside the package is changed. Modified atmosphere packaging (MAP) is a system of constituting air in the package by a prepared mixture of gases before sealing. Once the package is sealed, no further control is exercised over the composition of the in-package atmosphere [1e]. Details of MAP are given in Chapter-22.

a. Vacuum Packaging [8]

An alternative to controlling or modifying the atmosphere is vacuum packaging, where all of the gas in the package is removed. This can be a very effective way of retarding chemical changes such as development of oxidative rancidity, but care needs to be taken to prevent the growth of the pathogen, *C. botulinum*, which grows under anaerobic conditions. A specific pasteurisation process, referred to as the psychrotrophic botulinum process, is applied to the packaged food to reduce its numbers to commercially acceptable levels. By using vacuum packaging, and chilled storage, greatly extended shelf-lives have been achieved [12a].

Vacuum packing is a method of packaging that removes air from the package prior to sealing. This gets rid of oxygen which affects food in different ways, including: odor, color, taste and texture. This method involves the following steps [9]:

i. Placing items in a plastic film package.
ii. Removing air from inside.
iii. Sealing the package.

Shrink film is sometimes used to have a tight fit to the contents. The intent of vacuum packing is usually to remove oxygen from the container to extend the shelf life of foods and, with flexible package forms, to reduce the volume of the contents and package [10]. Vacuum packing reduces atmospheric oxygen, limiting the growth of aerobic bacteria or fungi, and preventing the evaporation of volatile components. Vacuum packing reduces freezer burn by preventing the food from exposure to the cold, dry air. Depending on the product, atmosphere, temperature, and the barrier properties of the package, vacuum packaging extends the shelf life of many foods [11].

- It is also commonly used to store dry foods over a long period of time, such as cereals, nuts, cured meats, cheese, smoked fish, coffee, and potato chips (crisps). The shelf life of meats can be extended by vacuum packaging, particularly when used with modified atmosphere packaging [12].
- Vacuum packing can also be used to store fresh foods, such as vegetables, meats, and liquids, because it inhibits bacterial growth.

The packages must have low permeability of oxygen. The following table represents how long some foods can last while using a vacuum sealer:

Sl. No.	*Freezer Storage*	*Regular*	*Food Saver*
1	Vegetables	8 Months	2-3 Years
2	Berries	1-6 days	1-2 weeks
3	Cheese	1-2 weeks	4-8 Months

Shelf Life of Vacuum Packaged Foods [12a]

Food	*Stored in*	*Normal Shelf-life*	*Shelf-life of Vacuum Packaged Product*
Cheese	Refrigerator	1-2 weeks	4-8 months
Cookies, crackers	Room temperature (periodically opening)	1-2 weeks	3-6 weeks
Flour, sugar, rice	Room temperature	6 months	1-2 years
Nuts	Room temperature	6 months	2 years
Oils with no preservatives, like safflower, canola, corn oil	Room temperature	5-6 months	1-1.5 years

To ensure the food safety, vacuum packing is often used in combination with other food processing techniques, such as retorting or refrigeration, to inhibit the growth of anaerobic organisms [13] because in an oxygen-depleted environment, anaerobic bacteria can proliferate, potentially causing food-safety issues. Some pathogens of concern in vacuum packed foods are spore-forming non-proteolytic *Clostridium botulinum*, *Yersenia enterocolitica*, and *Listeria monocytogenes* [13].

Types of Vacuum Sealers [8]

The following types of vacuum sealers are being used for vacuum packing of foods:

i. Single Vacuum Chamber Sealer

Single chamber sealers require the entire product to be placed within the machine. A plastic bag is typically used for packaging. Once the product is placed in the machine, the lid is closed and air is removed. Then, there is a heat seal inside the chamber that will seal the bag, after sealing the bag the chamber is refilled with air by the automatic opening of a vent to the outside.

ii. Double Vacuum Chamber Sealer

Double chamber sealers require the entire product to be placed in a plastic bag within the machine. Once the product is placed in the machine on the seal bar, the lid is closed and air is removed. Then a seal bar inside the chamber seals the product in the bag, after sealing the bag the chamber is refilled with air by the automatic opening of a vent to the outside. These are commonly used for the following products:

Fresh meat, Processed meat, Cheese (hard and soft), Candy and chocolate.

iii. Rotary Belt Type Vacuum Sealers

This type of vacuum packaging machine or vacuum sealer features the same function as the double chamber vacuum packaging machine as a 'vacuum bag sealer' but the rotary belt vacuum packaging machine is more convenient, as the belt rotates automatically while the bags are placed to the sealing bar and vacuum sealing process completed. The vacuumed and sealed bags are automatically unloaded. Rotary belt type packaging machines are commonly used for the following foods:

Fresh meat, Processed meat, Seafood, Pickles, Cheese (hard and soft), Candy and chocolate.

iv. Automatic Belt Vacuum Chamber Machine

This type of sealers require the entire product to be placed in a plastic bag or flow wrapped pouch within the machine. The product travels on the conveyor belt, it is automatically positioned in the machine on the seal bar, the lid is closed and air is removed. Then a seal bar inside the chamber seals the product in the bag. After sealing the bag, the chamber is refilled with air by the automatic opening of a vent to the outside. The lid is then opened and the product removed. Automatic belt vacuum chamber packaging machines are commonly used for the following foods:

Fresh meat (large portions), Processed meat, Large sausage logs, Cheese (hard and soft)

b. Post-Pasteurization Shrink Bag Packaging

Post-Pasteurization Shrink bag packaging promotes food safety by providing a skin-tight, high- barrier packaging that withstands the rigors of a post-packaging surface pasteurization process. It increases the shelf-life for ready-to-eat deli products like: poultry, pork and beef. It can withstand high temperatures for short amounts of time. The bags allow high-speed sealing and ink adhesion, as well as clarity and skin-tight shrinking of the bag.

High Barrier Shrink Vacuum Bags [8]

The amount of shelf life enhanced by a vacuum bag is dependent on the structure in the material.

- ☆ A standard vacuum bag is composed of a PA/PE structure where PA is for puncture resistance and PE is for sealing.

☆ The high barrier category includes the usage of more layers focused on the prevention of oxygen permeability and therefore shelf life protection. The following two materials are used in high barrier structures: Polyvinylidene chloride (PVDC) and Ethylene vinyl alcohol (EVOH).

Shelf life indication can be effectively measured by how many cubic centimeters of oxygen can permeate through 1 square meter of material over a 24-hour period. A standard PA/PE bag allows on average 100 cubic centimeters, PVDC allows on average over 10, and EVOH on average 1 cubic centimeter.

Multi-layer structures allow the ability to use strong oxygen-barrier materials for enhanced shelf life protection. The Premium Pack structure is a good example of EVOH based high barrier shrink material.

Sources

1. *Clemons, Rachel, "Extending the shelf life of fresh food."* CHOICE, CHOICE Australia, *5June2017, www.choice.com.au/food-and-drink/food-warnings-and-safety/food-safety/articles/extending-the-shelf-life-of-fresh-food.*

1a. *Soroka, 2002, Fundamentals of Packaging Technology, Institute of Packaging Professionals ISBN 1-930268-25-4.*

1b. *Shupeyko A. M.and Ishevsky A. L., Modern Food Packaging Technology, World Scientists, 2019.*

1c. *Prasad P. and Kochhar A., 2014, Active Packaging in food industry-a review. Journal of Environmental Science, Toxicology and Food Technology 8(5): 01-07.*

1d. *Wyrwa, J.and Barska, A., 2017, Innovations in the food packaging market: active packaging- A review article, European Food Research Technology 243: 1681-1692.*

2. *https: //en. wikipedia. org/wiki/Shelf_life.*

2a. *Wikipedia.*

2b. *Handbook of food safety engineering. Sun, Da-Wen. Oxford: Wiley-Blackwell. 2011. ISBN 978-1444333343.*

2c. *Ansari, I. A.; Datta, A. K., 2003, "An Overview of Sterilization Methods for Packaging Materials Used in Aseptic Packaging Systems". Food and Bioproducts Processing. 81 (1): 57-65.*

2d. *Willhoft, Edward, 1993, Aseptic Processing and Packaging of Particulate Foods (1st ed.), London: Blackie Academic and Professional, pp. 1-192.*

3. *The Difference Between Packaged and Processed, Valley Food Storage, 24 Aug.2016, valley food storage.com/difference-packaged-processed.*

4. *Smith, J. P., Bakery products, R. T. Parry, Principles and Applications of Modified Atmosphere Packaging of Foods, Springer, Boston, MA, 1993.*

5. *Smith, J. P., Ooraikul, B. Koersen, B., Jackson, E. D., Lawrence, E. D., Novel approach to oxygen control in modified atmosphere packaging of bakery products, Food Microbiol., 3 (4), 1986, pp. 315-320.*

6. *Settakorn Upasenhttps: //www.sciencedirect.com/ science/article/pii/S2405844018317821 - ! and Piyachat Wattanachai, Packaging to prolong shelf life of preservative-free white bread, Heliyon, Vol.4, Issue 9, 2018.*

6a. *Lesiow, T. and Kosiorowska, M, 2006, Opakow-aniaaktywneiin-teligentne w przetworstwiemiesa. Czêœæ I. Gospodarka-Miesna 3: 12 ¬ $ per cent V-18.*

6b. *Mahendra Pal, Mridula Devrani and Angesom Hadush, Recent developments in food packaging technologies, Researchgate, 2019, and Beverage and Food World -Vol.46- No. 1, Jan. 2019.*

7. *Ozdemir, M., and Floros, J. D., 2004, Active food packaging technologies. Critical Reviews in Food Science and Nutrition 44: 185-193.*

8. *Wikipedia.*

9. *Perdue, R, 2009, Vacuum Packaging, in Yam, K L (ed.), Encyclopedia of Packaging Technology, Wiley (published 2010), ISBN 978-0-470-08704-6.*

10. *Soroka, W. Illustrated Glossary of Packaging Terminology (Second ed.). Institute of Packaging Professionals.*

11. *Gorris, L G M, 1992, "Modified Atmosphere and Vacuum Packaging to Extend the Shelf Life of Respiring Food Products" (PDF). Hort Technology. 2 (3): 303-309. Retrieved August 14, 2018.*

12. *Seidman, S C, 1983, "Vacuum Packaging of Fresh Beef: A Review". Journal of Food Quality. 6 (1): 29-47.*

12a. *Vacuum pack and MAP, Lesson-23, http: //ecoursesonline.iasri.res.in/mod/page/view.*

13. *Mills, John., Donnison, Andrea; Brightwell, Gale (2014). "Factors affecting microbial spoilage and shelf-life of chilled vacuum-packed lamb transported to distant markets: A review", Meat Science. 98 (1): 71-80.*

4. Nanotechnology in Food Packaging [1]

The details on application of Nanotechnolgy in food processing are already discussed in Chapter-18. Here applications of Nano technologies in food packaging are discussed.

Nanopackaging is also highly recommendable since it releases different substances such as flavors, enzymes, antioxidants, antimicrobials, and nutraceuticals to extend the shelf life of the food products [2a]. Nanotechnology is new system which involves the manipulation of matter at a very small scale generally between 1 and 100 nanometres. Nanotechnology has diversified and wide applications in areas of packaging *etc.* Currently, over 400 global companies are using nanotechnology for food processing and packaging [2a, 2b]. Recent research on potential of Nanotechnologies in food packaging has highlighted the wide ranging food applications, including novel food packaging [2]. Potential benefits were reported particularly in food-package interaction and monitoring of quality and freshness [3].

Nanoparticles based sensors, array biosensors, nanocantilevers, nano-test strips, nanoparticles in solution, and electronic noses are the various types of nanosensors used in the food packaging plants [2a]. Nanosensor are extremely small device than can bind to thing which is to be detected and send back a signal. These tiny sensors are capable of detecting and responding to physico-chemical (sensors) and biological signal (biosensors), transferring that response into a signal or output that can be used by humans.

- ✰ Various nanosensors were developed for detection of internal and external conditions in food packaging [4]. Nano-sensors in package are important to trace the conditions of containers, and products and can detect gases in the food if there is food spoilage and package color change [5a].
- ✰ Nanosensors can be used to determine microbes, contaminants, pollutants, and ultimately the freshness of the food [5]. Film packed with silicate nanoparticles keeps freshness of food by reducing oxygen flow into the package and avoiding leaking of moisture out of package. This packaging technology detects the presence of microbial pathogens, toxins, pesticides, and other food spoiling agents in food within 2-7 days [2a, 5b].

A variety of nanomaterials are introduced as functional additives to food packaging, such as nanotitanium dioxide, titanium nitride nanoparticle, silver nanoparticle, nanoclay, and nano-zinc oxide [5a]. Nanoparticles can reduce the amount of material used and improve packaging function. It can also increase tensile strength up to forty percent and this decreases the amount of film used in pouch producing. Nanoparticles can increase thermal stability up to 350 percent and that keeps food fresh and can be reheated it in its original packaging. It can enhance barrier properties against gas permeations such as oxygen and carbon dioxide, ultraviolet permeation, moisture, and volatile compounds, develope active antimicrobial surfaces, and creates nano-biodegradable packaging materials [6, 7].

Food contact materials have been developed with improved flexibility, gas barrier properties, temperature control and moisture stability due to the inclusion of nanoscale fillers. The effect of nanopackaging has on a food depends on the composition of the nanomaterial that involved [2]. In addition to protection, nanofillers reduces the use of plastics as packaging materials and also required less energy for their production and biodegradation, these make them more friendly to environment [8]. Nanomaterials and nanoparticles may include any of the following nano forms [2, 5a]:

1. Nanoparticles
2. Nanotubes

3. Fullerenes
4. Nanofibres
5. Nanowhiskers
6. Nanosheets.

Advantages

Application of nanosensors in food packaging will assist consumers in followings:

1. Purchasing fresh and tasty products and also improve food safety [4].
2. Consumers rely on expiry date in food package provided by producers which is based on the assumption of the required storage and processing conditions. When either of these requirements is abused the product may likely deteriorate and this may not be known to consumer unless the food package is open. Nanosensors can offer solutions to this problem when apply to food package because they can detect presence of gasses, aromas, chemical contaminants, pathogens, and even changes in environmental conditions [2, 5].

Disadvantages/Toxicity

Despite the tremendous benefits of nanoparticles in food industry, there is great public concern regarding toxicity in human and the environment [7].

1. Lack of knowledge about the effectiveness and impact of nanoparticles on the environment and on human health is of great concern due to potential risk resulted from migration of nanoparticles into the food [9].
2. The migrants may likely react with oxygen and produce toxic substance [6].
3. There is a growing scientific evidence, which indicates that some free nanoparticles may cause harm to biological systems because of their ability to penetrate cellular barriers, and induce oxyradical generation that may cause oxidative damage to the cell [10].
4. There is an urgent need for regulation of nanomaterials before their incorporation into food processing, packaging, and food contact.
5. Before releasing any new product with the nanotechnology in food system, manufacturers must thoroughly assess its risk on human, animal and environmental health [6].

Sources

1. *Abdullahi Nura, Advances in food packaging technology-A review, Journal of Postharvest Technology, 2018,06(4): 55-64.*
2. *Cushen M., Kerry J., Morris M., Cruz-Romero M. and Cummins E., 2012, Nanotechnologies in the food industry- Recent developments, risks and regulation, Trends in Food Science and Technology, 24: 30-46.*

2a. *Mahendra Pal, Mridula Devrani and Angesom Hadush, Recent developments in food packaging technologies, Researchgate, 2019 and Beverage and Food World - Vol. 46 - No. 1, Jan. 2019.*

2b. *Neethirajan, S. and Jayas, D. S., 2011, Nanotechnology for the food and bioprocessing industries. Food Bioprocessing Technology 4: 39-47.*

3. *Amal M. M., 2015, Effects of nanocomposite based nano-silver and nano-titanium dioxideon food packaging materials, International Journal of Applied Science and Technology, 5(2): 26-35.*

4. *Omanovic - Miklicanin E. and Maksimovic M., 2016, Nanosensors applications in agriculture and food industry. Bulletin of the Chemists and Technologists of Bosnia and Herzegovina, 47: 59-70.*

5. *Joyner J. R. and Kumar D. V., 2015, Nano-sensors and their applications in food analysis: A review. The International Journal of Science and Technoledge, 3(4): 80-90.*

5a. *Pal, M, 2017, Nanotechnology: A new approach in food packaging. Journal of Food Microbiology Safety and Hygiene 2: 121.*

5b. *Wesley, S. J., Raja, P., Sunder Raj A. A. and Tiroutchelvamae, D., 2014, Review on nanotechnology applications in food packaging and safety. International Journal of Engineering Research 3: 645-651.*

6. *Abeer A. and Choudhary R., 2013, Current practices in bread packaging and possibility of improving bread shelf life by nanotechnology. International Journal of Food Science and Nutrition Engineering, 3(4): 55-60.*

7. *Mahmoud M. B., 2005, Nanotechnology in food industry; advances in food processing, packaging and food safety. Int. J. Curr. Microbiol. App. Sci.4(5): 345-357.*

8. *Sozer, N. and Kokini, J. L., 2009, Nanotechnology and its applications in the food sector. Trends in Biotechnology, 27(2): 82-89.*

9. *Amra B., Amra O. S. and Æatic, I. S., 2015, Application of polymer nanocomposite materials in food packaging. Croat. J. Food Sci. Technol, 7(2): 86-94.*

10. *Chaudhry Q., Castle L., Bradley E., Blackburn J., Aitken R. and Boxall A, 2008, Assessment of current and projected applications of nanotechnology for food contact materials in relation to consumer safety and regulatory implications. Food Standard Agency report, project A03063.*

5. Intelligent Packaging [1]

Intelligent packaging is another packaging system, which can help in food quality control. This technology incorporates nanosensors that provide information on the condition of the food inside the package. This packaging is happened with advance in time and facilities, such as temperature regulators, radiofrequency indicators, biosensors, and ripeness monitors [2]. This packaging performs sensing, detecting, recording, and communicating for decision making, and to extend storage time, and to enhance safety and quality. A package is called as "intelligent" as it has the above aforementioned extraordinary capacity to communicate with humans, and provide early warning to the consumer or food manufacturer [3]. In addition, it includes indicators that are used for quality control of packed food. The external attachment outside the package (time-temperature indicators), and the internal inside the package are the two indicators that are attached to the head-space or into the lid.

i. Time Temperature Indicator

It is a simple device that indicates about time-temperature change of a food product. The operation can be physical, chemical or biological irreversible change expressed by physical deformation including color change [4].

ii. Freshness Indicators

The fresh food products have following two types of changes:

a. Microbiological changes, such as formation of toxic compounds, unpleasant odor, and pH changes.
b. Lipid and pigments oxidation producing off flavors/odors, discoloration, and other adverse chemical/biological reactions.

A freshness indicator can directly reveal quality of a food product. Microbiological quality can be indicated due to the reaction between the indicator and the metabolites produced during growth of micro-organisms in the product. Such indicator in the package is used to ensure freshness of products during purchasing, and consumption. It can be used on packed perishable foods, such as milk, vegetables, fruits, cakes, and other chilled foods [5].

iii. Pathogen Indicators

This type of indicator helps to check the availability of pathogens by detecting immobilized antibodies. Commercially available Toxin Guard TM is a substance to produce polyethylene-based packaging material with this type of indicator. This works when the analyte (toxin, micro-organism), which is in contact with the material, and first bound to a specifically labeled antibody then to a capturing antibody printed as a specific pattern [1].

Sources

1. *Mahendra Pal, Mridula Devrani and Angesom Hadush, Recent developments in food packaging technologies, Researchgate, 2019 and Beverage and Food World - Vol. 46 - No. 1, Jan. 2019.*

2. *Peltzer, M., Wagner, J.and Jimenez, A, 2009, Migration study of carvacrol as a natural antioxidant in high- density polyethylene for active packaging. Food. Addit. Contam: Part A: Chem. Annal Contr Exposure Risk Assass 26: 938-946.*

3. *Kerry, J. P., O'Grady, M. N., Hogan, S. A., 2006, Past, current and potential utilization of active and intelligent packaging systems for meat and muscle based products: a review. Meat Science 74, 113-130.*

4. *Pavelkova, A., 2013, Time temperature indicators as devices intelligent packaging. Acta Universitatis Agriculturaeet Silviculturae Mendelianae Brunensis 61: 245-251.*

5. *Wilson, C., 2007, Frontiers of Intelligent and active packaging for fruits and vegetables. CRC Press, Boca Raton.*

Combination of Various Packagings [1]

The recent studies have revealed that combinations of various packaging techniques to prolong shelf life are more effective.

1. The study suggested [3] that even though using oxygen absorber sachets was very effective in terms of controlling the headspace oxygen, they could not slow down lipid oxidation of crackers when stored at high temperatures, attaining comparable sensory evaluation of crackers kept in absence of oxygen absorber sachets. An oxygen absorber sachet was found to extend the shelf life of pita bread and bakery products by impeding mould and yeast growth even in CO_2/N_2 MAP due to the fact that it absorbed oxygen trapped in food and in the air that permeated through the package [2].
2. As ethanol was well known as a disinfectant in the medical field, the ability to delay the microbial growth on rye bread depended on the size of ethanol emitter sachets [4].
3. The study [5] could not see any advantages of using a combination of an oxygen absorber sachet and an ethanol emitter sachet over to the use of an ethanol emitter sachet alone.
4. Several research studies have disclosed more effectiveness of sachet based AP or the combination of AP with a MAP over MAP alone [1, 2].
5. Scientists [6] added paper filters and a small plastic bag containing paper filters soaked with allyl, benzyl and phenyl isothiocyanates oil into bread plastic trays and found that these essential oils could reduce aflatoxins (AFs) and prevent microflora growth on the bread.
6. Researchers [7] prepared active polyolefin composites by melt extrusion of iron particles, kaolinite clay and high and linear low-density polyethylene (HDPE and LLDPE) and the nanocomposites films were subsequently prepared by compression moulding. The agglomeration of nanoclay was more pronounced in HDPE, resulting in less oxygen scavenger activity. Besides an active role, by incorporating the nanoclay into the films could reduce the oxygen and water permeability in passive action by 22-28 and 33 per cent, respectively [7, 8].

Prolonging of white bread shelf life was accomplished with the use of novel advanced packaging technologies [1].

- ✰ The combination of AP and MAP demonstrated a high potential for bakery product industries. It was illustrated that CO_2 was not a necessity in the MAP as long as the amount of O_2 was minimized by using N_2 as a balance gas.
- ✰ By incorporating an iron-based oxygen scavenger into the petroleum plastic films, the intermediate from the scavenging reaction acted as an antimicrobial agent in a similar way to CO_2.

☆ If the packaging film possesses a high barrier for O_2 transmission property, the usage of incorporating oxygen scavenger as well as the MAP is almost unnecessary. These technologies could be extended to other bakery products [1].

Sources

1. *Settakorn Upasenhttps: //www.sciencedirect.com/science/article/pii/S2405844018317821 - ! and Piyachat Wattanachai, Packaging to prolong shelf life of preservative-free white bread, Heliyon, Vol.4, Issue 9, 2018.*
2. *Smith, J. P., Ooraikul, B. Koersen, B., Jackson, E. D., Lawrence, E. D., Novel approach to oxygen control in modified atmosphere packaging of bakery products, Food Microbiol., 3 (4), 1986, pp. 315-320.*
3. *S. Berenzon, I. S. Saguy, Oxygen absorbers for extension of crackers shelf-life, LWT Food Sci. Technol., 31 (1), 1998, pp.1-5.*
4. *A. Salminen, K. Latva-Kala, K. Randell, E. Hurme, P. Linko, R. Ahvenainen, The effect of ethanol and oxygen absorption on the shelf-life of packed sliced rye bread. Packaging Technology and Science, LWT Food Sci. Technol., 9 (1), 1996, pp. 29-42.*
5. *E. Latou, S. F. Mexis, A. V. Badeka, M. G. Konto-minas, Shelf life extension of sliced wheat bread using either an ethanol emitter or an ethanol emitter combined with an oxygen absorber as alternatives to chemical preservatives, J. Cereal Sci., 52 (3), 2010, pp. 457-465.*
6. *F. Saladino, J. M. Quiles, F. B. Luciano, J. Mañes, M. Fernández-Franzón, G. Meca, Shelf life improvement of the loaf bread using allyl, phenyl and benzyl isothiocyanates against Aspergillus parasiticus, LWT Food Sci. Technol., 78, 2017, pp. 208-214.*
7. *M. A. Busolo, J. M. Lagaron, Oxygen scavenging polyolefin nanocomposite films containing an iron modified kaolinite of interest in active food packaging applications, Innovat. Food Sci. Emerg. Technol., 16, 2012, pp. 211-217.*
8. *M. J. Khalaj, H. Ahmadi, R. Lesan-khosh, G. Khalaj, Study of physical and mechanical properties of polypropylene nano-composites for food packaging application: nanoclay modified with iron nanoparticles, Trends Food Sci. Technol., 51, 2016 pp. 41-48.*

Advancement in Packing Solutions [1, 1b]

Antimicrobials/Plant based anti-microbials and novel additives are now being used to improve the packing of food hence shelf-life of foods. Sustainable packing (Biodegradable packing) is also in demand today. Some examples are discussed below:

1. Antimicrobial Packaging/Plant-based Antimicrobial Packaging

A. Antimicrobials Packaging [1b]

Antimicrobial packaging involves combination of food-packaging materials with antimicrobial substances such as the incorporation of antibacterial nanoparticles into polymer films to control microbial surface contamination of foods [2]. Antimicrobial food packaging systems have received considerable attention since these systems help control the growth of pathogenic and spoilage micro-organisms on food surfaces [3]. Antimicrobial packaging is performed in the form of solutions, which contain active component inside the packaging material [3a].

- ☆ During the plastic films production processes, antimicrobial substances are introduced [3a].
- ☆ Antimicrobial packaging can be done using compounds with natural antimicrobial activity, such as enzymes, and bacteriocins, which can be activated due to chemical or physical modifications [3b].
- ☆ Different studies on different nanoparticles incorporated into various polymeric matrices, indicate a great potential of antimicrobial nanocomposite system for prolonging the shelf-life and preservation of different food stuff.
  - i-. Nano-silver antimicrobial food packaging application is a novel approach toward the preservation of foods and the extension of their shelf-life [4].
  - ii. Intensive contact between the food product and packaging material is required in antimicrobial packaging, therefore, potential food applications include vacuum or skin-packaged products, *e.g.* vacuum-packaged meat, fish, poultry or cheese [2].

The following three basic categories of antimicrobial packaging systems are used [5]:

1. Incorporation of antimicrobial substances into a sachet connected to the package from which the volatile bioactive substance is released during further storage.
2. Direct incorporation of antimicrobial agent into the packaging film.
3. Coating of packaging with a matrix that acts as a carrier for the antimicrobial agent [5].

The antimicrobial effects of different nanoparticles used in different antimicrobial packaging system, are summarised below [6]:

Sl. No.	*Nano Particles*	*Polymer Matrix*	*Tested Micro-organism*
1	Ag/Chitosan [6]	PLA[1]	*Staphylococcus aureus* (ATCC6538) *Escherichia coli* (DSMZ 30083)
2	ZnO/Ag/Cu	PLA[1]/PEG[2]	*Lysteria monocytogenes, Salmonella typhimurium*
3	Ag/SiO_2/TiO_2	LDPE[3]	*Escherichia coli*
4	SiO_2	PBAT[4]	*Escherichia coli;*
5	ZnO	LDPE	*Bacillus subtilis; Enterobacter aerogenes*

Note: 1: Polylactic acid, 2: Polyethylene glycol, 3: Low density polyethylene, 4: Poly(butylene adipate co-terephatalate).

B. Plant-based Antimicrobial Packaging and Coating Solutions [1]

a. Scientists at Nanopack [7] have utilised the antimicrobial property of plant-based essential oils to develop an antimicrobial packaging. The packaging involves using halloysite nanotubes (HNTs) [8], hollow clay mineral fibres made up of aluminum and silicon atoms, capable of being infused with essential oils. While HNT absorbs ethylene, the ripening hormone, the infused essential oils [9] act as antimicrobial agents to prevent the food from degradation. The testing of the technical, industrial, and commercial viability of these smart antimicrobial surfaces for the application as an active food packaging solution, are currently being done.

b. Scientists from the Bhabha Atomic Research Center (BARC) have utilised the antimicrobial and antioxidant activity of mango peels [10] for incorporation into bio-based packaging films.

Both these packaging solutions are safe, durable, thermally-stable and even prevent water damage to food products. However, currently, they are expensive to use, and before widespread adoption, research on more food items would be required.

2. Novel Additive-based Packaging Solutions [1]

a. Sprouted potatoes have elevated levels α-solanine, which is toxic to consumers [11]. Scientists at the Chinese Academy of Sciences have devised a method to prevent sprouting in potatoes by using modified nanosilica [12] with amino silicone oil (ASO) to develop a hydrophobic nanosilica powder. Coating potatoes with nanosilica has been demonstrated by the researchers to be non-penetrating and safe for human consumption. It is also a cost-effective solution but consumers may be hesitant to use nanosilica-treated potatoes, even though it is safe.

b. Scientists at the Institute of Nano Science and Technology, India, are developing an inert graphene-oxide [13] based wrapper that releases salicylaldehyde (SA), *on-demand*. The release slows the ripening process of bananas and protects from spoilage-causing micro-organisms. This technology meets following three important criteria:

   i. Antimicrobial,

   ii. Controls ripening

   iii. Is not in direct contact with fruit. However, the wrapper may have limitations in the later stages of ripening, when the acid content reduces and the late polymerisation forms a lignin, as noted by the researchers and this challenge is need to be addressed.

3. Sustainable Packaging/Biodegradable Packaging [1a]

The use of biodegradable and environmentally friendly materials for food packaging is in demand as many packaging materials generates waste. Polylactide acid (PLA) plastics, sugar cane pulp, fiber composite, starch-based films [14], are some of the examples of biodegradable materials.

Sustainable packaging is a complex idea which must be applied with a systematic approach and critical thinking. The following three important issues to be considered in evaluation of packaging sustainability:

i. Entire lifecycle of the package from raw materials through to ultimate disposal to avoid problems being transferred from one part of the lifecycle to another.
ii. Interactions between the package and the product it contains so that the environmental impacts of the product-packaging system as a whole are minimised.
iii. There is also needs to consider 'triple bottom line' impacts of packaging: on the business, on people and on the natural environment.

Both biobased and biodegradable materials are being investigated as potential substitutes for petroleum-based plastics [15]. Biodegradable polymers can fulfill all the functions of food packaging without causing any threat to the environment.

The belief is that biodegradable polymer materials will reduce the need for synthetic polymer production (thus reducing pollution) at a low cost, thereby producing a positive effect both environmentally and economically [16].

The most acceptable definition of sustainable package is as:

i. It is beneficial, safe, and healthy for individuals and communities throughout its life cycle.
ii. It is sourced, manufactured, transported, and recycled using renewable energy.
iii. It meets market criteria for performance and cost.
iv. It maximizes the use of renewable or recycled source materials.
v. It is made from materials healthy in all probable end-of-life scenarios.
vi. It is manufactured using clean production technologies and best practices.
vii. It is designed to optimize materials and energy.
viii. It is recovered effectively and used in biological and/or industrial cradle-to-cradle cycles (SPC2011).

Sources

1. *Navneeta kaul and Paula Hock, Extending the shelf life of food and drink - far from an impossible dream, New Food, Nov2019.*

1a. *Abdullahi Nura, Advances in food packaging technology-A review, Journal of Postharvest Technology, 2018,06(4): 55-64.*

2. *Dele R., 2012, Application of plastics and paper as food packaging materials – An overview. Emir. J. Food Agric. 25(3): 177-188.*

3. *Amra B., Amra O. S. and Catic, I. S., 2015, Application of polymer nanocomposite materials in food packaging. Croat. J. Food Sci. Technol, 7(2): 86-94.*

3a. *Mahendra Pal, Mridula Devrani and Angesom Hadush, Recent developments in food packaging technologies, Researchgate, 2019 and Beverage and Food World - Vol. 46 - No. 1, Jan. 2019.*

3b. *Cierpiszewski, R (2016). Opakowaniaakty-wneiinteligentne [Active and intelligent packaging]. Wydawnictwo Uniwersytetu Ekonomicznego w Poznaniu, Poznañ, Chap 1, 2.*

4. *Amal M. M., 2015, Effects of nanocomposite based nano-silver and nano-titanium dioxideon food packaging materials, International Journal of Applied Science and Technology, 5(2): 26-35.*

5. *Sunil M., 2012, Antimicrobial food packaging to enhance food safety: current developments and future challenges. J Food Process Technol 3(5): 12.*

6. *Tanja I. R., Ivan S. R., Branka M. P. and Aleksandra R. N., 2016, Antimicrobial nanomaterials for food packaging applications, Food and Feed Research, 43(2): 119-126.*

7. *https:// www.nanopack. eu.*

8. *http: //ec.europa.eu/research/infocentre/article_en. cfm? and artid=49196 and caller=other.*

9. *http: //ec.europa.eu/research/info-centre/article_en. cfm? and artid= 49196 and caller=other.*

10. *https://onlinelibrary. wiley. com/ doi/full/ 10.1111/ jfs.12385,*

11. *https:// www.nature.com/articles/ cddis2015219,*

12. *https:// pubs. acs.org/doi/abs/10.1021/ acssuschemeng.8b01860,*

13. *https:// pubs.acs. org/ doi/10.1021/acsami.8b02048,*

14. *Nitaigour P. M., 2014, Advances in packaging methods, processes and systems. Challenges 5: 374-389.*

15. *Risch S. J., 2009, Food packaging history and innovations. J. Agric. Food Chem. 57: 8089-8092.*

16. *Pawar P. A. and Aachal H. P., 2013, Bioderadable polymers in food packaging. American Journal of Engineering Research, 2(5): 151-164.*

Chapter 20

Robotics Technology

Technological advancement is gradually finding applications in the agricultural and food products to meet the need of the growing population. Automation or use of automated system are replacing human operator. Presently, the technology is becoming more affordable and intelligent. Robotics automate many of the complex and repetitive tasks that are carried out in the dairy and food industry. **The field of robotics is both interdisciplinary and multidisciplinary as robots are amazingly complex systems comprising of mechanical, electrical and electronics hardware and software [1].**

Dairy and food industry has been lagging behind other industrial sectors in implementing robots but now the dairy and food industry is highly automated, from the raw material production to the processing and manufacture of products. The implementation of automation in the dairy and food sector offers great potential for improved safety, quality and profitability by optimizing process monitoring and control. The broad range of potential applications for robotics in dairy and food industries are given below **[1]**:

1. Grading of food products.
2. Pick and place operations.
3. Packaging and palletizing.
4. Meat processing.
5. Fruit and vegetable.
6. Milk and milk product.
7. Production and processing.

Automation technology is changing the way the milk is produced and processed. The applications of robotics and automation have been successfully achieved in a wide range of manufactured industries dealing with well -defined processes and products [2]. However there are following particular research challenges associated with the use of robots in the food industries [3]:

1. The objects being handled are variable in size, shape, weight and position, so that some form of intelligent sensing is required.
2. The objects to be handled are often delicate and covered with either slippery or viscous substances, and so the end effect or must be carefully designed if it is to handle the objects at high speed with secure lifting and without bruising.
3. The concern for hygiene, quality and consumer safety. The hygiene issue is becoming increasingly important for human health. But all the three challenges have been accepted by modern robots.

In the dairy and food industry most systems are also isolated, batch type operations that target a specific task. For automation to be successful, it must be integrated into the overall manufacturing system design and provide on-line, continuous control capability. India's first automated dairy plant with handling capacity of 1,000,000 LPD has been established at Gandhinagar near Ahmedabad in Western India. Dairy and food processing industries is highly labour-intensive, with sometimes labour costs at anything up to 50 per cent of the product cost. Improving productivity and reducing labour costs will therefore have a significant impact on profitability. Automating repetitive tasks will improve quality control and efficiency and reduce the high level of accidents. Today, the increasing technological development and sophistication of modern societies impose new quality and safety standards to the food producers. To assure the quality and safety of food products, automation can play a key role.

During milk and milk product processing, mostly fluid is transferred from one place to another through pumps, but still there are various operations where some form of solids has to be transferred repetitively. In dairy industry too, there is need for some movement of materials in hazardous atmosphere (as in ice store at -23 °C and in ice-cream hardening room at -30 °C). At this place, use of robot would be a good alternative. It can be expected that future dairies would be **'smart dairies'** employing such robots capable of processing and handling milk and milk products most economically with a through control on quality too.

One of the most important obstacles in the automation of food manufacturing is the biological variation in size, shape, and homogeneity of the raw materials. Dairy lends them readily to automatic processing because the milk(raw material) has to be handled in bulk. Accordingly, the dairy industry is among the most automated. The materials such as fruits, vegetables, meat, *etc.*, need to be handled on a more individual unit basis. This has hampered automation tremendously. Thus, food industry automation requires a level of flexibility uncommon to other mature industries [4].

In food industry, earlier use of robots was limited to packaging of food and palletizing in dairy, beverages, chocolates and food tins. In 1998, the launch of the Flex Picker robot revolutionized the food industry as it is the world's fastest pick and place robot [5].

India's first robotic dairy plant was opened in Virar, in suburban Mumbai by Amul Dairy, the largest food brand in India for milk and milk products. The Kaira District Co-operative Milk Producers Union Limited, known generally as Amul Dairy, launched the Virar plant, the sixth facility for Amul in the country, on the birth anniversary of Sardar Vallabhbhai Patel, India's first home minister. Built with a total investment of Rs 1.8 billion, the Virar plant, with an installed capacity of 1 million litre of milk processing per day, covers an area of 4.5 hectare and is the first across the globe with a fully automatic traffic management system for milk packing operations [6].

It is first in India to include a 50,000 litre/hour integrated milk reception facility, energy conservation by utilisation of hot water generator and water conservation through introduction of rain water harvesting system. Equipped with the largest ice cream manufacturing facility in the country(200,000 litres per day) [6].

Operations such as conveying of crates, packed pouches from high-speed packing machines and filling of pouches in crates are fully automated and controlled **through a centralised computer monitoring system**. The high-speed packing machines have been **integrated with online check weighers and robotic operations to avoid human touch of products [6].**

Advanced Automatic Robotic Dairy Plant [7]

Sources

1. *Meshram BD, Shaikh Adil and Suvartan Ranvir, Robotics: An emerging technology in dairy and food industry: Review, International Journal of Chemical Studies 2018, 6(2): 440-449.*
2. *Hurd SA, Carnegie DA, Brown NR, Gaynor PT. Development of an intelligent robotic system for the automation of a meat-processing task. Int J Intel Sys Technol Appl, 2005, 1: 32-48.*
3. *Peters R. Robotisation in food industry. 5th International Conference on the Food Factory for the Future. 2010, June 30 to July 2, Gothenburg, Sweden.*
4. *Judal A, Bhadania AG. Automation in Dairy and Food Processing Industry. International Conference of Advance Research and Innovation. 2015, 490-495.*
5. *Subhash Prasad, Application of Robotics in Dairy and Food Industries-A Review, International Journal of Science, Environment and Technology, Vol. 6, No 3, 2017, 1856 -1864,*
6. *India opens first robotic dairy plant, World Asia, Oct.2013.*
7. *India moves to Robotic Dairy plant, Milky world, An exclsive library for Hitech equipment 2015.*

Robot [1, 1b]

Robot, from the Czechoslovakian word, "robota" meaning forced labour. Robots are especially desirable for certain work functions because, unlike humans, they never get tired, they can work in physical conditions that are uncomfortable or even dangerous, they can operate in airless conditions, they do not get bored by repetition and they cannot be distracted from the task at hand. The robot is powerful, reliable and can be used in hot temperature area where a human after working for so long can become sick and exhausted [2, 3].

A robot can be defined as a programmable, self-controlled device consisting of electronic, electrical, or mechanical units. Robot is a machine that functions in place of a living agent. The definitions of robot given by different agencies are given below:

i. British Robot Association

"An industrial robot is a reprogrammable device designed both to manipulate and/or transport parts, tools, or specified manufacturing implements through variable programmed motions for the performance of specific manufacturing tasks."

ii. International Organization for Standardization (ISO)

A robot is "An automatically controlled, re-programmable, multi-purpose, manipulative machine with several degrees of freedom, which may be either fixed in place or mobile for use in industrial automation applications."

History of Robot [1]

The history of robot is tabulated on next page.

Year	*Development Activities*
1956	George Devol applied for a patent for the first programmable robot, later named 'Unimate'.
1961	First Unimate robot installed at General Motors, used for die casting and spot welding.
1986	Honda starts work on its first humanoid, robot named 'E0' (later to become ASIMO).
1988	SCAMP designed as the first robot pet with emotions.
1995	Robot used in packaging and palletisation line.
1997	Industrial Research Ltd. New Zealand, develop robot for sheep de-fleecing and cutting.
1999	GTRI develop a robot for deboning of meat.
1991	First Helpmate mobile autonomous robot used in hospitals.
1992	Development of robot for picking of citrus fruit in Spain.
2002	iRobot introduces Roomba, a personal robotic vacuum cleaner.
2004	Reed develop robot for harvesting of mushrooms.
2006	The world's first robotic rotary dairy was developed by Delaval.
2010	NASA and General Motors join forces to develop Robonaut-2, the new version of NASA's humanoid robot astronaut.

Parts of a Robot [1, 1b]

Robots are available in many shapes and sizes. Robots consist of following five components that work together [5]:

1. Controller,
2. Manipulator,
3. End effectors,
4. A power supply
5. A means for programming.

The relationship among these five components is illustrated below:

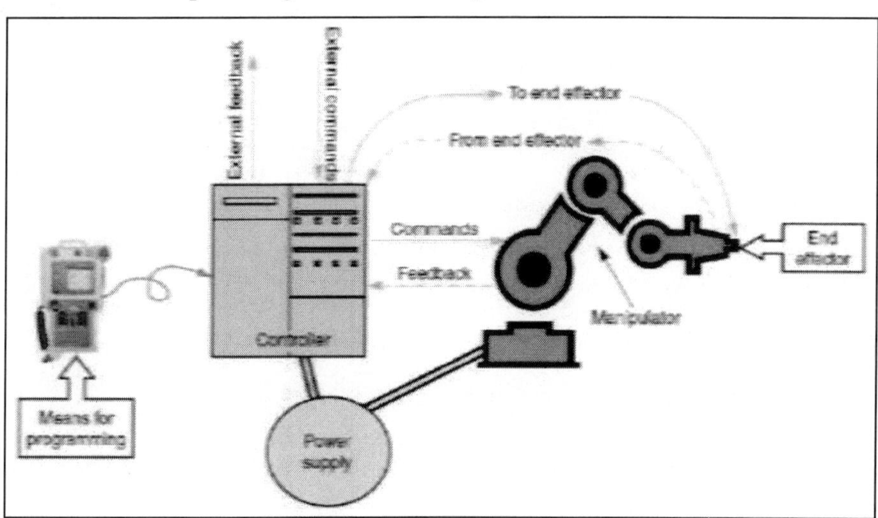

i. An Actuator

It is a motor or valve that converts power into robot movement. This movement is initiated by a series of instructions, called a program, stored in the controller's memory.

ii. Manipulator

It consists of segments that may be joined and that move about, allowing the robot to do work. The manipulator is the arm of the robot which must move materials, parts, tools, or special devices through various motions to provide useful work.

iii. End Effector

It is the robot hand, or the end-of-arm tooling on the robot. It is a device attached to the wrist of the manipulator for the purpose of grasping, lifting, transporting, maneuvering, or performing operations on a work piece.

iv. Power Supply

The power supply provides the energy to drive the controller and actuators. It may convert AC voltage to the DC voltage required by the robot's internal circuits, or it may be a pump or compressor providing hydraulic or pneumatic power. The three basic types of power supplies are electrical, hydraulic, and pneumatic.

v. The Means for Programming

It is used to record movements into the robot's memory. A robot may be programmed using any of several different methods. The teach pendant, also called a teach box or hand held programmer teaches a robot the movements required toper form a useful task. The operator uses a teach pendant to move the robot through the series of points that describe its desired path. The points are recorded by the controller for later use [6].

Types of Robot Used in Dairy and Food Industries [1, 1a]

The different types of robots were put into operation for several purposes under recent developments in the food industry. The old model SCARA robots for pick and place, spider robots for high speed picking and placing of light weight objects are the recent examples of robots used in the industry.

i. SCARA (Selective compliance assembly robot arm) Robots

SCARA robots are one of the types of stationery robots also known as horizontal articulated arm robots, with motions same as human arm. Its reliability for fast and repeatable movements make it fit for packaging palletizing, loading and unloading purposes [7].

ii. Delta Robots

This is also known as Parallel Link Robots are the category of modern day robotics. Delta robots are designed for high-speed handling of lightweight products and offer lower maintenance due to the elimination of cable harnesses and absent of multiple axis. Parallel robots are deployed into many food processing steps. Again they offer high speed transfer food stuffs, primary (unpacked) or secondary (packaged) through manufacturer lines and a multitude of processes.

Specifications for a Food Sector Robot [1, 1a]

A variety of robotic systems, developed for such sectors, are readily available from international companies and are often offered as manufacturing solutions for the food industry. Food sector robots should meet the following requirements:

i. Any design should meet the guidelines for food-handling equipment [8].
ii. Be ideally of stainless-steel grade AISI-304.
iii. Construction with ingress protection rating to IP67.
iv. All parts visible and accessible for inspection and manual cleaning.
v. Highest standards of hygienic design-The standards of hygienic design required will depend on the application, being higher in processing poultry and fish products than in processing dry-food products, such as biscuits or bread. A general purpose robot, which could potentially be used to process any type of product, should meet the highest standards of hygienic design.

Robotics [1, 1a]

Robotics is the branch of technology that deals with the design, construction, operation, and application of robots, as well as computer systems for their control, sensory feedback, and information processing. These technologies deal with automated machines that can take the place of humans in dangerous environments or manufacturing processes, or resemble humans in appearance, behaviour, and or cognition. The function of different component are given below [4]:

i. Processor

It is brain of the robot, calculates the motions and the velocity of the robot's joints, *etc.*

ii. Sensors

To collect information about the internal state of the robot or to communicate with the outside environment.

iii. Software

Operating system, robotic software and the collection of routines.

iv. Rover or Manipulator

Main body of robot (Links, Joints, other structural element of the robot).

v. Actuators

Muscles of the manipulators (servomotor, stepper motor, pneumatic and hydraulic cylinder).

vi. End Effector

The part that is connected to the last joint hand of a manipulator.

vii. Controller

Similar to cerebellum. It controls and coordinates the motion of the actuators.

Sources

1. *Meshram BD, Shaikh Adil and Suvartan Ranvir, Robotics: An emerging technology in dairy and food industry: Review, International Journal of Chemical Studies 2018, 6(2): 440-449.*

1a. *Subhash Prasad, Application of Robotics in Dairy and Food Industries-A Review, International Journal of Science, Environment and Technology, Vol. 6, No 3, 2017, 1856 -1864.*

1b. *Hamraj Singh, Robotics – An Emerging Technology in Dairy Industry International Journal of Mechanical and Production Engineering, ISSN(p): 2320-2092, ISSN(e): 2321-2071 Volume- 6, Issue-2, Feb.-2018.*

2. *Agrawal AK, Karthikeyan S, Goel BK, Khare A, Shrivastava AK, Mishra UK. Robotization of Indian Dairy Industry: An Indispensable Step in Futuristic Processing Plants. Published in proceeding of National Seminar on Paradigm shift in Indian Dairy Industry, 2010, 136-139.*

3. *Nayik GA, Muzaffar K, Gull A. Robotics and Food Technology: A Mini Review. J. Nutr. Food Sci. 2015, 5: 384.*

4. *Massey R, Gray J, Dodd T, Caldwell D. Guidelines for the design of low cost robots for the food industry, 2010, 37(6): 509-517.*

5. *Schilling, R. J. Fundamentals of Robotics: Analysis and Control. Prentice Hall, New Jersey, 1990.*

6. *Judal A, Bhadania AG. Automation in Dairy and Food Processing Industry, International Conference of Advance Research and Innovation, 2015, 490-495.*

7. *Brumson, B., Scara vs. Cartesian Robots: Selecting the right type for your applications, Robotics Online from http: //www.robotics.org/content-detail.cfm, 2011.*

8. *Lelieveld, H., Keener, L., 2007, Global Harmonization of Food Regulations and Legislation, EHEDG, Yearbook 2007, Trends in Food Science and Technology, 18(S1), 15-19.*

Benefits of Robots in Food Industry [1, 2]

The benefits of using robots in food processing are attractive, however, Robots can achieve consistent results more quickly than human employees can. Health and safety issues are also less of a problem with robots that can be designed to handle extreme conditions such as high temperatures. Especially in repetitive processes like cake decorating or chicken deboning, robots can help lower the cost of production through producing higher yields with no need for training or breaks [1]. Features and benefits of robots in food processing are tabulated on next page [2, 3]:

Sl. No.	Features	Benefits
1	Better process control	Easy to clean robot, minimum retention areas, connection protection.
2	High reliability, high speed	Increased productivity.
3	High dexterity, several mounting positions	Compact cell, less room required, simpler mechanical solution.
4	Cleanliness	Better hygiene.
5	Flexibility	Marketing innovative products and packaging.
6	Vision and conveyor tracking.	Product picked and controlled in process, in any position.

Sources

1. *Natasha Jane Chrisandina, Robotics in food manufacturing: Benefits and challenges, Prescouter Inc, Chicago, Aug.2018.*
2. *Meshram BD, Shaikh Adil and Suvartan Ranvir, Robotics: An emerging technology in dairy and food industry: Review, International Journal of Chemical Studies, 2018, 6(2): 440-449.*
3. *Anon, Robotics in Sweet Production, Lebensmiffelwirtrch., 1996, 7(4): 42-43.*

Robots in Food Processing [1]

Food processing can be divided into following two steps:

1. Primary Processing

In primary food processing, raw materials are extracted or transformed into food commodities. An example of such a process is milling wheat.

2. Secondary Processing

In secondary food processing, these food commodities are then turned into edible products. Robots are slowly being integrated into both steps with varying levels of success due to the varying levels of uniformity in the raw materials that each step starts with.

a. Robots in Primary Food Processing

In primary food processing, raw plant and animal products are extracted into commodities that can either go into secondary processing or be sold as-is. Because raw products have inconsistent dimensions, robots that require programming a default set of dimensions might not be well-suited to the task. Some examples of application of robots in primary food processing are given below:

i. Processing of Chicken Leg

One process where robots have successfully been integrated is in chicken leg deboning. The dimensions of a chicken leg have small variations, so many companies have developed an assembly-line system that can break down a piece of chicken leg into separate parts. With the growing popularity of chicken legs worldwide, the poultry industry has been able to keep up partly because of the widespread use of these machines.

ii. Beef Butchery

It is a process that robots cannot yet handle on their own. Beef carcasses are much more varied in shape and size than chicken legs, and human butchers rely on touch more than sight when butchering beef. JBS, a meat processing company based in Brazil, is investing in Scott Technology, a New Zealand robotics company, to build different robots that can together break down a beef carcass.

Robots in Secondary Food Processing

In secondary food processing, food commodities are turned into edible food products. As the raw materials in this process have already gone through primary food processing, there is more consistency in their dimensions and processes are easier to automate. Some examples of application of robots in secondary food processing are given below:

i. Cake Decorating

Mass-produced cakes have the same dimensions and similar decorations on them, so a robot can very easily be programmed to put icing on cakes in a specific configuration.

ii. Pizza-Making

Pizzas require frequent rotations in an oven, which may be dangerous for human workers especially if high production numbers are needed. Zume, a US-based startup, is designing some robots that can assist in the "low-level" tasks involved in pizza-making such as dough-mixing and in-oven rotating.

Sources

1. *Natasha Jane Chrisandina, Robotics in food manufacturing: Benefits and challenges, Prescouter Inc, Chicago, Aug.2018.*

Applications of Robotics

The details related to the applications of Robotics in different fields are given below:

1. Applications in Dairy Industry [1, 1b, 1c]

The application of robotics in the food industry has increased over recent years, particularly in the field of processing and packaging systems. Now the technology is becoming more affordable and the systems more intelligent, it may be feasible to automate many of the complex and repetitive tasks that are carried out in the Dairy/ food industry. The opportunity still exists to deliver significant benefits in terms of increased food shelf life, cost reductions and flexibility [2]. Dairy food products by virtue of their nature differ significantly in consistency and shape hence dairy industry has been lagging behind other industrial sectors in implementing robots.

However, there is a broad range of potential applications for robotics in dairy industries. The commercial application of robots in dairy industry is also widely spread at the end of processing lines like packaging and palletizing.

Robotic Milking [1]

Milking buffaloes and cows by machine, to replace the practice of milking by hand, has been known for more than century. Automatic milking systems (AMS) or milking robots are one of the most successful and important application of robotics in the dairy industry.

AMS also referred to as robotic milkers, were developed in Europe and became available there in 1992 and was introduced to the US in 2000. Robotic milking is a voluntary milking system, which allows the Buffalo/cow to set her own milking schedule. Each cow on a robotic milking platform is fitted with an electronic tag which allows the robot to identify her. When a cow enters her ID tag is read and she receives a feed reward customized to her level of production, the robot then cleans her teats, attaches the milk cups, and begins the milking processes when milking is complete, the cups disconnect as each quarter finishes milking and she exits [3, 4].

Milking Robot [4a].

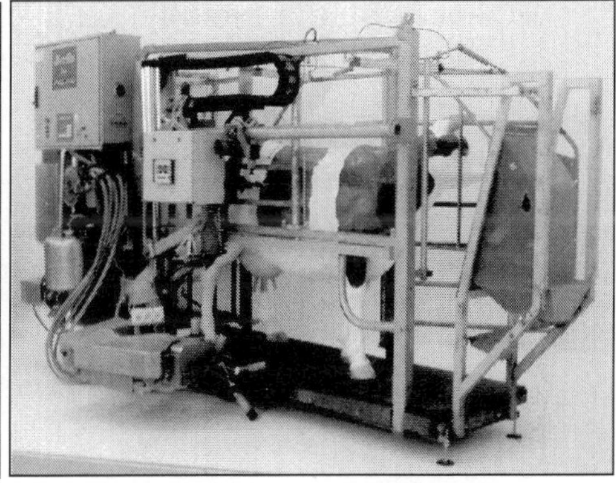

Automatic Milking [4b].

The world's first commercial robotic milking rotary has been unveiled by Swedish dairy equipment company DeLaval at a pilot farm at Quamby Brook, Tasmania, Australia. Featuring five robots, the rotary has a capacity to milk up to 90 cows per hour, enabling the robots to reach the cow from the side. With the use of laser technology, the robots focus a red light to determine the location of the cow's teats, clean them and attach the cups. The five robots work as per the following ways [5]:

1. The first two robots clean and prepare the teats for milking,
2. The second two attach the cups to the teats,
3. The last robot sprays the teats to disinfect them before the cows reaches the platform.

Once the milking is done, robotic liquid filling and finishing systems get the product ready for market. These robots handle many types of bottles, vials, bags, and pouches with precision filling from micro liter to multi liter. They accommodate a variety of products; and deal with fill volumes, dispensing profiles, containers and closure types, making them ideal for clinical trials, full scale production and contract manufacturing [6, 7, 8].

Advantages of Robotic Milking [1, 1b, 1c]

Milking by using robots has following various advantages :

1. **Economic benefit**: Labor flexibility and not needing to manage hired labor were the biggest advantages reported by current users of robotic milkers.
2. **Increased milking frequency**: Milking frequency may increase to three times per day. This may result in less stress on the udder and increased comfort for the cow. Higher frequency milking increases milk yield per cow.
3. **Management benefit**: Management of the herd can be made more efficient. For a farmer who's never managed his cows properly the robot computer will force him to do so. It informs him about blood in the milk, conductivity, and yield per quarter.
4. **Cow health and welfare benefits**: Milk Producers reported an improvement in cow health and a reduction in instances of mastitis following the transition to robotic milking. This was due to the less stress on the cows and to have better access to information on their cows. For example, benefits resulting from quarter-by quarter milking, which can help to reduce udder infections.

2. Robotics in Packaging [1, 1b, 1c]

There is a broad range of potential applications for robotics in food processing and packing.

1. In the meat industry, robots are used in slaughtering, deboning, cutting, sorting and packaging applications.
2. Robots can also be used for picking and placing items such as cookies, hamburgers, chocolate pralines, croissants, chicken fillets or pan cakes into primary packing. Additionally, robots are already used in baking lines to handle hot trays.
3. In the dairy industry, robots are used in cheese packaging, cheese slicing, and curd slicing *etc.* In cheese production, robots stir curds, transfer cheese moulds, and turn, cut, portion, package and palletize the cheeses. Integrated sensors and measuring systems enable the simple implementation of complex processes. Blocks of cheese arrive on wooden planks at the robot picking area. The special gripper allows the cheese blocks to be picked and placed onto a conveyor for further processing [9]. The picking and packing robots are shown in Figures given on next page:

Robotics allows for order picking and packaging to be much more efficient. For example, once an order has been placed online, the robot will be able to detect the location of the specific product, package aptly, and provide the customer with real-time information and status of the product being delivered [1d]. Packaging robots are able, but not limited to, carry out the following tasks [1d]:

1. Open,
2. Fill,
3. Pack,
4. Seal
5. Correctly label the package to be sent to the end-user.

For the packaging of food items, where speed, consistency or high levels of repetition are concerned, the **food robots** almost always win over humans in terms of efficiency. Robots are equipped with **intelligent vision systems** allowing the very specific placement of products on a belt to be done with incredible accuracy. **Vision system robots** are also utilized in sorting by color, shape or size [1e].

3. Application of Robotic in Agricultural [1]

An agricultural robot is a robot that is used for specific agricultural purposes. **Agricultural robots, sometimes known as agribots**, are seen as one of the key trends that will deeply influence agriculture. The main application with the agricultural industry would be regarded within the harvesting stage, monitoring the environment, and soil analysis. An advanced robotic systems will take care of and harvest plants, as well as carry out on-farm data collection, increasing crop yields [1d]. Many companies are utilising robots to increase their overall crop yields. These companies can do this by automatically analysing the crops, using a detailed sensor to gather data on the status of the crop within the external conditions to evaluate the best type of conditions to cultivate for crops. Robot can save farmers time and costs while also improving their costs and efficiency and now agricultural robots become a permanent fixture on many farms [1d].

Robots are being used in different activities (Sowing, Weeding, Spraying and Broad Acre Harvesting) related to agriculture. Some examples are given below [1]:

1. For spraying,0.5 m accuracy is usually sufficient. Here, demands of speed are at their most important, so there is a trade-off between a GPS with a once-per-second update with inertial assistance and the 5 Hz GPS that is becoming more common.
2. For yield monitoring, where the harvested yield might be apportioned into 5m squares, GPS with 0.5m precision is also sufficient. Precision GPS and inertial sensors have a clear lead. However, when a good furrow has been formed, planting can employ a simple mechanical furrow follower, a ball or wheel trailed from an arm projecting in front of the tractor.
3. When seedlings have emerged, machine vision offers great advantages over GPS guidance. While cultivating, any dynamic positional errors at the planting stage will be added to steering positional errors; the blades must be set further from the row if the risk of destroying plants is to be contained.
4. Vision can track the wander in the planted rows so that only one level of errors will be involved.
   a. When a cotton crop is ready for harvesting, simple methods can again be used. Mechanical stem feelers are ideal for guiding the harvester precisely along the rows and these have been successfully field-tested by the National Center for Engineering in Agriculture (NCEA).
   b. For wheat, however, vision could still be supreme.
5. A simpler robot already in widespread use is the center-pivot irrigation device. These systems are self-propelled, irrigating an area of up to 600 acres per pass. Add-ons such as GPS, moisture monitors, and even imaging devices add sensory input for decision making. In this way, water and fertilizer may be applied to specific areas of the field at specific rates dependent on conditions [10].

Sources

1. *Meshram BD, Shaikh Adil and Suvartan Ranvir, Robotics: An emerging technology in dairy and food industry: Review, International Journal of Chemical Studies, 2018, 6(2): 440-449.*

1b. *Hamraj Singh, Robotics - An Emerging Technology in Dairy Industry International Journal of Mechanical and Production Engineering, ISSN(p): 2320-2092, ISSN(e): 2321-2071 Volume- 6, Issue-2, Feb.-2018.*

1c. *Subhash Prasad, Application of Robotics in Dairy and Food Industries-A Review, International Journal of Science, Environment and Technology, Vol. 6, No 3, 2017, 1856 -1864.*

1d. *Robotics and automation in the food industry and its future, LAC Conveyors and automation, 2019.*

1e. *Advantages of Robots in Food Manufacturing, Robotworx, Scott company.*

2. *Wallin, P. J., 1997. Robotics in the food industry, Trends in Food Science and Technology. 8(5): 193-233.*

3. *Brogardh, T. 2007. Present and future robot control development -An industrial perspective. An Rev Control. 31: 69-79.*

4. *Butler, D., Holloway, L. and Bear, C. 2012. The impact of technological change in dairy farming: robotic milking systems and the changing role of the stockperson. J Royal Agric Soc, Eng. 173: 1-6.*

4a. *Rapid Growth of Milking Robots Market, 2019, Business Industry report, Open PR.*

4b. *Wikipedia.*

5. *Khodabandehlco, K.1994. Robot in Meat, Fish and Poultry Processing, Blackie. Legg, B., 1993. Hi-tech agricultural engineering-A contradiction in terms or the way forward, Mechanical Incorporated Engineer, pp. 86-90.*

6. *Butler, D., Holloway, L. and Bear, C., 2012, The impact of technological change in dairy farming: robotic milking systems and the changing role of the stockperson. J Royal Agric Soc, Eng. 173: 1-6.*

7. *Tedford, J. D. 1990. Developments in robot grippers for soft fruit packing in New Zealand. Robotica. 8: 279-283.*

8. *Yao, W., Cannella, F. and Dai, J. S., 2011, Automatic folding of cartons using a reconfigurable robotic system. Robotics and Computer-Integrated Manufacturing. 27: 604-613.*

9. *Kempthome, H. 1995. Robotic processing of carcasses, Food Technology in New Zealand, pp.1214.*

10. *Edan Y, Rogozin V. Robotic melon harvesting: Prototype and field tests, IEEE Trans. Robot. Autom., 2010, 16: 831-834.*

4. Robots in Freezers and Ovens [1]

When handling food, robots are located in freezers or near ovens, while these environments generally do not harm robots, some defence must be made to ensure the robots operate efficiently. The robots based inside freezers for palletizing operations preclude undesirable frosty condensation from building up on ice cream packages [2].

Sources

1. *Meshram BD, Shaikh Adil and Suvartan Ranvir, Robotics: An emerging technology in dairy and food industry: Review, International Journal of Chemical Studies, 2018, 6(2): 440-449.*

2. *Suganya P, Ramani S, Mathew AL, Elias S. Robotics in food processing, Beverage and Food World, 2011, 62-65.*

5. Application of Robotic in Food Processing Industries [1]

The food products are highly variable both in shape, sizes and structure which poses a major problem for the development of manipulators for its handling and for the application of robots. The commercial application of robots in food industry is widely spread at the end of processing lines like packaging and palletizing [1].

- ✰ However there is a broad range of potential applications for robotics in food processing: in the grading of food products, pick and place application, packaging and palletizing, meat and baking lines to handle hot trays [1].
- ✰ Food manufacturing robots are commonly used in the dispensing, feed placement, cutting, packaging or casing of food, pick-and-placing products into containers, and sorting [1b].
- ✰ Newer developments also allow for order picking, where an employee or customer is able to place an order electronically, and the food robot automatically fills that specific order [1b].

a. Grading of Food Products

The grading robot system made various effects like, labour substitution, objective grading operation without human subjective judgment, data accumulation for traceability and farming guidance to producer.

i. Scientists [2] developed a fruit harvesting robot for use in Japanese agriculture systems which commonly produce crops in greenhouses and in small fields.

ii. Researchers [3]developed an end-effect or for the delicate harvesting of mushrooms.

iii. A grading system using robots has been developed for use with deciduous fruits such as peaches, pears, and apples. System automatically picks fruit from containers and inspects all sides of the fruit [4].

b. Robotics in Fruit and Vegetable Processing Industry

Robot technology has proved able to handle agricultural products delicately and with a high degree of precision, and to gather information to create a database of products every season. Packing robots and palletizing robots have been a frequent feature in fruit grading facilities, while grading robots [4], which collect round-shaped fruits and inspect them using a machine vision system, are now being introduced in some East Asian countries. **Robotics.org** reports that a major advancement in robotic technology within food processing has been the introduction of more advanced grippers. These grippers allow for quick but gentle handling of sensitive food products like fruit and vegetables [1a].

c. Picking of Fruit and Vegetables

Brute force tree shakers might be used for picking some citrus fruit, but hand-picking is still common. Intelligent picking has presented a challenge to many robotics researchers [5]. At present, the automation consists of no more than conveyor belts extending each side of a tractor while hand-pickers walk the field, choosing which to cut, be they broccoli, rock melons, or cauliflowers.

The gathering of macadamia nuts is performed by a manually steered vehicle with a bristle roller which gathers up the nuts from the ground. What brings it to the attention of robotics is a localization task associated with selecting for varietal strains. Cameras inspect the rollers just before the nuts are stripped, so that the pick-up location is known precisely with respect to the vehicle, but that leaves the task of locating the vehicle.

GPS is unreliable under the tree canopy, so the system combines odometry with tree-trunk location using both sideways-looking visual streaming and radiofrequency identification (RFID) tagging [6].

Automatic Fruit Harvesting Method [1]

The details of some automatic fruits harvesting methods are given below:

a. Canopy Shaker

A canopy shaker was designed to clamp secondary limbs and to shake vertically. The shaker was extended into tree with a pantograph lift unit and shake always vertically. An excessive immature orange were removed during tests conducted. Two continuing canopy shakers were reported by Futch and Roka, one was self-propelled unit and another was tractor-drawn unit. These two units were used for juice processing plants.

b. Ultrasonic Sensing

This Technique can be illustrated by an example of an egg plant harvesting robot. A robot was designed to harvest egg plants trained on a V-shaped frame. The robot includes **sensors** (CCD camera and ultrasonic distance sensor), **a manipulator** with seven degrees of freedom, **an end-effector, a traveling carriage**. The sensors are attached to the end-effector.

The robot runs between rows and scans images of the egg plants on both sides through a combination of travel and manipulator control in the order. When the global sensor detects an egg plant, the end-effector approaches the fruit. At this point the fruit length is estimated by the visual sensor and the ultrasonic distance sensor. If the fruit is of marketable size, scissors hold and cut the peduncle.

c. Machine Vision Technique by Shape

Machine vision technique by shape can be most suitably used for citrus fruits. Researchers [7] have incorporated shape into their fruit detection algorithms. Citrus in general tends to have a round shape, whereas tree branches and leaves tend to have more straight and pointed shapes. Looking for round objects can be simple way to detect fruit, but as with colour detection, shape detection can also have several problems. The main problem is occlusion. Fruit occlusion can cause multiple fruit to appear as a single large fruit, unlike leaf occlusion where there is distinct contrast in colour between the leaf and the fruit. fruit detection algorithm that is economically feasible for mass harvesting will require a combination of both the colour and shape properties of citrus [8].

Sources

1. *Meshram BD, Shaikh Adil and Suvartan Ranvir, Robotics: An emerging technology in dairy and food industry: Review, International Journal of Chemical Studies, 2018, 6(2): 440-449.*

1a. *Robotics and automation in the food industry and its future, LAC Conveyors and automation, 2019.*

1b. *Advantages of Robots in Food Manufacturing, Robotworx, Scott company.*

2. *Kondo N, Monta M, Fujiura T. Fruit harvesting robots in Japan. Advances in Space Research, 1996; 18: 181-184.*

3. *Reed JN, Miles SJ, Butler J, Baldwin M, Noble R. Automatic mushroom harvester development. Journal of Agricultural Engineering Research, 2001, 78: 15-23.*

4. *Kondo N. Fruit grading robot. In: Proc. of IEEE/ASME International Conference on Advanced Intelligent Mechatronics, Kobe, Japan, 2003.*

5. *Edan Y, Rogozin V. Robotic melon harvesting: Prototype and field tests, IEEE Trans. Robot. Autom., 2010, 16: 831-834.*

6. *Dunn M, Billingsley J. Machine vision system for counting macadamia nuts, Australasian Conf. Robot. Autom., Brisbane, 2003.*

7. *Hannan M, Thomas Burks. Current Development In Automated Harvesting., 1987.*

8. *Shigehiko H, Tomohiko O, Kotaro K, Katsunobu G, Naoshi K. Robotic Harvesting Technology for Fruit Vegetables in Protected Horticultural Production, 1993.*

6. Online Egg Sorting [1]

Online poultry inspection by a multi-camera system can be employed to accurately detect and identify carcasses unfit for human consumption [2]. By automating this process, the level of accuracy in identifying defective eggs increases; and the rate of sorting is higher. The 1.4 Megapixel cameras are positioned in such a fashion as to capture images from every angle as the eggs roll down a conveyor belt. The cameras monitor the quality of eggs passing through the system and the images are analysed digitally.

7. Automatic Process Monitoring

Application of machine vision has been reported for controlling drying process of sliced apple. The vigilance of a drying process was provided due to online image analysis and correlation of image attributes (area, colour and texture) with physical parameters of drying (moisture and quality). Scientist [3] reported the use of digital imaging technology for the automatic monitoring of dry sugar granules and powders. This system provides particle size data to production line operators for process control and product quality improvement.

8. Livestock Inspection [1]

Poultry [4] is one application where the advances in image sensors and filters for particular spectra combine to provide automatic detection of disease and contaminants in a poultry processing line.

i. An application involving individual recognition of production animals is in full commercial use by Big Dutchman. In this circumstance, pigs are tagged with unique radiofrequency identification (RFID) ear tags. As each animal enters a feeding bay, the tag is read and an appropriate ration of food is delivered. This allows individualized diets and medication delivery to improve the bottom-line cost/benefit for pork producers.

ii. Another novel method of identifying and controlling animals is being developed in Southern Queens-land [5]. Using machine vision, each animal proceeding along a laneway towards water is classified at the species level. An automated gate then either allows or denies access to the watering point.

Sources

1. *Meshram BD, Shaikh Adil and Suvartan Ranvir, Robotics: An emerging technology in dairy and food industry: Review, International Journal of Chemical Studies, 2018, 6(2): 440-449.*

2. *Chen, Y, Chao K, Kim MS. Machine vision technology for agricultural Applications. Computers and Electronics in Agriculture, 2002, 36: 173-191.*

3. *Strickland ML. Online particle sizing aids in sugar production. Sugar Journal, 2000, 62: 14-20.*

4. *Big Dutchman: Callmatic 2. Last access 10 December, 2007, from www.bigdutchman.de/eng., 2005, 46: 42.*

5. *Vaughan R, Sumpter N, Frost A, Cameron S. Robot sheepdog project achieves automatic flock control, Fifth Int. Conf. Simulation Adapt. Behav, 1998,*

9. Meat Processing [1]

The main aim of using an industrial robot in meat processing is to reduce production costs and occupational injuries while improving process efficiency and hygiene. The strength of robotics, particularly in boning rooms where labour costs are inherently high, is in their ability to perform the required repetitive tasks more efficiently and consistently than is currently possible (Food Science Australia).

☆ Georgia Tech researchers have developed a system that uses advanced imaging technology and a robotic cutting arm to automatically **debone chicken** and other poultry products. This robotic system is used for the intelligent cutting and deboning of a chicken, as it prepares to slice through the shoulder joint of a chicken, cutting close to the bone to maximize breast meat yield and ensuring food safety by avoiding creation of bone chips [2].

- ☆ In beef production the first use of robotic equipment was in splitting complete carcass into carcass sides.
- ☆ The Meat Industry Research Institute of New Zealand (MIRINZ) has in particular been very active in automation of sheep and lamb slaughtering.
- ☆ The Danish company SFK-Danfotech has, in cooperation with the Danish Meat Research Institute (DMRI), has developed a series of dedicated robots for automation of pig slaughter line processes [3].

Vision-guided robots are speeding up these practices to make certain that the pieces are accurately portioned and cut, while packaging equipment is incorporating volumetric scanning systems.

a. Robots in Beef Processing [1]

The first area where robotic equipment exists is that for splitting of a complete carcass into carcass sides. The equipment for this is produced by a number of manufacturers. A chilled eviscerated carcass is mounted horizontally on an automatic guided vehicle and appraised using X-rays, 3D machine vision and ultra-sonic sensing. A robot is used to effect the separation with high pressure water, abrasive and air jets. Flesh is cut with the water jet while the air jets keep the severed meat clear of the cutting area. The abrasive jet is invoked when cuts are to be made through bone.

Research into automation of beef boning has taken place in Australia [4] and UK [5]. Both systems utilise force feedback systems to guide a reciprocating blades cutter along the surface of the bone.

b. Robots in Pork Processing [1]

Automated cutting systems for pork exist that separate a half carcass into fore, middle and hind sections. The process was developed in conjunction with DMRI and the equipment is produced by Attec in Denmark and Itec in Germany [6]. Carcasses hanging on a standard gambrel are pulled across a conveyer belt and the hind feet cut off.

Fully automatic systems for collecting and grading information and carcass stamping have been developed in Denmark and in Holland. Since 1989 all Danish export pork carcasses have been classified with the fully automatic Pig Carcass Classification Centre developed at the DMRI.

Sources

1. *Meshram BD, Shaikh Adil and Suvartan Ranvir, Robotics: An emerging technology in dairy and food industry: Review, International Journal of Chemical Studies, 2018, 6(2): 440-449.*
2. *Calderone L. Food processing without the human touch.http: //roboticstomorrow.com/ content.php?post_type=1851. Cambridge, 2013, 283-296.*

3. *Madsen KB, Nielsen JU. Automated meat processing. In Meat processing– improving quality, Kerry, J. P., Kerry, J. F., and Ledward, D. (Eds.), Wood head Publishing Ltd., 2002.*

4. *Clarke PT, Key SJ, Tang P. Development of robot technology for meat deboning. Conference Proceedings 19th International Symposium and Exposition on Industrial Robots (ISIR), Sydney, Australia, 1988, 427-439, IFS Publications.*

5. *Purnell G. Robotic equipment in the meat industry. MeatSci., 1998, 49S1: S297-307.*

6. *Folkman P. Fully automatic pig carcass cutting plant. Fieischwirtschaft, 1995, 75(9): 1127.*

Chapter 21

Block Chain Technology

The blockchain **[1, 2]** initially emerged as a way to manage enormous amounts of data transactions [1]. **Blockchain** is most simply defined as a decentralized, distributed ledger **technology** that records the provenance of a digital asset. A blockchain [2, 2a, 2b] originally block chain [2c], is a growing list of records, called **blocks,** that are linked using cryptography [2a]. Each block contains a cryptographic hash of the previous block, a timestamp and transaction data (generally represented as a Merkle tree). Blockchain was invented by Satoshi Nakamoto in 2008 to serve as the public transaction ledger of the cryptocurrency bitcoin [2a]. The invention of the blockchain for bitcoin made it the first digital currency to solve the double-spending problem without the need of a trusted authority or central server [2]. A blockchain is, in the simplest of terms, a time-stamped series of immutable records of data that is managed by a cluster of computers not owned by any single entity. Each of these blocks of data (*i.e.* block) is secured and bound to each other using cryptographic principles (*i.e.* chain) [1a].

In addition to financial companies, today, the technology is being widely adopted in other industries, including the food industry. In basic terms, a blockchain is a digital distributed ledger, maintained by a network of multiple computing machines. It is an online transaction ledger where information can be stored [1]. These blocks of information cannot be edited, adjusted or changed. New data is added to old in blocks, which forms a data chain [3]. Entities that wish to add data to a specific blockchain must go through a series of validation processes, reducing the chance that a hacker could alter the data. It is "an open, distributed ledger that can record transactions between two parties efficiently and in a verifiable and permanent way" [2]. Once recorded, the data in any given block cannot be altered retroactively without alteration of all subsequent blocks, which requires consensus of the network majority [2].

Blockchain technology has become popular because it's a decentralized way for many users to contribute to and access data in real time. This creates transparency, especially for those who want to see the data to validate anything that the data represents [1]. Blockchain technology has the potential for wide applications in the food industry and Ripe.io, a company at the forefront of creating blockchain technology in the food industry has started with a blockchain system in the tomato industry [4]. This is working with Dairy Farmers of America to create a system in the dairy industry [4]. **Blockchain** for the dairy industry gives **complete traceability** data to the consumers, which helps to increase trust in the product and confidence in food production.

Over the past year, numerous mainstream companies, such as Walmart and Bumble Bee Foods, have jumped into the blockchain space, harnessing the tech to improve their respective supply chains. Nestle was a "founding member" of IBM's Food Trust blockchain operation back in 2017, which seeks to improve food product traceability. In April 2019, Nestle and French retailer Carrefour teamed up to provide customers the ability to track Mousline Puree throughout the supply chain. In February 2019, Nestle also announced its intent to publicize data surrounding its top 15 commodities in an overall effort "to reach full supply chain transparency" [5].

It includes the secure handling and storing of administrative records and digital authentication to strengthen intellectual property rights and patent systems, as well as bring transparency throughout the supply chain, reduce **food** frauds and enhance **food** safety [6, 7]. Blockchain is an especially promising and revolutionary technology because it helps reduce risk, stamps out fraud and brings transparency in a scaleable way for myriad uses. There are several IT giants who have made different blockchain platforms or solutions to the food industry. A few examples include Food Trust and Watson platforms by IBM, Track and Trace, and Leonardo platforms by SAP, Track and Trace, and Internet of Things solutions by Oracle [6].

According to the 2018 policy paper by the Food and Agriculture Organization of the United Nations (FAO), blockchain could dramatically transform the food supply chain in India. As per the policy paper, blockchain technologies can help in building an immutable contract between the various players in supply chain, enabling further transparency in the system [6, 8].

Providing this transparency throughout the production chain also allows for enhanced food quality through better traceability, especially in the event of a food recall. Using the blockchain technology companies could get to the source of the food quality issue in a matter of hours rather than days [4].

Sources

1. *Mike Opperman, The Blockchain of Dairy, Farm journal's milk, 2018.*

1a. *Ameer Rosic, What is Blockchain Technology, Ask block chain community, 2016.*

2. *Wikipedia.*

2a. *Blockchains: The great chain of being sure about things. The Economist. 31 October 2015.*

2b. *Morris, David Z., 15 May 2016, "Leaderless, Blockchain-Based Venture Capital Fund Raises $100 Million, And Counting". Fortune. Archived from the original on 21 May 2016.*

2c. *Brito, Jerry; Castillo, Andrea, 2013, Bitcoin: A Primer for Policymakers (PDF) (Report). Fairfax, VA: Mercatus Center, George Mason University. Archived (PDF) from the original on 21 September 2013.*

3. *Arthur Linuma, President and co-founder of ISBX, a leading blockchain development agency.*
4. *Raja Ramachandran, CEO, Ripe.io.*
5. *Benjamin Pirus, Contributor Crypto and Blockchain, Nestle Tests Public Blockchains For Dairy Supply Chain, forbe, Jul 9, 2019.*
6. *Sanjeev Kumar Sharma and Vinay Singh, Applications of blockchain technology in the food industry, New Food, 2020.*
7. *A systematic review of consumer perceptions of food fraud and authenticity: A European perspective I Trends in Food Science and Technology 94, 2019, 79-90.*
8. *A review on blockchain applications in the agri-food sector I Journal of Science of Food Agriculture, 2019, 99: 6129-6138.*

History/Developmental Steps of Blockchain [1]

The history of Blockchain is tabulated below:

Year	*Developments*
2008	**Satoshi Nakamoto** publishes "Bitcoin: A Peer to Peer Electronic Cash System."
2009	The first successful Bitcoin (BTC) transaction occurs between computer scientist Hal Finney and the mysterious Satoshi Nakamoto.
2010	Florida-based programmer Laszlo Hanycez completes the first ever purchase using Bitcoin - two Papa John's pizzas. Hanycez transferred 10,000 BTC's, worth about $60 at the time.
2011	1 BTC = $1USD, giving the cryptocurrency parity with the US dollar. Electronic Frontier Foundation, Wikileaks and other organizations start accepting Bitcoin as donations
2012	Blockchain and cryptocurrency are mentioned in popular television shows like **The Good Wife**, injecting blockchain into pop culture. Bitcoin Magazine launched by early Bitcoin developer Vitalik Buterin.
2013	Buterin publishes "Ethereum Project" paper suggesting that blockchain has other possibilities besides Bitcoin (*e.g.*, smart contracts).
2014	**R3, a group** of over 200 blockchain firms, is formed to discover new ways blockchain can be implemented in technology. **PayPal announces** Bitcoin integration.
2015	**NASDAQ and San-Francisco** blockchain company Chain team up to test the technology for trading shares in private companies.
2016	Tech giant **IBM announces** a blockchain strategy for cloud-based business solutions. **Government of Japan recognizes** the legitimacy of blockchain and cryptocurrencies.
2017	Bitcoin reaches its all-time high at $19, 783.21/BTC. **Dubai announces** its government will be blockchain-powered by 2020.
2018	**Facebook** commits to starting a blockchain group and also hints at the possibility of creating its own cryptocurrency. **IBM** develops a blockchain-based banking platform with large banks like Citi and Barclays signing on.

Sources

1. *Blockchain, builtin.com.*

Basic Structure of Blockchain [1]

Blockchain is a process of passing information from stage one to stage two in a process, which takes places in a fully automated and safe environment.

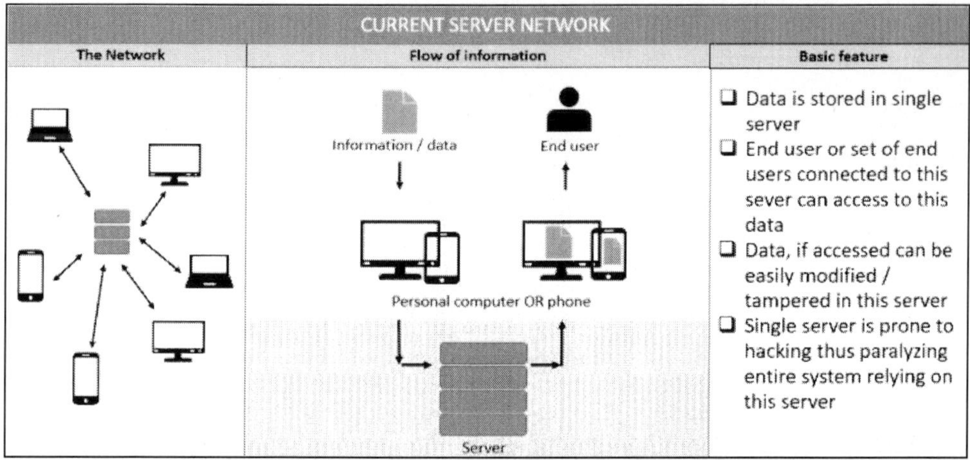

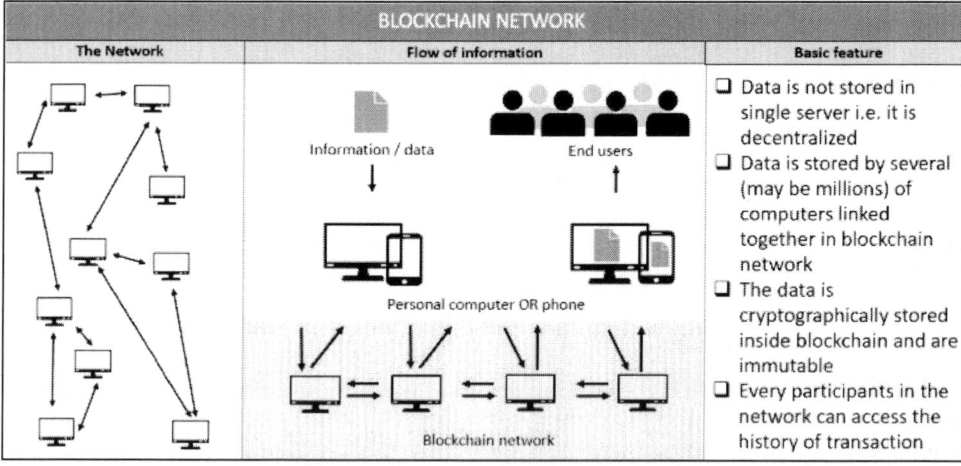

Within the food supply chain there are different participants involved in each step, from farmers, to processors, manufacturers, certifying agencies, government agencies, logistics, distributors, retailers and so on. Each of these participants shares critical information about the product that sits in their own local server. This information is not accessible to other participants and thus, there is an increased chance of food fraud in the system. Blockchain provides a secure environment where each of the participants in the blockchain network has access to each data and these data, once entered and verified, cannot be modified. For example, a farmer that provides an organic food certificate, which is verified by an authorised agency, cannot have that certificate tampered with at later date.

Source

1. *Sanjeev Kumar Sharma and Vinay Singh, Applications of blockchain technology in the food industry, New Food, 2020.*

Working of Blockchain Technology [1]

Blockchain consists of the following three important concepts:

Blocks, Miners and Nodes.

Blocks

Every chain consists of multiple blocks and each block has three basic elements:

1. The **data** in the block.
2. A 32-bit whole number called a **nonce.** The nonce is randomly generated when a block is created, which then generates a block header hash.
3. The **hash** is a 256-bit number wedded to the nonce. It must start with a huge number of zeroes.

When the first block of a chain is created, a nonce generates the cryptographic hash. The data in the block is considered signed and forever tied to the nonce and hash unless it is mined.

Miners

It create new blocks on the chain through a process called mining. In a blockchain every block has its own unique nonce and hash, but also references the hash of the previous block in the chain, so mining a block is not easy. Miners use special software to solve the incredibly complex math problem of finding a nonce that generates an accepted hash. There are roughly four billion possible nonce-hash combinations (32 nonce and 256 hash) that must be mined before the right one is found. When that happens miners are said to have found the "golden nonce" and their block is added to the chain.

Making a change to any block earlier in the chain requires re-mining not just the block with the change, but all of the blocks that come after. This is why it's extremely difficult to manipulate blockchain technology. When a block is successfully mined, the change is accepted by all of the nodes on the network and the miner is rewarded financially.

Nodes

Decentralisation is important concept in blockchain technology. No one computer or organization can own the chain. It is a distributed ledger via the nodes connected to the chain. Nodes can be any kind of electronic device that maintains copies of the blockchain and keeps the network functioning.

Every node has its own copy of the blockchain and the network must algorithmically approve any newly mined block for the chain to be updated, trusted and verified. Each participant is given a unique alphanumeric identification number that shows their transactions.

Sources

1. *Blockchain, builtin.com.*

Steps to Build a Blockchain Solution [1]

The following 8 steps are required to build a Blockchain solution:

Step 1: Identify a Suitable Use-case.

Step 2: Identify the Most Suitable Consensus Mechanism.

Step 3: Identify the Most Suitable Platform.

Step 4: Designing the Nodes.

Step 5: Design the Blockchain Instance.

Step 6: Building the APIs.

Step 7: Design the Admin and User Interface.

Step 8: Adding Future Tech.

Sources

1. *Rohas Nagpal, Chief Blockchain Architect, Primechain Technologies Pvt Ltd, , 2017*

Advantages/Benefits and Pre-requisites/Challenges of Block Chain Technology [1, 2]

Players Getting Benefits [1]

It is useful for producers, processing plants and consumers. The details are given below:

1. **Producers:** The growers of raw materials for the food products gain recognition in the commercial sphere with the implementation of blockchain in food supply chain. With good visibility, the growers/ producers have increased chances for bank loans to be sanctioned for business development.
2. **Processing Plants:** The plants that process raw materials into food products reap the benefit of establishing their brands across the world with blockchain integration. A popular brand obviously attracts more customers in the long run leading to massive returns.
3. **Consumers:** Blockchain presents the customers with quality food products hence improving the overall quality of lives making the planet a better place to live.

Benefits [2]

Blockchain technology is expected to bring multiple benefits to the multiple players in the food industry. The details are given below:

1. The supply chains can revitalize their management and handling, as they will have detailed information on member profiles, which provides a higher level of certainty over the safety of food. The proof of this to the consumer is by implementing QR codes and product labeling.
2. Consumers can access labelling information from their smart phones to know product information. On the other hand, supermarkets can access large amounts of data that travels through the supply chain.
3. In the agriculture sector, blockchain can help with inventory tracking and shipping processes, which helps reduce the associated costs that would otherwise go to a third party verifier. Blockchain technology can help combat food fraud and counterfeiters through its transparent and automated multi-ledger system.
4. The another challenge that manufacturers face, is to collect knowledge of origin of ingredients that go in to processed food. To build consumer trust, this information needs to be available to the consumer. Blockchain allows the grower and processor to share information with each other privately and securely.
5. Blockchain provides an open platform, and no third party is needed to authorize transactions. There are a set of rules that all participants must follow.
6. One of the biggest benefits of blockchain is its public nature. In a food supply chain, all of the transactions for one item of food can be seen and validated at any point of time.
7. The permanent record of transactions, provided by Blockchain cannot be altered or tampered with. Transactions are verified and approved by consensus among participants, making fraud more difficult.
8. The major advantage is that the time required to source any information regarding the supply chain and its components is reduced to a fraction of what it was before.
9. Because blockchain technology operates on a distributed (rather than centralized) platform. Information is updated for everyone at the same time. This empowers the whole supply chain to be more responsive to any fraud threats or food safety disasters.

Pre-requisites/Challenges [2]

In addition to the aforesaid benefits, this technology also has a set of pre-requisites and challenges. The blockchain technology has following pre-requisties:

1. The data at the source needs to be accurate.
2. All parties along the chain need to agree to adopt the technology.

3. Data would need to be collected from all the players in the value chain, which requires all of them to actively cooperate on what data to collect, how to capture and how to compile the information into meaningful product data that consumers can read.
4. So some effort would go into fixing the data before it is entered.
5. For food companies, the issue would be to carry out a proper analysis of the data to be uploaded.

Sources

1. *Trace food, Block chain does ot all.*

2. *Blockchain technology for Food Industry, Sourcetrace,*

Specialities and Pillars/Properties of Blockchain Technology [1, 2]

The specialities and pillars of Blockchain technology are discussed below:

1. Specialities [1]

i. It is not owned by a single entity, hence it is decentralized,
ii. The data is cryptographically stored inside,
iii. The blockchain is immutable, so no one can tamper with the data that is inside the blockchain
iv. The blockchain is transparent so one can track the data if they want to

2. Pillars/Unique features of Blockchain Technology [1, 2]

The three main properties of Blockchain Technology which have helped it gain widespread acclaim are as follows:

i. Decentralization

Unlike the current internet system, there is no single server where entire information is stored. These data are saved as copies in different computers (called nodes in the blockchain world) and there is no central authority supervising anything. These computers are connected to the blockchain network [2]. Before Bitcoin and BitTorrent came along, centralized services were used more but now decentralized system is used in blockchain technology. In a decentralized system, the information is not stored by one single entity. In fact, everyone in the network owns the information [1].

ii. Transparency

One of the most interesting and misunderstood concepts in blockchain is "transparency." Some users say that blockchain gives privacy while some say that it is transparent but in reality the person's real identity is secure, one will still see all the transactions that were done by their public address [1]. This should not be misinterpreted as violating privacy. The identity of parties involved in a transaction is hidden via complex cryptography and represented only by their public address (code) [2].

iii. Immutability

Immutability, in the context of the blockchain, means that once something has been entered into the blockchain, it cannot be tampered with [1]. Any data entered in blockchain cannot be modified/tampered with in the future. This immutability feature of blockchain has made it so popular that it is being explored in all the sectors where data integrity and data protection is of utmost importance [2].

iv. Trustless

Blockchain provides such a transparent environment that one need not to have a third-party to mediate or enhance trust between two parties involved in transaction. Trustless should not be confused with 'not trustworthy' [2].

Sources

1. *Ameer Rosic, What is Blockchain Technology, Ask block chain community, 2016.*
2. *Sanjeev Kumar Sharma and Vinay Singh, Applications of blockchain technology in the food industry, New Food, 2020.*

Food Safety, Food Tracebility and Prevention of Food Fraud by Blockchain Technology

This technology helps in maintaining required food safety and preventing food fraud. It also helps in Food tracebility.

a. Food Safety [1, 2]

Blockchain makes a supply chain more transparent at an all-new level. It also empowers the entire chain to be more responsive to any food safety disasters. Nestle, wallmart and Unilever are considering blockchain technologies for that very reason.

- ☆ By using blockchain, Walmart can provide all the information the consumer wants in 2.2 seconds. During an outbreak of disease or contamination, six days is an eternity. A company can save lives by using blockchain technologies.
- ☆ Blockchain also allows specific products to be traced at any given time, which would help to reduce food waste. For instance, contaminated products can be traced easily and quickly, while safe foods would remain on the shelves and not be sent to landfills.

b. Prevention of Fraud [1, 2]

However, it will work only if the data at the source is accurate, as current practices in the industry are much more open to human error.

- ☆ Much of the compliance data is audited by trusted third parties and stored either on paper or in a centralised database. These databases are highly vulnerable to informational inaccuracies, hacking, high operating costs, and intentional errors motivated by corruption and fraudulent behaviour.

- ☆ Blockchain operates anonymously, so mistakes would be traceable to individual culprits. Blockchain technology provides a method with which records are kept permanently.
- ☆ It facilitates data-sharing between disparate actors in a food value chain. Many retailers have sold fraudulent food products unknowingly. With the use of blockchain, it can be checked.

c. Food Traceability [3]

Food traceability has been at the center of recent food safety discussions, particularly with new advancements in blockchain applications. Due to the nature of perishable food, the food industry at whole is extremely vulnerable to making mistakes that would ultimately affect human lives. When foodborne diseases threaten public health, the first step to root-cause analysis is to track down the source of contamination and there is no tolerance for uncertainty. Consequently, traceability is critical for the food supply chain. The current communication framework within the food ecosystem makes traceability a time-consuming task since some involved parties are still tracking information on paper. The structure of blockchain ensures that each player along the food value chain would generate and securely share data points to create an accountable and traceable system.

Sources

1. *The Conversation Media Group.*
2. *Gul food manufacturing, UAE, 2020.*
3. *Elaine Qiu, How Blockchain will Transform the Food Supply Chain, Play and Plug.*

Applications of Blockchain Technology

Blockchain refers to a tamper-proof, distributed digital ledger that records transactions. It can be used in a number of ways across a range of different sectors, from **food** and drink to energy. Blockchain's capability of tracking ownership records and tamper-resistance can be used to solve urgent issues such as **food fraud, safety recalls, supply chain inefficiency and food traceability** in the current food system [1b].

Industry-wise Applications

The details related to the industry wise application of Blockchain technology are given below:

A. Application in Dairy

Dairy is ideal for blockchain technology because of the amount of data collected at each point of the supply chain, starting at the herd level. Data can be recorded at any point in the process from when the milk was produced on the farm all the way up to the time the milk bottle/milk pouch is on the store shelf [1, 4a].

- ☆ Blockchain could automate and sequence information by time stamp to create a full record for any item [4a].

- ✰ At the farm, data could be attached to a milk shipment, including when it was picked up and any lab work that goes with it. It's even feasible that production practices could be attached with the data, including any animal welfare activities, environmental practices, carbon footprint impacts, antibiotic applications and so forth.
- ✰ Once it leaves the dairy, each stop along the product production chain would have another set of data that could be attached with that specific load of milk.
- ✰ Once that product hits the final customer, that customer could have access to the complete ledger of data for that product [4a]. Individual customer brands have to use the blockchain process to bring value back through the supply chain.

i. *Trace Food Blockchain in Dairy Industry [1a]*

It assures to build the confidence in the product which ultimately raises the demand for the package. Tracefood allows you to trace each and every stage from the cattle shed to our table. **Blockchain for the dairy industry** traces and documents each and every stage involved from the shed to the fridge to present as the proof of the diary product's quality. **Dairy Supply chain blockchain** also tracks the milking and dairy processing management to exhibit the insight of the lab work to check the quality of the dairy product. **Blockchain for the dairy industry gives complete traceability** data to the consumers, which helps to increase trust in the product and confidence in food production. Blockchain's transport and storage transparency creates interest among the people which eventually increases the demand for the product. Fascinatingly, Blockchain allows anyone to compliment the cattle growers and the processors directly. This compliment from the consumers helps the cattle raisers to get loans and improvise their business.

ii. *Global Foods Giant Nestle*

It is now diving into public blockchain technology to enhance its dairy supply chain. Nestle will work with Australia-based startup OpenSC to bring its blockchain-supply chain work to the next level. World Wide Fund For Nature Australia (WWF-Australia) and Boston Consulting Group Digital Ventures founded the 2019 startup OpenSC. Nestle has planned to begin this technology in the dairy niche, tracking products from their New Zealand origin, to Nestle's Middle Eastern sites [2]. **Nestle** collaborated with OpenSc, a blockchain platform, to trace milk from farms and producers in New Zealand to Nestlé factories and warehouses in the Middle East. **Nestlé and** Carrefour collaborated to use IBM Food Trust platform for their GUIGOZ Bio 2 and 3 infant milk range [4].

iii. *Claru*

It is a Cork-based company using Blockchain technology to bring transparency to the food supply chain, specifically within the Irish dairy sector. Consumers appear more discerning, more willing to pay a premium for healthy, environmentally-friendly and ethically produced food. Transparency is predicted to become the largest differentiator for food brands in the near future [3].

iv. El Ordeño

It is one of the milk processors from Ecuador, adopted IBM Food Trust to bring transparency and reduce food waste. Through the QR code present in its TRU milk products. The consumers are able to get access to the information about the path that its TRU milk has travelled before it reaches them [4].

B. Application in Food Industries [4]

The several examples from different food industries where blockchain has been or is planning to be used in pilot-scale to bring transparency in the supply chain network and gain consumer trust, are given below:

a. Oil Industry

i. The food industry powerhouse also has plans to similarly track "palm oil in the Americas [2].

ii. The **Malaysian Palm Oil Council (MPOC)** in collaboration with BloomBloc has implemented blockchain technology to enhance accountability and traceability. It has developed a blockchain mobile app and web interface to trace palm oil throughout its supply chain. Each tree and related information, as well as other information of additional checkpoints, are uploaded into the system with the help of smartphone and this automatically creates an end-to-end digital database, enhancing transparency, accuracy and credibility for stakeholders and end customers [4].

b. Meat and Fish Industry [4]

i. **Shrimp: The Sustainable Shrimp Partnership (SSP)** has collaborated with IBM to use its Food Trust ecosystem to provide complete traceability of SSP shrimp for their consumers. Food Trust provides a safe and secure platform where data can be uploaded and shared.

ii. **Fish:** Cermaq salmon **and** Labeyrie **(Leading brand, Smoked salmon, France)** came together to utilise the IBM Food Trust platform to ensure traceability and transparency in their product supply chain. They are providing information about the fish- origin, hatching time and date, breeding conditions, length, time of its transfer to seawater, the kind of vaccination it was given, *etc.* Labeyrie is using blockchain technology for two of its Norwegian smoked salmon products. They are giving the consumers the opportunity to access all important information for building transparency. By scanning the QR code present on the package, the consumers can get access to the entire information of the product.

c. Organo Corporations

It is implementing IBM Food Trust, to selectively share key information about Organa products (Pure/Ultrapure water treatment equipments, water treatment chemicals, Filters, Energy, Medicine *etc.*) with consumers, supply partners, commerce partners, distribution partners, *etc.* The business wants to ensure transparency, but at the same time, ensure a highly secured environment. This transparency helps in building partner and consumer trust in every Organo product [4].

Sources

1. *Mike Opperman, The Blockchain of Dairy, Farm journal's milk, 2018.*

1a. *Trace food, Block chain does ot all.*

1b. *Elaine Qiu, How Blockchain will Transform the Food Supply Chain, Play and Plug.*

2. *Benjamin Pirus, Contributor Crypto and Blockchain, Nestle Tests Public Blockchains For Dairy Supply Chain, forbe, Jul 9, 2019.*

3. *John O'Reilly, Can Blockchain help dairy industry, Think business.ie, Ireland, 2020.*

4. *Sanjeev Kumar Sharma and Vinay Singh, Applications of blockchain technology in the food industry, New Food, 2020.*

4a. *Raja Ramachandran, CEO, Ripe.io.*

Chapter 22

Gas Packaging and Modified Atmosphere Package Design

MAP [1] is an advanced packaging technology that creates an optimal environment to extend produce shelf life and helps in preserving nutritional properties. The energy crisis, ecological awareness and demand for healthy and fresh food has created a need for a technology that allows distribution of fresh produce. It uses minimum processing and little energy and maximizes nutrition and flavour. Modified Atmosphere Packaging (MAP) theoretically, offers a possibility of meeting aforesaid requirements. **Modified atmosphere packaging (MAP) is defined as 'the packaging of a perishable product in an atmosphere which has been modified so that its composition is other than that of air.** MAP is the alteration of the gaseous environment produced as a result of respiration (passive MAP) or by the addition and removal of gases from food packages (active MAP) to manipulate the levels of O_2 and CO_2. MAP enables the fresh produce or perishable product to be packaged when it is fresh and then maintains it in that condition, thereby, reducing distribution costs and enhancing flavours and nutrition value for the consumer. Horticultural produce respires even after harvesting and the biological processes, which cause deterioration get accelerated which affects the nutritional value, flavour, texture and appearance. MAP slows the ongoing life processes not by changing the product but by adjusting its environment. Under the influence of MAP, the product is held in a state similar to that of animal hibernation. MAP maintains this state for a long period of time, during which deterioration is effectively stopped [1a]. In a modified atmosphere package, the product is exposed inside the pack to the normal atmospheric gases (oxygen, nitrogen, carbon dioxide and water vapour) but in concentrations which are different from those in the ambient air. The packaging consists of polymeric film pouch or plastic container with a specified gas permeability [1a]. Modified atmosphere is the practice of modifying the composition of the internal atmosphere of a package (food packages, *etc.*) in order to improve the shelf life [2].

The need for this technology for food arises from the short shelf life of food products (meat, fish, poultry, dairy) in the presence of oxygen. In food, oxygen is readily available for lipid oxidation reactions. Oxygen also helps maintain high respiration rates of fresh produce, which contribute to shortened shelf life [3]. Oxygen encourages the growth of aerobic spoilage micro-organisms [2]. Thus, the reduction of oxygen and its replacement with other gases can reduce or delay oxidation reactions and microbiological spoilage. Oxygen scavengers may also be used to reduce browning due to lipid oxidation by halting the auto-oxidative chemical process.

The modification process generally lowers the amount of oxygen (O_2) in the headspace of the package. Oxygen can be replaced with nitrogen (N_2), a comparatively inert gas, or carbon dioxide (CO_2) [2]. These gases can be applied individually or mixed according to specific ratios. CO_2, for instance, is most effective in inhibiting micro-organisms (such as mold and other common aerobic bacteria). It does this by dissolving into the food's liquid and fat phase, thereby reducing its pH value. It also penetrates biological membranes, causing changes in permeability and function [4]. A stable atmosphere of gases inside the packaging can be achieved using active techniques, such as gas flushing and compensated vacuum, or passively by designing "breathable" films.

MAP improves the shelf life and freshness of many foods. Modified atmosphere packaging allows natural and non processed food to be packaged in a way that extends the shelf life of the product. This increases the quality of the products image and texture, as well as the appeal of it's nutrition [4]. The MAP technique is useful for products that won't stand up to hard vacuuming or that tend to spoil in low oxygen environments. MAP can be used with fresh or processed foods, including meat, seafood, fruits, vegetables, various bakery products, snack foods and other dried foods.

Modified Atmosphere Packaging(MAP) or Controlled Atmosphere Packaging is making a big impact in produce packaging. This advanced process uses micro-perforation technology to alter the package atmosphere. The specifically engineered micro-perforations (Laser) control gas flow in and out of the package. This flow manages the oxygen and carbon dioxide levels inside the package and matches that flow to the products respiration rates. Oxygen flow into the package is reduced, slowing the ripening process and conserving moisture [1b, 1c]. The excess CO_2 that is produced by the produce is also able to diffuse. The controlled rate of exchange promoted by the MAP packaging allows the internal atmosphere of the package to reach an oxygen level lower than the regular air. The laser micro-perforations are more than 5 times smaller to standard mechanical micro-perforation. The ability to effectively manage respiration gasses helps to greatly improve produce shelf life [1b, 1c]. MAP packaging limits oxygen flow and moisture evaporation. The controlled gas exchange reduces respiration rates by limiting the oxygen available in the package [1b].

Sources

1. *Wikipedia.*

1a. *http: //icpe.in/icpefoodnpackaging/pdfs/24_modified.pdf.*

1b. *Clemons, Rachel, "Extending the shelf life of fresh food." CHOICE, CHOICE Australia, 5 June 2017, www.choice.com.au/food-and-drink/food-warnings- and-safety/food-safety/articles/extending-the-shelf-life-of-fresh-food.*

1c. *Extend Produce Shelf Life with Modified Atmosphere Packaging, Crawford Packaging.*

2. *Parry, R. T., 1993, Principles and applications of modified atmosphere packaging of foods, Boston, MA: Springer US. ISBN 978146 135 8923.*

3. *Boskou, D., Elmadfa, I., 2011, Frying of food: oxidation, nutrient and non-nutrient anti-oxidants, biologically active compounds and high temperatures (2nd ed.). Boca Raton: CRC Press. ISBN 9781439806821.*

4. *Nathan Dube, 2019, What is MAP, industrial packaging.com.*

History [1]

The historical development of MAP is summerised below:

- ☆ The first commercial application of modified gas atmospheres was for Controlled Atmosphere Storage (CAS) of fruits and vegetables.
- ☆ **1930-**Scientific investigations on the effect of gases on extending the shelf life of foods were conducted in 1930 on fresh meat.
- ☆ Fresh carcass meat was exported from New Zealand and Australia under CAS in the early 1930s.
- ☆ Numerous researchers have reported the increase of shelf life of meat products when they were stored in an atmosphere of CO_2.
- ☆ **1970-**Commercial retailing of fresh meat in MAP tray systems was introduced in European market in the early 1970s.
- ☆ In the past few years, there has been a considerable increase in the range of foods packed in modified atmospheres for retail sale including meat, poultry, fish, bacon, bread, cakes, and cheese products packed under MAP.

Sources

1. *http: //ecoursesonline.iasri.res.in/mod/page/view. Vacuum and MAP, Lesson- 23.*

Theory [1, 1a]

MAP has also been referred to as **"protective atmosphere packaging"** or when used in labelling as **"packaged in a protective atmosphere".** Four packaging techniques are included within this definition namely vacuum packaging, passive MAP, gas-flush MAP and MAP by active packaging. Passive modification is a slow process whereas active modification is faster and can be achieved by gas flushing, vacuum application or by using gas scavengers / emitters. Passive MAP is traditionally used with fresh and minimally processed fruits and vegetables and only with some selected cheese varieties.

Gas-flush MAP involves the establishment of a specific gas composition within the package in a single stage during the packaging operation, by flushing with the selected gas mixture before sealing. Depending on the desired residual O2, a vacuum operation may be needed prior to gas flushing.

Modified atmosphere can be created either passively by the product or intentionally by active packaging. Atmosphere within the package can be modified by these two ways [2]. The details of these two ways are given below:

i. Passively

In passive modification, the respiring product is placed in a polymeric package and sealed hermetically. Only the respiration of the product and the gas permeability of the film influence the change in gaseous composition of the environment surrounding the product. If the product's respiration characteristics are properly matched to the films permeability values, then a beneficial modified atmosphere can be passively created within a package [1a].

In passive modified atmosphere packaging (MAP), the high concentration of CO_2 and low O_2 levels in the package is achieved over time as a result of respiration of the product and gas transmission rates of the packaging film. This method is commonly used for fresh respiring fruits and vegetables. Reducing O_2 and increasing CO_2 slows down respiration rate, conserves stored energy, and therefore extended shelf life [1, 3].

ii. Actively

On the other hand, active MAP involves the use of active systems such as O_2 and CO_2 scavengers or emitters, moisture absorbers, ethylene scavengers, ethanol emitters and gas flushing in the packaging film or container to modify the atmosphere within the package [1, 3]. In case of active modification, two basic techniques are employed to replace air in MAP *i.e.* Gas flushing (the air is replaced by passing a stream of gas. The air gets diluted and the pack is sealed. The oxygen level in gas-flushed packs is up to 2 - 5 per cent) and Compensated Vacuum (vacuum is first applied to remove the air and then the desired gas or gas mixture is incorporated. Since it involves an extra step, this process is slower, but the residual oxygen is much less) [1a]. More details are given in Chapter-19.

Gases and Selection of Mixture of Gases in MA [1]

The common gases used in MA, are oxygen, nitrogen and carbon dioxide. Traces of carbon monoxide, nitrous oxide, ozone, argon, ethanol vapour and sulphur dioxide are also used. **Nitrogen** is an inert gas, has no anti-microbial activity and acts as a cushion, thereby preventing pack collapse. Since it displaces oxygen from the pack, oxidative rancidity is delayed. **Carbon dioxide** is responsible for the bacteriostatic and fungistatic effect in MA packaged food. It retards the growth of moulds and aerobic bacteria. The inhibitory effect of carbon dioxide to micro-organisms is increased as the temperature is lowered because of increased solubility [1a]. Basically, carbon dioxide is used in 20 per cent concentration and it is characterized by a strong inhibitory property, which slows down the development of bacteria [1b]. The selection of mixture of gases for a MA package depends on followings [1]:

1. The type of product,
2. The packaging materials.
3. The storage temperature.

The atmosphere in an MA package consists mainly of adjusted amounts of N_2, O_2, and CO_2 [2, 4]. Reduction of O_2 promotes delay in deteriorative reactions in foods such as lipid oxidation, browning reactions and growth of spoilage organisms [2, 5].

- ☆ Low O_2 levels of 3-5 per cent are used to slow down respiration rate in fruits and vegetables [2].
- ☆ In the case of red meat, however, high levels of O_2 (<"80 per cent) are used to reduce oxidation of myoglobin and maintain an attractive bright red color of the meat [6]. Meat color enhancement is not required for pork, poultry and cooked meats, therefore, a higher concentration of CO_2 is used to extend the shelf life [4].
- ☆ Levels higher than 10 per cent of CO_2 are phytotoxic for fruit and vegetables, so CO_2 is maintained below this level.
- ☆ N_2 is mostly used as a filler gas to prevent pack collapse [4, 5]. In addition, it is also used to prevent oxidative rancidity in packaged products such as snack foods by displacing atmospheric air, especially oxygen, therefore extending shelf life [4, 5].
- ☆ The use of noble gases such as Helium (He), Argon (Ar) and Xenon (Xe) to replace N_2 as the balancing gas in MAP can also be used to preserve and extend the shelf life of fresh and minimally processed fruits and vegetables. Their beneficial effects are due to their higher solubility and diffusivity in water, making them more effective in displacing O_2 from cellular sites and enzymatic O_2 receptors [7].

There has been a debate regarding the use of carbon monoxide (CO) in the packaging of red meat due to its possible toxic effect on packaging workers [6]. Its use results in a more stable red color of carboxymyoglobin in meat, which leads to another concern that it can mask evidence of spoilage in the product [5, 6].

Sources

1. *Wikipedia.*

1a. *http: //icpe.in/icpefoodnpackaging/pdfs/24_modified.pdf.*

1b. *Shupeyko A. M.and Ishevsky A. L., Moddern Food Packaging Technology, World Scientists, 2019.*

2. *Robertson, G. L., 2006, Food Packaging Principles and Practice (2nd ed). Florida: CRC Press. pp. 313-330, ISBN 978-0-8493-3775-8.*

3. *Brody, A. L., Zhuang, H., Han, J. H., 2011, Modified atmosphere packaging for fresh-cut fruits and vegetables. West Sussex, UK: Blackwell Publishing Ltd. pp. 57-67. ISBN 978-0-8138-1274-8.*

4. *Fellows, P. J, 2017, Food processing technology: principles and practice (4th ed). Duxford, UK: Woodhead Publishing. pp. 992-1001. ISBN 978-0-08-101907-8.*

5. *Blakistone, B. A., 1998, Principles and applications of modified atmosphere packaging of foods (2nd ed.). London: Blackie Academic and Professional, pp. 1-38, ISBN 978-0751403602.*

6. *Djenane, D., Roncales, P., 2018, "Carbon monoxide in meat and fish packaging: advantages and limits". Foods,* **7** *(2): 12. doi: 10.3390/foods7020012.*

7. *Ghidelli, C, Perez-Gago, M. B, 2018, "Recent advances in modified atmosphere packaging and edible coatings to maintain quality of fresh-cut fruits and vegetables". Critical Reviews in Food Science and Nutrition. 58 (4): 662-679. doi: 10.1080/104 083 98.2016. 121 1 087.*

Effects of Gases on Micro-organism [1]

Different gases being used in MAP have important effects on micro-organism [1]:

Carbon Dioxide

1. Lactic acid bacteria are resistant to the gas, and may replace aerobic spoilage bacteria in modified atmosphere packaged meat.
2. Most of the yeasts are also resistant to carbon dioxide.
3. Anaerobic bacteria, including food poisoning organisms, are little affected by carbon dioxide. Consequently, there is a potential health hazard in MAP products from these micro-organisms.
4. Molds and some Gram negative, aerobic bacteria are inhibited by carbon dioxide concentrations in the range 5-50 per cent [2].
5. Too high CO_2 concentrations in the atmosphere can lead to an increased drip loss during storage. This can be explained by the pH drop induced by CO_2 dissolving in the water phase of the product, causing a decrease in the water binding capacity of the proteins.

Nitrogen

It has only a role to inhibit the oxidation of fats by replacing oxygen. It is used as a bulking material to avid collapse of MAP packages when the carbon dioxide dissolves in the food since its solubility in water is low [1].

Sources

1. *Mahendra Pal, Mridula Devrani and Angesom Hadush, Recent developments in food packaging technologies, Researchgate, 2019, and Beverage and Food World -Vol.46-No. 1, Jan. 2019.*

2. *Brennan, J. G., 2006, Food Processing Handbook, Wiley-Vch Verlag GmbH and Co., Weinheim.*

Equipment [1, 6]

The details related to the equipments for MAP and quality assurance of products are given below:

a. The Equipment for MAP [1]

FFS packing machine, Chamber machine and snorkel machines are being used. The details regarding these types of machine are given below:

i. Form-Fill-Seal Packing Machine

In using form-fill-seal packaging machines, the main function is to place the food product in a flexible pouch suitable for the desired characteristics of the final product. These pouches can either be pre-formed or thermoformed. The food is introduced into the pouch, the composition of the headspace atmosphere is changed within the package and then it is heat sealed [2]. These types of machines are typically called pillow-wrap, which horizontally or vertically form, fill and seal the product [3]. Form-fill-seal packaging machines are usually used for large scale operations.

ii. Chamber Machines

These are used for batch processes. A pre-formed wrap is filled with the food product and introduced into a cavity. The cavity is closed and vacuum is then pulled on the chamber and the modified atmosphere is inserted as desired. Sealing of the package is done through heated sealing bars, and the product is then removed. Although this batch process is labor intensive and time consuming but it is relatively cheaper than packaging machines which are automated [2].

iii. Snorkel Machines

These are used to modify the atmosphere within a package after the food has been filled. The food product is placed in the packaging material and positioned into the machine without the need of a chamber. A nozzle, which is the snorkel, is then inserted into the packaging material. It pulls a vacuum and then flushes the modified atmosphere into the package. The nozzle is removed and the package is heat sealed. This method is suitable for bulk and large operations [2].

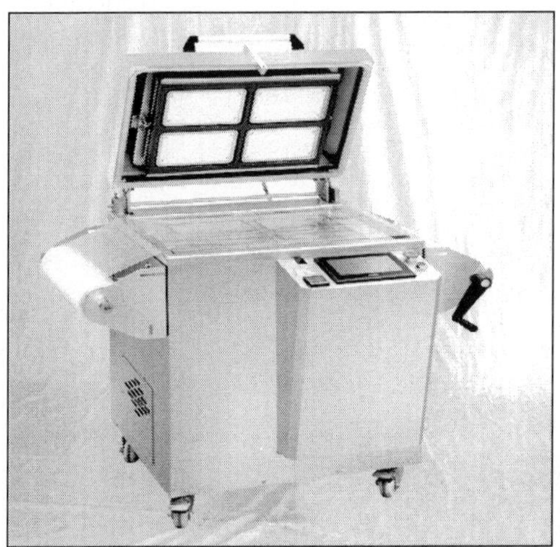

MAP Packing Machine Semi Automatic [4].

MAP Machine [5]

MAP [6]

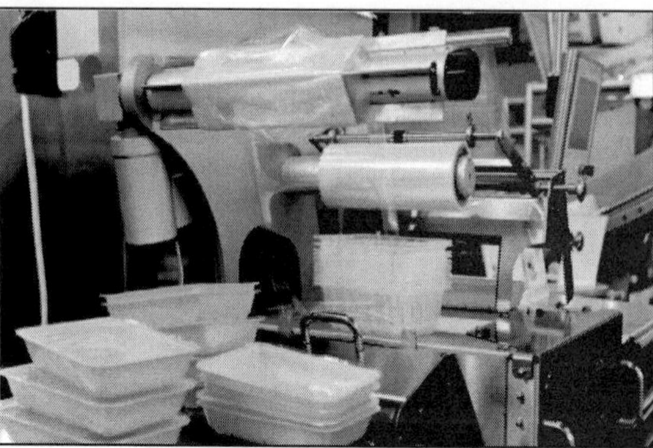

MAP [7]

b. The Equipment for Quality Assurance [6]

To assure the quality of modified atmosphere packaging system, the following effective support is required:

i. Gas Mixers

Gas mixing systems for 2 or 3 defined gases, designed for a variety of applications, particularly for all areas with constant or varying mixed gas flow demands. Electronic flow control system for modified atmosphere packaging (MAP) to suit flow pack machines in the food industry and for room atmospheres *e.g.* for the storage of fruit and vegetables.

ii. Gas Analysers

There is a cordless hand held Gas Analyser for checking modified atmosphere packs (MAP). It is an ideal instrument for portable, faster and accurate sample tests at the packaging machine, in stores or in laboratories. Compact Gas Analyser is for checking and/or monitoring modified atmosphere packs/packaging (MAP). For continuous analysis (in-line) and also intermittent sampling via a needle at the packaging machine, in stores or in laboratories.

iii. Leak Detection

Bubble leak detection systems for packages. The Leak detection system (MASTER® EASY) features the detection of the smallest of leaks without the need for trace gases. This is a Leak detection system for packages containing CO_2. Another leak detection system (LEAK-MASTER®) features non-destructive testing of the smallest of leaks without the need for costly helium.

iv. Ambient Air Monitoring

Gas monitoring system for the depletion of oxygen (O_2), carbon dioxide (CO_2) and other gases.

Sources

1. *Wikipedia.*

2. *Mullan, Michael; McDowell, Derek, 2011, Food and Beverage Packaging Technology. Oxford, UK: Wiley-Blackwell. pp. 263-294. doi: 10.1002/9781444392180.ch10. ISBN 9781444392180.*
3. *Blakistone, B. A., 1998, Principles and applications of modified atmosphere packaging of foods (2nd ed.). London: Blackie Academic and Professional. pp. 1-38. ISBN 978-0751403602.*
4. *Shenzhen Penglai Industrial corporation Ltd..*
5. *UTOPLAS Brand MAP.*
6. *www.wittgs.com.*
7. *Co_2meter.com.*

Packing Materials for MAP [1]

The main components of the material for the production of packages are laminate of PTE/PE, films from VDC copolymers, the laminate of cellophane with polyethylene, PET and PA films in the form of sleeves, from which air is removed by thermal shrinkage of the product being packaged [1a]. Other packaging materials that are in use for MAP are, Ethylene vinyl alcohol (EVOH), Polypropylene (PP), Polystyrene (PS), Polyvinyl chloride (PVC) and Polyvinylidene chloride (PVDC) [2].

MAP packages are either thermoformed trays with heat-sealed lids or pouches. These trays, and pouches need to be made of materials with low permeability to gases (CO_2, N_2, and O_2) [1b]. The details related to thermoformed packaging and Flexible film are given below [1]:

i. Flexible Films

These are commonly used for products such as fresh produce, meats, fish and bread as these films provide suitable permeability for gases and water vapor to reach the desired atmosphere. Pre-formed trays are formed and sent to the food packaging facility where they are filled. The package headspace then undergoes modification and sealing. Pre-formed trays are usually more flexible and allow for a broader range of sizes as opposed to thermoformed packaging materials as different tray sizes and colors can be handled without the risk of damaging the package [2a].

ii. Thermoformed Packaging

This packing however is received in the food packaging facility as a roll of sheets. Each sheet is subjected to heat and pressure, and is formed at the packaging station. Following the forming, the package is filled with the product, and then sealed [3]. The advantages that thermoformed packaging materials have over pre-formed trays are mainly cost-related, thermoformed packaging uses 30 per cent to 50 per cent less material.

When selecting packaging films for MAP of fruits and vegetables the main characteristics to consider are as follows [4]:

1. Gas permeability.
2. Water vapour transmission rate.

3. Mechanical properties.
4. Transparency.
5. Type of package.
6. Sealing reliability.

- ✰ Traditionally used packaging films like LDPE (low-density polyethylene), PVC (polyvinyl chloride), EVA (ethylene-vinyl acetate) and OPP (oriented poly-propylene) are not permeable enough for highly respiring products like fresh-cut produce, mushrooms and broccoli.
- ✰ As fruits and vegetables are respiring products, there is a need to transmit gases through the film. Films designed with these properties are called permeable films.
- ✰ Other films, called barrier films, are designed to prevent the exchange of gases and are mainly used with non-respiring products like meat and fish.
- ✰ MAP films developed to control the humidity level as well as the gas composition in the sealed package are beneficial for the prolonged storage of fresh fruits, vegetables and herbs that are sensitive to moisture. These films are commonly referred to as modified atmosphere/modified humidity packaging (MA/MH)films.

Selection of Packaging Materials [2]

The selection of these plastic packaging materials are based on the following attributes:

1. Food contact approval.
2. Gas and vapour barrier properties.
3. Optical properties.
4. Antifogging properties.
5. Mechanical properties.
6. Heat sealing properties.

Sources

1. *Wikipedia.*

1a. *Shupeyko A. M. and Ishevsky A. L., Modern Food Packaging Technology, World Scientists, 2019.*

1b. *Lesiow, T. and Kosiorowska, M, 2006, Opakowaniaaktywneiin-teligentne w przetworstwiemiesa. Czêœæ I. Gospodarka-Miesna 3: 12¬$ per cent V-18.*

2. *http: //ecoursesonline.iasri.res.in/mod/page/view, Vacuum and MAP, Lesson- 23.*

2a. *Mullan, Michael; McDowell, Derek, 2011, Food and Beverage Packaging Technology. Oxford, UK: Wiley-Blackwell. pp. 263-294. doi: 10.1002/9781444392180.ch10. ISBN 9781444392180.*

3. *Schmidt, F, 2003, "Modelling of infrared heating of thermoplastic sheet used in thermoforming process", Journal of Materials Processing Technology, 143-144, 225-231, doi: 10. 1016/s0924-0136(03)00291-7.*

4. *Robertson, G. L., 2006, Food Packaging Principles and Practice (2nd ed). Florida: CRC Press. pp. 313-330, ISBN 978-0-8493-3775-8.*

Advantages and Disadvantages of MAP Technology [1]

The benefits of MAP technology to the manufacturer, retailer as well as consumer far outweigh the drawbacks.

a. Advantages [1, 1a]

i. Increased shelf-life allowing lesser frequency of loading of retail display shelves,
ii. Improved presentation – clear view of the product and all round visibility,
iii. Hygienic stackable pack, sealed and free from product drip and odour,
iv. Potential shelf-life increases by 50 to 400 per cent,
v. Reduction in production and storage costs due to better utilisation of labour, space and equipment,
vi. Fresh appearance,
vii. Product can be distributed to long distances,
viii. High quality product
ix. User friendly.

MAP under different conditions have different advantages. The details are summarised in following table [2]:

Sl. No.	*Duration and Temperature*	
	19 Days and 45°F	*7 Days and 55° F*
1	Preservs natural sweetness.	Reduce produce waste.
2	Extends shelf-life while maintaining flavor.	Maintain saleable weight.
3	Maintains quality and freshness without additives and Preservatives.	Maintain colour, texture and visual appeal.
4	No salt, sugar and syrup needed.	—
5	Allow to pick closer to peak ripeness.	—

b. Disadvantages [1, 1a]

i. Capital cost of gas packaging machinery,
ii. Cost of gases and packaging materials,
iii. Increased pack volume increases transport costs and retail display space,
iv. Potential growth of food borne pathogens due to non-maintenance of required storage temperature by retailers and consumers,

v. Temperature control is necessary during storage.
vi. Special equipment is required.

Sources

1. *http: //icpe.in/icpefoodnpackaging/pdfs/24_modified.pdf.*

1a. *http: //icpe.in/ecourseonline.iasru.res.in., Vacuum and MAP, Lesson- 23.*

2. *Extend Produce Shelf Life with Modified Atmosphere Packaging, Crawford Packaging*

Packing and Storage Requirements

a. Packaging Requirements for Extended Shelf-life [1]

The following three packaging requirements must be met to meet extended shelf-life objectives:

i. Choose the gas mix that is conducive to maintaining or extending desired shelf life,
ii. Select materials that deliver the right barrier for a specific product and shelf-life requirements
iii. Monitor seal integrity throughout the manufacturing process.

b. Requirements for Storing a Product in MAP [2]

The following requirements are imposed on packaging for storing a product in a gas-modified environment:

i. Complete tightness,
ii. Chemical inertness,
iii. Environmental safety,
iv. Food tolerances,
v. High barrier properties (no transmission of oxygen, moisture, corrosive media inside, the gas modified environment from the inside to the outside).

The success of the MAP depends on following various factors:

i. Good initial quality of the product,
ii. Good hygiene practices during fishing,
iii. Selection of the right packaging material,
iv. A safe packing equipment,
v. Good maintenance and control of temperature and a proper gas mixture for the product and the gas/product ratio.

Sources

1. *Advanced packaging technology helps extend snack and bakery product shelf life, Snack food and wholesale Bakery.*

2. *Shupeyko A. M.and Ishevsky A. L., Modern Food Packaging Technology, World Scientists, 2019.*

Application of MAP in Foods [1]

The defining the optimal gas atmosphere for a food product in a specific packaging design , is an important consideration in MAP . This optimal atmosphere depends on the intrinsic parameters of the food product (pH, water activity, fat content, type of fat) and the gas / product volume ratio in the chosen package type. The intrinsic parameters will determine the sensitivity of the product for specific microbial, chemical and enzymatic degradation reactions.

Many food products such as red meat, seafood, minimally processed fruits and vegetables, salads, pasta, cheese, bakery goods, poultry, cooked and cured meats, ready meals and dried foods are packaged under MA [2]. A summary of optimal gas mixtures for MA products are shown in the following table:

MAP for different Food Products and Optimal Gas Mixtures [3]

Sl. No.	*Product*	*Oxygen per cent*	*Carbon Dioxide per cent*	*Nitrogen Per cent*
1	Red Meat	80-85	15	—
2	Poultry	—	25	75
3	Fish	—	60	40
4	Cheeses	—	100	—
5	Bread	—	70	30
6	Fresh Pasta	—	—	100
7	Fruits and Vegetables	3-5	3-5	85-95

MAP in Dairy Products [5a]

MAP has the potential to increase the shelf life of a number of dairy products like fat-filled milk powders, fat spreads and cheeses.

i. Whole Milk Powder [5a, 9]

It is particularly susceptible to the development of off-flavours due to fat oxidation. Commercially, the air in the pack is removed under vacuum and replaced with 100 per cent N_2 or N_2/CO_2 mix and the powder is hermetically sealed in metal cans where in product will have low levels of residual O_2 (<1 per cent). Use of O_2 scavengers may also be used [5a].

Milk powder, above all for use in baby food, is a highly sensitive product. It is especially important to ensure that oxygen is displaced from the packaging in order to extend the shelf life. In practice, packaging is carried out in pure nitrogen with as low a residual oxygen content as possible [9].

ii. Cheeses [5a, 9]

MAP is used to enhance the shelf-life of chesses.

a. *Cheddar Cheese [5a]*

It is traditionally vacuum packed while MAP is being used with high concentration of CO_2 gas mixes and it is important to maintain the balance of correct CO_2 level in the gas mix so as to avoid excessive pressure being put on the pack seal.

b. *The Cottage Cheese [5a]*

It is a high-moisture, low-fat product that is susceptible to a number of spoilage organisms including *Pseudomonas* spp. Use of N_2/CO_2 atmospheres showed significant extension of the shelf life of cottage cheese. Gas mixtures containing CO_2: N_2 in the proportion of 40: 60 can increase the shelf life of cottage cheese significantly.

c. *Soft Cheese [9]*

It can quickly become rancid. This problem can also be tackled with CO_2 modified atmospheres. However, as soft cheese absorbs CO_2 to a significantly higher extent, there is a risk of the packaging collapsing. A correspondingly lower CO_2 content should therefore be chosen.

d. *Grated and Sliced Cheeses [10]*

Grated or sliced cheddar are also packaged in modified atmospheres. Grated cheese is normally packed in an atmosphere of 70 per cent nitrogen and 30 per cent CO_2. By limiting CO_2 to 30 per cent, manufacturers can avoid package collapse.

iii. Cream [10]

Cream and dairy products containing cream rapidly turn sour in oxygen or air atmospheres. By replacing oxygen with nitrogen, however, manufacturers can prevent rancidity and the growth of aerobic bacteria.

iv. Milk [9]

In the case of milk products such as yoghurt or cream, there is a risk of the products absorbing too much CO_2 and becoming sour. A lower CO_2 content should therefore be chosen.

Sl. No.	*Dairy Products*	*Shelf-life*	
		In All	*With MAP*
1	Butter, yoghurt, cream *etc.*	1–4 weeks	2–12 week
2	Milk powder	4–8 months	1–2 years
3	Soft cheese	4–14 days	1–3 weeks
4	Hard cheese	2–3 weeks	4–10 weeks

MAP in Fish and Seafoods [9]

Fish and seafood are some of the most sensitive foods. They are at risk of rapidly declining in quality and spoiling even shortly after the catch. The most important element for a longer shelf life is cooling close to 0° Celsius. Modified atmospheres with minimum 20 per cent CO_2 also retard the growth of bacteria. CO_2 components around 50 per cent are frequently used. Higher CO_2 concentrations can lead to undesirable side effects such as liquid loss or a sour taste. In the case of low-fat fish and shellfish, O_2 is also used in the packaging. This prevents a fading or loss of the colour. Shelf-life in MAP of some fishes is tabulated below [9]:

Sl. No.	*Fish and Seafood Products*	*Shelf-life*	
		In All	*With MAP*
1	Raw low fat fish and crustaceans	2–4 days	4–9 days
2	Raw high fat fish and crustaceans	2–3 days	4–6 days
3	Cooked fish	2–4 days	3–4 week
4	Smoked and refined fish and seafood products	5–10 days	7–21 days

Fish Packing [6].

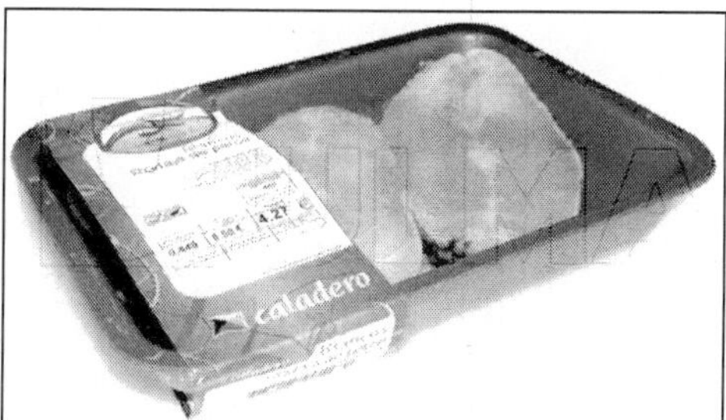

Fish Packing [7].

Fresh Salmon [7].

Vegetable Packing [8].

MAP in Meat/Poultry [9]

Meat and sausage products, above all raw meat, are very prone to spoiling due to microbial growth on account of their high moisture and nutrient content. No matter whether beef, pork or poultry, spoilage begins from the moment of slaughter and especially all butchering. Besides high hygiene standards and permanent cooling, modified atmospheres can significantly extend the shelf life of meat and sausage products. CO_2 is the most important among the protective gases. At concentrations above 20 per cent, CO_2 can considerably reduce microbial growth.

Sl. No.	Meat Products	Shelf-life	
		In all	With MAP
1	Raw red meat	2–4 days	5–8 days
2	Sliced cooked meat	2–4 days	2–5 weeks
3	Sausages	2–4 days	2–5 weeks
4	Raw poultry	4–7 days	16–21 days
5	Cooked, smoked and refined poultry products	5–10 days	7–21 days

Poultry is sensitive to rapid spoilage and is therefore subject to higher requirements for permanent cooling. In poultry, a modified atmosphere with CO_2 content will extend the shelf life. A high oxygen content is also used for poultry without skin so as to retain the colour of the meat. The CO_2 can partly be absorbed by the foods. To prevent the packaging from collapsing, nitrogen is used as a supporting gas.

MAP in Fruit and Vegetables [9]

Modified atmospheres in packaging(MAP) make it possible to offer consumers fresh and untreated products *e.g.* fresh fruit and vegetables with a long shelf life. Fruit and vegetables are subject to very special requirements in regard to the nature of the packaging and atmosphere because unlike other food, fruit and vegetables continue breathing after the harvest and consequently require an oxygen content in the packaging and that why the packaging film does not have to be fully tight.

- ☆ By taking the product's breathing and the permeability of the film, typically via micro-perforation, into account, the composition of carbon dioxide, nitrogen and low amounts of oxygen ideal for the product can be maintained.
- ☆ An EMA (equilibrium modified atmosphere) term is used here. The gas composition is individually adapted to the corresponding product. Thorough cleaning along with hygienic processing are the fundamental preconditions for long-lasting freshness.
- ☆ Modified atmospheres, in conjunction with corresponding cooling, can be used to extend the shelf life of fresh produce, while achieving an attractive packaging design at the point of sale.

Sl. No.	Fruit and Vegetables	Shelf-life	
		In all	With MAP
1	Fresh cut salad or fruit	2–5 days	5–10 days
2	Cooked vegetables	3–14 days	7–21 days

MAP in Pasta and Ready-made Meals [9]

The nature and composition of fresh pasta and, in particular, readymade meals are very different. In the majority of cases, modified atmospheres can significantly extend the shelf life without using oxygen. Mixtures of CO_2 and nitrogen are used here. The concentration of the gases is oriented to the content of the product:

1. If, there is a risk that large volumes of CO_2 will be absorbed by the product,
2. The nitrogen content should be chosen higher to prevent the packaging from collapsing.

Sl. No.	*Pasta and Readymade Meals*	*Shelf-life*	
		In all	*With MAP*
1	Pizza	7–10 days	2–4 weeks
2	Fresh pasta	1–2 weeks	3–4 weeks
3	Combination products (sandwiches, pies, meat and fish in batter *etc.*)	2–7 days	3–21 days
4	Chilled fruit and vegetable juices	1 week	2–3 week

MAP in Bakery [4, 9]

The shelf life of bakery products (bread, cake and biscuits) is primarily affected by potential mould formation. Packaging involving a modified atmosphere with CO_2 and without oxygen extensively prevents the products from becoming mouldy and extends the shelf life. The collapsing of packing due to CO_2 absorption by the products is observed and this can be prevented by using nitrogen as a supporting gas in many cases [9].

Sl. No.	*Bakery Products*	*Shelf-life*	
		In all	*With MAP*
1	Cakes	several weeks	up to one year
2	Bread	some days	2 weeks
3	Pre baked bread	5 days	20 days

Most MAP for bakery products is mainly high CO_2 contents with the elimination of O_2 due to the fact that CO_2 is considered as an antimicrobial agent. In high water content food like bakery products, it is found that CO_2 can dissolve in water to form carbonic acid, lowering the pH. This acidification of the cell contents causes the death of bacteria. It is concluded [5] that mould growth could not be prevented, but could be delayed by N_2 and/or CO_2 up to 5-10 days. The only possibility to prevent mould growth was to maintain the level of O_2 to be below 0.4 per cent [4].

MAP in Snacks and Nuts [9]

Snack products (potato crisps or peanuts), primarily involve problems associated with the fat content of the food. There is a risk of oxidation and if the packaging is not optimal, products can quickly become rancid. Modified atmospheres with 100 per cent nitrogen are frequently used and a premature spoilage can be prevented, while these atmospheres also provide protection from mechanical damage to sensitive products, *e.g.* potato crisps in conventional packets.

Sl. No.	*Snacks and Nuts*	*Shelf-life*	
		In all	*With MAP*
1	Dried foods (snacks, coffee, dried milk, herb *etc.*	4–8 months	1–2 years

MAP in Coffee [9]

Coffee (dried product) is relatively insensitive to spoilage by micro-organisms. To prevent the oxidation of the fatty acids and making the product rancid is greater, coffee is packaged with the exclusion of oxygen. Instead, a modified atmosphere comprising pure nitrogen is frequently used in coffee sachets or capsules.

MAP in Wine [9]

Gases are often used to protect wine in the different phases of its production process and to retain the quaility of the product. The tank headspace is replaced with an inert gas or a gas mixture, for example of CO_2, N_2 or Ar. The composition of the gases is chosen according to the type of wine.

Sources

1. *Wikipedia.*
2. *Kirtil, E and Oztop, M. H., 2016, "Controlled and modified atmosphere packaging", Reference Module in Food Science. doi: 10.1016/B978-0-08-100596-5.03376-X. ISBN 9780081005965.*
3. *Parry, R. T., 1993, Principles and applications of modified atmosphere packaging of foods, Boston, MA: Springer US. ISBN 9781461358923.*
4. *Advanced packaging technology helps extend snack and bakery product shelf life, Snack food and wholesale Bakery.*
5. *Smith, J. P., Ooraikul, B. Koersen, B., Jackson, E. D., Lawrence, E. D., Novel approach to oxygen control in modified atmosphere packaging of bakery products, Food Microbiol., 3 (4), 1986, pp. 315-320.*

5a. *http://ecourseonline.iasri.res.in., Vacuum and MAP, Lesson- 23.*

6. *leybold.com.*
7. *ULMA Packing.*
8. *Ken Research, 2020.*
9. *www.wittgas.com.*
10. *MAPAX for Dairy, Industrial gases, www.lindegas.com.*

Index

I

K

L

M

N

O

P